HILL & MESSENT
CMR: CONTRACTS FOR THE INTERNATIONAL CARRIAGE OF GOODS BY ROAD
(THIRD EDITION)

LLOYD'S SHIPPING LAW LIBRARY

Series editors: M. T. Wilford and T. G. Coghlin

LLOYD'S SHIPPING LAW LIBRARY

The Ratification of Maritime Conventions
edited by The Institute of Maritime Law,
University of Southampton
(1990) (looseleaf)

*Contracts for the Carriage of Goods
by Land, Sea and Air*
by David Yates *et al.*
(1993) (looseleaf)

Marine Environment Law
by John H. Bates and Charles Benson
(1993)

Voyage Charters
by Julian Cooke,
Timothy Young, Andrew Taylor
John D. Kimball, David Martowski
and LeRoy Lambert
(1993)

Marine War Risks
second edition
and Supplement to
the second edition
by Michael D. Miller
(1994)

Merchant Shipping Legislation
by Aengus R. M. Fogarty
(1994) (looseleaf)

Multimodal Transport
by Ralph De Wit
(1995)

*CMR: Contracts for the International
Carriage of Goods by Road*
third edition
by Andrew Messent
and David Glass
(2000)

Time Charters
fourth edition
by Michael Wilford, Terence Coghlin
and John D. Kimball
(1995)

The Law of Shipbuilding Contracts
second edition
by Simon Curtis
(1996)

The Law of Tug and Tow
by Simon Rainey
(1996)

Enforcement of Maritime Claims
second edition
by D. C. Jackson
(1996)

Laytime and Demurrage
fourth edition
by John Schofield
(2000)

London Maritime Arbitration
by Clare Ambrose
and Karen Maxwell
(1996)

Berlingieri on Arrest of Ships
third edition
by Francesco Berlingieri
(2000)

EC Shipping Law
second edition
by Vincent Power
(1998)

*Limitation of Liability
for Maritime Claims*
third edition
by Patrick Griggs and Richard Williams
(1998)

Ship Sale and Purchase
third edition
by Iain Goldrein QC *et al.*
(1998)

*Shipping and the Environment—
Law and Practice*
by Colin de la Rue and
Charles B. Anderson
(1998)

P. & I. Clubs: Law and Practice
third edition
by Steven J. Hazelwood
(2000)

Admiralty Jurisdiction and Practice
second edition
by Nigel Meeson
(2000)

Hill & Messent
CMR: Contracts for the International Carriage of Goods by Road

THIRD EDITION

By

ANDREW MESSENT
MA Cantab
Partner, Holmes Hardingham, London

AND

DAVID A. GLASS
LL.M.
Lecturer in Law, Cardiff Law School

|L|L|P|

LONDON   HONG KONG
2000

LLP Professional Publishing
(a trading division of Informa Publishing Group Ltd)
69–77 Paul Street
London
EC2A 4LQ

EAST ASIA
LLP Asia
Sixth Floor, Hollywood Centre,
233 Hollywood Road
Hong Kong

First Published 1984
Second Edition 1995
Third Edition 2000

British Library Cataloguing in Publication Data
A catalogue record
for this book is available
from the British Library

ISBN 1–85978 593–X

Text set in 10 on 12pt Times by
Interactive Sciences Ltd.
Gloucester.
Printed in Great Britain by
MPG Books, Bodmin,
Cornwall

Preface to the Third Edition

Another five years has passed since the last edition, a relatively rapid succession compared with the period of eleven years between the first two editions. Once again, much credit for this and for the new edition in general, must go to David Glass, in particular in relation to the ever increasing wealth of cases and jurisprudence from other contracting countries.

When the first edition of this book was published in 1984, there were 23 signatories to the Convention. By 1995 when the second edition was published, there were 33. As this edition goes to print, there are 42 states which are party to the Convention. This reflects the growing importance and influence of the Convention, which is also particularly evident in countries such as Austria and Belgium which have adopted the provisions of the Convention directly into their national law for application to purely domestic operations of carriage of goods by road.

Although it will be appreciated that in a work such as this we cannot provide a comprehensive guide as to the interpretation of the Convention throughout the varied jurisdictions which apply its provisions, we have endeavoured to keep abreast of developments in other contracting countries, in particular the decisions from the main jurisdictions (Belgium, France, Germany and Holland), where there is a constant and heavy flow of decisions concerning the Convention.

Of particular interest are, for example, decisions from the German Supreme Court concerning issues of proof, and in respect of rights of disposal, where an irregular CMR Consignment Note has been issued. There are also important decisions from the same court concerning the rights of action of a consignee who has refused to take the delivery of the goods, and on the basis for drawing a distinction between partial and total depreciation of the consignment. Numerous decisions across different jurisdictions have added to the growing body of law on the defence of "unavoidable circumstances" under Article 17(2) of the Convention, including an important statement on the Dutch understanding of this defence from the Dutch Supreme Court.

The problems associated in recent times with making a secure delivery in the Russian Federation are reflected in several decisions across different jurisdictions.

We have endeavoured also to reflect cases and developments which impact on the application of the Convention in particular countries. The recent reform of the German domestic transport law through the Transport Law Reform Act is an important example.

This Act has implications for the application of the CMR rules on wilful misconduct in that country, and consequently there is an expanded treatment of the German approach to this aspect. Of more general interest is the fact that this reform brings rules which have been influenced by the principles contained in the CMR Convention into the operation of German domestic transport law, thus reflecting the approach in countries such as Austria and Belgium which is referred to above.

So far as developments in this country are concerned, here too there have been a number of cases on both the Article 17(2) defence and on wilful misconduct. In addition, perhaps potentially the most significant development since the last edition is the first instance decision in the case of *Gefco UK Ltd* v *John Mason*, in the light of which previous assumptions as to the scope for the operation of national law in the context of CMR cases may need to be re-examined. In view of the wide-ranging potential of that case, it is to be hoped that it finds its way into the law reports.

Our considerable thanks are due to Mrs Araya Stevenson for her assistance with the translation of numerous German decisions, and once again we are grateful to our respective families for their indulgence in respect of the time involved in the preparation of the new edition. As with the previous editions, the IRU and FTA/RHA/SITPRO forms of consignment note are reproduced by kind permission of those bodies, the Carriage of Goods by Road Act 1965 is reproduced by kind permission of Her Majesty's Stationery Office and the French text of the Convention is reproduced by kind permission of the United Nations.

London, August 2000   A. D. M.

Preface to the Second Edition

More than a decade has now elapsed since the first edition of this book. In that time, there has been a considerable amount of additional case law on the Convention, both in this country and elsewhere. In the same period, I moved back into practice, the demands of which have delayed this second edition much longer than either I or, I am sure, Lloyd's of London Press, would regard as desirable.

In view of the difficulties I was encountering in making progress with the new edition, I am indebted to David Glass, who very kindly agreed to assist, and whose very considerable efforts have been invaluable. His input on the Continental case law in particular has gone far beyond what I could ever have hoped to achieve on my own. Beyond that, however, he has contributed countless improvements and amendments. To single out one area for special mention, he undertook the heroic task of reformulating the treatment of Article 17(4)(c) into what is, we both hope, a more useful analysis in the light of developments since the last edition.

Inevitably, the length of the text has increased. With a view both to keeping the size of the book within manageable proportions and to avoiding undue disruption of the structure, we have incorporated much of the additional material, in particular the Continental case law, by way of footnote. However, because we appreciate that the Continental case law is not readily accessible, we have sought to give some detail of the content of the more important decisions referred to. On occasion, as will be seen, this has resulted in some fairly substantial footnotes, but in our view this represented the best way of presenting the material without the text becoming inordinately long and unwieldy.

Also with reference to the question of the accessibility of the relevant case law, it is both surprising and regrettable, in particular in view of the relatively small number of cases, that a larger proportion of the English cases are not reported in any of the readily accessible series of reports. Again, for that reason some of the English cases have been set out in some detail.

We have endeavoured to include all the English cases on the Convention of which we are aware to date. There are, however, two very recent county court judgments on wilful misconduct which we have not included partly because both were too late to be readily incorporated in the text but more importantly because both are currently the subject of appeal.

Finally, I must on behalf of David and myself proffer our considerable thanks to Janet Stevenson, Jill Green and Gabriela Kutting for their invaluable assistance in translating numerous German decisions. Our thanks, too, to our respective families for the disruption over the gestation of the new edition. My thanks must also go to my secretary, Sandra, who managed to cope with the enormous volume of amendments to the text without any noticeable disruption to her normal prodigious output. As with the first edition, the IRU and FTA/RHA/SITPRO forms of consignment note are reproduced by kind permission of these bodies, the Carriage of Goods by Road Act 1965 is reproduced by kind permission of Her Majesty's Stationery Office, and the French text of the Convention is reproduced by kind permission of the United Nations.

London, July 1995   A. D. M.

Preface to the First Edition

As related in the introduction to this book, the somewhat inelegantly named Convention on the Contract for the International Carriage of Goods by Road (the acronym CMR by which it is known being drawn from its French title: *Convention Relative au Contrat de Transport International de Marchandises par Route*) came into force in this country in 1967. Although a slow starter in this country in terms of decided cases involving its provisions, the rate at which such cases have come before the courts has increased steadily in recent years. Dr Hill's untimely death prevented him from finishing this book on the Convention, which is based largely on his prodigious research into its operation, both in this country and abroad. It has been my privilege to have the opportunity of completing the text, a privilege for which I am most grateful to Colina Hill.

Dr Hill had conceived of the book as an article by article approach to the provisions of the Convention, and although this has necessitated considerable amounts of cross-referencing in the discussion of particular provisions, it seemed to me that such an approach is to be preferred for the purposes of a book such as this.

I am regretfully unable to give such acknowledgements as Dr Hill may have thought appropriate, but for myself I must primarily thank Adrian Hardingham, who has with great patience read over everything I have passed to him, contributed countless helpful ideas and suggestions, and given graciously of his time in discussion of many detailed points considered in the book. I am also indebted to him for his encouragement. My considerable thanks are also due to John Hodgson, who at short notice was able to provide much assistance with the first six chapters; his linguistic abilities have been particularly appreciated. I must also thank Lloyd's of London Press and Colina Hill for their considerable but much tried patience in awaiting the finished text.

As for the typing, my thanks are due to all those who assisted, including my wife, Trudie, to whom I am also indebted for bearing with me generally throughout. I should also like to record my thanks to Richard Morris of Trent Polytechnic Library for his frequent assistance in tracing and obtaining materials. Finally, the IRU and the FTA/RHA/SITPRO forms of CMR Consignment Note are reproduced by kind permission of those bodies respectively, the Carriage of Goods by Road Act 1965 is reproduced by kind permission of Her Majesty's Stationery Office, and the French text of the Convention is reproduced by kind permission of the United Nations.

A.D.M.
Nottingham,
March 1984.

Contents

CHAPTER 7   DELAY, NON-DELIVERY AND CASH ON DELIVERY   171

CHAPTER 8   DANGEROUS GOODS   187

CHAPTER 9   COMPENSATION   193

Table of Cases

Table of European Cases

Table of Legislation

(Paragraph numbers printed in **bold** indicate where text is reproduced)

Table of Conventions, etc.

(Paragraph numbers printed in **bold** indicate where text is reproduced)

Table of Abbreviations

Hof	District Appeal Court, Holland, and Hof van Beroep (District Appeal court, Belgium)
H.R.	Hoge Raad (Supreme Court, Holland)
HVC.	Hof van Cassatie (Supreme Court, Belgium)
I.C.L.Q.	International and Comparative Law Quarterly
IMF	International Monetary Fund
IRU	International Road Transport Union
I.T.L.Q.	International Trade Law Quarterly
J.P.A.	Jurisprudence du Port D'Anvers (Belgium)
J.T.	Journal des Tribunaux (Belgium)
Jung	Christian Jung, CMR—Survey, Analysis and Trends of Recent German Case Law (1997) U.L.R. 148–168
K.B.	King's Bench Reports
KG.	Kantongerecht (District Appeal Court, Holland)
Krings	Ernest Krings, CMR: Examen de la jurisprudence des Cours de Cassation de France et de Belgique, et du Hage Raad des Pays–Bas (1999) U.L.R. 140–163, 767–795
Lamy	Lamy Transport (updated annually, references are to 1999 edn., and vol. 1 except where otherwise indicated), Paris
LG.	Landgericht (District Court, Germany)
L.J.Ch.	Law Journal, Chancery
Ll.L.Rep.	Lloyd's List Law Reports (to 1950)
Lloyd's Rep.	Lloyd's Reports (post 1950)
LMCLQ	Lloyd's Maritime and Commercial Law Quarterly
Loewe	Commentary on the Convention of 19th May 1956 on the Contract for the International Carriage of Goods by Road (CMR) (1976) 11 E.T.L. 311
L.R.C.P.	Law Reports, Common Pleas
Mayer	Carl Ulrich Mayer, CMR—Aperçu et réflexions sur la jurisprudence autrichienne en dernier ressort (1994–1996), (1997) U.L.R. 169
Mercadal	Barthélémy Mercadal, Droit des transports terrestres et aériens, 1996, Paris
M.L.R.	Modern Law Review
M. & W.	Meeson & Welsby's Reports
Nickel Lanz	Mari-Carmen Nickel Lanz La Convention relative au contrat de transport international de marchandises par route (CMR), 1976, Hamburg
N.J.	Nederlandse Jurisprudentie (Dutch law reports)
N.J.W.	Neue Juristische Wochenschrift (German law reports)
N.L.J.	New Law Journal
Öest.Jur.	Österreichischer Juristenzeittung (Austrian law reports)
OGH.	Oberster Gerichtshof (Austrian Supreme Court)
O.J.	Official Journal of the E.E.C.
OLG.	Oberlandesgericht (German and Austrian provincial Appeal Court)
Palmer	Norman E. Palmer, Bailment, 2nd edn., 1991, London
Putzeys	Jacques Putzeys, Le Contrat de Transport Routier de Marchandises, 1981, Brussels
Q.B.	Queen's Bench Reports
RB.	Rechtbank van Koophandel (Belgian and Dutch District Commercial Court)
Rev. Trim. Dr. Comm.	Revue Trimestrielle de Droit Commercial
RHA	Road Haulage Association
RID	International Regulations concerning the Carriage of Dangerous Goods by Road

RIW	Recht der Internationalen Wirtschaft
Rodière	The Convention on Road Transport (1970) 5 E.T.L. 620–636, (1971) 6 E.T.L. 3–18, 306–324, 574–593
RW	Rechtskundig Weekblad
S. & S.	Schip en Schade (Dutch law reports)
Sc.L.R.	Scottish Law Reports
SITPRO	Simplification of International Trade Procedures Board
Stra Güv	Strassengüterverkehr
Theunis	Jan Theunis (ed), International Carriage of Goods by Road CMR, 1987, London.
Thume	Karl–Heinz Thume, Kommentar Zur CMR, 1995, Heidelberg
TranspR	Transportrecht
UfR	Ugeskrift for Retsvoesen (Danish law reports)
U.L.C.	Uniform Law Cases (since 1973, incorporated into Uniform Law Review)
U.S.C.A.	United States Court of Appeal
U.S.D.C.	United States District Court
VersR	Versicherungsrecht
W.L.R.	Weekly Law Reports
Yates	David Yates (ed), Contracts For the Carriage of Goods by Land, Sea and Air, London, 1993 (Looseleaf update).

INTRODUCTION

Background

0.1 The history and origins of the Convention have been well documented elsewhere,[1] but in brief it arose from concern at the problems posed by the increasing growth in the international carriage of goods by road after the Second World War. The nature of that concern is apparent from the preamble to the Convention:—

"The Contracting Parties, having recognised the desirability of standardising the conditions governing the contract for the international carriage of goods by road, particularly with respect to the documents used for such carriage and to the carrier's liability . . . ."

The impetus towards such standardisation originally came from the International Institute for the Unification of Private Law (UNIDROIT) in Rome, which in 1948 agreed with the International Road Transport Union (IRU) and the International Chamber of Commerce (ICC) to set up a Committee of Experts to prepare a preliminary draft Convention on the subject. This work was taken over eventually by the United Nations Economic Commission for Europe, acting through a Working Party on Legal Questions which was a subsidiary of the E.C.E. Inland Transport Committee. Ultimately, the Convention was opened for signature at a special session of the Inland Transport Committee on 19 May 1956. With the deposit of the fifth instrument of ratification it entered into force on 2 July 1961, and to date 43 States have become parties to it.[2] The Convention became part of the law of the United Kingdom by virtue of the Carriage of Goods by Road Act 1965, although by section 14(4) of the Act an Order in Council was required to bring it into operation. This Order was made on 5 June 1967, and the United Kingdom Instrument of Accession was deposited in Geneva on 21 July 1967. The Convention therefore came into force 90 days later, on 19 October 1967.

0.2 Not surprisingly, a number of difficulties and shortcomings in relation to the Convention have emerged over the period of its operation. Although the Convention itself provides for the convening of a review conference at the request of a contracting State,[3] as yet there have been no proposals for such a conference.[4] However, in one area where

---

1. Donald, "CMR—An Outline and Its History." [1975] LMCLQ 420; Loewe, paras 1–8. Clarke, para. 1; Haak, pp. 8–10.
2. See para. 1.1 fn. 1 *infra*.
3. Article 49.
4. See Loewe, paras 299–300. The Inland Transport Committee of the E.C.E. did meet in 1972 to review the Convention, but decided that no changes were necessary; see Clarke, p. 2. In 1983, a report was issued by the FIATA ad hoc Working Group on CMR Revision which recommended a considerable number of changes. However, no steps have been taken so far to implement these recommendations. Among the more important recommendations are that the current strict liability of the carrier should be replaced with a lesser liability based

particular difficulties were experienced in practice—namely the appropriate rate of conversion for the limit of compensation laid down by the Convention—the Inland Transport Committee of the E.C.E. adopted a Protocol to the Convention.[5] That Protocol has been ratified by this country, and was brought into force in this country by the Carriage by Air and Road Act 1979, which made the necessary alterations and additions to the Convention. Those changes are discussed in detail in Chapter 9.

0.3 The other principal problem to have beset the overall scheme of the Convention is that it was drawn up prior to the vast increase in the use of containers as a means of transporting goods, and is so drafted as to exclude any contract which would otherwise be within the scope of the Convention but where containerised goods are separated from the road vehicle and transported for part of the journey by another mode of transport.[6] The Protocol of Signature to the Convention did contain an undertaking to negotiate a separate convention on combined transport operations, but it was not until 1980 that a Convention on International Multimodal Transport of Goods was adopted by a United Nations Conference.[7] That Convention lays down a detailed régime for international multimodal transport,[8] and when in force will operate to govern those multimodal transport operations which do not fall within the scope of CMR.[9] However, it will not affect those multimodal operations which do fall within CMR,[10] and in any event it is not anticipated that it will enter into force for some considerable time.[11]

Interpretation of the Convention

Uniformity

0.4 The CMR Convention is an international Convention aimed at standardisation,[12] and as such the English courts have consistently drawn attention to the desirability of a uniformity of approach to its provisions in the various jurisdictions which might be called upon to interpret it.[13] Thus, for example, in *Ulster-Swift Ltd.* v. *Taunton Meat Haulage Ltd.*, Megaw, L.J., observed:

on a duty to take reasonable care (as in Article 5 of the Hamburg Rules) and that the rules relating to successive carriage be abolished entirely.

5. Cmnd. 7480, H.M.S.O.

6. See *infra*, paras 2.8–2.11.

7. For the background to the Convention, see Chrispeels, "The United Nations Convention on International Multimodal Transport of Goods: A Background Note", (1980) 15 E.T.L. 355. The text of the Convention (UNCTAD document TD/MT/CONF 6) is reproduced in (1980) 15 E.T.L. 87.

8. Defined in Article 1 of the U.N. Convention.

9. See generally Mankabady, "The Multimodal Transport of Goods Convention: A Challenge to Unimodal Transport Conventions", (1983) 32 I.C.L.Q. 120.

10. Article 4(1), U.N. Convention.

11. By Article 36, ratification by 30 States is required, and to date only nine States have become contracting parties. So far as this country is concerned, the Convention is one in relation to which the EC has competence, so there has to be accession by the Community as a whole, and it would seem that there is no likelihood of this in the near future.

12. See *supra*, para. 0.1.

13. The Convention does not contain a provision of the sort to be found in some international conventions, such as Article 3 of the U.N. Convention on the Carriage of Goods by Sea, Hamburg, 1978: "In the interpretation and application of the provisions of this Convention regard shall be had to its international character and to the need to promote uniformity." As Haak points out, however, p. 14, a similar function is assumed by the preamble to the Convention. Interestingly, the preamble was not included with the text of the Convention as scheduled to the 1965 Act.

"But as it is an international Convention, we must do our best to interpret it, so far as we can, with a view to promoting the objective of uniformity in its interpretation and application in the Courts of States which are parties to the Convention."[14]

Equally consistently, however, the courts have found the desired uniformity elusive of achievement in practice. It is necessary therefore, before embarking on a discussion of the detailed provisions of the Convention, to give some consideration at this stage to the approach which the judiciary in this country adopts in relation to its interpretation, and the sources to which it is legitimate to have regard for that purpose.

The proper approach to interpretation

0.5 Traditionally, the English judiciary has adopted a narrow approach to the interpretation of statutes. Judges have confined themselves more or less exclusively to the words of the statute, seeking therein the intentions of the legislature but only so far as disclosed by the wording of the statute itself. The words there found were to be given their literal and ordinary meaning, however absurd the result. Only in cases of ambiguity was it regarded as permissible to start considering such matters as the "mischief" the Act was aimed at correcting in interpreting its provisions.[15] However, recent years have seen a general liberalisation of the approach of the English courts. Thus, for example it has become possible for Lord Griffiths to remark that "the object of the court in interpreting legislation is to give effect so far as the language permits to the intention of the legislature . . . . The days have long since passed when the courts adopted a strict constructionist view of interpretation which required them to adopt the literal meaning of the language. The courts now adopt a purposive approach which seeks to give effect to the true purpose of the legislation . . . ".[16] So too in *Sidhu* v. *British Airways*,[17] Lord Hope observed that "it is now well established that a purposive approach should be taken to the interpretation of international conventions which have the force of law in this country".

0.6 It is, therefore, now well accepted that the approach used in the interpretation of domestic legislation may not be appropriate in the interpretation of international conventions. As Bingham, J., in *A.B. Bofors UVA* v. *A.B. Skandia Transport* observed:

" . . . it is clear from authority that I am not to approach the matter with the blinkered gaze traditionally attributed to English judges but am both permitted and required to look at the matter in a slightly broader light and to take full account of the international intentions underlying the Convention."[18]

As indicated by Bingham, J., the essence of this broader approach is that regard is to be had to the objects and intents of the Convention—the schematic, teleological, or purposive approach—rather than a strict literal approach.[19] The reasons why such a broader

---

14. [1977] 1 Lloyd's Rep. 346, 348.
15. See generally Cross, *Statutory Interpretation*.
16. *Pepper (Inspector of Taxes)* v. *Hart* [1993] 1 All E.R. 42, 50.
17. [1997] 2 Lloyd's Rep. 76, 80. The judgment of Lord Hope contains an informative review of the permissible aids to interpretation.
18. [1982] 1 Lloyd's Rep. 410, 412.
19. Haak, at p. 17, describes the function of the teleological approach as being to control the interpretation arrived at by the other means so as to safeguard an interpretation which conforms with the purport and nature of the particular provisions concerned.

approach has commended itself to the English courts is indicated in the observation of Viscount Dilhorne in *Buchanan* v. *Babco* that

" . . . it is proper and indeed right, in my opinion, to have regard to the fact that Conventions are apt to be more loosely worded than Acts of Parliament. To construe a Convention as strictly as an Act may indeed lead to a wrong interpretation being given to it."[20]

The correct approach therefore is to adopt

" . . . a normal manner . . . unconstrained by technical rules of English law, or by legal precedent but on broad principles of general acceptation."[21]

0.7 The dangers of the purposive approach were, however, pointed out by Megaw, L.J., in the *Ulster-Swift* case,[22] namely that judges become legislators with potentially widely differing views of the policy which was or ought to have been behind the legislation. In short, the result of the purposive approach might be the opposite of the desired uniformity.[23] These inherent dangers are well illustrated by the judgment of Lord Denning, M.R., in the *Buchanan* case. Having discerned a gap in the Convention, he set about filling it by reference to his perception of the intentions behind the provision in question. His approach was unanimously rejected by the House of Lords both on the facts of the case[24] and in principle. The judgment of Lord Salmon is particularly apposite:

"For a court to construe a statute is one thing but to graft a provision on to it on the ground that the court thinks it reasonable to do so would bring the law into chaos and introduce a like chaos into the business of international carriers and those who contract with them . . . ."[25]

In any event, the views of Lord Denning, M.R., would appear to be based on a misconception as to European methods of interpretation. The nature of that misconception is apparent from his reference[26] to the practice of the European Court in support of his approach; he was, in essence, seeking to expand the approach to Community law to an entirely different context, and one which does not have the safeguard of one supreme court to supervise the national courts involved in its interpretation.[27] This point was noted by

20. [1978] 1 Lloyd's Rep. 119, 125. See also Mustill, J., in *J.J. Silber Ltd.* v. *Islander Trucking Ltd.* [1985] 2 Lloyd's Rep. 243, 245.

21. Lord Wilberforce in *Buchanan* at p. 122, citing with approval the much approved dictum of Lord Macmillan in *Stag Line Ltd.* v. *Foscolo Mango & Co. Ltd.* (1931) 41 Ll.L.Rep. 165, 174.

22. [1977] 1 Lloyd's Rep. 346, 351, quoted with approval by Lord Edmund-Davies in *Buchanan* v. *Babco* at p. 133.

23. See too Rodière, paras 149–150.

24. The gap was not apparent to any of the other judges in the Court of Appeal or the House of Lords.

25. At p. 128. See also the comments of Lord Wilberforce at p. 123, Viscount Dilhorne at p. 125 and Lord Edmund-Davies at p. 131. Hirst, J., in *Eastern Kayam Carpets Ltd.* v. *Eastern United Freight Ltd.*, Queen's Bench Division, 6.12.83, unreported except on LEXIS, rejected an argument that the House of Lords' disapproval of Lord Denning's view was based on the specific wording of the relevant provision and not of general application. A distinction is sometimes made between intended and unintended gaps (see Haak, p. 18), nevertheless, as Hirst, J., pointed out, to totally disregard the words used in a provision would trespass beyond interpretation, however liberally applied. Cf. Saville, J., in *Shell Chemicals U.K. Ltd.* v. *P. & O. Roadtanks Ltd.* [1993] 1 Lloyd's Rep. 114, 116, referred to *infra*, para. 0.26.

26. At p. 237.

27. See Megaw, L.J., in the *Ulster-Swift* case at p. 348 and, generally, Munday, "The Uniform Interpretation of International Conventions", (1978) 27 I.C.L.Q. 450 and Sacks and Harlow, "Interpretation, European-Style", (1977) 40 M.L.R. 578. See further Haak, pp. 10–38, Haak, Theunis, Chapter 15, Clarke, paras 3–7 and, generally, Giles, *Uniform Commercial Law*, Leyden, 1970, Jacobs and Roberts, *The Effect of Treaties in Domestic Law*, London, 1987 and Mann, "Uniform Statutes in English Law", (1983) 99 L.Q.R. 384.

Morritt, L.J., in *Gefco U.K. Ltd.* v. *Mason*.[28] Invited to imply an additional requirement to the terms of Article 1(1) of the Convention, he observed as follows[29]:

" . . . I do not think that the implication contended for would be consistent with the overall intention of the authors of the Convention to provide for harmonization across national boundaries. Given that there is no procedure for obtaining consistent application of the Convention in all countries to which it applies such as exists for European legislation or for some other international Conventions . . . in my view this Court should be slow to imply conditions for the application of the Convention. Otherwise discrepancies will arise between the effect of the Convention in the different states party to it."

0.8 The purposive approach to international conventions is perhaps best illustrated by reference to the House of Lords' decision in *Fothergill* v. *Monarch Airlines Ltd.*,[30] a case under the Warsaw Convention. The court was required to interpret the words "damage to baggage" for the purposes of the requirement of a written notification of claim to the airline. The defendant airline argued that these words included partial loss, and, not having received such notification within the specified time limit, the plaintiff's claim should be barred.

0.9 Following the traditional common law approach, literally the defendant's argument had to fail: "Damage is damage and loss is loss."[31] Nevertheless, turning to a purposive approach, it was clear that for the purposes of a time limit for claiming from the airline, there was no reason for any difference of approach between damage to baggage and partial loss of baggage. Having discerned an ambiguity in the Convention by reference to the French text,[32] and confirmed in its view by a consensus of academic opinion,[33] the House held that "damage" included partial loss and that the plaintiff's claim was accordingly barred.

0.10 This approach does not necessarily produce any clear answer, however. Thus in *A.B. Bofors-UVA* v. *A.B. Skandia Transport*,[34] Bingham, J., found himself with two opposite purposive interpretations of Article 33, and was compelled in effect to make a constructive choice between the two. Also in *Buchanan* v. *Babco*,[35] the House of Lords, adopting the purposive approach, concluded that a particular item of expense was recoverable under Article 23(4). However, as is indicated by the fact that the decision was reached by a bare majority, it may be questioned whether the decision there reached really does accord with the intentions behind the Convention. In particular, as is indicated in the discussion below, the width of expression used by the majority may paradoxically operate to exclude matters intended to be included.

0.11 The limits of the purposive approach must also be noted. In *Cummins Engine Co. Ltd.* v. *Davis Freight Forwarding (Hull) Ltd.*,[36] the Court of Appeal did not let the fact that the conclusion it reached produced inconvenience distort what the court otherwise regarded to be the clear meaning of Article 39(2).

28. [1998] 2 Lloyd's Rep. 585.
29. At p. 590.
30. [1980] 2 Lloyd's Rep. 295.
31. Lord Scarman at p. 310, quoting with approval from the American case *Schwimmer* v. *Air France* (1976) 14 Avi. 17,466, 17,466.
32. See *infra*, para. 0.14.
33. See *infra*, para. 0.22.
34. [1982] 1 Lloyd's Rep. 410, discussed *infra*, para. 10.123.
35. See *infra*, paras 9.25–9.33.
36. [1981] 2 Lloyd's Rep. 402, discussed *infra*, para. 11.121.

0.12 Finally, mention should be made of the Vienna Convention on the Law of Treaties[37] which is now in force. Article 31(1) of that Convention provides:

"A treaty shall be interpreted in good faith in accordance with the ordinary meaning to be given to the terms of the treaty and in their context and in the light of its object and purpose . . . ."

Although the Convention applies only to treaties concluded after its operation,[38] clearly the principles set out in Article 31(1) closely reflect the approach of the English judiciary as outlined above. Furthermore, in *Fothergill*,[39] Lord Diplock pointed out[40] that what this Convention says in Articles 31 and 32 about the interpretation of treaties does no more than codify existing public international law.[41]

0.13 In conclusion therefore, the purposive approach should be seen as a natural part of the process of coming to the correct interpretation of the provision contained in an International Convention, which is intended to provide uniform law.[42]

The French text

0.14 The text of the CMR Convention is in two versions of equal status, in English and in French. However, the Carriage of Goods by Road Act 1965, which brought the Convention into force in this country, provides that the Convention "*as set out in the Schedule to this Act* shall have the force of law in the United Kingdom",[43] and then only includes the English text in the Schedule. In cases where there is some evidence of a divergence of approach between the English and the French texts, the courts have therefore had to consider how far it is permissible to have regard to the French text in interpreting the Convention, most notably in *Buchanan* v. *Babco*. In the Court of Appeal,[44] Roskill and Lawton, L.JJ., both felt able to derive assistance from the French text, to which they regarded it as permissible to have regard in cases of ambiguity.

0.15 The difficulties inherent in the interpretation of international conventions are well illustrated by the fact that in the House of Lords[45] none of their Lordships thought that the French text was, on the facts of the case, of any assistance. However, they did consider

---

37. Cmnd. 4818. See generally Sinclair, *The Vienna Convention on the Law of Treaties*, 2nd edn. (1984).
38. But see Lowe and Williams, "A Shirt, A Pair of Sandals, A Cardigan", (1981) 44 M.L.R. 452.
39. *Supra*, fn. 30.
40. At p. 304.
41. On this point see Gardiner, *Treaty Interpretation in English Courts* [1994] LMCLQ 184, 185. See, further, Roth and Happ, "Interpretation of Uniform Law Instruments According to Principles of International Law", (1997) U.L.R. 700, who present arguments in favour of using principles of international law in respect of the interpretation of rules of uniform law; cf. De Meij, "Interpretatie van verdragen van uniform (vervoer) recht", (1998) 33 E.T.L. 607 (with summary in English at pp. 644–647).
42. Cf. Gardiner, *supra*, fn. 41, at p. 187. Lord Wilberforce, however, at p. 298 said: "I start by considering the purpose of Article 26 and I do not think that in doing so I am infringing any 'golden rule'. Consideration of the purpose of an enactment is always a legitimate part of the process of interpretation, and if it is usual, and indeed correct, to look first for a clear meaning of the words used it is certain in the present case, both on a first look at the relevant text and from the judgments in the courts below that no 'golden rule' meaning can be ascribed." See also Bingham, J., in *Swiss Bank Corp.* v. *Brink's-MAT Ltd.* [1986] Q.B. 853, 857 in respect of the Warsaw Convention, and Lord Diplock in *The Morviken* [1983] 1 Lloyd's Rep. 1, 5 in the context of the Hague-Visby Rules. In *Microfine Minerals & Chemicals Ltd.* v. *Transferry Shipping Co.* [1991] 2 Lloyd's Rep. 630, 632, Judge Kershaw felt the need, however, for ambiguity in the CMR provision before being constrained to look for a purposive construction to resolve it.
43. Section 1(1), italics added.
44. [1977] 1 Lloyd's Rep. 234.
45. [1978] 1 Lloyd's Rep. 119.

the extent to which it was legitimate to have regard to the French text. Lord Wilberforce[46] thought it legitimate to consider the French text even where there is no apparent ambiguity, and expressed himself unwilling to lay down rules as to how that reference should be made. Lord Dilhorne[47] expressed some doubt as to the propriety of doing so where only the English text is scheduled to the Act. Lord Salmon[48] thought it permissible where "there is some doubt as to its meaning in one language", and Lord Edmund-Davies,[49] with whose dissenting judgment Lord Fraser of Tullybelton agreed, expressed misgivings but in the event agreed with Lord Wilberforce, save that he was of the view that there must be an initial ambiguity.

0.16 The question of the proper approach to a French text again came before the House of Lords in *Fothergill* v. *Monarch Airlines Ltd.*,[50] but on that occasion the court was required, not just permitted, to have regard to the French text by the terms of the enacting statute. Again, however, the court did not lay down any rules as to how such reference is to be made, save to expressly reject any rule that evidence from qualified experts should be required. Again, however, their Lordships were unable to derive any positive guidance from the French text.

0.17 It can safely be assumed therefore, at least in cases of ambiguity,[51] that it is permissible to have regard to the French text of the Convention. As to the manner of that regard, there are no hard and fast rules: it would seem that the judge can take cognisance of his own linguistic knowledge,[52] that recourse can be had to dictionaries, academic writers and judicial decisions in other countries; the evidence of a language expert may

---

46. At p. 122.
47. At p. 126.
48. At p. 128.
49. At p. 132.
50. [1980] 2 Lloyd's Rep. 295.
51. In *Fothergill* v. *Monarch Airlines Ltd.*, Lord Roskill (at p. 315) referred to Lord Wilberforce in *Buchanan* as laying down that such reference was only permitted in cases of ambiguity. As indicated above, that is not what Lord Wilberforce in fact said. Hirst, J., in *Eastern Kayam Carpets Ltd.* v. *Eastern United Freight Ltd.* (*supra*, fn. 25) seems not to have required a preliminary test of ambiguity before considering argument based on the French text and reaching a conclusion contrary to counsel as to the support to be obtained from it. Article 33 of the Vienna Convention on the Law of Treaties, which deals with multi-lingual texts, has a different focus. Relevant parts of this provide as follows: (1) When a treaty has been authenticated in two or more languages, the text is equally authoritative in each language, unless the treaty provides, or the parties agree that, in the case of diversions, a particular text shall prevail; (2) The terms of the treaty are presumed to have the same meaning in each authentic text; . . . (4) Except where a particular text prevails in accordance with paragraph (1), when a comparison of the authentic text discloses a difference of meaning which the application of Articles 31 and 32 does not remove, the meaning which best reconciles the texts, having regard to the object and purpose of the treaty, shall be adopted. See further, Kuner, "The Interpretation of Multi-lingual Treaties: comparison of Texts versus the Presumption of Similar Meaning" (1991) 40 I.C.L.Q. 953. Similarly to the views of Lord Diplock (*supra*, para. 0.12), the Dutch Supreme Court has expressed the view that Article 33(4) of the Vienna Convention represents a codification of existing international law (29.6.90 (1990) 25 E.T.L. 589, 629), and has applied this to a perceived difference between the English and the French text in respect of Article 2 of CMR, see *infra*, para. 2.36, fn. 85 (see also the later decision of the court, 14.6.96, *infra*, para. 2.28 fn. 57). The view of Czapski, "Responsabilité du transporteur routier lors du transroulage et du ferroutage", (1990) 25 E.T.L. 172, 179, that preference should be given to the more expansive language seems incorrect (at least as an unqualified rule, although clearly the author sees it as an application of the Vienna Convention principles, see further, "Interpretation de la Convention CMR à la lumière du droit international public", (1998) 33 E.T.L. 461, 490–493). Likewise the view that gives a preference for the supposedly clearer text, or the French text as being the language used in creating the convention; see Bombeeck, Hamer, Verhaegen, "La responsabilité du transporteur routier dans le transport par car-ferries", (1990) 25 E.T.L. 110, who, at p. 129, consider also that the requirement of ambiguity, which appears from the majority of expressions of opinion in *Buchanan*, is not grounded sufficiently in international law.
52. But see the doubts of Lord Edmund-Davies (p. 131) and Viscount Dilhorne (at p. 126) in *Buchanan*.

assist, but is not obligatory. In *ICI PLC* v. *MAT Transport Ltd.*,[53] Staughton, J., who had not been referred to a French-English dictionary, nor assisted by experts and who admitted that his knowledge of French was partial, referred to Larousse in support of his view that the word *"reclamation"* includes not only a claim but also a complaint or protest.

Case law

0.18 English courts have also consistently been prepared to have regard to judicial decisions from other jurisdictions as guidance in their approach to the Convention when faced with difficulties of interpretation. However, whenever such guidance has been sought, the tendency has been for the results to be less than conclusive. Thus, for example, in *Buchanan* v. *Babco*,[54] the House of Lords was faced with two decisions, one favouring each side of the question at issue. Equally in the *Ulster-Swift*[55] case, the Court of Appeal was referred to the several approaches apparent from European decisions on Articles 17 and 18. In *Buchanan*, Lord Salmon said[56]: "If a corpus of law had grown up overseas which laid down the meaning of Article 23, our courts would no doubt follow it for the sake of uniformity which is the object of the Convention to establish. But no such uniformity exists." Most influential, therefore, is any such corpus, or "coherent body of authority"[57] or "impressive consensus".[58] However, despite the absence of any such corpus, Mustill, J., in *Silber*[59] was able to derive assistance from two decisions which he considered "useful in that they show how the present problem has appeared to experienced lawyers in other jurisdictions, and they put this court on its guard against too readily dismissing certain interpretations as untenable". In *Eastern Kayam Carpets Ltd.* v. *Eastern United Freight Ltd.*[60] Hirst, J., felt able to accept and adopt a Dutch decision, although ultimately distinguished by him in the case before him.

0.19 Attempts to rely on foreign case law thus also demonstrate well the difficulties involved in achieving a uniform approach to the interpretation of the Convention. Nor should this be regarded as surprising: judicial divergence of opinion is to be expected internationally as it is nationally. But having regard to each others' decisions over a period of time does at least represent a pragmatic approach by the judiciary to an inevitable problem.[61,]

0.20 Such reference should not be made without caution, however.[62] As was pointed out by Lord Wilberforce in *Fothergill* v. *Monarch Airlines*,[63] the decisions may not be of the highest courts; the process of law reporting varies in different countries so that those decisions available might not represent a balanced picture; and, as he put it:

---

53. [1987] 1 Lloyd's Rep. 354, 361. See *infra*, para. 10.99.
54. *Infra*, paras 9.34–9.35.
55. *Infra*, para. 6.114.
56. At p. 128.
57. Per Mustill, J., in *J.J. Silber Ltd.* v. *Islander Trucking Ltd.* [1985] 2 Lloyd's Rep. 243, 246.
58. Per Staughton, J., in *ICI PLC* v. *MAT Transport Ltd.* [1987] 1 Lloyd's Rep. 354, 359.
59. *Supra*, at p. 246.
60. Queen's Bench Division, 6.12.83, unreported except on LEXIS.
61. Cf., however, Czapski, "Interpretation de la Convention CMR a la lumière du droit international public", (1998) 33 E.T.L. 461, 494. The resolution of divergence through the creation of a supranational court (see Loewe, "Le CMR à 40 ans", (1996) U.L.R. 429 (see English summary, p. 441) is, perhaps, too much to hope for.
62. *Tetroc Ltd.* v. *Cross-Con (International) Ltd.* [1981] 1 Lloyd's Rep. 192, 195.
63. [1980] 2 Lloyd's Rep. 299, 300–301.

" . . . it was not beyond argument when the facts of each case were carefully examined on which side the preponderance in quantity, or quality, lay."

Other sources

0.21 In *Fothergill* v. *Monarch Airlines*[64] the House of Lords also considered the admissibility of other materials in the interpretation of international conventions in cases of ambiguity. Specifically, the court was concerned with the question whether or not the *travaux préparatoires* of a convention can be taken into account, and indicated that because of the relative availability in published form of the relevant material, cautious reference might be permitted. However, there are no *travaux préparatoires* available in relation to the CMR Convention—at least, none in the sense indicated in *Fothergill*, where the relevant minutes had been published by I.C.A.O. and made available for sale by H.M.S.O., and this was clearly central to their Lordships' view in that case. Lord Wilberforce[65] noted two conditions before *travaux préparatoires* can be taken into account. First, the material involved must be public and accessible, and secondly, it must clearly and indisputably point to a definite legislative intention.[66] The English headnote of one German case in *European Transport Law* seems to indicate that reference may be made to preliminary studies.[67] None, however, are specified.

0.22 In *Fothergill*, the court also considered the question of how far the views of foreign jurists could be taken into account. On the question with which they were there concerned, there was a considerable consensus of foreign academic opinion in favour of one particular interpretation, and, having decided that there was ambiguity in the Convention, the court did regard it as legitimate to take that into account.[68] In relation to the CMR Convention, English courts have several times mentioned Loewe's "Commentary on the Convention"[69] as being helpful in general terms,[70] but on no occasion has it been specifically referred to as being of assistance in reaching any particular conclusion. The German case referred to in the preceding paragraph[71] does refer to his Commentary as confirming the court in its view, but there are no observations in the English cases to like

---

64. *Supra.* See too *Sidhu* v. *British Airways* [1997] 2 Lloyd's Rep. 76, 80–81.

65. At p. 302.

66. Cf. the criteria adopted in *Pepper (Inspector of Taxes)* v. *Hart* [1993] 1 All E.R. 42, in respect of the more liberal approach now adopted by the English courts in respect of domestic legislation. Article 32 of the Vienna Convention confirms the use of such material as a *"supplementary"* means of interpretation (see Gardiner, *supra*, fn. 41, at p. 188) to which recourse may be had in order to confirm the meaning resulting from the application of Article 31, or to determine the meaning when the interpretation according to Article 31 leaves the meaning ambiguous or obscure, or leads to a result which is manifestly absurd or unreasonable. The Dutch Supreme Court has expressed the view (29.6.90 (1990) 25 E.T.L. 589, 627) that it is not possible to have regard to the drafting history of the Convention since no *travaux préparatoires* have been published or made available for public consultation.

67. BGH., 6.7.79 (1980) 15 E.T.L. 863.

68. Lord Diplock's remarks (at p. 306) would indicate his lack of enthusiasm in this respect: "The persuasive effect of learned commentaries . . . will depend on the cogency of their reasoning. Those to which your Lordships have been referred contain perhaps rather more assertion than ratiocination, but for the most part support the construction favoured by your Lordships."

69. (1976) 11 E.T.L. 311.

70. E.g. *Thermo Engineers Ltd.* v. *Ferrymasters Ltd.* [1981] 1 Lloyd's Rep. 200, 203; *Buchanan* v. *Babco* [1977] 1 Lloyd's Rep. 234, 238; *Michael Galley Footwear Ltd.* v. *Dominic Iaboni* [1982] 2 All E.R. 200, 205. In *J.J. Silber Ltd.* v. *Islander Trucking Ltd.* [1985] 2 Lloyd's Rep. 243, 245, Mustill, J., made reference to this and other writers on the Convention.

71. Fn. 67.

effect. His views on the burden of proof in Article 18 were specifically disapproved in the *Ulster-Swift* case.[72]

0.23 The CMR Convention shares some similarity to the CIM Convention, which was of much earlier origin,[73] and provided a model for the creators of the CMR Convention.[74] It may be legitimate therefore for the courts to make comparison with the provisions of this Convention as is the practice of the courts on the Continent.[75] Comparison with other carriage conventions and the approach towards them taken by the courts may sometimes be brought in aid by the courts at least in respect of more general matters of method and principle.[76] However, a more cautious approach is needed when making comparison with specific provisions. So, for example, in *Swiss Bank Corp.* v. *Brink's-MAT Ltd.*[77] Bingham, J., in dealing with a provision of the Warsaw Convention, and having been referred to the practice of the English courts in respect of the Hague Rules, said[78]:

> "It does not, however, appear that this practice is uniformly followed, and M. Drion points out that in, for example, the Netherlands there is a different practice. It does not seem to me that the English maritime practice is something which should be of very strong persuasive authority in the construction of this convention in an entirely different field."

National law

0.24 As will become apparent, there are a number of areas in relation to which the Convention does not provide a ready answer to particular circumstances. Loewe was of the view that the Convention does not represent a comprehensive code on the international carriage of goods by road[79]—it merely regulates certain aspects of the contractual relations between the various parties involved.

0.25 In some areas, the Convention itself gives guidance as to the law to be applied.[80] However, inevitably there are many matters on which the Convention offers no guidance, so the question arises as to the correct approach in such areas. The better view was thought to be that the answer was to be sought in the national law. In *Eastern Kayam Carpets Ltd.* v. *Eastern United Freight Ltd.*[81] Hirst, J., applied the common law to found the liability

---

72. [1977] 1 Lloyd's Rep. 346, 352.

73. The original version was 1890, and the version in force at the time the CMR Convention was drafted was the fourth revision of 1933. The current version is the CIM Rules which forms Appendix A of the COTIF Convention 1980, which is given effect in this country by the International Transport Conventions Act 1983. See Clarke, Yates, part 4.

74. E.g. Haak notes at p. 9 the decision taken to adopt Article 27 of CIM as the basis of the liability regime of the CMR Convention. CIM is also said to have been the model for many of the provisions of the Warsaw Convention. See Gardiner, "Treaty Interpretation in English Courts" [1994] LMCLQ 184, 187.

75. See, e.g., BGH., 11.12.81 (1983) 18 E.T.L. 63, 69, and further examples given by Clarke, p. 13 fn. 69.

76. As seen *supra*. A further example is *Eastern Kayam Carpets Ltd.* v. *Eastern United Freight Ltd.*, Queen's Bench Division, 6.12.83, unreported except on LEXIS, where Hirst, J., drew support from the case of *Stag Line Ltd.* v. *Foscolo Mango & Co.* [1932] A.C. 328 in respect of the Carriage of Goods by Sea Act 1924, in deciding that local law can apply where the CMR Convention does not. See *infra*, paras 7.32–7.35. See too *Gefco (U.K.) Ltd.* v. *John Mason*, Q.B.D., 22.1.00, where the Court had regard to a decision of the Court of Appeal concerning the Warsaw Convention, *Sidhu* v. *British Airways PLC* [1997] 2 Lloyd's Rep. 76, in considering the application of the CMR Convention.

77. [1986] Q.B. 853.

78. At p. 860

79. Loewe, para. 9.

80. E.g. the law of the place where the goods are situated in Article 20(4), and the law of the court or tribunal seised of the case in Article 32(1) and (3).

81. Queen's Bench Division, 6.12.83, unreported except on LEXIS.

of the carrier, having found that the issue was not regulated by CMR.[82] He also pointed out that unless local law always underlay the Convention, it would have been unnecessary to limit tort liability under Article 28 and also to exclude any recourse other than those specifically stated in Article 23(4).[83]

0.26 So too in *Shell Chemicals U.K. Ltd.* v. *P. & O. Roadtanks Ltd.*[84] where the CMR carrier caused damage to the plaintiff by delivering the wrong goods: although the contract was regulated by the CMR Convention, no provision was made for the particular liability alleged. However, this did not mean that the defendants were not liable. Since CMR did not operate so as to exclude or limit the liability of the defendants arising from their breach of contract, in such circumstances, the defendants were accordingly held to be fully liable.[85] Saville, J., observed as follows[86]:

"The conclusion that I have reached does not entail that there is any 'gap' in the Convention. That proposition could only be maintained if it is first assumed that those who framed the Convention intended to set up a code which exhaustively covered all the rights and obligations that could arise out of the carriage of goods by road or contracts for such carrier."

0.27 However, a new direction may be signalled by the first instance decision in the case of *Gefco (U.K.) Ltd.* v. *John Mason.*[87] This case concerned the claimants' claim for freight charges under the terms of an "umbrella contract" for the carriage of the goods of a particular retailer. As a result of alleged breaches of the umbrella contract, the defendant claimed to have lost their own contract for the carriage of the retailers' goods, and they accordingly counterclaimed for damages. It was held by the Court of Appeal as a preliminary issue, that the Convention applied to the "umbrella contract". Subsequently, the matter reverted to the High Court for the determination of two further preliminary issues, namely (a) whether the defendants' counterclaim, a claim for pure economic loss consequent on the claimants' alleged breaches of contract, was unrecoverable under the Convention since it fell outside the scope of Article 17 and Article 23, and (b) whether, as a delay claim, it was time-barred under Article 30. It is the former issue which is relevant in the present context.

0.28 The claimant argued that the defendants' claim was based on liability for delay, which is governed by Article 17 of the Convention. However, such a claim is limited to the amount of the freight charges by Article 23(5), and further, no claim could be brought in any event because of the absence of a reservation under Article 30(3). The defendant argued that these provisions had no application to the particular circumstances, on the

82. He cited the decision of LG. Bremen, 6.5.65 (1966) 1 E.T.L. 691 where local law was applied in respect of the failure of the carrier to collect the goods. He also quoted Clarke (first edn. p. 5, now 3rd edn. pp. 16–17): "If a gap is found at the level of the CMR, that gap will be treated as a hollow rather than a void: below the CMR there remains the substratum of existing common law, to which the courts may turn without usurping the function of the legislature and without the necessity to guess what the draftsmen of the CMR would have said, if the question had been put to them."

83. To similar effect, see e.g. Arrond. Breda, 16.12.69 (1970) 5 E.T.L. 67; OGH., 13.6.86 (1988) 23 E.T.L. 198; BGH., 24.10.91 (1992) 27 E.T.L. 839; Cass., 17.3.92 (1992) B.T. 253. The only doubt in this respect stems from the judgment of Lord Denning, M.R., in the *Buchanan* case (*supra*) where, as has been seen, having detected a gap in the Convention he set about filling it by reference to his view of the intention of the Convention. As has been explained already (*supra*, para. 0.7), this approach was rejected by the House of Lords on the facts of the case, and considerable doubt was also cast on the legitimacy of the breadth of Lord Denning's approach.

84. [1993] 1 Lloyd's Rep. 114. See further *infra*, paras 10.53–10.56.

85. The decision of the court in LG. Bremen, 6.5.65 (1966) 1 E.T.L. 691 is to similar effect.

86. At p. 116.

87. Q.B.D., (Birmingham District Registry), 21.2.2000, unreported.

basis that the duties imposed under Article 17 are owed to the cargo interests or to the sender, not to another carrier, the relationship between carriers being governed by Chapter VI. They also relied on repudiatory breach on the part of the claimants.

0.29 The claimants relied strongly on the decision of the House of Lords in *Sidhu* v. *British Airways*,[88] where it was held that the Warsaw Convention represented the exclusive cause of action for passengers claiming for loss, injury and damage sustained in respect of an international carriage by air. That case concerned a claim by passengers on a British Airways flight which landed at Kuwait after Iraq had invaded in the course of the Gulf War. The passengers were detained as a result. Claims for personal injury were brought at common law, because there was no remedy under the Convention, whereby it is only possible to claim in respect of personal injury if sustained on board the aircraft or in the course of embarking or disembarking. It was held by the House of Lords that there was no remedy under the national law. As Lord Hope put it[89]:

"The Convention does not purport to deal with all matters relating to contracts of international carriage by air. But in those areas with which it deals—and the liability of the carrier is one of them—the code is intended to be uniform and to be exclusive also of any resort to the rules of domestic law."

0.30 In the *Gefco* case, it was held that, as with the Warsaw Convention, under the CMR Convention the carriers surrender their freedom to limit or exclude liability in exchange for the limits contained in the Convention itself. It was accepted that both the Warsaw and the CMR Conventions were intended to function in the same way, and the judge said that, on its wording, the CMR Convention was intended to constitute a comprehensive code regulating the liability of a carrier. However, he went on to hold that the wording of Article 23(5) in particular and Chapter IV of the Convention in general was wide enough to cover the counterclaim for persistent delay. Accordingly, the claim was limited to the freight charges in relation to each individual movement and was time-barred in any event, by reason of the failure to lodge written reservations as required in relation to claims for delay under the Convention.

0.31 On the renunciatory breach point, although the judge refused the defendants leave to amend their pleadings, he went on at the request of the parties to consider the position as if renunciatory breach by the claimant had been established. The defendants relied heavily in this context on the decision in the *Shell Chemicals* case referred to above, in particular the concession made by the parties and approved by the judge in that case that the Convention would not apply where there was a complete failure to perform. They argued that, by the same reasoning, damages for renunciatory breach fall outside the scope of the Convention. The claimants argued that in the light of the *Sidhu* case, the concession was wrongly made and the *Shell* case was no longer good law.

0.32 In the event, Judge Gibbs avoided deciding whether or not the *Shell* case was correctly decided. He held that since the relief sought by the defendants fell within the scope of the Convention, the fact that it was framed in common law terms did not of itself take the claim outside its scope. The claim remained within Article 17(1), so the Convention applied in any event. As he observed:

88. [1997] 2 Lloyd's Rep. 76.
89. At p. 87.

"Under those circumstances a finding that it was exempt from the provisions of the CMR scheme regulating the liability of carriers would be to undermine the Convention by allowing the domestic court to provide a remedy under its own law."

0.33 So far as the *Shell* case, was concerned, Judge Gibbs regarded it as based on the view of the judge, that there was a limit to the rights and obligations intended to be covered by the Convention, and that the circumstances in that case fell outside that limit. He did not regard it as inconsistent with *Sidhu*, because in that case the claim fell within the broad category of damage provided for by the Convention, i.e. personal injury, but in circumstances in which no remedy was provided. He regarded the case before him as broadly parallel with *Sidhu*, whereas in the *Shell* case, the facts were distinguishable because they fell outside the range of circumstances to which the limits of liability applied.

0.34 However, clearly there is considerable room for a difference of view as to which side of this line a particular case may fall. As will be apparent from the quotation from the judgment of Lord Hope set out above, in describing the areas with which the Convention deals, he spoke broadly in terms of "the liability of the carrier". On the basis of that breadth of analysis, it is not easy to see circumstances involving the liability of the carrier where there is room for the operation of national law. It accordingly remains to be seen how this issue will be developed by the courts. Although Judge Gibbs in the *Gefco* case was able to avoid a finding that the *Shell* case was wrongly decided, it should probably be viewed with some caution. Equally, it was held in the case of *Eastern Kayam Carpets Ltd. v. Eastern United Freight Ltd.* that CAD instructions to the carrier fell outside Article 21 so that national law applied.[90] That case too may have to be regarded with some caution.

0.35 To the extent that it might seem unjust that the defendants in the *Gefco* case were deprived of a remedy, Judge Gibbs observed as follows:

" . . . that injustice has to be weighed against the benefits of a predictable commercial regime, enabling all parties to CMR contracts to regulate and plan their affairs sensibly and prudently. Such a regime facilitates a wider form of justice, against which apparent unfairness in individual situations should be weighed."

0.36 There have been instances in which the national courts have been over-willing to hold the point at issue beyond the scope of the Convention and thus subject to national law. Thus in *Moto Vespa S.A. v. MAT (Brittania Express) Ltd.*,[91] it is argued below[92] that Mocatta, J., was rather too ready to fall back on the national law. As Hirst, J., pointed out in the *Eastern Kayam Carpets* case:

" . . . this court should not lightly hold that a given set of facts within the scope of the Convention is not covered by any Article."[93]

90. See *infra*, paras 7.32–7.35.
91. [1979] 1 Lloyd's Rep. 175.
92. See *infra*, para. 10.77.
93. In Germany, in the decision of BGH., 6.7.79 (1980) 15 E.T.L. 863, it was said that the CMR, being an international agreement, must be interpreted primarily according to the principles expressed in it, or, if need be by reference to preliminary studies for it. See *supra*, para. 0.21, fn. 67.

0.37 Ultimately, it is a matter of the correct interpretation of the CMR Convention as to whether a matter is within the scope of its autonomous operation[94] or is to be referred to local law in accordance with the principles of conflicts of law.[95] This is true not only in respect of the scope of application of the provisions of CMR but also as to whether, in determining the meaning of the words and concepts used by CMR, it is permissible to refer to local law for assistance.[96] The first duty of the court is to strive for a uniform meaning.[97]

94. Haak, p. 10, Pesce, Theunis, p. 2. See Roth and Happ, "Interpretation of Uniform Law Instruments According to Principles of International Law", (1997) U.L.R. 700, 702, for a distinction between autonomous interpretation, which is the correct approach to adopt in respect of rules of uniform law, and autonomous methods of interpretation, which should give way to the application of principles of international law, see *supra*, para. 0.12 fn. 41.

95. In *A.B. Bofors UVA, C.A.V. Ltd.* v. *A.B. Skandia Transport* [1982] 1 Lloyd's Rep. 410, Bingham, J., applied Swedish law, as the proper law, to the construction of the arbitration clause in the contract. The further question as to whether this clause satisfied the relevant provisions of CMR (*infra*, para. 10.123) was not a matter for the proper law but for "English law as the appropriate governing law for interpreting the Schedule to an English Act". *Ibid.*, at p. 412. Within its sphere, CMR applies without resort to the rules of the conflict of laws, see Clarke, *International Encyclopedia of Comparative Law, Volume III, Private International Law, Chapter 25 Transport by Rail and by Road*, 1996, para. 24.

96. Cf. the view of Pesce, "Uniform Law on International Carriage of Goods: A System of Special Rules", (1997) 32 E.T.L. 503, which sees little scope for the application of national law even in respect of perceived "gaps."

97. In OLG. Düsseldorf, 27.3.80 (1983) 18 E.T.L. 89, application of the principles of conflicts of law was deemed inappropriate in interpreting the meaning of "unavoidable circumstances" in Article 17(2) (as to which, see *infra*, para. 6.28 ff.). To similar effect: Cass., 27.1.81 (1981) B.T. 219. In OGH., 27.4.87 (1988) 23 E.T.L. 193, it was held that the presence of indications based on international practice according to which the particular provision (Article 32(2)) is to be interpreted meant that a different solution could not be adopted by reference to national law. See also OLG. Munich, 23.7.96 (1997) TranspR 33 (1998) U.L.R. 206, *infra*, para. 1.53 fn. 144. It is possible that in respect of some of the terms used in the Convention, the intention of the draftsmen was not to anticipate a uniform meaning derived from its context but to permit application of the local law either to give content to the word or to provide illustrations of possibilities within a sphere of meaning. Thus it may be possible to interpret a word such as "cash" in Article 21 as meaning that which is considered as cash according to the law of the court seised of the case, or by the proper law, since there is little in the context of CMR to enable a uniform meaning to be expounded (cf. *Eastern Kayam Carpets Ltd.* v. *Eastern United Freight Ltd.*, Queen's Bench Division, 6.12.83, unreported except on LEXIS, cf. App. Paris, 31.1.83 (1983) B.T. 183, BGH., 10.2.82 (1983) 18 E.T.L. 32). Even if a uniform meaning of a word or phrase can be derived, it may depend on the local law to determine how it is to be applied to the facts. See, e.g., the phrase "action is pending" in Article 31(2), discussed *infra*, para. 10.38 ff., and cf. *Dresser U.K. Ltd.* v. *Falcongate Freight Management Ltd.* [1992] 2 All E.R. 450.

CHAPTER 1

Scope of Application of the Convention

1.1 The scope of the Convention's application is governed by Chapter I, comprising Articles 1 and 2. Article 2, however, is concerned with carriage involving combined transport, namely where for part of the journey the vehicle is carried by sea, rail, inland waterway or air, and is considered in detail in the next chapter. The present chapter is therefore concerned with Article 1, which lays down the general prerequisites for the application of the Convention. The principal provision is Article 1(1), which provides as follows:—

"This Convention shall apply to every contract for the carriage of goods by road in vehicles for reward, when the place of taking over the goods and the place designated for delivery, as specified in the contract, are situated in two different countries, of which at least one is a Contracting country,[1] irrespective of the place of residence and the nationality of the parties."

The contract of carriage of goods

1.2 It will be noted that Article 1(1) applies the Convention to "every *contract* for the carriage of goods". Likewise, the English title of the Convention is "Convention on the *Contract* for the International Carriage of Goods by Road". The significance of this somewhat awkward phraseology is crucial to an understanding of the scope of the Convention, since it makes it plain that what is covered is not the international carriage of goods by road as such, but only a contract for such carriage. Thus, by way of example, if goods are carried by road from London to Dover under one contract, shipped across the Channel under another contract, then carried by road from Calais to Paris under a third contract, this is in one sense an international carriage of goods by road. It would not,

---

1. By Order in Council, the Contracting states are certified to be: Austria*, Belgium*, Bulgaria, Czech Republic, Denmark*, Finland*, France*, Germany*, Greece*, Hungary*, Italy*, Luxembourg*, The Netherlands*, Norway*, Poland, Portugal*, Romania*, Spain*, Sweden*, Switzerland* and Yugoslavia: see S.I. 1980 No. 697. Until superseded, this Order in Council is, by s.2(2) of the Carriage of Goods by Road Act, stated to be conclusive evidence of the matters so certified. Since that time, however, the following states have become contracting parties: Belarus, Bosnia-Herzegovina, Croatia, Estonia*, Georgia*, Iran*, Ireland*, Latvia*, Lithuania*, Macedonia*, Moldova, Morocco, Russian Federation, Slovak Republic, Slovenia, Tajikistan, Tunisia*, Turkmenistan*, Ukraine and Uzbekistan*. (The countries marked with an asterisk have also acceded to the SDR Protocol. See *infra*, para. 9.9.) In *Chloride Industrial Batteries Ltd.* v. *F. & W. Freight Ltd.* [1989] 3 All E.R. 86, the Court of Appeal held that the word "country" refers to a Sovereign State which is competent to enter into an international convention on its own behalf. Since Jersey is not so competent, road transport between England and Jersey was held not to be subject to CMR. Section 9 of the Act provides for the power to extend CMR to (a) the Isle of Man; (b) any of the Channel Islands; (c) any Colony and (d) any state or territory which is for the time being a protectorate or protected state for the purposes of the British Nationality Act 1948. This power has been exercised in respect of Gibraltar, Guernsey and the Isle of Man. It has not been exercised in respect of Jersey, and so CMR is not part of the law of Jersey.

however, be governed by the Convention, since it is not within the wording of Article 1(1): it is not a "contract for carriage of goods by road . . . when the place of taking over the goods and the place designated for delivery are situated in two different countries".[2] In order to comply with this requirement it would be necessary for there to be one contract governing the entire carriage,[3] or at least one contract which involved carriage by road between two countries. So, to vary the facts of the above example, if the first contract was for the carriage of the goods from London to a depot on the outskirts of Calais, the segments either side of the Channel being carried out by road, and the second carriage was for onward by road to Paris, the first contract would be subject to the Convention,[4] but the second would not since it has no international content.[5]

1.3 On the continent, a distinction has been drawn between a contract of carriage and a contract of hire which, falling outside Article 1, is not governed by the Convention.[6]

2. Though it may be open to the court to construe it as a single contract of carriage if, e.g., the same parties are involved at each stage. See Comm. Verviers, 7.4.79 (1979) 14 E.T.L. 664 and Comm. Anvers, 3.4.77 (1977) 12 E.T.L. 411. Putzeys (para. 236) cites the latter case in the context of suggesting that the issue of two consignment notes in such circumstances may be considered a fraud on the law which will be ignored by the courts (cf. BGH. 17.4.97 (1998) 33 E.T.L. 427, (1998) TranspR 21, (1998) U.L.R. 206). Cf. App. Paris, 27.1.86 (1986) B.T. 742 where the goods, having travelled from Antwerp, arrived at Roubaix in France which was the place indicated in the CMR consignment note as the destination. They were unloaded from the vehicle, stored and cleared through Customs and then despatched to Attiches in France where they were finally delivered to the consignee. The court held that this further transport constituted a fresh contract of carriage governed by French law and not CMR even though the same operator was involved (cf., however, Cass., 6.6.95 (1995) B.T. 474). In the context of a groupage operation, the Court of Appeal of Versailles, 15.1.86 (1986) B.T. 742 (criticised by Chao, (1986) B.T. 733, 734) held that the theft of goods from the road vehicle which was in the course of collecting other goods destined for the same destination was not subject to CMR. At that point the goods had not begun their international journey since the driver was making his way to the operator's depot (cf. BGH., 27.1.82 (1985) 20 E.T.L. 349) cf. however, Comm. Bobigny, 20.1.2000 (2000) B.T. 192.

3. This was not the case in Cass., 25.3.97 (1997) B.T. 276, (1997) U.L.R. 128. A first carrier carried the goods solely in France and the sender arranged for a second carrier to transport the goods to Spain, the goods being discharged and exchanged at the frontier. The first stage was not subject to CMR, it not being the common intention of the parties that the whole transport should be seen as part of a single contract for international carriage. Naturally, the consignment note provides evidence of the extent of the agreed carriage. Where an operator signed a CMR consignment note for a journey from France to Tunisia, he found himself responsible under CMR, notwithstanding that he also issued an internal consignment note for the domestic section performed by him and that another operator was also named as carrier on the CMR consignment note, cf. Cass., 21.11.95 (1995) B.T. 831. See, however, BGH., 17.4.97 (1998) 33 E.T.L. 427, (1998) TranspR 21, (1998) U.L.R. 206, where the consignment note was not signed by the sender.

4. Subject to Article 2, *infra*, Chapter 2. In a case where a reference to Tokyo appeared on the consignment note, the carriage could not be regarded as international since the goods were in containers to be transhipped from road vehicle to ship: App. Bordeaux, 21.12.94 (1995) B.T. 74, (1996) U.L.R. 400.

5. See, e.g., RB. Dordrecht, 7.2.79 (1980) 8 E.L.D. 145: onward carriage within one country at request of consignee after international carriage had been completed; CMR did not apply. (CMR applied, however, to an onward movement under the instructions of the sender after the original consignee refused the goods, OLG. Munich, 12.4.91 (1991) TranspR 298, (1991) R.I.W. 676). Cf. Cass., 22.9.83 (1984) 19 E.T.L. 111, where after an international carriage performed by a sub-contractor, the national on-carriage by the contracting carrier was regarded as a separate contract. This was in respect of the relationship between the sub-contractor and the contracting carrier (cf. *infra*, Chapter 11, in respect of successive carriage). It did not affect the result as there was proof that the damage occurred during the international stage. The important consideration at this point is to determine the extent of the transport agreed between the carrier and the sender, and the fact that different stages are performed by various sub-contractors will not, in itself, determine whether the total transport has been split into different contracts of carriage. Nevertheless the fact that the subcontractor issues a fresh consignment note may combine with other facts to confirm that a fresh carriage is being undertaken, as in a case where the sender ordered the carrier to temporarily store the goods at a depot in the country of destination before ordering their despatch to a new place of delivery. App. Versailles, 3.2.97 (1997) B.T. 491, cf., App. Versailles, 23.11.95 (1996) B.T. 257.

6. See generally Haak, p. 78; Clarke, paras. 10(c) and 10(d). See e.g., Cass., 17.12.96 (1997) B.T. 19, (1997) U.L.R. 630. See also OLG. Innsbruck, 20.6.95 (1997) TranspR 343. A further question which has arisen is

This distinction does not relate to the simple lending of a vehicle by an owner to another,[7] but rather to where a driver is also supplied with the vehicle. The distinction is not made by the Convention itself,[8] but it is always important to consider whether a contract in a particular case can truly be defined as being one of carriage.[9] While CMR emphasises an agreement for carriage between certain points this is not, in this context, the central factor. Rather the fact that international carriage has taken place which has been agreed to by the lender creates the possibility of responsibility under CMR.[10] The crucial issue is whether there is an agreement to carry.[11]

1.4 In France there has been considerable development of this issue, where the criteria for the distinction which are employed for the purposes of domestic carriage[12] are also considered to be relevant[13] for the purposes of CMR.[14] There is a presumption in favour of carriage in cases of doubt, since a contract of hire is considered to be the exception.[15] The courts look to the nature and use of the documentation[16] and the remuneration,[17]

---

whether the parties are sufficiently independent for there to be truly a contract of carriage, as in LG. Münster, 18.6.98 (1998) TranspR 471, where the court was faced with an "owner-driver" contract considered as too close to an employment contract to be governed by CMR.

7. Merely providing e.g. a trailer does not, in itself, make the owner a carrier: Hof. Antwerpen, 15.3.89 (1989) 24 E.T.L. 574, 588. Such supplier may, however, be a transport operator who has engaged in a contract to carry and as such has become a CMR carrier. See the discussion *infra*, para. 11.49 ff.

8. LG. Bremen, 6.5.65 (1966) 1 E.T.L. 691, 696; see, however, Haak, p. 82 n. 225.

9. See also *infra*, para. 1.13 ff. in respect of forwarding operations. Cf. companies operating under the UIRR Conditions of the International Road-Rail Federation who agree to load road vehicles received at a terminal onto railway wagons and receive them at the destination terminal for re-delivery to the haulier. Their contracts are better viewed as contracts for handling and carriage of the vehicle and goods by rail rather than contracts for carriage by road involving combined transport within Article 2 (*infra*, Chapter 2). See further, Yates, para. 6.2.2.4. A further issue is whether the element of carriage forms only an incidental part of a wider contract; see Libouton, (1982) J.T. para. 18, who cites a Belgian case (Bruxelles, 9.11.77, R.G.A.R., 1979, no. 10079) where the transport by the seller of goods to his buyer was merely incidental to the sale and not a distinct contract of carriage. Cf. the distinction between a contract of sale and a contract of services (Atiyah, *Sale of Goods*, 9th edn., p. 19).

10. Where a vehicle is lent for a period of time, the fact that the lender agrees to be subject to the orders of the customer means that where the vehicle is used for an international operation the owner has thereby agreed to this use and has, in consequence, agreed to international carriage by road (so long as he is, in fact, otherwise properly regarded as a carrier). Similarly, where a carrier makes a standing offer to receive instructions for the carriage. These are just as much agreements to carry between the points instructed by the customer as a definite agreement to carry between those points. The position is analogous to certain types of combined transport, see *infra*, para. 2.7 fn. 10.

11. The more usual factual context is in respect of sub-contracting and successive carriage, see *infra*, Chapter 11.

12. In France, there is both regulation of the hire of heavy vehicles (Lamy, para. 785 et seq.) as well as a "*contrat type*" decreed to operate in the absence of contrary agreement (Lamy, para. 563). See also Chao & Fos Colette, (1986) B.T. 340. An operator held to be a CMR carrier may well have a recourse action against the lender, see e.g. App. Paris, 5.4.95 (1995) B.T. 811.

13. They were also relevant to determine whether the carrier was in breach of a clause controlling sub-contracting in a case concerned with the application of CMR: Comm. Paris, 12.5.98 (1998) B.T. 688.

14. A different approach, adopted in Cass., 22.10.96 (1996) B.T. 758, cuts across the distinction. The court saw the driver of the vehicle as being under the control of the hirer (Le Patron provisoire du conducteur) so that the lender was not liable for the driver's failure to complete the hirer's (impossible-to-achieve) instruction to deliver by a certain time. Cf. Cass., 17.6.97 (1997) B.T. 596, where the distinction was sidestepped by a finding that the accident was due to the brake of the semi-trailer supplied by the hirer so that he, rather than the tractor operator, took on the entire CMR responsibility to the claimant.

15. App. Paris, 21.3.84 (1984) B.T. 336, Cass., 5.2.85 (1986) B.T. 350, Lamy, para. 568; App. Paris, 16.2.96 (1996) B.T. 463, this appears to be a matter of evidence rather than a presumption of law; see Tilche, (1992) B.T. 798.

16. I.e. whether a transport or a hire document has been issued by the alleged lender.

17. Remuneration by distance or on a daily rate points to hire, whereas by journey (see, however, App. Paris, 3.10.96 (1996) B.T. 799) or by weight points to transport; see Chao & Fos Colette, op. cit., p. 341. See also App.

among other factors.[18] A further factor is whether the lender has insured his liability for the goods.[19] Naturally in the context of CMR, the issue and contents of the consignment note will be important, although not conclusive. Thus the fact that the lender is mentioned as carrier in the consignment note does not prohibit him from showing that he is not a carrier.[20] Nevertheless the fact that another party is named as carrier will help to support a claim that the owner is only a hirer.[21] In essence, the courts look to see who is truly in control of the transport operations to determine whether, as between two parties, one of them is lender or carrier.[22]

1.5 In England, the issue is likely to be determined by a consideration of whether the lender of a vehicle with a driver has become a bailee of the goods. As long as there is no transfer of the lender's servant so as to affect the question of vicarious liability, the control that the servant has over the vehicle and, in consequence, the goods in it or drawn by it will normally be sufficient to found responsibility as possessor of the goods in the lender.[23] It is possible that the degree of control that the hirer has over the movement of the goods is such that, although there is no transfer of the servant, the hirer rather than the lender is in possession of the vehicle and, in consequence, in possession of the goods.[24] In general, however, a temporary transfer of the servant will need to be shown. The issue therefore is not likely to be viewed in identical terms in England as on the continent, given the different juridical structure that underpins the law of carriage.[25] The same analysis may also apply to the even more rarefied question of whether a lender of a tractor with driver to pull the trailer of the customer is hire or carriage. On the Continent, it seems possible to say that the factors discussed above will apply equally. So for example in Belgium, a

Paris, 30.9.83 (1984) B.T. 335, App. Paris, 21.3.84 (1984) B.T. 336, Comm. Bordeaux, 13.3.91 (1992) B.T. 119, App. Paris, 1.12.92 (1993) B.T. 115, cf. App. Paris, 20.5.92 (1992) B.T. 805.

18. Tilche, op. cit., p. 799 notes the movement towards appreciating the degree of dependence, with the courts taking into account several matters in addition: whether the vehicle is regularly garaged with the hirer; whether it is in the hirer's colours (cf., however, App. Paris, 22.2.80 (1980) B.T. 239); whether a servant of the hirer travels on the vehicle; whether the lender is registered as such; whether the contract refers to the vehicle being put at the *disposition of* the hirer (although in itself this may be considered as ambiguous: App. Paris, 21.3.84 (1984) B.T. 336). See further App. Paris, 17.1.96 (1996) B.T. 134.

19. Comm. Bordeaux, 13.3.91 (1992) B.T. 119. The fact that he is insured for hiring out the vehicle with a driver may not contradict his acting as a carrier, the CMR consignment note having come into his hands: App. Paris, 29.9.95 (1995) B.T. 836.

20. Cass., 1.12.92 (1992) B.T. 790, (1993) 28 F.T.L. 745, in light of Article 9 of CMR, see *infra*, para. 4.24 ff. See also in Austria, OGH., 8.9.83 (1985) 20 E.T.L. 282. Cf. Cass., 5.1.88 (1988) II U.L.R. 732 where, after a theft of goods from the hired vehicle, the owners of the vehicle issued a CMR consignment note in respect of the forwarding of goods which had been recovered. This meant that they were also considered as carrier for the remaining goods which had not been recovered, since it was not possible for the same vehicle to be used both for hire by the hirer and for carriage by the lender at the same time.

21. App. Paris, 1.12.92 (1993) B.T. 115. Cf. App. Paris, Rouen, 1.6.95 (1996) B.T. 322.

22. *"La Maîtrise des Operations de Transport"*, see Comm. Colmar, 20.9.91 (1992) B.T. 119, Cass., 5.1.88, (1988) II U.L.R. 732, Comm. Bobigny, 26.5.92 (1992) B.T. 805. Likewise, in Austria, OGH., 8.9.83 (1985) 20 E.T.L. 282. A "Lohnfuhrverträge" involves a manned vehicle being placed at the disposal of the hirer for any goods and journey: OLG. Innsbruck. 20.6.95 (1997) TranspR 343, cf. Herber/Piper, Art. 1. Rdn. 42.

23. *L. Harris (Harella) Ltd.* v. *Continental Express Ltd.* [1961] 1 Lloyd's Rep. 251. See generally Palmer, p. 494 *et seq.*

24. See the example given by Palmer at p. 500.

25. A basis of the distinction in France is the idea that the hirer borrows the vehicle, and borrows the driver to drive rather than to carry. This is reflected in the fact that liability of the lender is a liability for defects in the vehicle and for the fault of the driver in respect of the operations particularly relevant to driving the vehicle (see Chao and Fos Colette, (1986) B.T. 340, 343, and App. Paris, 7.4.94 (1994) B.T. 352) as opposed to the strict liability of a carrier. This cuts across the notion of possession which underpins liability at common law. The interpretation of contractual terms, however, may involve the recognition of more subtle distinctions e.g. *Gillespie Bros. Ltd.* v. *Roy Bowles Transport Ltd.* [1973] 1 Lloyd's Rep. 10, 19.

tractor operator who was instructed to pull a semi-trailer, load a container, drive from Antwerp to Boussois in France, where a consignment of glass was to be loaded into the container, and then forward the container and its load to Antwerp was held to be a CMR carrier and not merely a haulier.[26]

1.6 This last issue is often linked with the question as to what constitutes goods for the purposes of Article 1. The Convention provides no definition, but Loewe[27] has put the position as follows:

"The term 'marchandises' in the French text should not be interpreted narrowly as meaning goods which are carried from one place to another for the purposes of sale. If that had been the author's intention, the exceptions provided for in Article 1, paragraph 4,[28] would not have been necessary, since the operations mentioned in that paragraph do not involve goods in the sense which has just been indicated. The term should rather be construed as meaning tangible goods in general. However, a passenger's luggage is not 'marchandises'."[29]

Continental cases have displayed a difference of opinion as to whether a trailer can be goods distinct from the goods contained within it. So long as the contract can be viewed as one of carriage, as opposed to hire, there does not seem to be any reason why even a mode of transport cannot be "goods" for the purposes of CMR.[30]

1.7 Another issue in relation to the scope of the Convention arose before the Court of Appeal in the case of *Gefco U.K. Ltd.* v. *Mason*.[31] Under a long-term contract ("the umbrella contract"), Mason had sub-contracted the carriage of Marks & Spencer goods to Gefco. The umbrella contract contained various terms as to such matters as reporting and in particular the time within which delivery of the individual consignments was to be effected. As a result of alleged breaches by Gefco, the contract with Marks & Spencer was lost. Gefco claimed outstanding freight from Mason, and Mason counterclaimed damages for loss of profit in relation to the umbrella contract.

1.8 It was alleged by Gefco that the umbrella contract was subject to the CMR Convention, whereas Mason argued that, although the individual movements were subject to the Convention, the umbrella contract itself was not. In particular, it was argued that the requirement for a consignment note could not be applied to a long term contract for multiple movements of unascertained goods.

1.9 On the hearing of a preliminary issue as to whether the Convention applied to the umbrella contract, it was held by the Court of Appeal that it did. Although it was argued

---

26. Cass., 2.2.90 (1990) II U.L.R. 434; (1990) 25. E.T.L. 319.

27. Para. 25.

28. See *infra*, para. 1.55.

29. Presumably because this is covered by a sister convention to CMR, the Convention on the Contract for the International Carriage of Passengers by Road (CVR) 1973 (not yet in force), see Glass and Cashmore, p. 137.

30. In Holland (RB. Amsterdam, 16.4.75 (1975) S. & S. 18.1, reported by Haak, pp. 83–4) a laden trailer was held to be goods. There are conflicting Belgian decisions: RB. Antwerpen, 27.10.71 (1982) 17 E.T.L. 1054 held a tanker trailer not to be goods (the damage to the trailer occurring after delivery of the goods carried in the trailer), Cass., 2.2.90 (as reported in (1992) B.T. 141) suggests that the loaded semi-trailer was the transported goods (Tilche, (1992) B.T. 142, suggests that this goes against the French jurisprudence). See also RB. Antwerpen, 7.11.86 (1986) 22 E.T.L. 453 and RB Antwerpen, 1.2.96 (1996) 31 E.T.L. 579 where damage to a trailer during haulage following movement of the goods was held to be governed by CMR. In Germany, a carrier was liable under CMR for the loss of the sender's trucks which were being carried by means of the sender's transporter, but under national law for loss of the transporter, OLG. Düsseldorf, 25.10.94 (1996) TranspR 152 (see Jung, p. 150). Cf., in respect of a container, a decision in respect of the RHA Conditions, *Acme Transport* v. *Betts* [1981] 1 Lloyd's Rep. 131.

31. [1998] 2 Lloyd's Rep. 585.

for Mason that Articles 4 and 6, which require the insertion of various details into the consignment note, could not be complied with in the context of the umbrella contract, the Court of Appeal was of the view that the umbrella contract nevertheless fell squarely within Article 1(1) of the Convention. Accordingly, it was a contract to which the Convention applied unless there was some further limit to be applied to the scope of its application. The Court of Appeal was of the view that any such implication was inappropriate.

1.10 Four further general points should be noted with regard to the scope of the Convention as provided for by Article 1(1). First, the place of residence or nationality of the parties is irrelevant for the purposes of deciding whether or not the Convention applies[32]: its scope is determined entirely with reference to where the goods are taken over for carriage and where they are delivered.[33] Secondly, under the terms of Article 1(1), only carriage for reward is within the scope of the Convention. English law does not, however, concern itself with the adequacy of consideration,[34] and it would seem that any consideration which is sufficient to support a contract for the purposes of normal English common law rules[35] will equally be sufficient to render the contract a carriage for reward for the purposes of Article 1(1). Thirdly, the Convention does not at present apply to carriage between the United Kingdom and Ireland.[36] Fourthly, the Convention will apply to a contract for international carriage of goods by road even if the consignment never reaches the frontier of the country where the goods are taken over for carriage; the *contract* is nevertheless for the international carriage of goods by road,[37] so long as a place is designated for delivery in another country.[38]

The parties to the contract

1.11 Having established that the contract of carriage is subject to the Convention, it is next essential to establish who the parties to the contract are for the purposes of the application of the Convention. The Convention refers throughout to the sender, the carrier, and the consignee, but does not offer any assistance in establishing their exact status, none of them being further defined. Section 14(2) of the Carriage of Goods by Road Act 1965

32. Residence is, however, relevant for the purposes of jurisdiction: Article 31, *infra*, para. 10.30.
33. OLG. Karlsruhe, 24.5.67 (1967) U.L.C. 289. OLG. Düsseldorf, 18.11.71 (1973) II U.L.C. 428, (1973) 8 E.T.L. 510, OLG. Celle, 13.1.75 (1975) 10 E.T.L. 410, (1975) 3 E.L.D. 582.
34. See, e.g., *Thomas* v. *Thomas* (1842) 2 Q.B. 851.
35. See generally *Halsbury's Laws*, 4th edn., Vol. 9, paras. 316–328. See further *infra*, para. 1.12.
36. See Protocol of Signature to the Convention, Appendix A. However, since Ireland is now party to the Convention, as soon as the necessary legislation has been amended, the Convention will apply to carriage between the two countries. Unfortunately, there is no immediate sign that the necessary Parliamentary time for the necessary amendments is likely to be found in the near future. For the present, the Convention will therefore only apply to such movements if contractually incorporated.
37. RB. Antwerpen, 9.12.77 (1978) 13 E.T.L. 110: CMR applied even though defendant's participation in the carriage was limited to the domestic section. See also Cass., 18.3.86 (1986) B.T. 607.
38. In Germany, in a case concerned with jurisdiction (see *infra*, Chapter 10, para. 10.21 ff.), the place of delivery was provisionally agreed subject to further determination by the sender. The place ultimately so determined was the place designated for delivery: OLG. Hamburg, 7.4.94 (1995) TranspR 115 (see Jung, p. 151 and also LG. Munich 19.7.94 (1995) TranspR 116). Note also the view of Tilche, (1996) B.T. 168, 169 that an instruction changing the point of delivery back to the original country might result in the application of national law, particularly if there is a change of consignee and consequent possible novation of the contract.

provides that each is to be treated as a person concerned in the carriage, but this takes matters no further.[39] In *M. Bardiger Ltd.* v. *Halberg Spedition APS*[40] Evans, J., considered that section 14(2) deems the sender to be a statutory party to the contract of carriage so that there was no need for the person he identified as the sender to be in privity of contract with the carrier in order to be able to exercise rights under the Convention. However, in *Aqualon (U.K.) Ltd.* v. *Vallana Shipping Corp.*[41] Mance, J., had difficulty in accepting this view of the role of section 14(2), which is to be read with section 1 which gives the force of law to the provisions of the Convention. The pre-condition to the application of the Convention is that there should be a contract for the carriage of goods by road, but neither section 1 nor section 14(2) is concerned to effect a statutory expansion or definition to the parties to any such contract which would otherwise exist under the applicable national law.[42]

1.12 The common[43] assumption therefore is that the contract contemplated by CMR is the normal consensual contract, as opposed to some "real" or "formal" contract, which, apart from such amendments as are made by CMR itself, is to be found by the court in the normal way, including the question of the identity of the parties to that contract. The initial opposing parties to the contract will normally be the sender and the carrier. It has been said that the first definition of the concept of sender to be used in the history of the Convention is "he who in his own name contracts with the carrier".[44] Several decisions have involved the finding that the party who contracts with the carrier is the sender and have denied this status to, for example, the seller from whose premises the carrier has collected the goods.[45]

---

39. See further the discussion of section 14(2) *infra*, para. 11.15, fn. 26.

40. Queen's Bench Division, 26.10.90, unreported except on LEXIS.

41. [1994] 1 Lloyd's Rep. 669.

42. In contrast, as the learned judge pointed out, to the fact that within the text itself there are specific Articles expanding the class of persons who may be treated as party to a contract governed by the Convention (e.g. Articles 13 (see *infra*, para. 5.18 ff.) and 34 to 36 (see *infra*, Chapter 11) and other Articles which, in certain circumstances where there is such a contract, allow reliance on the provisions of the Convention against or by persons not so party (e.g. Article 28, see *infra*, para. 9.65). The approach of Mance, J., is closer to that expressed by Diplock, L.J., in *Post Office* v. *Estuary Radio* [1968] 2 Q.B. 740, 757.

43. Haak, p. 44, L.G. Bremen, 6.5.65 (1966) 1 E.T.L. 693, 697, Hof. Antwerpen, 23.2.93 (1993) 28 E.T.L. 934.

44. Haak, p. 253, quoting from E/ECE/TRANS/SCI/79, E/ECE/Trans/WP9/13, 1.5.50, p. 2.

45. In France: App. Paris, 6.1.71 (1971) U.L.C. 149, a seller of goods collected by the carrier from his premises, who was named as sender on the consignment note but who had not signed it, was held not liable to pay the freight charges. He had acted only as an intermediary in finding a carrier for his buyer and had made it clear to the carrier that he was not a party to the contract. In App. Douai, 11.3.82 (1982) B.T. 199, a buyer who gave an order to a carrier to collect goods from a seller was held to be the sender and consequently the proper person to be sued under Article 22 in respect of dangerous goods (see *infra*, Chapter 8), cf. App. Colmar, 12.12.93 (1995) B.T. 793. The court stated that the sender is the person who concludes the contract, who treats with the carrier and not necessarily the person from whose premises the goods are taken over; see also App. Aix-en-Provence, 25.11.82 (1983) B.T. 591 to similar effect in respect of the right to sue the carrier. In both of these last two cases, the seller was named as the sender on the consignment note, but it was not signed by him (on the significance of this, see Article 5, *infra*, para. 4.11). In the latter case, the buyer's agent in fact signed the consignment note, this party having participated in the operations concerning the loading of the goods (cf. Trib. gde. inst. Metz, 7.9.82 (1983) B.T. 246, where the intermediate seller was not named as sender or consignee in the consignment note, and although he was presented as having ordered the "taking in charge", had not given sufficient justification for his right to sue). To similar effect in Germany is the statement of the BGH., 28.4.88 (1989) 24 E.T.L. 89, 92. In Belgium, in Comm. Bruxelles, 12.4.72 (1972) 7 E.T.L. 1046, the name of the forwarder appeared as sender on the consignment note but the court held the forwarder's principal liable to the carrier under Article 10 of the Convention. The forwarder was considered to be the agent of the principal who had thereby concluded the contract with the carrier and was therefore the sender. Contrast Putzeys, para. 111.

Other decisions rely more heavily on what is stated in the consignment note.[46] It may well be, however, that the Convention must be read in the light of the applicable local law, so that allowance must be made for variations as to which parties will be considered as contracting parties. This in turn raises the possibility that more than one person may take the status of sender. This is possible in France,[47] and in English law it is conceivable that the principles of undisclosed agency[48] can be applied to render both principal and agent co-contractors of the carrier and thereby take the status of joint sender towards the carrier.[49]

1.13 An even greater degree of uncertainty surrounds the question of who has the status of carrier where a forwarder intervenes as an intermediary between the shipper of the goods and the carrier who performs the actual transport. Three sets of facts can be distinguished:

(a) The forwarder undertakes the carriage of the goods and therefore the obligations of the carrier under the Convention. If any part of the carriage is then sub-contracted, the sub-contractor will be a successive carrier, if the requirements of Article 34 are satisfied, and/or a person for whom the carrier is responsible within the meaning of Article 3.

(b) The forwarder is treated as the sender. In this case the shipper will not be a party to the CMR contract of carriage and will have neither rights nor liabilities thereunder unless he is designated as consignee or becomes a person to whom the latter, in exercising his right of disposal under Article 12, has ordered delivery of the goods, or unless the consignee's rights under the contract have been assigned to him.[50]

(c) The forwarder is treated as agent of one or other party to the contract of carriage, being merely an intermediary and not himself a party to the contract. In practice, while it is relatively common for the small forwarding company to act as agent for its client the shipper, it would be rare (given current trade usages) for a forwarder to act as agent for a carrier.

1.14 In practice it may prove difficult to establish which factual analysis is appropriate to a particular transaction. This difficulty stems from the ambiguity of the position of the

---

46. BG. Switzerland, 2.6.81 (1995) 30 E.T.L. 668. In Holland, in HR., 7.12.73 (1974) 9 E.T.L. 724, it was held that the party who concluded the contract with the carrier had a contractual right of action against the carrier, notwithstanding the fact that his supplier appeared as sender on the consignment note, and his purchaser appeared as consignee. Cf. further App. Paris, 8.11.95 (1996) B.T. 132 where a seller was held to be the sender having signed the consignment note without making it clear that he was acting on behalf of the person designated as sender. The court ignored the terms of sale to which the carrier was not a party. Note further that in Hof. Gent, 28.6.96 (1996) 31 E.T.L. 843, CMR was held to be inapplicable to an existing relationship between a carrier and a purchaser of goods where the latter, in relation to the carrier, was neither directly nor indirectly interested in the contract of carriage. This relationship was governed by national law.

47. Article 101 of the *Code de Commerce* grants contractual rights to a party against a carrier when acting through a *commissionnaire de transport* in addition to the contractual rights of the *commissionnaire*. He thereby becomes the *"expéditeur réel"*, Lamy, para. 555, *infra*, para. 5.20. Both parties can therefore exercise rights as sender. For the *"destinataire réel"*, see *infra*, para. 5.20.

48. See, generally, *Bowstead on Agency*, 16th edn.

49. See, *infra*, the view of Judge Martin in *Tetroc Ltd.* v. *Cross-Con (International) Ltd.* [1981] 1 Lloyd's Rep. 192, 195 in respect of the continued application of the principles relevant to the relationship of principal and agent.

50. Cf. Carriage of Goods by Road Act 1965, section 14(2)(c).

freight forwarder.[51] Where, for example, a forwarder consolidates his clients' shipments into a single load to be forwarded on one vehicle he may in fact be the sender under the CMR, especially if he signs the consignment note in that capacity.[52] Similarly, where the forwarder operates his own transport fleet and himself carries the goods for some part of the journey, he may be categorised as the CMR carrier. Perhaps the most difficult case is that where the forwarder offers a regular freight service on a given route, but either operates no vehicles himself or at most operates a local collection or delivery service.

1.15 The question of the status of the forwarder for the purposes of the Convention is essentially the same question as to whether, for the purposes of English domestic law, a party constitutes a carrier or a forwarding agent.[53] In determining either question, it is necessary to look beyond the form of the contract and to analyse the obligations actually undertaken by the forwarder. If he in fact contracts to undertake the international carriage of goods by road, he is likely to be treated as the carrier for CMR purposes, whether he likes it or not, and whether or not the actual carriage of the goods is sub-contracted. A forwarder who is in fact a carrier cannot avoid his liability under CMR by seeking to avail himself of the provisions of his contract, since the Convention itself provides that if a contract is within the scope of its application no party can exempt himself from his liability thereunder.[54] Where a forwarder offers a comprehensive range of services on the short sea crossings across the Channel and the North Sea, it will not be easy for him to establish his status under the CMR in general terms. Indeed it can be suggested that the status of the parties to a contract which is potentially governed by the CMR can only be

---

51. See generally Hill, *Freight Forwarders*, Chapter 3. The 1981 edition of the Standard Trading Conditions of the Institute of Freight Forwarders (now the British International Freight Association) specified that the forwarder contracts as forwarding agent only (clause 2). Such description cannot be regarded as conclusive (see, e.g., *Ex parte White, Re Nevill* (1871) 6 Ch. App. 397, affd. sub. nom. *John Towle & Co. v. White* (1873) 29 L.T. 78; *Michelin Tyre Co. v. Macfarlane* (1917) 55 Sc. L.R. 35; *Lamb v. Goring* [1932] 1 K.B. 710; *Harris (Harella) Ltd. v. Continental Express Ltd.* [1961] 1 Lloyd's Rep. 251) as was in effect conceded by reference in clause 2 to section III of the Conditions, which regulated the position of the forwarder where deemed to be carrier. The current Standard Trading Conditions of BIFA (2000) expressly make the distinction discussed in the text between the forwarder acting as an agent and acting as a principal, although clause 4 seeks to influence the criteria used to make the distinction. It should be noted, however, that where printed conditions clearly indicate that only an agency service is offered, this may more readily take effect (unless directly contradicted), as in *Victoria Fur Traders Ltd. v. Roadline (U.K.) Ltd.* [1981] 1 Lloyd's Rep. 570, where a British Airways Waybill stated that "The ... Carrier ... in re-forwarding the goods shall do so only as agents of the ... Owner ...". Cf. BIFA Conditions clauses 11 and 12.

52. See Article 5, *infra*, paras 4.11–4.14. The fact that the forwarder has, or has not, signed the consignment note is not conclusive, however. See Articles 4 and 9(1) *infra* (para. 4.2 ff. and 4.24 ff.) respectively, and *Tetroc Ltd. v. Cross-Con (International) Ltd.* [1981] 1 Lloyd's Rep. 192, 196, discussed generally *infra*, para. 1.18 ff. In *Kala Ltd. v. International Freight Services (U.K.) Ltd.*, Queen's Bench Division, 7.6.88, unreported except on LEXIS, the forwarders were named as the sender on the CMR consignment note and their agent was named as consignee. It was accepted by the plaintiffs that they had no contract of carriage with the forwarders. It is noteworthy that the goods had been called forward to the forwarders' handling facility for consolidation into trailer loads and that the forwarders issued a "through Bill of Lading" which stated: (1) that they were "authorised to enter into contracts with carriers ... subject to the latter's usual terms and conditions"; (2) that the forwarders do not act as carriers but as forwarding agents and are only responsible for careful selection of third parties; and that (3) the forwarders will, on request, subrogate their claims against carriers and other parties. These terms are similar to those used by continental forwarders, see *infra*, para. 1.46 ff. and also the clauses of the UIRR conditions.

53. See, e.g., *Davis (Chas.) (Metal Brokers) Ltd. v. Gilyott and Scott Ltd.* [1975] 2 Lloyd's Rep. 422. *Marston Excelsior Ltd. v. Arbuckle Smith & Co. Ltd.* [1971] 2 Lloyd's Rep. 306. *The "Maheno"* [1977] 1 Lloyd's Rep. 81 (New Zealand Supreme Court). *Salsi v. Jetspeed Air Services Ltd.* [1977] 2 Lloyd's Rep. 57. See generally Yates, Part 7, Glass and Cashmore, Chapter 2, Bugden, *Freight Forwarding and Goods in Transit*, Chapter 3.

54. See Article 41, *infra*, Chapter 12.

established *ex post facto* and *pro hac vice*. If the forwarding agent retains his status as such (i.e. he does not fall to be treated as the carrier) his relationship to his client will fall to be governed by normal agency principles and not by the CMR.[55] In such cases, the forwarder's principal will be the CMR sender or consignee, and the party who has undertaken the carriage will be the carrier for the purposes of the Convention.

English case law

1.16 Turning to the English case law on the matter, the leading case is that of *Ulster-Swift Ltd.* v. *Taunton Meat Haulage Ltd.*[56] In that case the consignors arranged for the carriage by road of pig carcases from Northern Ireland to Switzerland. They contracted in the first instance with Taunton, who subsequently sub-contracted the entire carriage to a Dutch carrier, Fransen, who carried the carcases to Switzerland. On arrival it was found that the carcases had deteriorated. The consignors accordingly sued Taunton, who in turn joined Fransen as third party. The third party contended, *inter alia*, that they were not successive carriers within the meaning of Article 34 on the basis that Taunton were not carriers in the first place, not having performed any part of the carriage. It was therefore necessary for the court to consider the status of Taunton in relation to Article 1 of the Convention. The case is unusual in that it was here the third party arguing that the defendants were forwarding agents, rather than the defendants themselves, but this makes no difference to the legal issues involved.

1.17 In the event, the Court of Appeal held that Taunton were carriers notwithstanding the fact that they had performed no part of the carriage:

"In our judgment, a person who contracts to carry goods is a 'carrier' even if he sub-contracts the actual performance of the whole of the carriage to someone else."[57]

The court was confirmed in this view by Article 3, which, as will be seen,[58] makes the carrier responsible for the acts of "his agents and servants and of any other persons of whose services he makes use for the performance of the carriage . . . ":

"In our view, this article entitles a person who has contracted to carry to perform the contract by means of a sub-contractor, though he himself remains liable under the Convention as a 'carrier'."[59]

In effect therefore, the Court of Appeal held that because Taunton had contracted to carry the goods, they were carriers within the meaning of the Convention. Implicit in this is the assumption that if, as distinct from having contracted to carry, they had contracted to *arrange* the carriage, they would have been forwarders rather than carriers.

1.18 Further support for such an analysis can be drawn from the decision of Judge Martin in *Tetroc Ltd.* v. *Cross-Con (International) Ltd.*,[60] from which further guidance can be drawn as to how the distinction referred to above is to be applied in practice. In that case the plaintiffs contracted with the defendants for the carriage by road of machine tools

55. In *Tetroc Ltd.* v. *Cross-Con (International) Ltd.* [1981] 1 Lloyd's Rep. 192, discussed *infra*, para. 1.18, Judge Martin said at p. 195 that in considering European decisions on this issue "the English law governing the relationship of principal and agent is in no way superseded by them".
56. [1977] 1 Lloyd's Rep. 346.
57. Ibid., at p. 359, per Megaw, L.J.
58. *Infra*, Chapter 3.
59. [1977] 1 Lloyd's Rep. 346, 359, per Megaw, L.J.
60. [1981] 1 Lloyd's Rep. 192.

from Copenhagen to this country. The defendants, however, performed no part of the carriage, having sub-contracted its performance to other carriers not party to the proceedings. The defendants argued that they were forwarders only.

1.19 Judge Martin commenced by observing that, following the *Ulster-Swift* case, the fact that the defendants had not performed any part of the carriage did not mean that they were not carriers within the meaning of Article 1. Then, after a review of the documentation and other evidence before him, he concluded:

" . . . in the end as was said by Mr. Justice Bean in the case of *Hair & Skin Trading Co. Ltd.* v. *Norman Air Freight Carriers and World Transport Agencies Ltd.*—' . . . when a Judge has to decide whether a party is acting as principal or an agent, it is very much a matter of impression, what impression the evidence forms'.[61] In this case I have come firmly to the conclusion the defendants contracted as principals, that they were carriers within the meaning of CMR . . . "[62]

Clearly this is not very helpful in establishing any reliable guide as to a party's status as carrier under the Convention, Judge Martin giving no guidance as to which factors in particular he relied on in reaching this conclusion. It is instructive, however, to look at the various matters which he had considered in arriving at his conclusion.

1.20 First, he noted that the defendants had invoiced the plaintiffs, in effect, with an all-in charge in respect of the carriage, commenting:

" . . . one notes that this was not a case where the price quoted by [the sub-contractor] had been set out and then a commission added on."[63]

In considering this aspect of the case, Judge Martin referred to a Dutch case on this point, where the fact that the defendant had invoiced an all-in charge meant that he was held to be a carrier for the purposes of CMR, even though a third party had executed the carriage.[64] He said that he took this case as a guide, though not a conclusive one.

1.21 Secondly, he referred to a number of letters from previous transactions containing the plaintiffs' instructions to the defendants which included the words: "Would you kindly arrange onward transport to . . . ." Judge Martin expressly said he did not attach much importance to the use of the word "arrange" used in a business context:

---

61. [1974] 1 Lloyd's Rep. 443, 445.
62. [1981] 1 Lloyd's Rep. 192, 198.
63. At p. 195. See also *Electronska Industrija Oour TVA* v. *Transped Oour Kintinentalna Spedicna* [1986] 1 Lloyd's Rep. 49 discussed *infra*, para. 1.29. The method of charging used by the forwarder has long been an important factor in determining the status of the forwarder quite apart from CMR, see *Colley* v. *Brewer's Wharf and Transport Ltd.* (1921) 9 Ll.L.Rep. 5, 7 and *Harris (Harella) Ltd.* v. *Continental Express Ltd.* [1961] 1 Lloyd's Rep. 251, 259. In the non-CMR case of *Marston Excelsior Ltd.* v. *Arbuckle Smith & Co. Ltd.* [1971] 2 Lloyd's Rep. 306, Denning, M.R., at p. 310, was troubled by the fact that the forwarder added 10% in sending on the performing carrier's quotation, but did not feel that this altered the fundamental nature of the arrangement between the parties as one of agency. See further, *infra*, para. 1.40 fn. 94.
64. Arrond. Rotterdam, 10.11.70 (1971) 6 E.T.L. 273. (The reference cited by Judge Martin, at least as it appears in the report, is incorrect.) See too *Naarden International (U.K.) Ltd.* v. *Gelders (U.K.) Transport Ltd.* (1980) unreported, where the fact that an overall rate had been negotiated for an anticipated series of transactions was held indicative of the defendant's status as carriers. It was also indicated that the defendants would have been estopped from disputing their status as carriers by their conduct after the claim arose. See also *Texas Instruments Ltd.* v. *Nason (Europe) Ltd.*, discussed *infra*, para. 1.33.

"I agree it could mean that the defendants should arrange for other people to affect [sic] onward transport. On the other hand, it could mean that the defendants should arrange to do it themselves."[65]

1.22 Thirdly, the defendants had relied on the fact that the sub-contractors rather than themselves appeared as carriers in the consignment note. Judge Martin answered this by reference to Article 4[66] and 9(1),[67] without further comment.

1.23 Fourthly, he referred to a letter from the defendants to the sub-contractors in which the defendants described themselves as freight forwarders, noting that he had been told that these are words sometimes used to describe a forwarding agent. In the same letter the defendants referred to the plaintiffs on a number of occasions as their principals, but Judge Martin did not think that this was intended by the writer to connote any precise legal relationship, but was intended merely to describe the concern which had instructed them without the need to name them.[68]

1.24 Much of the remaining evidence was regarded by Judge Martin as the parties to the action simply stating their respective positions after the damage had occurred, but he did note that one of the defendants' managers had accepted in cross-examination that, in negotiating the contract, the plaintiffs would have been interested only in an all-in figure for the carriage, and that a price would have been given with no reference to the quotation from the sub-contractor.

1.25 Finally, Judge Martin noted that the defendants had not told the plaintiffs what arrangements had been made on their behalf. Referring to another Continental decision,[69] he commented that this was not conclusive, but that it could be taken into account.

1.26 As already stated, Judge Martin gave no indication at the end of the day as to which of these matters he relied on in reaching his decision that the defendants had contracted as carriers. Nevertheless, the only factors in favour of the plaintiffs' contention which he expressly mentioned as factors to be taken into account were the method of charging adopted by the defendants, and the fact that the defendants had not informed the plaintiffs of the arrangements made on their behalf. It seems reasonable to conclude, therefore, that he was relying on those two factors in particular in reaching his decision.[70]

1.27 The same problem, albeit in a slightly different context, arose in *Moto Vespa S.A. v. MAT (Brittania Express) Ltd.*[71] In that case lathes were purchased by the Spanish plaintiffs from a Birmingham manufacturer, the contract having been negotiated through

---

65. [1981] 1 Lloyd's Rep. 192, 196. Cf. *Harris* at p. 259, where Paull, J., in deciding that forwarders were not agents, took note of the fact that the words used in the consignment note were "please collect" not "please arrange for the collection of".

66. *Infra*, para. 4.2 ff.

67. *Infra*, para. 4.24 ff. At p. 196 Judge Martin also referred to an earlier transaction where, in the bundle of correspondence relating to it, the forwarders referred to the performing carrier as "the carriers". This does not mean that it followed that the forwarders themselves were not carriers.

68. So too in the *Naarden* case (*supra*), the use of the words "import agents" was not regarded as affecting the position. See, however, the *Elektronska* case, discussed *infra*, para. 1.29.

69. Hof. Brussel, 19.10.72 (1974) 9 E.T.L. 608. The same case is reported in (1973) 8 E.T.L. 503, where the point referred to by Judge Martin appears in the English headnote.

70. The evidence also revealed that the forwarder's manager, who had not regarded his company as being a carrier, agreed that the company had only one vehicle, which had been obtained for one particular job. Contrast *Marston Excelsior, supra*, para. 1.15 fn. 53, where Phillimore, L.J. said at p. 311: "It would of course be a very odd thing if a firm which owned no vehicles . . . to carry goods . . . should . . . never having done this sort of thing before . . . enter into a contract as carriers. But no doubt it is possible."

71. [1979] 1 Lloyd's Rep. 175.

E.S., the manufacturer's agent in Spain. Subsequently E.S. contracted with the defendants for the carriage of the lathes from England to Spain. The lathes were damaged in transit, and in the subsequent action the plaintiffs claimed that the contract of carriage with the defendants was made by E.S. on their behalf as undisclosed principals, thus entitling them to sue the defendants on the contract. The issue therefore was essentially the same as that considered above, namely: was E.S. contracting as agent for the plaintiffs?

1.28 The evidence before the court was that there had been a long course of dealing between E.S. and the defendants in which E.S. had entered into contracts for the provision of carriage by the defendants. Those contracts were made orally and informally over the telephone, the defendants always looking to E.S. to meet their carriage charges. On the occasion in question, the evidence was that the plaintiffs would have sought different quotations for the carriage and insurance, then asked E.S. to arrange the contract on their behalf, which, the evidence was, he then did, as usual over the telephone and with nothing in writing. As regards remuneration, E.S. received commission from the manufacturers on the sale and also a commission on the insurance from the insurance company involved.[72] There was in the documents an invoice from E.S. to the plaintiffs for insurance charges, which the court took as indicative of agency. Also among the documents was a form of instructions sent by E.S. to the defendants some time after the formation of the oral contract to deliver the lathes to Moto Vespa "by order of the last named", thus making it clear for whom he was acting. Mocatta, J., concluded that E.S. was acting as forwarding agent only, and accordingly the plaintiffs were entitled to sue the defendants under the Convention.[73]

1.29 Two further cases provide useful illustrations of how this issue is likely to be treated by the courts, whereas the most recent cases suggest new possibilities of analysis.[74] In *Elektronska Industrija Oour TVA* v. *Transped Oour Kintinentalna Spedicna*[75] the plaintiffs sued in respect of a consignment of cartons of television sets which were damaged in the course of transport from the former Yugoslavia, and which would have arrived in South Wales but for the accident. The driver was employed by a company against whom an unsatisfied judgment existed and the plaintiffs sought to attack the forwarder through whom the transport arrangements had been made following instructions given to them by the plaintiffs' agent. As in *Tetroc Ltd.* v. *Cross-Con (International) Ltd.*,[76] Hobhouse, J., began[77] with the usual reference to *Ulster-Swift*[78] and the decision that the word "carrier" under CMR is not confined to the person who performs the carriage. He reviewed the relevant contrasting arguments as being between the defendants acting either as agents for the plaintiffs and sub-agents of the plaintiffs' agent or the defendants making their contract with the plaintiffs as principals and, as principals, making a sub-contract with the performing carrier. He noted that it was agreed that the question had to be decided on the facts having regard to all the evidence, and he

---

72. He therefore could not himself have been the CMR carrier, since there was no carriage for reward.

73. The actual carriers issued a waybill which named the plaintiff as consignee. Although it does not appear that this was the CMR consignment note, the issue of title to sue might have been resolved by the application of Article 13. See *infra*, para. 5.18, and cf. the *Texas Instruments* case, discussed *infra*, para. 1.33.

74. In particular *Bardiger Ltd.* v. *Halberg Spedition APS*, Queen's Bench Division, 26.10.90, unreported except on LEXIS, *infra*, para. 1.39.

75. [1986] 1 Lloyd's Rep. 49.

76. *Supra*, para. 1.18.

77. At p. 50.

78. See *supra*, para. 1.16.

considered that it was not possible for the court to approach the question as a pure matter of the legal construction of the communications between the plaintiffs' own agent and the forwarder.[79]

1.30 In reviewing this evidence, Hobhouse, J., treated some factors as having only limited weight in shifting the balance. Firstly, he noted that it had been conceded that the CMR consignment note, although highly relevant evidence,[80] is in no way conclusive. In this case, he considered that, although it was signed by the driver who named his employer as the carrier, it was of low evidential value since it only came into existence after the contracts setting the carriage arrangements in place had been made. Furthermore, the driver would not have been likely to have been directing his mind to the legal antecedent factual position even if he was aware of it.[81]

1.31 Secondly, the defendants' description of themselves as "forwarders", which had been argued as being equivalent to "forwarding agents", did not carry with it any implication that the defendants were not prepared to make contracts for carriage as principals. As he pointed out[82]:

"to carry on the business of 'agent' is not the same as saying that you are contracting in the capacity of agent for a principal."

On the other hand, the evidence from the plaintiffs' agents showed that they did not care whether the defendants performed the transport themselves or engaged someone else to do it. In substance, they requested that the defendants transport rather than arrange to transport. The defendants described themselves as "specialised trading services" and as "full load and groupage specialists to Yugoslavia". In a previous transaction with the plaintiffs' agent, they had sometimes used their own vehicles. Although the defendants did, on occasion, perform various services on an agency basis, their normal way of conducting business involving the carriage of goods by road, when not carrying themselves, was to make a profit on the freight charged to the customer. Invoices from previous transactions revealed no element of agency. Furthermore, had an invoice been raised for this transaction, freight would have been charged in the same way. The contract with the performing carrier gave no indication that the defendants were only agents.

1.32 The course of dealing thus described reinforced the view that the true position was that the defendants and their customers preferred to deal with each other in relation to carriage by road as principals. The defendants' normal way of doing business was to contract to carry and then decide whether or not to sub-contract, it being a matter of chance and convenience whether they used one of their own lorries.[83] As the learned judge pointedly remarked[84]:

---

79. Cf. Steyn, J., in *Singer* v. *Tees & Hartlepool Port Authority* [1988] 2 Lloyd's Rep. 164 who, at p. 167, stated that "On this aspect I heard no oral evidence from the individuals who concluded the contract. This is an issue to be resolved on the documents, and in the light of objective features of the transaction".

80. See Article 9, *infra*, Chapter 4.

81. Arguably the more relevant question to ask is what authority the driver has to sign the consignment note so as to bind the person he represents as the carrier, see *infra*, para. 1.36, fn. 88 and cf. OLG. Nuremberg, 23.2. 94 (1994) TranspR 288 (see Jung p. 151), OGH., 19.1.94 (1994) TranspR 437, (1996) B.T. 84 (see Mayer, p. 170).

82. At p. 52, cf., *supra*, fn. 49.

83. Similarly in *Singer* v. *Tees & Hartlepool Port Authority, supra*, fn. 79, a lump sum payment agreed with the forwarder "for crating and delivering to the U.K. port" indicated that the customer engaged the forwarder for a complete package of services, leaving it to the forwarder to sub-contract where necessary.

84. At p. 53.

"They were enjoying the advantages of a principal contractor, for example, the freedom to make a profit for themselves. They must accept a corollary of liability."

1.33 A contrast is provided by the decision in *Texas Instruments Ltd.* v. *Nason (Europe) Ltd.*[85] A consignment of machine calculators was stolen in the course of carriage from Italy to the United Kingdom. The plaintiffs[86] took the precaution of suing all those connected with the carriage but at the trial attention was focused on the liability of the second defendants who, as lessees of a trailer, had supplied it for the carriage of the goods. The tractor had been supplied by the third defendants, who were the employer of the driver. The plaintiffs' case depended on their establishing that the second defendants were either the first carrier or the last or performing carrier.[87]

1.34 Part of the plaintiffs' case was that the fourth defendants were forwarding agents, who had contracted for the carriage with the second defendants via the English first defendants, acting as their sub-agent.

1.35 The second defendants sought to defend the claim against them by asserting that the forwarders were really carriers, with the consequence that they were not burdened with the status of first carrier. The plaintiffs' case depended heavily on the consignment note which, translated from Italian, indicated that the fourth defendants were the sender by order of the plaintiffs, and their name was stamped in the relevant box. The second defendants were named as "transporters" and the box for the transporter's signature and stamp contained the driver's signature along with a stamp identifying the second defendants as carrier.

1.36 Inevitably, the view of Hobhouse, J., in the *Elektronska* case was used in support of the defence, which sought to weaken the usual presumption to be drawn from the consignment note. It was argued that the driver could not have appreciated the contractual arrangements between the fourth defendants and their groupage customers. Tudor Evans, J., accepted that the value of the consignment note depends on all the evidence and the circumstances, but he did not think that Hobhouse, J., was intending to hold that the fact that the CMR consignment note post-dated the contract always inevitably detracted from its weight. He accepted, however, the direct evidence of the driver that he was authorised to sign on behalf of the second defendants as the carrier, which he took to be strong evidence that they were the first carrier.[88]

---

85. [1991] 1 Lloyd's Rep. 146. Cf. the decision in Cass. (Belg.), 17.9.87 (1988) 23 E.T.L. 201, *infra*, para. 1.51 fn. 133.
86. An issue in the case was which party had title to sue, see para. 5.22.
87. See Articles 34 and 36, *infra*, Chapter 11.
88. It might be thought that it does not follow that the forwarders were not also carriers, see Judge Martin's comments in *Tetroc* v. *Cross-Con*, *supra*, fn. 66. More pertinent, perhaps, might have been the reason why the indication of the sender took the form that it did. In *Aqualon (U.K.) Ltd.* v. *Vallana Shipping Corporation* [1994] 1 Lloyd's Rep. 669 the defendant forwarders questioned whether the consignment note, signed by the driver (though made out by the plaintiff), and indicating that the forwarders were the carrier, could assist as to the identity of the contracting parties and submitted that Article 9 does not make it *prima facie* evidence of the identity of the parties. Mance, J., found it difficult to see why the consignment note should not be *prima facie* evidence of the identity of the carrier when this is stated on the face of the note. It was important that the dealings with the consignment note in this case were a major factor in substantiating the belief of the plaintiffs that they were dealing with the forwarders as carriers. Note that an unsigned consignment note is not *prima facie* proof, *City Vintages Ltd.* v. *SCAC Transport International*, Queen's Bench Division, 1.12.87, unreported except on LEXIS, see further *infra*, para. 4.12. Arguably, a consignment note naming a party as carrier and signed by the driver should not be proof of the identity of the carrier (i.e. proof that the named carrier has accepted responsibility as such) if he can successfully impugn the authority of the driver to sign on his behalf. Whether it can automatically be assumed that a driver has the authority of his employer to sign the consignment note is

1.37 Tudor Evans, J., also considered evidence led by the plaintiffs of communications after the loss between the second defendants' loss adjuster and the first defendants, in which the second defendants were described as "principal carriers". Used in the context of the CMR Convention, he treated this as an admission that the second defendants were the first carriers.

1.38 Further evidence revealed that the plaintiffs had indicated that they viewed the fourth defendants as an "international carriage company" and that it was left up to them how they effected carriage. He considered much of this latter evidence to be ambiguous, including the fact that the fourth defendants had been given an order "for the provision of carriage". It was also a fact that the fourth defendants charged a flat rate fee, but he considered that this was only one factor to be considered with others, and was out-weighed by the evidence from the consignment note and the admission referred to above. Consequently, the second defendants were held to be the first carriers.[89]

1.39 The cases considered above have treated the issue as one to be resolved by determining an answer to the question whether the forwarder is acting as principal or as agent. If the forwarder takes the position of principal in relation to the customer, it is assumed that he has thereby contracted to carry.[90] However, in *M. Bardiger Ltd.* v. *Halberg Spedition APS*[91] Evans, J., took the view that there was nothing which required the conclusion that unless a forwarder makes a contract of carriage, he must be considered as an agent and accordingly subject to the duties of an agent in respect of remuneration and account. He was of the view that it is possible for a forwarder to contract as principal, but without thereby contracting as a carrier.

1.40 In this case, no less than seven parties were connected with the carriage of raw mink skins which were stolen in the latter stages of a transit from Copenhagen to London. The plaintiffs had bought the skins at an auction at a fur centre, which had an arrangement with certain forwarders who were resident at that centre and who contracted on standard charges and to whom all instructions for carriage of skins purchased at auctions had to be directed. The plaintiffs duly passed their instructions to one of these forwarders, who became the first defendants. The order was passed down a chain of parties until it reached

a question which has been raised in Germany (see Jung, p. 151). Arguably, any implication of authority should depend upon the contract made by the employer.

89. A further factor which indicated to the judge that the second defendants were the first carrier was their supply of the trailer for the carriage. The availability and supply of a vehicle has been considered relevant to the issue of whether a party has contracted to carry (see *supra*, fn. 69), but this seems to extend it as a factor in considering the chain of carriers. A central question relevant at this point of the judgment was the second defendants' claim that they were not carriers at all since they had sub-let the vehicle to the third defendants. This was rejected for failure to plead it (see *infra*, p. 152) and because there was no evidence of it. The evidence pointed the other way having regard to the degree of control that the second defendants had over the movement.

90. Said to be the difference between arranging carriage and effecting carriage, see e.g. per Beattie, J., in *The Maheno* [1977] 1 Lloyd's Rep. 81, 86 (New Zealand Supreme Court). In the latter role, they may well be expressed to be acting as transportation contractors undertaking an obligation personally to procure that the goods are carried, see *Salsi* v. *Jetspeed Air Services* [1977] 2 Lloyd's Rep. 57, 60, *Landauer & Co.* v. *Smits & Co.* (1921) 6 Ll.L.Rep. 577, 579, *Lynch Bros.* v. *Edwards & Fase* (1921) 6 Ll.L.Rep. 371, *J. Evans & Son (Portsmouth) Ltd.* v. *Andrea Merzario Ltd.* [1976] 1 W.L.R. 1078, 1082, [1976] 2 Lloyd's Rep. 166, 168 and [1975] 1 Lloyd's Rep. 162 (Q.B.D.) at p. 168; see further Hill, "Forwarding and Sub-Contracting Complications", [1976] LMCLQ 63 and Holloway, "Troubled Waters: The Liability of a Freight Forwarder as a Principal under Anglo-Canadian law", (1986) 17 J.M. L.C. 243. Despite the expression of a role as "transportation contractor", the possibility that a party might take such a role and yet not be a CMR carrier has not been explored, cf. *infra* para. 150 fn. 118. This is a role distinct from that accepted by the court in the *Bardiger* case.

91. Queen's Bench Division, 26.10.90, unreported except on LEXIS.

the fourth defendants, who were the first to admit the status of carrier and thereafter the goods passed, physically or vicariously, through the hands of several carriers before they reached the last carrier. The fourth defendants made out the consignment note, showing the fur centre as sender and the plaintiffs as consignees.[92] The judge accepted on the evidence that the second and third defendants were carriers, in that they had been approached as trucking companies offering a service to the United Kingdom which they would provide either themselves or, as happened in the event, by sub-contractors.[93] With regard to the first defendants, however, Evans, J., accepted that whilst they took the position of principals towards the customer, they had not undertaken to carry. Having acknowledged the ambiguity in the description of a forwarder as "forwarding agent" and in an undertaking to "arrange" carriage, he considered that the nature of the business carried on by that contracting party, to the knowledge of the other party, is of greater weight. The fact that a flat rate was charged was significant but not conclusive.[94] Unlike the second and third defendants, no-one contemplated that they would carry the goods themselves nor could they have done so. They were employed because of their residence at and close links with the fur centre, and because the plaintiffs had no choice but to employ one of the forwarders there. Consequently, they were held not to have undertaken to carry.[95]

1.41 Whilst the *Bardiger* case recognises the possibility of the forwarder taking an intermediate position, clearly this will not be a common finding. In *Aqualon (UK) Ltd.* v. *Vallana Shipping Corporation*[96] Mance, J., warned that such an interpretation would frequently conflict with the expectations of a goods owner consigning goods to the continent.[97] In his view, a finding that a contracting party was intended to fulfil an

---

92. None of the alleged carriers were named as carriers, the spaces for "carrier" and "subsequent carrier" having been left blank. Cf. *Aqualon* v. *Vallana*, *infra*, where the forwarder was named as carrier on the CMR consignment note made out by the sender. The forwarder would subsequently alter the consignment note to indicate a different carrier and stamped the box for the sender's instructions along with the words "as agents only". This alteration was held not to alter the status of the forwarder as a contracting carrier since the plaintiffs had never been alerted to their error in making out the consignment note in the way which reflected their belief that they were contracting with a CMR carrier.

93. The attempts of these defendants to show that they were not expected to carry since they were small companies or known to possess no vehicles were contradicted either by evidence of their actual scale or the belief of others in the chain that they were carriers.

94. At an earlier point in the judgment he noted the artificiality of finding the defendants to be agents since they took a trading profit. True agents would be under a duty as to remuneration and account. This also concerned Lord Denning, M.R., in the *Marston* case, *supra*, fn. 63, but cf. *Cohen* v. *Brantford Corporation*, Court of Appeal, 8.12.87, unreported except on LEXIS, where a forwarder's practice of including an element of profit when performing a service did not prevent their characterisation as agents in respect of the supply of a container. In *Bardiger*, the forwarders' trading conditions sought to remove the method of charging from the relevant criteria, as did clause 5(b) of the previous version of the BIFA Standard Trading Conditions 1989. It might be thought possible for a customer to be taken to have agreed to an implied exclusion of the duty to account where it is clear that a flat rate is charged (cf. the decision of the Belgian Supreme Court cited by Libouton, (1973) 8 E.T.L. 9, n. 11), the real difficulty being caused where the agent purports to be passing on an outgoing which does not represent the true cost to him as in the *Marston* case. Where a forwarder contracts as agent, he may nevertheless be free to perform the carriage himself but he will not be entitled to charge so as to make a profit over and above the least cost to get the work done: *Immediate Transportation Company Ltd.* v. *Speller, Willis & Co.* (1920) 2 W.L.R. 645.

95. This did not, it seems, make the forwarders the sender. Rather, the judge considered that the effect of section 14(2) has the effect of deeming the plaintiff to be party to the contract as sender. See further, however, *supra*, para. 1.11, and *infra*, para. 11.15 fn. 26.

96. [1994] 1 Lloyd's Rep. 669.

97. And in this case, unlike in the *Bardiger* case, would have resulted in there being no CMR carriage, at least in relation to the whole of the transport of the goods.

intermediate role of this nature will need to be supported by evidence demonstrating clearly that this was the intention.[98]

1.42 Another instance is provided by the case of *Lacey's Footwear (Wholesale) Ltd.* v. *Bowler International Freight Ltd.*,[99] where it was contended on behalf of the defendants that they had contracted only as the English handling agents for the actual carriers. Reliance was placed in part on the fact that their standard terms stipulated that they contracted solely as agents and not as carriers, and that they did not make or purport to make any contract of carriage. However, the terms also provided that they were entitled to perform the contract themselves. Further, in the course of attempts to solicit the plaintiffs' business, Bowler had described themselves in terms of being carriers. Their Annual Report described the principal business of the company as being that of international haulage contractors, and the freight account was submitted to the plaintiffs and not to the actual carriers. In addition, a letter sent after the event from Bowler to the actual carriers stated "we contracted this consignment to you for delivery . . . ". It was held that Bowler had contracted as carriers.

1.43 On the other hand in *Inco Europe Ltd.* v. *First Choice Distribution*,[100] the most recent case in this country on the issue, the Court of Appeal held that the plaintiffs had failed to make out a good arguable case that the forwarder had contracted as carrier. The forwarder had contracted subject to the FENEX[101] conditions, the tenor of which the court regarded as being to the effect that the forwarder acts as agent; the forwarder had been contracted to "arrange" the transport, which was described as indicative but not conclusive; the CMR note described the forward as sender; the invoice which referred to "truck-freight" was described as neutral in the following terms:

"It is as consistent with Steinweg being a forwarder as with its being a carrier. As a forwarder and consignor Steinweg would be liable for the freight to the carrier and would be entitled to charge [the claimants] for that sum, which is what occurred."[102]

However, unfortunately the court appears to have overlooked the fact that, presumably, the freight charged to Steinweg by the carrier, and that charged by Steinweg to the claimant, was not the same amount.[103]

Burden of proof

1.44 On principle, he who alleges should prove. In practice, this will mean that the onus of proof will be on the plaintiff, who will commonly be seeking to establish the

---

98. In *Kala Ltd.* v. *International Freight Services (UK) Ltd.*, Queen's Bench Division, 7.6.88, unreported except on LEXIS, the practice of a forwarder was explained as issuing a "Bill of Lading" to their customer which stated that they "do not act as carriers but as forwarding agents. In consequence they are only responsible for the careful selection of third parties instructed by them". The forwarder would then sign the CMR consignment note as sender, consigning the goods to their own agent. It was accepted by the parties that no contract of carriage existed between the forwarder and the customer. The court accepted that in detaining the goods the forwarder was acting under rights granted by the customer and confirmed by CMR. The "Bill of Lading" made it clear that the forwarder would subrogate claims against the actual carrier on request. This is the common position with Continental forwarders, see *infra*, para. 1.46 ff. In *Aqualon*, it was suggested that a forwarder in such an intermediate position would owe a duty to enforce the carriage contract for the benefit of the consignor. Unless this is expressed, it is hard to see how it can be implied. In *Bardiger* such a duty was not referred to and would have been unnecessary on the view of the judge that the goods owner had a direct statutory right as "sender".

99. [1997] 2 Lloyd's Rep. 369.

100. [1999] 1 All E.R. 820.

101. The General Conditions of the Netherlands Association for Forwarding and Logistics.

102. Hobhouse, L.J, at p. 830.

103. This was an interlocutory application only and the extent of the evidence before the court is not apparent from the report.

responsibility of the forwarder as carrier. However, it may be suggested that there should be a reversal of the burden of proof, as in the case of a claim for loss or damage against a carrier,[104] on the grounds that the relevant evidence is more often in the hands of the forwarder than his customer.[105] In essence, the search should be for the intention of the parties objectively ascertained, and often evidence of objective intention may only be provided by facts drawn from the relationship of the forwarder with the performing carrier than with the customer. Nevertheless, the customer will normally be seeking to present his view of the contract that he believes himself to have made in the particular case.[106] In the standard case where there is some element of doubt, it would seem unlikely that the plaintiff will have much difficulty in at least setting up a *prima facie* case.[107]

1.45 In Belgium, a line of authority, noted in the first edition of this book, indicated that in that country, the burden of proof rested on the party interested in the goods. An opposite tendency in these courts has been noted.[108] A similar approach has also been taken in France in respect of whether a carriage or hire contract has been made,[109] but has been criticised.[110] This line of authority, however, places emphasis on a requirement for the forwarder engaged in a commission involving transport to be clear that no obligation for carriage is being undertaken.[111] It may be that an English claimant will find himself far advanced into establishing a *prima facie* case if he can show that no clear indicators of agency were present in the contract he made with the forwarder.

Continental case law

1.46 Before attempting to deduce any general guidance from these cases, it is useful to have regard to the Continental case law on the same issue. However, before doing so, it is important to realise that, as distinct from English law, where the precise status of the forwarder is largely undefined, under Continental legal systems the forwarder is given a specific legal status under the commercial law provisions of the various legal codes.[112] In particular, the difficult point for the English common lawyer to grasp is that the forwarder under most Continental systems, while remaining liable to his client as an agent, does for some purposes contract with the actual carrier as a principal. So under English law, if the forwarder is contracting as an agent rather than as carrier, he will create privity of contract

---

104. See Palmer, *Bailment*, 2nd edn., p. 49.

105. A view stated only to be rejected by Clarke, p. 30.

106. And, no doubt, can rely on facts indicating a holding out that a particular service is being offered, see *Luigi Riva di Fernando* v. *Simon Smits Co. Ltd.* (1920) 2 Ll.L.Rep. 279, *Claridge, Holt & Co.* v. *King & Ramsey* (1920) 3 Ll.L.Rep. 197, *Landauer* v. *Smits* (1921) 6 Ll.L.Rep. 577, 578.

107. The *Texas Instruments* case, discussed *supra*, paras 1.33–1.38, may well reflect the difficulty a carrier may have in displacing a *prima facie* case where there is evidence of his status as a carrier and his need is to displace that evidence by proving the equivalent status on the part of another party in the face of the plaintiff customer's contention that he had indeed contracted as carrier.

108. Libouton, (1982) J.T. 694 No. 4, applied by Comm. Liège, 27.6.85 (1985) 20 E.T.L. 572, 576. The decision of RB. Antwerpen, 27.2.67 (1968) 3 E.T.L. 1244, noted in the first edition of this book, is said (by Haak, at p. 64 n. 124) to have been quashed by the decision of Hof. Brussel, 24.1.69 (1969) 4 E.T.L. 937. The same tendency in favour of the forwarder's customer has been noted in relation to the Dutch courts, see Haak, p. 67.

109. Cass., 5.2.85 (1986) B.T. 350, cf. App. Nimes, 26.5.82 (1982) B.T. 407, and Lamy, para. 568.

110. By Clarke, loc. cit., cf. Haak, p. 64.

111. As in Hof. Brussel, 26.4.83 (1983) 18 E.T.L. 511.

112. See generally Hill, op. cit., p. 50 ff. See also "Les auxiliaires de transport dans les pays du marché commun", IDIT, 1977.

between his principal and the carrier. His principal will therefore be the consignor or consignee for the purposes of the Convention and have direct rights of action against the carrier under the Convention.

1.47 In contrast, under the Continental systems, the forwarder is normally considered to contract under a contract of commission, and unlike other mandataries, contracts with the carrier in his own name and does not create privity of contract between his customer and the carrier.[113]

1.48 The forwarder will therefore take the position of sender, and only if he assigns his rights to his principal will the principal have a direct right of action against the Convention carrier. It is common, however, for the trading conditions of Continental forwarders to give the forwarders' customer a right to such an assignment or to require the forwarder to otherwise take action or assist the customer in making a claim against the person responsible.[114] In Holland, the right to an assignment is given statutory recognition under the New Civil Code.[115] The only contractual redress will thus be against the forwarder, and because that contract is not a contract of carriage, it will not be subject to CMR.

1.49 Broadly, however, the Continental approach to the distinction between the contract of forwarding and the contract of carriage is similar to that as outlined above in relation to this country. It must be stressed that the *consequences* of that distinction may be different, as outlined above, but the distinction itself is concerned, in a like manner to that discussed above in relation to English law, with the difference between a contract to *arrange for* the execution of the carriage, and to a contract to *effect* that carriage. Consequently, similar criteria can be found to operate in the cases. A consideration of the factors used in the Continental courts may provide further guidance in respect of this issue and are therefore outlined below. This is directed primarily to the position in France, Belgium, Holland and Germany where issues of the relationship between forwarding operations and CMR have undergone considerable development. Relevant cases can, of course, be found in other jurisdictions.[116]

1.50 A caveat must be drawn, however, in respect of French law. In other Continental countries the contract of commission involves a liability towards the principal similar to that of an agent in English law. This is not the case in France, where the *commissionnaire de transport* is subject to a strict liability.[117] This liability is subject to freedom of contract

---

113. Exceptionally under the French *Code de Commerce* (Article 101) the "*lettre de voiture*" is deemed to create a contract between the shipper and the carrier, or between the shipper, the forwarder and the carrier. The shipper therefore acquires direct rights against the carrier under the contract. See Hill, op. cit., p. 54. In such a case the forwarder's customer can take the status of sender (*expéditeur réel*) under CMR notwithstanding that he is not named as sender in the consignment note, Lamy, para. 555, Cass., 27.10.75 (1975) B.T. 426. The forwarder may also maintain a recourse action against the carrier, likewise governed by CMR, Lamy, vol. 2. paras. 61 and 66; see e.g. Comm. Nantes, 10.9.98 (1998) B.T. 780.

114. Examples are: clauses 51(a) and 52(a) of the German Forwarders' Standard Terms and Conditions (ADSp), 1993, see clause 26 of the conditions of the Confederation des Expediteurs de Belgique 1980, clause 11(5) of the Dutch Forwarding Conditions (FENEX) 1992, cf. also *Kala Ltd.* v. *International Freight Services (U.K.) Ltd.*, *supra*, where the "Bill of Lading" issued by the forwarder contained a clause stating that "the undersigned forwarding agents will, on request, subrogate their claims against carriers and other parties".

115. N.B.W., section 8.2.3.3–2, reproduced in clause 16(4) of the FENEX Conditions 1992.

116. E.g. Portugal: the issue of a forwarding agent's certificate of receipt did not prevent the issuer from being regarded as the "effective" carrier and, in consequence, subject to CMR, Supremo Tribunal da Justiça, 17.11.94 (1997) U.L.R. 200.

117. An *obligation de résultat*, which means a responsibility for the proper execution of the transport by those whose services he procures as well as for his own personal fault: Articles 97–99 of the *Code de Commerce*.

and does not engage the forwarder in liability under CMR.[118] Nevertheless, it reflects a role closer to that of a transport operator. This means that some factors which in England and in other countries would be relevant to distinguishing the forwarder from the carrier are of more importance in France in distinguishing a *commissionnaire de transport* from a simple *transitaire* whose role and consequent liability is that of an agent.[119]

1.51 In France therefore, the fact that the carriage has been sub-contracted has traditionally been a strong indicator of status as *commissionnaire de transport*. The modern view, however, is to consider the common intention of the parties,[120] and a clear indication of this common intention will be overriding.[121] Thus an operator whose consignor's shipping instruction (*feuille d'expedition*) mentioned that he was chosen as transporter with an order to transport was held to be a CMR carrier notwithstanding that he sub-contracted the whole of the carriage.[122] Of central importance will be the transport document itself[123] and, in particular, whether the operator is named as transporter.[124] In one case, where a forwarder was involved in a transport from Nimes to Hull, the goods being consigned under a groupage arrangement to London and then on-carried to Hull, the forwarder was held to be a *commissionnaire de transport* and not a CMR carrier. The CMR consignment note issued by the groupage carrier named the forwarder as sender and the on-carrier as

118. Nor can the forwarder rely on it except to the extent permitted by French national law, cf. Cass. 25.6.96 (1996) B.T. 507. Normally, however, the *commissionnaire de transport* can rely on the regime of liability which applies to the carrier, except in the case of personal fault, and can rely also on the domestic rules relating to limitation of action, see App. Paris, 18.12.96 (1997) B.T. 570, Cass., 12.11.97 (1997) B.T. 820, see further, Lamy, para. 452.

119. Such as the method of charging. An all-in charge is more indicative of status as a *commissionnaire de transport* than as a *transitaire*, Lamy, vol. 2, para. 19, and see App. Paris, 28.10.81 (1982) B.T. 311, App. Paris, 15.3.83 (1983) B.T. 305 and App. Paris, 25.4.84 (1985) B.T. 61. See further App. Paris, 27.11.96 (1997) B.T. 405. Haak, p. 62, criticises the approach of French law since the fact that the *commissionnaire de transport* assumes a responsibility to ensure that the transport is effected should be sufficient to engage a responsibility under CMR. The French approach thus narrows the scope of CMR in a way which is not conducive to the interests of uniformity.

120. The changing view of the courts on this is traced by Brunat, (1982) B.T. 422; see also Comm. Paris, 14.3.78 (1978) 13 E.T.L. 742.

121. Lamy, vol. 2, para. 6. A lower court which fails to mention the background documents and elements on which it bases its decision is liable to have its decision referred back on cassation, since the mere fact of sub-contracting is not sufficient necessarily to imply status as a *commissionnaire de transport*, Cass., 25.6.79 (1980) 15 E.T.L. 79.

122. App. Paris, 17.11.83 (1984) B.T. 390. See also the non-CMR decision, Cass., 2.2.88 (1988) 23 E.T.L. 223, (1988) B.T. 490 and the comments of Chao (1988) B.T. 477.

123. Cf. Cass., 29.10.91 (1991) Bull. IV No. 326 (Krings, p. 157) where two consignment notes were issued indicating different persons as carriers. For the court it was necessary for the lower court to have considered whether the issue of the second consignment note exonerated the "carrier" indicated in the first.

124. Lamy, vol. 2, para. 6 citing, among others, Cass., 13.10.80 (1980) B.T. 598 and Cass., 17.12.80 (1981) B.T. 155 (these cases were in fact concerned with the issue of whether the relevant operator was a hirer or carrier, see *supra*, paras 1.3–1.4; see also Comm. Fleurs, 6.11.81 (1982) B.T. 297, Cass. 3.3.98 (1998) B.T. 231, (1998) U.L.R. 897, App. Paris, 18.12.96 (1997) B.T. 570. Putting a stamp and signing the CMR consignment note in the "signature of carrier" box signified that the forwarder was the carrier, notwithstanding that another operator had rubber-stamped the "carrier" box, in App. Paris, 23.11.95 (1996) B.T. 257, but being named as carrier in the sender's headed note was insufficient in the absence of the forwarder's signature in App. Paris, 26.3.97 (1997) B.T. 764. Cf. App. Paris, 22.4.77 (1978) 13 E.T.L. 407 where the forwarder was named as sender on the CMR consignment note and received, in effect, an order to organise the carriage and was considered therefore not to be a CMR carrier. As between the *commissionnaire* and the actual road carrier CMR can apply whether or not the *commissionnaire* is named as sender in the consignment note: Cass., 25.6.91 (1991) Bull. IV No. 243 (Krings, p. 148). When seeking recourse it is important that the *commissionnaire* keeps within the time limit prescribed by Article 32, see e.g. the difficulty faced by the *commissionnaire* in App. Versailles, 20.5.99 (1999) B.T. 866.

consignee.[125] Apart from the transport documents, there may be other evidence of intention: for example, the issue of an invoice for *"traction"* in one case led to classification as a CMR carrier.[126] In Belgium, the distinction is drawn between a *commissionnaire de transport*, whose legal responsibility is assimilated to that of a carrier,[127] and a *commissionnaire-expéditeur* who has the responsibility of an agent under a contract of commission. The position in respect of the burden of proof has been noted above.[128] An operator given an instruction which used the words: "transport of machine to U.S.S.R." and "please transport by special truck", and who did not give the sender information about the transport agreement which the operator claimed to have concluded on his behalf, was held to be a CMR carrier.[129] This was despite the facts that he did not have his own vehicles,[130] he had previously contracted as a forwarding agent only, and that he had referred in correspondence to conditions of trading which were inconsistent with his being a carrier (there being no proof that these conditions had been accepted by the sender).[131] In another case, acceptance of an instruction to deliver "by your groupage lorry" was held to indicate a contract of carriage,[132] whilst an instruction to *"expédier"* indicated forward-

125. Cass., 21.6.82 (1982) B.T. 427, cf. App. Paris, 1.7.81 (1981) B.T. 499, App. Paris, 15.3.83 (1983) B.T. 305, App. Nimes, 26.5.82 (1982) B.T. 407. The court characterised the contract as one whereby the forwarder contracted with freedom to organise the transport *"de bout en bout"*, cf., Cass. 6.10.92 (1992) Bull. IV No. 298 (Krings, p. 147). The more complex the movement (the role of the *commissionnaire* being one of co-ordination) the more likely the contract will be one of commission rather than transport, and vice versa, cf. Cass, 13.2.78 (1978) B.T. 210, (1979) II U.L.R. 239 (see further App. Aix-en-Provence, 13.2.97 (1997) B.T. 570 in the context of a ro-ro transport from Morocco to Marseille where the *commissionnaire* performed the local delivery but was not classified as the CMR carrier). See also Brunat, (1982) B.T. 422, 423, who comments on the fact that the sender was not informed of the actual carriage arrangements; cf. the *Tetroc* case, *supra*, para. 1.18. Where the *commissionnaire* does not have freedom of organisation because the choice of carrier has been imposed on him he does not have to answer for the carrier: App. Riom, 11.3.98 (1998) B.T. 362.

126. App. Paris, 20.10.81 (1982) B.T. 38. In App. Paris 5.4.95 (1995) B.T. 811 the fact that the defendant hired a vehicle and driver for the journey indicated that he was simply a transporter, but see the comment to the case. Other factors include whether the operator is registered as a *commissionnaire* and not insured for transport activities: App. Dijon, 31.1.95 (1995) B.T. 183.

127. Formerly by Articles 2 and 4 of the Law of 25 August 1891 as incorporated in Titre VII bis, Livre 1 of the *Code de Commerce*, see further Libouton, *"Les auxiliaires de transport dans les pays du marché commun"*; see further Haak, p. 63, Putzeys, p. 35, Clarke, pp. 28–29, Libouton, (1973) 8 E.T.L. 2, 6, (1974) J.T. 505, (1982) J.T. 693. This assimilation is retained under the new law of 3.5.1999 (Article 38(5)) which now makes CMR applicable to domestic carriage.

128. See *supra*, para. 1.44. It is necessary to be astute to the attitude taken by the relevant court to this, since this will affect the weight given to particular factors.

129. Hof. Brussel, 19.10.72 (1973) 8 E.T.L. 503, (1974) 9 E.T.L. 608. It has been said of this case that the court took a more cautious approach (than, e.g., Hof. Brussel, 23.12.71 (1972) 7 E.T.L. 865, see *infra*, fn. 130) and that, while it took the words used by the parties at face value, it was ready to look behind the appearances if the common intention of the parties was different, see Clarke, p. 28; see also Libouton, (1974) J.T. 505, para. 3, who cites App. Bruxelles, 25.5.72 (1972) J.P.A. 219, where the statement *"Nous mettons à votre disposition des marchandises pour expedition par camion à telle date à telle destination"* was judged reconcilable with transport rather than *expedition*, save where other factors indicate a contrary intention.

130. Similarly, in respect of this factor, Comm. Verviers, 7.4.79 (1979) 14 E.T.L. 664 where a clear order for transport was accepted by the forwarder who wrote to the claimant that the goods were sent "by our lorry".

131. To similar effect: Hof. Brussel, 5.12.68 (1969) 4 E.T.L. 958. See also Civ. Charleroi, 1.10.68 (1969) J.P.A. 129, cited by Libouton, (1973) 8 E.T.L. 9, where the forwarder wrote a letter saying *"we can undertake the carriage of the machines for the sum of … "*; similarly, Comm. Bruxelles, 26.10.72 (1973) 8 E.T.L. 516, where the firm wrote an offer to the customer to effect the transport.

132. Hof. Brussel, 23.12.71 (1972) 7 E.T.L. 865. Libouton states, at (1982) J.T. 694, that "groupers" are, in effect, generally considered as transporters, citing Comm. Anvers, 26.9.80, unpublished, cf. App. Bruxelles, 4.12.69 (1969) J.P.A. 440 in Libouton, (1973) 8 E.T.L. 2, 10.

ing rather than transport,[133] and it has also been held that the indication of a party as carrier on the CMR consignment note will justify a finding that they have the status of a carrier.[134] Again, when an operator does not carry out the transport himself, but issues the freight invoice,[135] or submits an account for "carriage" and describes the actual transporter as an "agent" or "branch", he will be treated as the CMR carrier.[136] The method of charging may also provide guidance. If an "all-in" charge is made, that is, a charge for freight which does not indicate any profit or commission element, this militates in favour of the operator being regarded as a CMR carrier.[137] The court may also take indications of a transport role from the operator's trading conditions,[138] or from its business description or commercial stationery.[139]

1.52 In Holland, the introduction of Book 8 of the New Civil Code has added a new dimension to the liability of a forwarder which could have implications for his classification for the purposes of CMR. If the forwarder carries himself, he is deemed to be the carrier.[140] If he employs a carrier, and loss or damage occurs, he must inform the customer of what arrangements have been made and disclose all documents relevant to a claim. Should he fail to do so, he is liable to the customer to the same extent as if he had made the contract himself.[141] This would seem to provide a significant foundation for the protection of the forwarder's customer.[142]

133. Cass., 5.11.36 (1937) J.P.A. 70, and *"re-expédier"*, Hof. Brussel, 26.6.74 (1975) 10 E.T.L. 235, affirming RB. Antwerpen, 5.5.71 (1975) 10 E.T.L. 230. Cf. Cass., 9.4.76 (1976) Pas. 889, see Libouton, (1982) J.T. 694.

134. Cass., 17.9.87 (1988) 23 E.T.L. 201 (on this point a similar decision to the *Texas Instruments* case, *supra*, para. 1.33). See also Libouton, (1982) J.T. 694 and the case there cited: Comm. Tongres, 29.6.79. Also cited are cases which have considered that a party may still contract to carry notwithstanding that the consignment note is issued by the sub-contractor (App. Anvers, 9.5.79) or names the operator as sender (Comm. Anvers, 26.9.80).

135. Comm. Bruxelles, 12.2.77 (1978) 13 E.T.L. 285.

136. Hof. Brussel, 21.1.69 (1969) 4 E.T.L. 943. Cf., however, Comm. Bruxelles, 4.2.72 (1972) 7 E.T.L. 573 (see Libouton, (1974) J.T. 505) where the forwarder's reference to the performing carrier as his sub-contractor did not displace the finding that he had not contracted to carry.

137. Comm. Bruxelles, 28.9.77 and other cases cited by Libouton, (1982) J.T. 693, 694, also App. Bruxelles, 24.1.69 (1969) 4 E.T.L. 937, (1970) U.L.C. 17; App. Bruxelles, 26.4.83 (1987) R.D.C.B. 145. Contrasting cases can be found: Hof. Brussel, 23.6.66 (1967) 2 E.T.L. 1006, and see Libouton, (1973) 8 E.T.L. 2, 9.

138. App. Bruxelles, 24.1.69 (1969) 4 E.T.L. 937, (1970) U.L.C. 17: "the organiser of daily international transportation must be held to be a carrier and not a forwarding agent when the firm's General Conditions emphasise the carriage procedure rather than mere forwarding instructions, when the order is accepted without the firm's making quite clear that it is not undertaking any service of carriage in its own identity and when a bill is rendered for 'road freight' at the same time as for various expenses such as insurance, Customs formalities, telephone", Libouton, (1973) 8 E.T.L. 2, 9). Cf. App. Bruxelles, 26.4.83, *supra*, fn. 136.

139. Comm. Bruxelles, 12.2.77 (Libouton, (1982) J.T. 694), mentioning transport as an activity in correspondence; similarly, *"Transport Européens"* in Comm. Bruxelles, 24.1.73 (1973) J.C. 201; Comm. Courtrai, 13.7.78, (Libouton, loc. cit.), description of the company as *"société Belge de transport"*; Hof. Brussel, 23.12.71 (1972) 7 E.T.L. 865, *"consortium de transports Européens"*; cf. Comm. Bruxelles, 4.2.72 (1972) 7 E.T.L. 573, where the description *"union de transporteurs par autocamions"* was indicative of the organisational role appropriate to commission rather than to transport. Other cases have treated such terminology to be of less significance, emphasising rather the particular contract made by the operator whose general activities may not be confined to a single type of operation: App. Bruxelles, 25.5.72 (1972) J.P.A. 219 (Libouton, (1974) J.T. 505).

140. N.B.W., Section 8.2.3.2, see Article 16(1) of the FENEX Conditions 1992.

141. N.B.W., Section 8.2.3.3, see also Article 16(2) and (5) of the FENEX Conditions 1992.

142. It is, as yet, unclear how far the provisions of the Civil Code will be considered relevant to classification of international road carriage operations. Decisions of the Dutch courts prior to the adoption of these provisions utilised similar factors (and conflicting decisions) to those indicated *supra*, (see further, Haak, p. 67) e.g. Arrond. Rotterdam, 10.11.70 (1971) 6 E.T.L. 273, Hof. s'Gravenhage, 11.11.77 (1978) 6 E.L.D. 423 ("all-in" charge); cf. GH. Amsterdam, 6.1.66 (1969) 4 E.T.L. 151 where charging a flat rate was displaced by the request for and

1.53 In Germany, provisions of the Commercial Code specifically relate to forwarding and indicate circumstances where the forwarder is placed in a position equivalent to that of a carrier.[143] These provisions have been utilised to assimilate the position of the forwarder to that of a carrier both in domestic law and in respect of CMR cases.[144] However, the factors indicated in the Commercial Code are not the only ones to which the German courts have made reference in determining the status of a forwarder. So for example in the context of air carriage, a forwarder named as carrier in the air waybill was held to have contracted as carrier.[145]

Conclusions

1.54 It will be observed therefore that there is considerable similarity between the approach adopted by the English courts and that indicated by the Continental decisions. The search ultimately is for the objective understanding of the parties of their relationship, the crucial distinction being, as already stated, between the forwarder contracting to arrange the carriage and contracting to undertake responsibility for the performance of the carriage. Some tentative conclusions may be drawn:

issue of a forwarding agent's receipt; RB. Amsterdam, 8.4.64 (1964) U.L.C. 169 ("our liners", "our vehicles" does not necessarily imply carriage); Hof. The Hague, 9.1.70 (1970) S. & S. No. 97 ("both transports were properly carried out by us" refers to carriage). Previous dealings between the parties have been considered relevant in Holland, Hof. The Hague, 11.11.77 (1978) S. & S. No. 58, RB. Breda, 7.7.81 (1982) S. & S. No. 96. Haak has noted, critically, at p. 77 the decision in RB. Amsterdam, 5.12.79 (1980) S. & S. No. 96 whereby a party was held to be a transport operator but not subject to CMR (but subject nevertheless to Article 91 of the former commercial code in the absence of incorporation of the FENEX Conditions) contrary to the decision in RB. Almelo, 22.8.79 (1980) S. & S. No. 61.
143. Both under the former provisions of the Commercial Code and the new provisions introduced by the Transport Law Reform Act which came into force on 1.7.1998 (see Herber, "The New German Transport Legislation", (1998) 33 E.T.L. 591, Looks, Kraft and Bird, "The Reform of German Transport Law", [1999] I.T.L.R. 93). The former provisions were ostensibly concerned with regulating the forwarder's right of commission, but the new provisions more clearly assimilate the forwarder's position to that of a carrier. They apply where the forwarder carries the goods himself (*Selbsteintritt*), charges a fixed sum for transportation (*Spediteur zu festen kosten*) or deals with groupage (*Samelladungverkehr*). Similar rules are applicable in Austria: OGH., 14.7.93 (1994) TranspR 189, OGH., 17.3.98 (1998) TranspR 361, LG. Salzburg, 1.9.93 (1994) TranspR 120.
144. This is traced by Haak, p. 68 *et seq*. The former Article 413(2) was applied in the context of the German long distance haulage regulations (KVO) to subject a forwarder's groupage operation to these mandatory rules, BGH., 3.3.72 (1972) VersR 431 (cf. BGH., 13.1.78 (1978) N.J.W. 1160, (1978) 13 E.T.L. 402, where the forwarder was held not to have the liability of a carrier in respect of damage occurring in the warehouse prior to despatch with the actual carrier). This broad interpretation of the commercial code was then applied in respect of CMR in the decision of the BGH., 18.2.72 (1972) VersR 729 whereby the Supreme Court held a forwarder who had charged a fixed sum to be a CMR carrier (also BGH., 21.11.75 (1976) VersR 36, (1976) N.J.W. 1029). The KVO was subsequently amended (in 1978 by insertion of Article 1(5)) to provide that a forwarder assimilated to a carrier by reason of Articles 412–413 was subject to the compulsory liability rules of the KVO only if he carries with his own vehicles. This did not, however, affect the approach previously taken in respect of CMR as confirmed by BGH., 10.2.82 (1982) VersR 543, (1983) 18 E.T.L. 32, applied in respect of groupage from the point that the goods are consigned to the actual carrier: OLG. Hamburg, 18.6.92 (1992) TranspR 421 up to the time of delivery to the person specified in the forwarding contract, BGH, 25.10.95 (1996) TranspR 118 (1997) U.L.R. 414. The position of the forwarder is now regulated by the provisions of the new commercial code but without change to the position under CMR.
145. OLG. Hamburg, 10.9.74 (1975) VersR 660, noted by Schoner, "The freight forwarder as an air carrier" (1980) 5 Air Law 2, similarly, in a case applying CMR, LG. Munich, 22.1.97 (1998) TranspR 166. Cf. however, OLG. Oldenburg, 25.9.96 (1997) TranspR 229. The emphasis on the criteria in the commercial code has been criticised as contrary to the spirit of uniformity, Haak, p. 73; see further Arnade, (1992) TranspR 341. In the context of CMR, see also OLG. Munich, 5.7.89 (1990) TranspR 16 and OLG. Munich, 23.7.96 (1997) TranspR 33, (1998) U.L.R. 206. The Munich court sees the question as an autonomous one divorced from national law and dependant more on the practice of, and interpretation by, the courts of other States party to the Convention. See further OLG. Munich, 8.7.97 (1998) TranspR 353.

(a) To avoid the difficulties outlined above, it is important for the parties to spell out in the contractual documents exactly what their mutual rights and obligations are. The courts will, however, look to the substance of the transaction and what the parties call themselves will not be conclusive, although potentially relevant.

(b) An operator can be treated as a carrier despite the fact he performs no part of the carriage himself. The mere fact that he owns no vehicles is not conclusive, although the customer's knowledge of that fact may militate against an allegation that the customer understood him to be undertaking responsibility as a carrier.

(c) An operator who apparently accepts responsibility for the whole of the transport arrangements, without specifying that he is arranging rather than effecting the transport, will generally be held to be a CMR carrier. This may be reinforced by the way the service or the scale of the operator has been represented. Any previous course of dealing between the parties or well-known course of dealing of the operator can be relevant to confirm or negate the alleged role.

(d) An operator who charges for the whole operation, or specifically charges for carriage, will generally be held to be a CMR carrier.

(e) If the operator has full control over the performance of the carriage, without the involvement of the shipper in the arrangements made, this may well indicate that he is a CMR carrier.

(f) The indications made on the CMR consignment note will be important where the alleged carrier is named as such on it and where, given the circumstances surrounding the making of the contract, this is likely to reflect the understanding of the customer.

At the end of the day, however, the question is one of fact in every case, so it must be stressed that none of the above can be regarded as conclusive in individual cases. One is left with the words of Bean, J., quoted with approval by Judge Martin[146]:

" . . . it is very much a matter of impression, what impression the evidence forms."

Supplementary provisions

1.55 Article 1(1) therefore contains the general rule as to the application of the Convention. The remainder of Article 1 is concerned with miscellaneous supplementary points:

> "2. For the purposes of this Convention, 'vehicles' means motor vehicles, articulated vehicles, trailers and semi-trailers as defined in article 4 of the Convention on Road Traffic dated 19th September 1949.
> 3. This Convention shall apply also where carriage coming within its scope is carried out by States or by governmental institutions or organizations.
> 4. This Convention shall not apply:
>> (a) to carriage performed under the terms of any international postal convention;
>> (b) to funeral consignments;
>> (c) to furniture removal.
> 5. The Contracting Parties agree not to vary any of the provisions of this Convention by special agreements between two or more of them, except to make it inapplicable to their frontier traffic or to authorise the use in transport operations entirely confined to their territory of consignment notes representing a title to the goods."

146. See *supra*, para. 1.18.

These provisions can be shortly dealt with, no points of real substance being involved.

1.56 First, with reference to the definition of "vehicles", this is set out in Appendix A by way of a note to Article 1(2). It is suggested elsewhere[147] that reference to the definitions contained in Article 1 of the Convention on Road Traffic of 1968 should be read into CMR in place of the reference to the definitions in the 1949 Convention, but it is doubtful whether this can be done in this country in the absence of specific legislative enactment, and in any event the respective definitions do not significantly differ for the purposes of CMR, so the question seems entirely academic. Also in connection with the definition of vehicles, it has been held in Denmark that a special ship's trailer, used for the purpose of the ocean crossing, is not a vehicle for the purposes of CMR.[148]

1.57 Article 1(3) applies the Convention to carriage performed by States or governmental institutions, but the requirement of Article 1(1) that the carriage be for reward is likely to exclude anything other than commercial operations. Article 1(3) refers to carriage by States or governmental institutions, but plainly carriage *for* such bodies will equally be covered by the Convention, provided it otherwise complies with Article 1.

1.58 So far as Article 1(4) is concerned, one possible difficulty is where mixed consignments are involved, but there seems no reason why the Convention should not apply to those parts of the consignment which do not fall within the exceptions listed. The other difficulty is that neither "funeral consignments" nor "furniture removal" are defined by the Convention. In relation to the former, Loewe distinguishes between transport operations directly related to funerals—giving the example of flowers and wreaths accompanying the coffin containing the "mortal remains"—and the carriage of objects merely intended for funerals.[149] He is doubtless correct.

1.59 Rather more difficulty, and rather more practical importance, might attach to the definition of "furniture removal". On its strict wording, a distinction might be drawn between furniture and other household effects, but it seems unlikely that such was intended.[150] This would seem to be confirmed by reference to the French text, in which the equivalent exclusion is in respect of "*transports de déménagement*". It would also seem from a Belgian decision that a distinction is to be drawn between furniture removals and a contract for the carriage of furniture. The former involves the removal of furniture from premises, packing them up, and the reverse procedure at the other end of the transit. CMR was therefore held to apply to a contract for the carriage of a container already packed with furniture.[151]

---

147. Loewe, paras 30–33.

148. Supreme Court, Denmark, 28.4.89 (1989) 24 E.T.L. 345, 357. Cf. OLG. Hamburg, 13.3.93 (1994) TranspR 193, where it was held that a swap-body (a form of trailer without wheels) is a vehicle in the sense of Article 1 of CMR, but, *quaere*, see further Basedow, (1994) TranspR 338 and OLG. Hamm, 2.2.95 (1996) VesR 1304.

149. Para. 35. The exception concerns funerals involving human remains, see Pesce, Theunis, p. 12, n. 49. There is a European Convention of 26.10.73 on the carriage of human remains.

150. It was intended that the whole question of furniture removals be the subject of a separate Convention, but subsequent negotiations were eventually abandoned. See Loewe, para. 36. The carriage of exhibition goods is not furniture removal: RB. Rotterdam, 12.2.82, affirmed by Hof. 's Gravenhage, 4.5.84 (1984) S. & S. No. 116, cited by Haak, p. 51 n. 59.

151. RB. Antwerpen, 1.4.80 (1980) 15 E.T.L. 461. Cf. Hof. Antwerpen, 16.10.85 (1986) 21 E.T.L. 377. A different distinction suggests removal as involving a change of establishment of goods in the same ownership. On that basis, a transfer of goods being the subject of a sale would not be a "removal": OLG. Hamburg, 28.2.85 (1985) TranspR 436. See Pesce, Theunis, p. 12 n. 50 and Clarke, p. 36. See further, Herber/Piper, Art. 1 Rdn. 60 ff, Koller, Art. 1 Rdn. 10.

Voluntary adoption of the Convention

1.60 There is nothing to prevent the parties to a contract of carriage from voluntarily adopting the Convention in relation to a contract for carriage which would not otherwise be subject to it.[152] Indeed, where the standard form IRU consignment note is used, the Convention is likely to be incorporated in any event regardless of whether it would otherwise apply to the carriage, by virtue of the provision it bears on its face that "This carriage is subject, notwithstanding any clause to the contrary, to the Convention on the Contract for the International Carriage of Goods by Road (CMR)".

1.61 One difficulty that arises in relation to voluntary adoption is in cases where the Convention itself provides that it shall not apply. The type of situation envisaged is, for example, voluntary adoption in the context of a contract for the carriage of a coffin (containing the "mortal remains"[153]): can the Convention apply when the Convention itself provides that it does not? A problem of rather wider application is the voluntary adoption of CMR in relation to carriage within one State: again, Article 1(1) provides that it shall not so apply. The answer lies in the general contractual test of the presumed intention of the parties: implicit in the agreement that the Convention apply must be the exclusion of those provisions of the Convention which would prevent it so applying.[154] Thus in *Princes Buitoni Ltd.* v. *Hapag-Lloyd Aktiengesellschaft*,[155] a contract for multi-modal transport from Vancouver to Liverpool provided that CMR was to apply during the road carriage in Europe. It was held that this meant that the provisions of CMR would be applied to the road carriage which took place between Felixstowe and Liverpool. There were no grounds for limiting the clause so that it read "road carriage in Europe between two different countries in Europe". So too in *Ken Bell (International) Ltd.* v. *Eric Bacon Transport Services Ltd.*[156] it was held that the CMR Convention applied contractually to one part of the consolidated shipment which was being carried only within Italy. The apparent difficulty arising from the contradictory provision of Article 1(1) was not even addressed in the judgment.

1.62 A further problem in relation to voluntary adoption of the Convention concerns the application of the Unfair Contract Terms Act 1977 which negates exclusion clauses deemed to be unreasonable within the meaning of that Act.[157] That Act does not apply to cases governed by CMR by virtue of section 29(1)(a) which excludes the application of the Act to any contractual provision "authorised or required by the express terms or

---

152. So, for example, the Convention might be voluntarily adopted in cases of container shipments: see *infra*, para. 2.8 ff. and see also *Buchanan (James) & Co. Ltd.* v. *Babco Forwarding & Shipping (U.K.) Ltd.* [1978] 1 Lloyd's Rep. 119.

153. See *supra*.

154. See *Adamastos Shipping Co. Ltd.* v. *Anglo-Saxon Petroleum Ltd.* [1958] 1 Lloyd's Rep. 73. Cf. Hof. Antwerpen, 16.10.85 (1986) 21 E.T.L. 377 cf. however App. Riom, 11.3.98 (1998) B.T. 362

155. [1991] 2 Lloyd's Rep. 383.

156. C.L.C.C. (B.L.) 22.7.98, unreported.

157. See generally Yates, *Exclusion Clauses in Contracts*, 2nd edn. Cf. in respect of mandatory laws applicable in other countries, e.g. Holland, HR., 26.5.89 (1990) E.T.L. 717, (1990) II U.L.R. 438. See further, Haak and Swart, *Road Carriers Liability in Europe, Part 2*, 1995, p. 87 and Van de Laarschot, "De toepasselijkheid van de CMR op binnenlands wegvervoer", (1997) Tijdschrift Vervoer & Recht, 9, discussing RB. Rotterdam, 1.8.96 (1997) S. & S. No. 22 and RB. Rotterdam, 12.9.96 (1997) S. & S. See also Krings, p. 145. For Belgium see supra, fn. 153 and Hof. Antwerpen, 8.3.93 (1994) 29 E.T.L. 104. In France, a reference to CMR can displace the "*contrat type*" otherwise implied into a domestic contract of carriage: Cass., 1.7.97 (1997) B.T. 537. This is subject to those rules of national law which are of compulsory application, see Tilche (1997) B.T. 531, Bernadeau, "La CMR en tant que règle des transports intérieurs", (1998) 33 E.T.L. 785, cf., App. Bordeaux, 21.12.94 (1995) B.T. 74, (1996) U.L.R. 400.

necessary implication of an enactment". Where CMR is adopted voluntarily, however, the CMR exclusions and limitations are not required nor, strictly, are they authorised. However, even in cases of voluntary adoption, it may be that the Unfair Contract Terms Act is excluded because the contract is an "international supply contract" within the meaning of section 26 of that Act. But even if that section is inapplicable, as, for example, in the case of internal carriage, it would seem extremely unlikely that an English court would be willing to declare voluntary contractual reliance on the CMR exclusions and limits of liability to be unreasonable.[158]

1.63 Finally, in relation to voluntary adoption, it will be seen that Article 28[159] overrides normal common law rules of privity of contract by permitting persons not party to the contract of carriage to rely on the exclusions and limitations of liability. This would seem to be by virtue of the stipulation in section 1 of the Carriage of Goods by Road Act 1965 that the provisions of the Convention "shall have the force of law" in relation to a contract to which the Convention applies. Where the Convention is adopted voluntarily, it must be doubted whether normal rules of privity are similarly overridden so as to affect the rights and liabilities of persons not party to the contract of carriage.[160]

1.64 A different kind of adoption which is becoming a significant trend, is the incorporation of CMR into rules of national law. This can take the form of direct incorporation, as in Austria[161] and Belgium,[162] through the adoption of rules influenced by CMR as in Germany[163] and Holland[164] or through partial incorporation as contract terms imposed in the absence of terms adopted by the parties, as in Spain.[165] The prospect in the longer term may be the increasing use of the principles of CMR as a basis for national laws governing the carriage of goods by road and the gradual harmonisation of such laws throughout Europe.[166]

---

158. Apart from the Act, the EC Council Directive on Unfair Contract Terms in Consumer Contracts (93/13/EEC, 5.4.1993) is now applicable to carriage contracts made with consumers by regulations which came into force on 1.7.1995 (S.I. 1994 No. 3159).

159. *Infra*, Chapter 9, paras 9.65–9.71.

160. The party seeking to rely on them may be able to do so if a "Himalaya clause" in the contract with the sender is available for their benefit (*New Zealand Shipping Co. Ltd.* v. *A.M. Satterthwaite & Co. Ltd., The Eurymedon* [1975] A.C. 154), or if they can establish a bailment on CMR terms, *The Pioneer Container* [1994] 2 All E.R. 250.

161. By Article 1 of the National Carriage of Goods by Road Act 1990.

162. Law of 3.5.1999.

163. Under the Transport Law Reform Act, in force from 1.7.1998.

164. Through Title 13 of Book 8 of the New Civil Code, in force from 1991 and formerly through the *Wet Overeenkomst Wegvervoer*, in force from 1983.

165. See Sanchez-Gamborino, (1997) 32 E.T.L. 345.

166. A symposium entitled "CMR—A National Law" was held in Rouen on May 14, 1998, under the auspices of *l'Institut du Droit International des Transport*, where the prospects for this development were considered, see further Clarke, "Contracts for the Carriage of Goods by Road: Cabotage in the United Kingdom", (1998) J.B.L. 591 and "the Transport of Goods in Europe: patterns and problems of uniform law", [1999] L.M.C.L.Q. 36, Herber, "CMR: Unidroit Should Not Let This Child Go!", (1998) U.L.R. 475.

CHAPTER 2

Combined Transport

Scope

2.1 Combined transport, sometimes referred to as intermodal or multimodal transport, occurs where goods are transported successively by two or more different methods of transport. This may be either by transferring the goods from one method of transport to another, or alternatively by placing the one means of transport on to another for part of the transit. CMR only makes partial provision for combined transport operations in Article 2, since this provision only covers the latter variety of combined transport, so-called "piggy-back" operations. Partly as a result of this, Article 2 is not a satisfactory provision and is the source of considerable problems. These provisions are, of course, of particular significance for this country since, unlike other European countries, all shipments of goods to and from the United Kingdom must be subject to a combined transport operation as the existence of the Channel and the North Sea means that all goods must be transported by sea, air or rail for at least some part of the transit.[1] A more comprehensive régime for combined transport is established by the United Nations Convention on International Multimodal Transport of Goods,[2] but this is not yet in force and in any event it will not apply to such contracts for international carriage as are subject to the CMR Convention by virtue of Article 2.[3]

2.2 Article 2 provides as follows:—

"1. Where the vehicle containing the goods is carried over part of the journey by sea, rail, inland waterways or air, and, except where the provisions of article 14 are applicable, the goods are not unloaded from the vehicle, this Convention shall nevertheless apply to the whole of the carriage. Provided that to the extent that it is proved that any loss, damage or delay in delivery of the goods which occurs during the carriage by the other means of transport was not caused by an act or omission of the carrier by road, but by some event which could only have occurred in the course of and by reason of the carriage by that other means of transport, the liability of the carrier by road shall be determined not by this Convention but in the manner in which the liability of the carrier by the other means of transport would have been determined if a contract for the carriage of the goods alone had been made by the sender with the carrier by the other means of transport in accordance with the conditions prescribed by law for the carriage of goods by that means of transport. If, however, there are no such prescribed conditions, the liability of the carrier by road shall be determined by this Convention.

---

1. It seems that it was at the instigation of the English delegation that Article 2 was included in the Convention: Haak p. 94.
2. See *supra*, para. 0.3.
3. Article 30(4) U.N. Convention.

2. If the carrier by road is also himself the carrier by the other means of transport, his liability shall also be determined in accordance with the provisions of paragraph 1 of this article, but as if, in his capacities as carrier by road and as carrier by the other means of transport, he were two separate persons."

2.3 It should first be noted that Article 2 will apply irrespective of whether it is the first, the last or the middle part of the carriage in respect of which the road vehicle is carried on another mode of transport. Indeed, it is possible that what might otherwise merely be domestic carriage of goods could become subject to CMR by virtue of Article 2. For instance, if goods are to be carried by road from Birmingham to Dover and thence by ship on the road vehicle to Boulogne without being unloaded, the transit would be subject to CMR even if the vehicle never left the port area in Boulogne. Equally, it will make no difference whether the carriage by the other mode takes place during the road carriage by the first or any successive carrier.

2.4 On the face of it a carrier by another mode will not himself be a successive carrier within the meaning of Article 34, since that Article is restricted to road carriers.[4] However it may be that his taking over of the goods and consignment note, in accordance with the provisions of Article 34, amounts to an acceptance of responsibility in respect of the whole of the carriage by road and not merely for the period of carriage by sea.[5]

2.5 Turning to the detailed provisions of Article 2, it will be observed that Article 2(1) falls into two parts: the general rule, and the proviso to that rule. Each will be considered in turn.

The general rule

2.6 The general rule under Article 2(1) is that where goods are carried by another mode of transport and the goods are not unloaded from the vehicle, then the whole carriage will be governed by the Convention, notwithstanding the fact of that different mode of transport. So far as this aspect of Article 2(1) is concerned there are few difficulties. One, however, is that there may be room for argument as to what is meant by "unloaded from the vehicle" for the purposes of Article 2(1). In particular, in the French text the corresponding expression is "*rupture de charge*", and it is possible that courts applying the French text may come to different conclusions from those which apply the English text, as the two expressions would seem to carry different nuances of meaning.[6]

2.7 First of all, since the word "vehicle" is defined in Article 1(1) to include trailers and semi-trailers,[7] it is clear that where a trailer or semi-trailer is detached from its tractor unit for the purposes of onward carriage by another mode, this will not amount to unloading from the vehicle for the purposes of Article 2(1).[8] However, where actual unloading of the goods has taken place this would convert the operation into one of combined or intermodal

---

4. As to successive carriage generally, see *infra*, Chapter 11.

5. See per Hobhouse, J., *Dresser UK Ltd.* v. *Falcongate Freight Management Ltd.* (1991) 26 E.T.L. 798, 804.

6. As to the relevance of the French text, see *supra*, para. 0.14 ff.

7. *Supra*, para. 1.55.

8. Thus in *Thermo Engineers Ltd.* v. *Ferrymasters Ltd.* [1981] 1 Lloyd's Rep. 200, discussed *infra*, para. 2.20 ff., there seems to have been no dispute that separation of a trailer from its tractor unit did not amount to unloading. See also Arrond. Rotterdam, 10.11.70 (1971) 6 E.T.L. 273.

transport which would not be covered by the Convention.[9] For example, if goods are unloaded from the road vehicle on to the quay and then loaded into the ship on their own, whether accompanied by the road vehicle on the sea transit or not, then the operation will be outside Article 2. The operation of the CMR rules will therefore be excluded from that moment or else be precluded from applying at all if the road vehicle has not yet crossed an international frontier, unless a subsequent stage of the carriage involves an international carriage by road, in which case CMR would apply to that part of the carriage.[10]

Carriage by container

2.8 Turning to the question of shipment by containers in circumstances where the container is separated from its carrying vehicle for the purposes of onward shipment by another mode, this would appear to be outside Article 2(1): for in such circumstances, as the container itself is not within the Convention's definition of a "vehicle",[11] it must be regarded as "goods" and therefore, having been "unloaded from the vehicle", Article 2(1) cannot apply.[12] Given the vast increase in the use of containers since the Convention was drafted, this exclusion from the scope of the Convention is of obvious significance.[13] It also has the more far reaching consequence in terms of the road transit either side of the transit by the other mode. If, for example, a container is to be carried by road from London to Paris, but is unloaded from the vehicle for the purposes of the sea transit, not only is the sea transit excluded from the Convention, but the road transit either side of the Channel is also excluded since there is no "international contract for the carriage of goods by road".[14] Only where the shipment is subsequently carried by road across another national frontier can it then be said that the onward transmission from the Channel port amounts to an international contract for the carriage of goods by road so as to bring that segment of the carriage within the Convention.[15] The carriage within this country and across the Channel would still be outside the terms of the Convention.

2.9 If this interpretation be correct, which in view of the wording of Article 2(1) would seem inescapable, this would mean that a considerable volume of short sea traffic will not

9. Though in such circumstances the carriage may be subject to the U.N. Convention on International Multimodal Transport of Goods, when in force. See *supra*, para. 0.3.

10. Thus in RB. Antwerpen, 23.9.75 (1976) 11 E.T.L. 279, where there was a combined transport by sea from Hong Kong to Rotterdam and thence by road to Antwerp, it was held that CMR applied to the road transit from Holland to Belgium. The court rejected the argument that the carrier had not contracted to carry by road but had contracted for combined transport which might or might not involve carriage by road. The international carriage by road was therefore not subsumed into some wider contract. See Ramberg, Theunis, p. 23, Cf. Arrond. Rotterdam, 28.10.99 (2000) S & S No. 35.

11. Article 1(2), *supra*, para. 1.55. In respect of a swap-body, however, see OLG. Hamburg, 13.3.93 (1994) TranspR 193, *supra*, para. 1.56, fn. 147.

12. RB. Antwerpen, 7.9.73 (1973) 8 E.T.L. 754. See too RB. Rotterdam, 2.1.76 (1978) 6 E.L.D. 232 where it was made clear that before road/rail combined transport would fall within Article 2(1), the container and the road vehicle carrying it must not be separated for the purpose of the carriage by rail, cf., Hof. Gent, 17.9.92 (1993) International Vervoersrecht over Land en Zee, 293. However, see also the case referred to *infra*, para. 11.49 fn. 97.

13. That significance is, however, reduced where the damage occurs in circumstances covered by the proviso to Article 2(1), *infra*, para. 2.17 ff., since then the appropriate prescribed conditions would apply in either case—subject to what is said below with reference to *Thermo Engineers Ltd.* v. *Ferrymasters Ltd.* [1981] 1 Lloyd's Rep. 200.

14. Cf., HR., 24.3.95 (1995) S. & S. No. 72, (1996) N.J. No. 317, (1996) U.L.R. 760.

15. RB. Antwerpen, 23.9.75 (1976) 11 E.T.L. 279 *supra*, fn. 10. Where periods of transport covered by CMR are preceded or followed by carriage involving other modes of transport, difficult issues of demarcation can arise, see eg *infra*, para. 6.8, fn. 23 and para. 6.10, fn. 31.

be subject to the Convention at all.[16] It may therefore be desirable to arrange that in such cases the contract of carriage should expressly incorporate the terms of CMR so as to avoid any uncertainty as to what rules as to liability are to apply.[17] In the absence of such express incorporation, in the event of the carriage being outside the Convention, then it would be open to the carrier to rely on his standard trading conditions, assuming, of course, that those are incorporated into the contract.

2.10 In an important Danish decision, a carrier was held able to rely on his trading conditions in relation to container transport between Denmark and England.[18] A freight forwarder claimed an indemnity for loss of the goods from the carrier, whose rates differed depending on whether a trailer or a container was used, the choice being determined by the carrier. CMR terms were offered in respect of transport by trailer, but if a container was used, terms derogating from CMR were to apply. The day before the transport took place, the forwarder was informed of the number of the transport unit, from which he could tell whether a trailer or a container was to be used, although the implications of the choice were not spelled out. In the event, a container was used and the court held that CMR was not applicable to the transport since the container was not carried continuously by road. Consequently, the carrier was free to rely on his standard trading terms. In a dissenting judgment, one of the judges considered that the carrier should have stated his choice and made sure that the forwarder knew the consequences of it. On his view therefore, the contract should have been treated as one for carriage by road notwithstanding the use of a container which, in fact, involved unloading from a road vehicle.[19]

2.11 The difficulties which arise where unloading of the goods involves a breach of contract are considered below. However, where unloading is expressly or impliedly permitted, the interpretation so far suggested is that the effect of Article 2 is to take the contract outside of CMR altogether in circumstances where no continuous international road carriage occurs or, at least, is not contemplated by the carrier.[20] An alternative interpretation might be to regard Article 2 as subject to Article 1 so that, notwithstanding the fact there there is a change of mode which is permitted by the terms of the contract, if the contract is nevertheless properly to be regarded as a contract of carriage by road then CMR is compulsorily applicable to it.[21] This view can be supported by a recent Dutch decision[22] where a mobile crane was carried by road from Cairo to Alexandria, unloaded from the truck, carried by sea under a Bill of Lading to Antwerp and then finally by road to a destination in Holland. The crane was damaged during the first stage of the journey

---

16. Though again such carriage may be subject to the U.N. Convention on International Multimodal Transport of Goods, when in force, which expressly defines "goods" as including containers (Article 1(7)).

17. On voluntary incorporation of CMR, see *supra*, Chapter 1, paras. 1.60–1.64.

18. Supreme Court of Denmark, 28.4.89 (1989) 24 E.T.L. 345.

19. The majority of the court pointed out that the forwarder was a professional who had carried out a number of transports with the carrier and must have known the conditions, cf. Clarke, pp. 38–39. The forwarder had himself contracted with the goods owner, supplied printed CMR consignment notes to him and accepted liability under CMR.

20. Cf. App. Bordeaux, 21.12.94 (1995) B.T. 74, (1996) U.L.R. 400.

21. Cf. Comm. Bobigny, 24.11.95 (1996) B.T. 38 where the Warsaw Convention governing carriage by air was applied to goods destroyed during a section of international carriage by road. The carrier was exercising a permission to use an alternative mode of transport in order to put them on a flight from London in fulfilment of an agreement to carry them from France to Nigeria under cover of an air waybill which made the Warsaw Convention applicable.

22. RB. Rotterdam, 24.1.92 (1993) S. & S. No. 89, see Cleton (1992) II U.L.R. 186, 191. The decision is criticised by van Beelen, "CMR en overlading" in *Vergelijkend Zeerecht*, 1994, p. 48.

and the carrier sought to apply his trading conditions to that stage. The court held that CMR applied and displaced the conditions which derogated from it. The court acccepted an interpretation whereby, although CMR did not apply to the sea leg, it did apply to the road leg since the conditions of Article 1 had been fulfilled.[23]

Unloading from the vehicle

2.11A Article 2 does not itself provide that where goods are unloaded and transferred on to another mode of transport, CMR no longer applies. This, however, necessarily follows from the fact that the goods are no longer being carried by road. It is the purpose of Article 2 to preserve the operation of the Convention in the limited circumstances therein provided for; in the absence of such circumstances, transfer of the goods on to another means of transport will mean (subject to the express or implied terms of the contract[24]) either that CMR ceases to apply, or, if a national frontier has not at that stage yet been crossed, that CMR is prevented from coming into operation at all.[25]

2.12 It should be noted, however, that an unloading will only be relevant for the purposes of Article 2(1) if it occurs immediately before, during or immediately after transportation by another mode of transport. Unloading of the goods at any other stage of the road transport will be of no relevance for the purposes of Article 2. Indeed it would seem, in the absence of any provision in the Convention, that unloading the goods otherwise than in connection with transportation by another mode will not have any significance at all on the application of CMR to the performance of the contract in question.[26] Arguably, however, this is a matter of fact and degree. So, for example, if goods are unloaded from a vehicle into a warehouse as part of a normal transhipment pending loading onto another vehicle, these operations would presumably be covered by the Convention. If, on the other hand, the warehousing was for a more prolonged period of storage by the carrier in between two stages of an international road carriage, then arguably that part of the contract, being essentially a contract for storage rather than for the carriage of goods by road, would not be governed by CMR.[27]

2.13 Unloading the goods from the vehicle will not affect the operation of Article 2(1), however, where the unloading takes place in circumstances covered by Article 14. Article 14 is discussed in detail in Chapter 5 but in brief it makes special provision for circumstances where it is or becomes impossible to carry out the contract of carriage in accordance with the terms of the consignment note before the goods arrive at the place designated for delivery. In such cases, Article 14 provides for the carrier to obtain instructions from the person entitled to dispose of the goods; where the carrier is unable to obtain instructions from that person, he is entitled to take such steps as appear to him to be in the best interests of the person entitled to dispose of the goods. Where, therefore,

---

23. Cf., however, App. Grenoble, 29.4.98 (1998) B.T. 419 where the indication to the carrier that the goods were to be sent from St. Julien c.i.f. Le Pirée meant that a maritime transport was intended, so that the pre-carriage by road within France was not subject to CMR.
24. See, however, *supra*, para. 2.11.
25. In both cases this will be subject to the operation of Article 14. See *infra*, Chapter 5, paras 5.25–5.30.
26. See, e.g., *Moto Vespa S.A.* v. *MAT (Brittania Express) Ltd.* [1979] 1 Lloyd's Rep. 175.
27. Cf. a case in respect of carriage by sea: *Mayhew Foods Ltd.* v. *O.C.L.* [1984] 1 Lloyd's Rep. 317, see further Yates, para. 6.4.2.6.

Article 14 applies and the goods are unloaded, either as a result of instructions received from the cargo interests or as a result of steps taken by the carrier where unable to obtain those instructions, Article 2(1) will still operate to bring any subsequent carriage by another mode of transport within the orbit of the Convention.

2.14 One difficulty that arises in relation to the general rule in Article 2(1) is as to whether it can apply where the transport by another mode or the unloading is carried out on the initiative of the road carrier rather than where envisaged either expressly or impliedly by the original contract. Where carriage from this country is concerned, clearly carriage by another mode must at least be impliedly envisaged. This is not so, however, either in the case of unloading or for European transport operations, and in the case of both transport from this country and European transport operations, carriage by one alternative mode might be envisaged by the contract, whereas the road carrier in fact arranges carriage by another mode. It might be argued that for the purposes of the general rule under Article 2(1) the use of a different mode of transport causes no real problem, since the carriage will in any event be subject to CMR, but such is to ignore that such carriage might take the entire performance of the contract outside the cargo interests' insurance policy. Further, if the carrier can unilaterally decide to vary the means of transport but is still able to rely on the CMR defences, the cargo interests might be severely prejudiced. Conversely, where the proviso to Article 2(1) comes into operation, the carriage may not be governed by CMR, and the cargo interests will in such circumstances find that their claim is subject to an unexpected régime, in respect of which they will probably not have made appropriate insurance arrangements. In cases where the goods have been unloaded by the carrier, again CMR would not apply in circumstances where the cargo interests were expecting the carriage to be governed by it.

2.15 On the one hand, Article 2(1) makes no express provision whereby the unloading or the use of the other mode of transport must be agreed, either expressly or impliedly, with the cargo interests. Further, it is provided by Article 6(2)(a) that, where applicable, the consignment note is to contain a statement that transhipment is not allowed. It can be argued from this that, in the absence of such a statement, transhipment is permitted. Although transhipment is not defined, it would seem wide enough to cover carriage by another mode, whether or not the goods are unloaded. On the other hand, the prejudicial nature of the factors put forward above would seem to militate strongly against the road carrier unilaterally deciding to unload the goods or utilise transportation by another mode, and Loewe is of the view that the carrier cannot arrange a different mode of carriage than that initially provided for.[28] This view can be supported by a Belgian decision where it was held that a carrier who had undertaken an international road transport operation and then on his own initiative had unloaded the goods for carriage by sea without the consent of the shipper or the consignee could not rely on that unloading to release himself from the terms of CMR.[29] On the other hand, in a Dutch decision[30] it was held that the plaintiff could not maintain that the agreement was to carry solely by road, since ro-ro transport was a mode

---

28. Loewe, para. 57.
29. RB. Antwerpen, 9.12.77 (1978) 13 E.T.L. 110.
30. RB. Amsterdam, 18.11.87, a decision rejected on different grounds by the Hoge Raad. Both decisions are reported at (1990) 25 E.T.L. 589. Cf. Hof. Gent, 30.4.97 (1997) 32 E.T.L. 606: in the absence of contrary instructions, and to the extent that he demonstrates normal and habitual diligence, a carrier who performs a superposed carriage subject to Article 2 is free to choose the port of shipment.

of transport in general use and the plaintiff was obviously aware of it, having previously complained of trailers being carried on deck.[31]

2.16 It would seem, therefore, that the better view is that unless the carriage by another mode, or the unloading, has been done with the prior consent of the cargo interests, then the carrier cannot evade his CMR responsibilities. From this it follows that the carrier is not permitted by Article 2 unilaterally to vary the mode of performance of the contract, so such variation would amount to breach of contract, and in the event of the cargo interests suffering losses by reason of that breach which are unrecoverable under CMR, such losses should be recoverable under the national law for that breach of contract. To the extent that the carrier attempts to rely on his CMR defences and limitations of liability in such circumstances, it can further be argued that the principle of deviation, well recognised in carriage by sea cases,[32] should operate to deprive him of any such protection. Although the scope of this principle in anything other than carriage by sea cases must be a matter of some doubt since the decision of the House of Lords in *Photo Production Ltd. v. Securicor*,[33] it would seem that the reasoning behind the carriage by sea cases, namely that there has been a change of adventure and therefore a change of risk,[34] is equally applicable to the present circumstances. Such reasoning would, at least, seem to be applicable to the defences in Article 17.[35] On the other hand, application of this reasoning may only be permissible if consistent with the proper interpretation of the provisions of CMR, bearing in mind that CMR is an instrument of uniform law and should not be construed in the light of domestic concepts such as the principle of deviation.[36] The possibility that the breach of contract by the carrier may also be considered as wilful misconduct would provide considerable protection for the cargo interests without the need to resort to the principle of deviation.[37]

31. Cf. App. Paris, 15.2.82 (1982) B.T. 141 where the court, taking into account the CMR consignment note which named only the road carrier as carrier, and the customs document which indicated transport by road and a frontier point of exit which was consistent with a road transport to destination, considered that the road carrier was in breach by shipping the lorry by sea for part of the journey, which resulted in the need for additional unexpected documentation. In the Dutch case referred to in the text, it was held that the plaintiff could not maintain an action based on the carriage on deck since they had not maintained their objection, and in view of the frequency with which they had made use of the carrier's services. See also RB. Gent, 19.6.90 (1991) 26 E.T.L. 377, where the court held that a road carrier who places a closed vehicle which can resist normal weather conditions on board a ferry as deck cargo is not in default. Cf. Comm. Marseille, 4.2.86 (1990) 25 E.T.L. 223, where the road carrier was held negligent in loading the semi-trailer on deck, having been informed that the goods "must be kept in conditions of perfect impermeability". See also App. Aix-en-Provence, 16.1.90 (1991) B.T. 65.

32. See *Scrutton on Charterparties*, 18th edn., p. 257 ff.

33. [1980] 1 Lloyd's Rep. 545; see in particular the comments of Lord Wilberforce at p. 550 where he indicates that the deviation cases are to be regarded as *sui generis*; he does not make it clear, however, whether he is referring to deviation cases arising in carriage by sea cases only, or whether he is including those cases where the principle has been extended to other forms of bailment (e.g. *Alexander* v. *Railway Executive* [1951] 2 K.B. 882) including carriage by rail, e.g., *London and North Western Rwy.* v. *Neilson* [1922] A.C. 263.

34. *Hain Steamship Co.* v. *Tate and Lyle Ltd.* (1936) 55 Ll.L.Rep. 159, 178.

35. See RB. Te Leuven, 23.10.90 (1992) 27 E.T.L. 272, the English headnote for which states: "The rules of this Convention, and in particular Article 17, only concern risks arising during carriage and during essential loading and unloading operations. The rules do not apply where the damage occurred during a trans-shipment of the goods which was not provided for in the contract and had not been agreed."

36. Cf. Clarke, pp. 88–89.

37. See *infra*, paras 9.72–9.106 and 10.58. In the Dutch decision noted *supra* at fn. 35, the court applied Article 29 in respect of wilful misconduct to deprive the carrier of his right to limitation of liability since the unauthorised transhipment constituted an augmentation of risk. Cf. Cass., 30.11.82 (1983) B.T. 129 where a carrier by road in the course of carrying a mechanical digger loaded on to his semi-trailer found himself unable to continue the journey by road due to the breakdown of his vehicle, so he arranged for the trailer and the goods

The proviso to Article 2(1)

2.17 The greater part of the text of Article 2(1) is concerned with the position where the general rule, as outlined above, does not apply, so that the carriage by the other means will not be governed by CMR. In order to understand this part of Article 2(1) it is essential to realise that the relations between the road carrier and the carrier by the other mode will in most cases be governed by a legal régime which differs from the CMR Convention, applicable as between the cargo interests and the road carrier. The amount recoverable under that regime may be greater or less than under CMR according to the circumstances, but more importantly the defences available under that régime have been specifically designed to take account of risks particular to that method of transport. This may well mean, if the road carrier is forced to rely on CMR, that the CMR defences are inappropriate to protect him against a claim by the cargo interests but in circumstances where he cannot effect recovery from the other carrier under the régime applicable to that mode of transport. The basic purpose of this part of Article 2(1) would seem to be to avoid such difficulties for the road carrier in circumstances where loss, damage or delay has occurred during carriage by that other mode and without fault on his part.

2.18 There are thus two linked features: first, to bring the relationship between the cargo interests and the road carrier into line with the special rules developed for particular modes of transport, and secondly, to enable the road carrier's recourse to be as close as possible to his own liability.[38] A balance is struck between the interests of the two parties. The cargo owner's recovery may be more limited than under CMR, but this is justified by the fact that there is no reason why he should recover more than if he himself had contracted with the carrier by the other means.[39] On the other hand, the cargo owner is protected by the fact that his recovery is not to be subjected totally to the carrier's need to match his recourse possibilities since CMR will only relinquish its hold as a mandatory régime to another mandatory régime.

2.19 With this in mind, if reference is made to the relevant part of Article 2(1), it will be observed that its scope can be broken down into three separate but cumulative requirements. Before it can apply,[40] the loss, damage or delay must have occurred (a) during the carriage by the other means of transport; (b) without fault on the part of the road carrier; and (c) due to an event which could only have happened "in the course of and by reason of" the other mode of transport. Each of these requirements must be considered in turn.

to be transported by rail, during which an accident occurred. The accident was largely due to the negligence of the sender in preparing the goods for transport, who was thereby held to be liable to the railway and unable to obtain full indemnity from the carrier by road. The fact that the carrier had not sought to obtain instructions under Article 14 did not affect the position, since the change in the mode of transport played no causal role in the damage and the sender suffered no prejudice by reason of the change of mode. At the very least, however, a carrier should not be permitted to benefit from an unauthorised transhipment, for example by the application of the proviso to Article 2, see *infra*, and cf. BGH., 17.5.89 (1990) 25 E.T.L. 76 (concerning carriage by air).

38. Bombeeck, Hamer and Verhaegen, "La Responsabilité du Transporteur Routier dans le Transport par Car-Ferries", (1990) 25 E.T.L. 110, para. 51.

39. Op. cit., para. 8.

40. Before it can apply it must, at least, be invoked by one of the parties: RB. Antwerpen, 17.1.97 (1998) 33 E.T.L. 686.

"during the carriage by the other means of transport"

2.20 As to the first of these requirements, this raises no particular problem of interpretation, but rather one of proof,[41] since it might be difficult to prove at which stage of a journey any particular loss or damage occurred.[42] This aspect of Article 2(1) did, however, fall to be considered by an English court in *Thermo Engineers Ltd.* v. *Ferrymasters Ltd.*,[43] where a consignment of machinery which projected over the top of the trailer on which it was loaded was damaged when it struck the deckhead above the lower cargo deck when being loaded into a vessel for shipment across the Channel. Neill, J., indicated his agreement with the view that CMR must be viewed as having been intended to be consistent with other transport Conventions, so that as a general rule the question of when sea carriage begins would be determined according to the Hague Rules. Since under the Hague Rules carriage by sea includes loading operations,[44] and as on the facts the trailer was across the line of the stern of the vessel at the time the damage occurred, he concluded that this amounted to damage occurring during carriage by sea for the purposes of Article 2(1). The plaintiffs' contention that carriage by road only came to an end when the vehicle was secured on board the vessel, so that the load's effective means of carriage had ceased to be the wheels of the trailer, was rejected.

"not caused by an act or omission of the carrier by road"

2.21 The main problem in relation to this second requirement concerns the application of Article 3. As will be seen, Article 3 renders the CMR carrier responsible for the acts and omissions of his servants, agents, and sub-contractors.[45] The difficulty that arises is as to whether it is permissible to have regard to Article 3 in determining whether loss, damage or delay is due to "an act or omission of the carrier by road" for the purposes of Article 2(1). This question was discussed by Neill, J., in *Thermo Engineers Ltd.* v. *Ferrymasters Ltd.*[46] Having turned to consider whether the damage to the machinery was due to the fault of the road carrier, he observed that it was necessary to have regard to Article 3, and that as a general rule this meant that the carrier would be liable for the acts of sub-carriers who perform carriage by another mode. However, he continued:

"I consider, however, that in art. 2 the words 'carrier by road' have to be construed in such a way as to impose a narrower responsibility. A construction which imposed a wide responsibility on the carrier by road would, in my view, be contrary to the purposes which par. 1 of art. 2 seeks to achieve. It would also be inconsistent with par. 2 of that article."[47]

---

41. The burden will naturally fall on the party seeking to benefit from the proviso, normally the road carrier, who will seek to obtain such benefit as he can from the presumptions to be derived from the issue of clean transport documents by the carrier by the other mode, cf. App. Versailles, 15.10.92 (1993) B.T. 239 where the presumption derived from a clean bill of lading was displaced by other evidence (such as the absence of seawater) suggesting that the damage was not caused during a storm at sea. See also App. Paris, 13.10.86 (1986) B.T. 689 (reversed on other grounds by Cass., 5.7.88 (1990) 25 E.T.L. 221).
42. This will be a question of fact in every case (RB. Rotterdam, 9.1.79 (1979) 7 E.L.D. 279), but the road carrier taking over the goods from a carrier by another mode will be well advised to ensure that any appropriate reservations are made at that stage.
43. [1981] 1 Lloyd's Rep. 200.
44. Article II and *Pyrene Co. Ltd.* v. *Scindia Steam Navigation Co. Ltd.* [1954] 1 Lloyd's Rep. 321. Provided, of course, that the carrier by sea has assumed contractual responsibility for that operation, see *Balli Trading Ltd.* v. *Afalona Shipping Co. Ltd., The Coral* [1993] 1 Lloyd's Rep. 1.
45. See generally, *infra*, Chapter 3.
46. [1981] 1 Lloyd's Rep. 200, *supra*.
47. At p. 204. As to Article 2(2), see *infra*, para. 2.41 ff.

2.22 Unfortunately no further guidance was given on this particular aspect of the case, but the substance of what he seems to have been saying was that once it has been decided that the carrier is at fault, so as to bring the situation within the general rule in Article 2(1) and thus subject to the Convention, then it is permissible to have regard to Article 3 in determining the liability of the road carrier; in determining the initial question of whether the proviso to Article 2(1) applies, however, a narrower approach must be adopted. It must surely be correct that Article 3 cannot be relied on to determine the initial question of fault on the part of the carrier for to hold otherwise would in effect deprive the proviso of any real substance: the carrier would almost inevitably be at fault either personally or under Article 3 where the damage is due to the fault of a servant, agent, or sub-contractor. The only situation where CMR would not apply would be where the damage is caused by some independent third party, such as a collision at sea caused through the negligence of another vessel.

2.23 On the other hand, it might be objected that this approach deprives the general rule in Article 2(1) as to the application of the Convention of much of its substance, excluding from its ambit what must be the cause of the majority of situations where goods are damaged while in transit by another mode of transport, namely the negligence of the carrier by that mode. However, such is to ignore the overall purpose of the proviso to Article 2(1) outlined above, and the fact that CMR is only excluded by the proviso to Article 2(1) where there are "conditions prescribed by law" in respect of carriage by that other mode: otherwise CMR will apply.[48]

2.24 Neill, J., gave no guidance as to what he meant by "a narrower responsibility", but he did effectively concede that the reference in Article 2(1) to "an act or omission of the carrier by road" was not restricted to purely personal fault on the part of the road carrier, since he went on to consider the plaintiffs' argument that the road carriers were at fault on another basis altogether. What had happened was that the machinery had been delivered to a dock company, and the damage occurred while the machinery was being loaded into the ship by the dock company under the supervision of the ship's officers. The plaintiffs first alleged negligence on the part of the defendants themselves in failing to warn the dock company and the ship's officers of the excess height of the load. This argument, however, which clearly would have been sufficient personal fault for the purposes of Article 2(1), was rejected on the facts since the protrusion of the load above the top of the trailer was inescapably obvious, so that there was no need for any warning.

2.25 The substance of the plaintiffs' second argument in this respect consisted of the allegation that the dock company were at fault in their capacity as agents for the road carriers in failing to pass on information as to the height of the load, both within their own organisation and to the ship's officers. In other words, the plaintiffs' argument was that the dock company were fulfilling a dual role: first, as agents for the road carrier and second, as stevedores, and that they were at fault in their former role, this being sufficient to fix the road carrier with liability. This argument was again rejected due to the obvious nature of the load. But it was only rejected on its facts, thus admitting of the possibility, which must surely be right, that where there is fault on the part of an agent (which must necessarily include a servant on basic principles of vicarious liability) which occurs while fulfilling his duties as agent for the carrier by road in connection with the carriage by road,

---

48. Article 2(1) *in fine*. See *infra*, para. 2.29 ff.

rather than while acting in any other capacity, this will be sufficient to amount to fault of the carrier for the purposes of Article 2(1).

"some event which could only have occurred in the course of and by reason of the carriage by that other means of transport"

2.26 It is clear that this requirement incorporates two distinct elements: not only must the damage occur in the course of the transport by the other mode,[49] but it must also be due to that other mode. In straightforward cases, this is unlikely to produce any difficulty. Thus, for example, if a shipment of goods is loaded onto a trailer which in turn is transported on board a ship across the Channel, and the load is damaged owing to the rolling of the ship, there are two possibilities. First, if the damage results from the fact that the load has not been properly secured within the trailer, then the proviso will not apply,[50] since the damage results from the fault of the road carrier or those otherwise responsible for the loading of the goods[51]; although it is damage occurring "in the course of" carriage by sea, it cannot be considered as having occurred "by reason of" such carriage since where a road vehicle is to travel by sea such factors must be taken into account by those responsible for loading. Where, on the other hand, the damage to the goods results from the trailer itself being improperly secured on board the ship, then the damage is both "in the course of" and "by reason of" the carriage by sea.

2.27 Difficulty does arise, however, from the use of the words "could only have occurred" in this part of Article 2(1). Thus in *Thermo Engineers Ltd.* v. *Ferrymasters Ltd.*[52] the plaintiffs argued that since the damage to the load could have occurred on land, by striking any overhead obstruction such as a low bridge, it could not be said that the damage "could *only* have occurred" during and by reason of sea carriage. The substance of their argument, therefore, was that this requirement was directed at risks exclusive to that particular mode of transport such as, in case of carriage by sea, wetting by sea water. Neill, J., however, refused to allow this argument in the following terms:

"One is concerned to consider not whether the loss or damage could only have occurred in the course of the other means of transport but whether *the event* could only have so occurred. It seems to me that any adequate description of the relevant events in this case would have to include a statement to the effect that a collision with the bulkhead of a ship had taken place in the course of loading the ship. Such an event could only have occurred in the course of, and by reason of, the carriage by sea."[53]

In other words, one must look to the description of the events which have occurred in order to determine this question. Particularly in the context of the overall purpose of Article 2(1) such an approach produces a satisfactory result: one is concerned with the question of whether the road carrier has a claim against the carrier by the other mode[54]; that question can only be answered by a precise description of the events resulting in the

49. To this extent it therefore would appear to duplicate the earlier requirement that the loss, etc., must occur during the transport by the other mode.

50. See, e.g., App. Paris, 23.3.88 (1990) 25 E.T.L. 215.

51. As to responsibility for loading, see *infra*, para. 6.75 ff.

52. [1981] 1 Lloyd's Rep. 200.

53. At p. 205, italics added.

54. Since if so, and that claim is governed by another legal régime, then the purpose of Article 2(1) is to ensure that where the road carrier is not at fault the same régime governs the road carrier's liability to the cargo interests.

damage which has occurred. To adopt the narrow view put forward by the plaintiffs potentially would be to prejudice the road carrier, leaving him without adequate rights of recovery from the carrier by the other mode, in circumstances where the road carrier was not at fault.

2.28 A different conclusion was reached by a Dutch court,[55] specifically disagreeing with the decision in *Thermo Engineers*, where it was held in a similar case that when a trailer turned over in the course of being unloaded by means of a tugmaster, the mere fact that it overturned whilst on board the ship did not necessarily imply that such an accident could not have happened ashore so as to be considered as an event which could only have occurred in relation to the transport by sea. Although the view has been expressed elsewhere that the approach adopted in the *Thermo Engineers* case renders this condition largely redundant,[56] the narrower approach adopted by the Dutch court would overly restrict the possibility of matching the régimes for the purposes of recourse, since most of the exceptions recognised by, for example, the Hague Rules could equally occur on land. Recently, however, the Dutch Supreme Court, in a case where a fire broke out on board ship, resulting in the loss of the ship and cargo, with these purposes in mind, rejected the narrow approach taken in the earlier decision.[57] The wider approach adopted in the *Thermo Engineers* case does still leave room for the condition to apply: for example, a breakdown of a refrigerated vehicle which results in damage occurring during the sea stage and not due to an act or omission of the carrier by road.[58]

"conditions prescribed by law"

2.29 Once the three requirements outlined above have been fulfilled, then the claim is not regulated by CMR but by such "conditions prescribed by law for the carriage by that means of transport" as would have applied had the cargo interests contracted directly with the carrier by the other mode.[59] In such circumstances, therefore, the road carrier remains liable for the loss, but is subject to an alien régime, and will not, for example, be able to rely on the protection of Article 17(4)[60] but will be able to rely on any protection afforded by that régime. If, however, there are no such prescribed conditions,[61] his liability will still be regulated by CMR. It should be remembered that only the "*liability*" of the carrier is subjected to these prescribed conditions, CMR regulates the liability of the carrier in the provisions of Chapter IV.[62] It is not to be supposed that the provisions in other chapters which might touch on liability, such as in Chapter V,[63] are to be similarly subjected to the alien régime.[64]

55. Hof. The Hague (1990) I U.L.R. 242. Cf. Hof. Gent. 30.4.97 (1997) 32 E.T.L. 606.
56. See Clarke, pp. 45–46.
57. H.R., 14.6.96 (1996) S. & S. No. 86, (1996) 31 E.T.L. 558, (1997) N.J. No. 703, Cleton, "Digest of Recent Decisions of Netherlands Courts concerning International Transport Law Conventions," 755, 760.
58. See Czapski, "Responsabilité du Transporteur Routier lors du Transroulage et du Ferroutage", (1990) 25 E.T.L. 172, 173.
59. Such a contract would have a different subject matter, in that the cargo interests would contract only for the carriage of his goods, whereas the road carrier is contracting for the carriage of goods plus vehicle. This difference is taken into account by the use of the word "alone" in Article 2(1).
60. As to which, see *infra*, Chapter 6.
61. As in Arrond. Arnhem, 18.7.96, (1997) S. & S. No. 33, which concerned live animals carried by sea, live animals being excluded from the potentially applicable regime of carriage by sea.
62. See *infra*, Chapters 6–9.
63. See *infra*, Chapter 10.
64. Hof. The Hague (1990) I U.L.R. 242, App. Aix-en-Provence, 30.5.91 (1992) B.T. 281.

2.30 The difficulty, of course, is in deciding what is meant by "conditions prescribed by law".[65] Given the purpose of Article 2 one could with some confidence have presumed that for countries which are party to such Conventions it means the Hague or Hague-Visby Rules for carriage by sea, CIM for carriage by rail, and the Warsaw Convention (as amended) in respect of carriage by air. The decision of Neill, J., in *Thermo Engineers Ltd. v. Ferrymasters Ltd.*[66] casts some doubt on this. As we have already seen, it was decided in that case that if there were any "conditions prescribed by law", these would have applied to the plaintiffs' claim. The defendants therefore argued that liability should be governed by the Hague Rules.[67] This, however, was rejected by Neill, J., largely on the basis of Article V of the Hague Rules,[68] which permits the sea carrier to increase his responsibilities and liabilities under the Hague Rules as long as such an increase is included in the Bill of Lading. This led Neill, J., to the conclusion that since the sender and the carrier by sea could have entered into terms different from those contained in the Hague Rules,

" . . . the compensation payable to the plaintiffs is to be calculated in accordance with such conditions as they could, and would, have agreed with a carrier by sea in November, 1975, if a separate contract for the carriage of the [machinery] . . . had been made."[69]

2.31 Although one can appreciate Neill, J.'s difficulty in applying the Hague Rules where those rules themselves provide for variation, such is to ignore the fact that the only variation permitted is an increase in the carrier's liability. It is true that had the plaintiffs concluded a contract directly with the sea carrier they might have negotiated an increase in that liability. It is also true that Article 2(1) does express itself in terms of liability being determined in accordance with the terms which would have governed the contract in the event of the cargo interests contracting directly with the sea carrier. It is nevertheless submitted that there are serious objections to Neill, J.'s approach. First, on the wording of Article 2(1) itself, this does not set out to deduce the terms of a fictional contract between the cargo interests and the sea carrier and then apply them, but in effect simply asks the question: would such a contract be governed by prescribed conditions? To that question, the answer is indisputably positive: the Hague Rules would have applied. As for the fact that those conditions might have been varied by the parties, this does not affect the fact that, in the absence of such variation, the Hague Rules would apply. Article 2(1) does not refer to "conditions prescribed by law subject to any permitted variations".

---

65. It has been suggested that the English text can be interpreted to mean any statutory rules relevant to the carriage by the other mode and not just mandatory rules which is the sense more strongly suggested by the French text *"dispositions impératives de la loi"* see, e.g., Ramberg, Theunis and Baranyai, note to RB. Gent, 19.6.90 (1991) 26 E.T.L. 377, 383. The Dutch Supreme Court, 29.6.90 (1990) 25 E.T.L. 589 (discussed *infra*) felt that the two versions could be harmonised by consideration of the purpose of the provision (see *supra*), so that the intended meaning is to be taken as a reference to uniform rules laid down by international convention in the realm of transport. The English version was felt to be particularly consistent with an intention to refer over to a convention such as the Hague Rules despite the possibility that it might be possible for the parties to a contract to derogate from those rules (pp. 628–629). See further, *infra*, paras 2.36–2.37.

66. [1981] 1 Lloyd's Rep. 200.

67. The facts of the case arose in 1975, before the Hague-Visby Rules were in operation, but the latter make no amendments to Article V.

68. Neill, J., also referred to some clauses in the bill of lading, unfortunately without specifying their content.

69. [1981] 1 Lloyd's Rep. 200, 206.

2.32 It may be further objected that such an interpretation is clearly inconsistent with one of the overall purposes of the Convention, namely uniformity, and further conflicts with the general aim of any legal régime, namely certainty and consistency. What could be less certain than terms the plaintiff might have agreed with a sea carrier some six years previously? Even conceding that this might be ascertainable in individual cases, where, for example, reference can be made to a previous course of dealing, the ability of the cargo interests to negotiate better terms than those contained in the Hague Rules would clearly vary with the circumstances of every case.[70]

2.33 For these reasons it is to be hoped that Neill, J.'s decision in this respect will not be followed, and that in such circumstances the Hague Rules will be applied. It may, moreover, further be objected that even if it be right that the Hague Rules do not apply in such circumstances, it cannot be right that permitted variations of the Hague Rules amount to "conditions prescribed by law": they may be permitted but they are certainly not prescribed. The logical result of a finding that the unvaried Hague Rules are not prescribed should have been that no conditions are prescribed, with the result that under Article 2(1) itself CMR should be applied.

2.34 It should also be appreciated that a similar difficulty arises in respect of the application of the Hague Rules in the period prior to the loading of the goods on to the ship and after the discharge from the ship. Article VII of these Rules grants freedom of contract in respect of those periods, notwithstanding that the shipowner may have the goods in his care. In a Dutch case,[71] the court rejected a reference over to the Hague Rules since the event causing the loss (non-shipment) occurred prior to shipment, presumably viewing the rules as not mandatorily applicable at that point. Although it would seem inescapable that the Hague Rules are not "prescribed by law" in relation to these periods, it does produce an unfortunate hiatus from the point of view of the road carrier. His liability under the CMR régime will continue in circumstances where there is no equivalent compulsory régime applicable to the carrier by sea.[72]

2.35 Furthermore in the *Thermo Engineers* case, it was clear that the contract of carriage by sea was covered by a bill of lading,[73] so the actual carriage by sea clearly fell within the scope of the application of the Hague Rules. The question arises as to the position in circumstances where it is uncertain whether the Rules would have been

---

70. Cf. Glass & Cashmore, para. 3.13.

71. Hof. The Hague, 19.12.79 (1980) S. & S. No. 33 (reported by Haak, p. 102, and see also Clarke, p. 46 n. 7).

72. Clarke is critical of this in line with his view that Article II of the Hague Rules does not provide for a sequential and exclusive series of duties restricted to the period between loading and discharge, see [1989] LMCLQ 394, 395, referred to by Gleeson, C.J., in *P.S. Chellaram & Co. Ltd v. China Ocean Shipping Co, The Zhi Jiang Kou* [1991] 1 Lloyd's Rep. 493, 498 (Supreme Court of New South Wales Court of Appeal), and see Kirby, P., at p. 516. Cf. however, Bingham, L.J. in *The Captain Gregos* [1990] 1 Lloyd's Rep. 310, 315 with reference to Article 1(e). On the one hand, it is arguable that Article VII would not have been necessary if a purely temporal approach to the sequence of responsibilities in Article II and III(2) had been intended. On the other hand, Article VII may be intended to confirm the limitation on the scope of application of the Rules as a whole. It begins "Nothing herein contained shall prevent . . . " whereas Article 5 begins "a carrier shall be at liberty . . . ". Arguably it is only during the period indicated that mandatory rules are imposed in respect of the precise duties of the carrier which thereby regulate the basis of his liability. The basis of liability outside this period is not dictated by the Rules, but derives from the general principles of law relevant to it. It does not necessarily follow that all of the Rules are excluded from applying to events occurring outside this period (e.g. limitation of action). It is rather a matter of interpretation of the particular rule and its place in the totality of the scheme.

73. [1981] 1 Lloyd's Rep. 200, 205.

applicable had the sender himself contracted for the carriage of goods by sea. If no bill of lading had been issued, or if the goods had been carried on deck and stated to have been so carried,[74] would it still have been possible to view the Rules as the prescribed conditions? The more fundamental question therefore is how a court is to determine whether there are applicable conditions prescribed by law. Whilst this question has not arisen for consideration in England, it has been the subject of differing approaches on the Continent and differing opinions among commentators.[75]

2.36 In France and Belgium, the issue has been resolved by considering that if the contract between the carrier by road and the operator of the other means of transport is subject to mandatory rules, then it can be assumed that the contract which would have been made by the sender would also have been mandatory, but not otherwise. Thus the Antwerp Appeal Court considered[76] that the issue by the sea carrier to the road carrier of a non-negotiable transportation document meant that there was no binding legal provisions so that no provisions could be invoked by the road carrier to deviate from the régime laid down by CMR. In France, the Supreme Court held[77] the Hague-Visby Rules not to apply to the carriage of goods on deck since they had been made applicable only by virtue of a paramount clause and were thus not mandatorily applicable to the transport.[78] It is but a short step from this to suggest that, since the application of the Hague-Visby Rules always depends ultimately on the willingness of the carrier by sea to comply with the conditions for their application,[79] they are never truly mandatory and therefore never to be applied.[80] Since those responsible for drafting CMR are particularly likely to have had the Hague Rules in mind in drafting Article 2, it is hardly likely that such interpretation could

74. Article 1(c) of the Hague-Visby Rules.

75. Clarke, para. 15; Ramberg, Theunis, Theunis, Chapter 2; Theunis, Chapter 16 (see also a slightly amended version of this in German, "Die Haftung des Straßenfrachtführers bei der Ro/Ro-Beförderung", (1990) TranspR 263); Czapski, "Responsabilité du Transporteur Routier Lors du Transroulage et du Ferroutage", (1990) 25 E.T.L. 172; Bombeeck, Hamer, Verhaegen, "La Responsabilité du Transporteur Routier dans le Transport par Car-Ferries", (1990) 25 E.T.L. 110, Herber, "Haftung beim Ro/Ro-Verkehr", (1994) TranspR 375, van Beelen, "De aansprakelijkheid van de wegvervoerder bij stapelvervoer conform Article 2 CMR", (1991) 26 E.T.L. 743. Especially useful is the report by the Advocate-General W.E. Haak to the Dutch Supreme Court, reported in French (1990) 25 E.T.L. 589, 606, where three possible interpretations are set out.

76. Hof. Antwerpen, 15.3.89 (1989) 24 E.T.L. 574, 589. See also RB. Gent, 19.6.90 (1991) 26 E.T.L. 377 and OLG. Hamburg, 15.9.83 (1984) VersR 534. Cf., however, Hof. Antwerpen, 22.12.97 (1998) 33 E.T.L. 399, 410–414.

77. Cass., 5.7.88, (1989) 24 E.T.L. 49, (1989) B.T. 449, see Chao (1989) B.T. 439.

78. Reversing the lower court decision of App. Paris, 13.10.86 (1986) B.T. 689. The case has been criticised (see Clarke, p. 50 and Achard, (1989) D.M.F. 219) for failing to recognise that once a case falls within Article 10 of the Rules, then even by the operation of a paramount clause (see Article 10(c)) they become applicable mandatorily and not merely as a matter of contract. The court also failed to recognise that since there was no agreement to carry the goods on deck, the Rules were mandatorily applicable anyway. Both of these errors were recognised by the decision in App. Paris, 23.3.88 (1990) 25 E.T.L. 215, discussed by Clarke and by Achard. The latter error also appears to have been made in Belgium in RB. Gent, 19.6.90 (1991) E.T.L. 377, where the court followed the decision of the French Supreme Court.

79. In *Browner International Ltd.* v. *Monarch Shipping, The European Enterprise* [1989] 2 Lloyd's Rep. 185, where the Hague-Visby Rules were held not to apply compulsorily where the carrier by sea had not issued a bill of lading but merely a waybill, Steyn, J., said, at p. 188, "It follows that shipowners, if they are in a strong enough bargaining position, can escape the application of the rules by issuing a notice to shippers that no bills of lading will be issued by them in a particular trade. Subject to the limited restriction introduced by the Unfair Contract Terms Act 1977, in favour of carriage for consumers . . . the position is that freedom of contract prevails".

80. Cf. OLG. Celle, 4.7.86 (1987) VersR 1087, (1987) TranspR 275, see Herber, "Die CMR und der Roll-on/Roll-off Verkehr", (1988) VersR 645, 646.

possibly have been intended.[81] The Dutch Supreme Court has come to an opposite view which gives more weight to the fact that Article 2 employs a fiction which does not depend in all respects on the actual contract made by the parties to the contract of carriage by sea.[82] In this case the CMR carrier shipped the semi-trailer containing the goods on the deck of a ship under a non-negotiable waybill. The lower court applied CMR rather than the Hague-Visby Rules.[83] The Supreme Court held that the tribunal should have applied the Hague-Visby Rules: the application of Article 2 is not to be determined by the actual contract made between the road carrier and the non-road operator, since the sender is not party to the negotiations between them; in order to protect the sender, it is necessary that the contents of the fictional contract indicated by the proviso be construed objectively. In reaching its decision, the court considered that the drafters of CMR had uniform law in mind.[84] CMR itself being a uniform law which seeks to balance the interest of the parties, the security afforded by this is maintained only by reference to other uniform laws, clearly in mind at the time of drafting, which are adapted to the nature and risks of the particular modes of transport concerned.[85]

2.37 In line with this reasoning, it is not necessary that the uniform law should be as strict as CMR. Thus the fact that the Hague-Visby Rules are more liberal in permitting a degree of derogation[86] does not deprive them of their nature for this purpose since the drafters of CMR could not be understood as having wished to omit reference to rules so clearly valuable to maritime transport.[87] Finally, in order to know which uniform law to apply, the court indicated[88] that the starting point is the same journey by the other means of transport and to apply the Convention relevant to it. If more than one Convention applies, then that to which the country of shipment by the other means of transport is party is applied, and if this Convention is treated differently in different countries, then the version contained in the law of the country of shipment is similarly applied. In effect therefore, starting from an objective basis, the background facts of the transport in issue are used to resolve the difficulties involved in applying an abstract solution, which in effect applies the uniform rules without regard to the specific rules governing their application, in particular, the requirement that a bill of lading be issued.[89] It remains to be

---

81. Since roll-on/roll-off transport and the more frequent use of waybills rather than bills of lading was a later development: Advocate-General W.E. Haak in his report to the Dutch Supreme Court, (1990) 25 E.T.L. 589, 616, para. 5.3.1.

82. H.R., 29.6.90 (1990) 25 E.T.L. 589 (also (1990) 25 E.T.L. 232), rejecting the conclusions of the Advocate-General.

83. RB. Amsterdam, 18.11.87, see (1990) 25 E.T.L. 589, 600. The parties made use of a leap-frog procedure to reach the Supreme Court.

84. In France, the courts have applied their own internal law where this is mandatory, see App. Aix-en-Provence, 30.5.91 (1992) B.T. 281.

85. (1990) 25 E.T.L. 589, 628. On the basis of this objective, the perceived difference between the English and French versions of CMR (see *supra*) could be reconciled.

86. As in Article 5 of the Rules.

87. (1990) 25 E.T.L. 589, 629, the court drawing some support from the English version.

88. Ibid.

89. See Clarke, pp. 52–53. Cf. the criteria suggested by Bombeeck *et al.*, op. cit., fn. 75, paras. 49 and 55. The Dutch court saw the Hague-Visby Rules as applicable, presumably applying the version in force in Sweden as the country of shipment. Had that country not been a party to any carriage by sea convention nor applied any version of the Rules by its domestic legislation, the court would have been unable to choose between the Hague and the Hague-Visby Rules (and today the Hamburg Rules) and presumably would have been forced back to CMR. Presumably also, that is the only extent to which the legal reality of the situation can impinge. The court would not appear to envisage a search for the practical possibility of the sender being able to send his goods under the terms of a uniform régime, cf. Czapski, op. cit., fn. 75, p. 176.

seen whether an English court would be prepared to adopt a similarly purposive approach.[90] Furthermore, recently a Dutch court has adopted an approach which suggests a return to the more direct application of the hypothetical contract by giving the road carrier the opportunity to adduce evidence that had the sender made a direct contract with the non-road contractor, that contract would have been subject to compulsory provisions.[91]

The Channel Tunnel

2.38 There are two services provided as a result of the opening of the Channel Tunnel. Firstly, there is "Le Shuttle", which operates as a roll-on/roll-off service between Folkestone and Calais. Since the goods are not unloaded from the road vehicle, CMR will continue to apply so far as the road haulier is concerned. However, since the service is not a listed railway for the purposes of the CIM, it has been free to settle its own terms and conditions which will apply as between the haulier and Eurotunnel. Since those terms are not "prescribed by law", clearly the road haulier will not be able to rely on those conditions under Article 2 of the CMR Convention. Liability under the current conditions is excluded if loss or damage is due to "circumstances beyond its reasonable control or not within its power or ability to remedy", which is clearly wider in scope than the equivalent defence available to the CMR carrier under Article 17(2). Although the limit of liability under the conditions is 8.33 SDRs per kg, which corresponds with the CMR limit, there is a maximum liability of 70,000 SDRs per consignment, which is defined so as to include the tractor and trailer unit.

2.39 As distinct from the roll-on/roll-off operation described above, there is also a through-rail container/swap-body operation from inland terminals in this country to inland terminals in mainland Europe. In this country, the trains are operated by Railfreight Distribution, which is a wholly owned subsidiary of British Rail, and in France they are operated by SNCF. However, the marketing of space on the through-rail operation is delegated to independent commercial concerns, who market the space to other freight operators, utilising their own trading conditions. For the purposes of a claim against the railways, they will operate under the CIM rules in COTIF. Unless there is a subsequent international road carriage, the CMR Convention will not apply to the through movement, because the goods will be "unloaded" for the purposes of the international transit.[92]

Apportionment of liability

2.40 The inclusion in the proviso to Article 2(1) of the words "to the extent that it is proved" will permit the court to apportion liability where it was partly incurred subject to CMR and partly subject to "the conditions prescribed by law". Thus, for example, where the road carrier has damaged a load which he is carrying and the carrier by the other mode

---

90. An English court, if it were to be convinced that the issue is to be determined by objective criteria, might well apply whatever uniform rules are applicable to that mode of transport under English law, reading "conditions prescribed by law" as those conditions in force in the country seised of the case.

91. RB. Rotterdam, 1.7.94 (1995) S. & S. No. 99, interlocutory judgment, concluded 10.4.97, (1999) S. & S. No. 19. The road carrier was held to have failed to adduce the required evidence. See further, for criticism, van Beelen, "Stapelvervoer: Duke of Yare en Gabriële Wehr", (1997) Tijdschrift Vervoer & Recht 15, 18).

92. See *supra*, paras. 2.11–2.16.

then loses part of the damaged load, liability can be divided, presumably on a causal basis, CMR applying to the initial damage and any "prescribed conditions" to the subsequent loss.

"Two separate persons"

2.41 Article 2(2) provides that where the carrier by road is also the carrier by the other mode, Article 2(1) applies in the way explained above, but as though he were two separate persons for the purposes of the different modes of carriage. As a preliminary point, it would seem unlikely that this provision will be of much significance, since where an undertaking carries out different modes of carriage, it will normally in any event establish separate corporate entities in respect of each operation.

2.42 However, where relevant, Article 2(2) gives rise to the following implications. If the loss occurs during the actual road carriage, no problem arises as the responsible operator is in any event a CMR carrier by road. Similarly, if it is not known where the loss occurred there will be no difficulty because, as it cannot be proved that the loss occurred during carriage by the other mode, then under Article 2 the Convention would in any case apply to the whole carriage. If, however, the loss can be shown to have occurred during carriage by the other mode then under Article 2(2) one starts by applying Article 2(1) in the normal way. Where, therefore, the damage has occurred during transport by the other means of transport without having been unloaded from the road vehicle, then CMR will apply. It is, however, in relation to the proviso to Article 2(1) that the fictional separation of entities becomes of significance. Where damage has occurred in the course of and by reason of transit by the other mode of transport without fault in the part of the carrier *qua* road carrier, then the claim will be regulated by the relevant "conditions prescribed by law", and only by CMR in the absence of such conditions. Where, on the other hand, damage has occurred which is due to fault on the part of the carrier as road carrier, the general rule in Article 2(1) will apply and CMR will nevertheless apply to the claim.[93] By way of example, if he was responsible for loading the vehicle and damage occurred during sea transport as a result of defective loading of the vehicle, then there is fault on the part of the carrier as road carrier, and CMR applies. If, however, the damage results from negligent stowage by the carrier as sea carrier, so that the proviso to Article 2(1) applies, the claim will be regulated by any "conditions prescribed by law", and only by CMR in the absence of such conditions.[94]

Direct action against the other carrier

2.43 Article 2 does not preclude the claimant from suing the carrier by the other mode direct. Such an action, however, will not be subject to CMR, but to the national laws of

---

93. Support for the view expressed in the text can be gained from the view of Neill, J., in *Thermo Engineers Ltd.* v. *Ferrymasters Ltd.* [1981] 1 Lloyd's Rep. 200, 204 that Article 3 could not apply so as to render the road carrier liable for the fault of a sub-contractor carrying out the other means of transport, since such would be inconsistent with Article 2(2) (see *supra*, paras 2.21–2.25). Plainly if the road carrier is to be regarded as at fault by virtue of Article 3 for the purposes of the application of the proviso where damage is caused through the fault of the carrier by the other mode, there is no point in the fictional separation of entities.

94. With some operations, however, it may be particularly difficult to determine whether they are being performed by the carrier as road carrier or as carrier by sea, e.g. the operation of loading the vehicle on to the ship. The Hague Rules will regulate this operation only where the carrier by sea has undertaken responsibility for it. If he is the carrier both by sea and road there is no-one with whom he can agree.

the country concerned.[95] Such an action will be extra-contractual, that is in tort or delict.[96] A direct claim against the carrier by the other mode may be a useful alternative where the road carrier has become insolvent, but such an action will not, of course, assist the claimant in respect of loss which has occurred while the goods are being carried by road.

Compatibility with Guadalajara Convention

2.44 The question was raised by certain governments some time after the conclusion of CMR as to whether Article 2 is compatible with Article 2 of the Guadalajara Convention 1961, which dealt with the problem of international carriage by air which is performed by a person other than the contracting carrier. Further discussion is outside the scope of this text; suffice it to say that it has tacitly been agreed at governmental level to leave the question in abeyance until the question of conflict does in fact arise.[97] In this country, section 8 of the Carriage of Goods by Road Act 1965[98] was included specifically so as to enable any conflict with the Guadalajara Convention which might become apparent to be resolved by Order in Council.[99] Loewe, however, is of the view that there is no such conflict.[100]

---

95. Article 28 (*infra*, Chapter 9, paras 9.65–9.71) will not assist the defendant since it is not "carriage under this, Convention".

96. But note that the Hague-Visby Rules are applicable to claims both in contract and in tort: Article IV *Bis*. See, however, *The Captain Gregos* [1990] 1 Lloyd's Rep. 310.

97. See E.C.E. Inland Transport Committee document W/TRANS/SCI/355 Add. 6, Sept. 18, 1969, which discusses the problem at length.

98. Set out in Appendix A.

99. Hansard, Vol. 715, pp. 1066–7.

100. Loewe, para. 59.

CHAPTER 3

Carrier's Liability for Third Parties

3.1 As has already been seen in relation to Article 1, it will often be the case that the party with whom the cargo interests contract for the carriage will arrange for a third party to carry out all or part of the carriage. Further, there may be ancillary services in relation to the road carriage, such as loading the goods or dealing with Customs formalities, which the carrier arranges for a third party to perform. The general aim of Article 3 is to ensure that the carrier will remain personally liable where he has delegated the performance of any of his duties to a third party.

3.2 Article 3 provides as follows:—

> "For the purposes of this Convention the carrier shall be responsible for the acts and omissions of his agents and servants and of any other persons of whose services he makes use for the performance of the carriage, when such agents, servants or other persons are acting within the scope of their employment, as if such acts or omissions were his own."

"The carrier"

3.3 Before Article 3 can come into operation it must first be shown that the party against whom it is invoked is "the carrier" for the purposes of the Convention. This initial stage may well pose problems where the performance of the entire carriage has been sub-contracted to another carrier, for in such circumstances, as has been seen, the party with whom the cargo interests first contracted may argue that he was contracting as a forwarder only rather than as a carrier, so as to be outside the operation of Article 3. On this question, and on the question of status as carrier generally, reference should be made to the discussion in Chapter 1.

3.4 Once it is established that a particular party is a carrier, which will include both the first carrier and any successive carrier within Article 34,[1] then that party cannot avoid liability for any loss or damage by claiming that it has resulted from the act or omission of a third party where the latter falls into one of the two categories laid down in Article 3.

"Agents and servants of the carrier"

3.5 First, the carrier will be liable for the acts of his "agents and servants".[2] At common law a carrier would similarly be liable for the actions of his servants as a question of

---

1. On Article 34 generally, *see infra*, Chapter 11.
2. By way of examples: GH. Dordrecht, 18.5.66 (1968) 3 E.T.L. 416, (1968) U.L.C. 155—a carrier was liable where his driver fell asleep while driving. App. Colmar, 11.1.72 (1972) B.T. 90, (1973) I U.L.C. 276—a carrier was liable for the negligence of his driver for giving a written undertaking to the sender as to the condition of a tank trailer.

vicarious liability,[3] and on general agency principles would also be liable for the actions of his agent acting on his behalf in relation to transport operations.

"Any other persons"

3.6 Secondly, the carrier is responsible for "any other persons of whose services he makes use[4] *for the performance of the carriage"*. Although in most cases it is unlikely that this phrase will cause any difficulty, the italicised words do raise a possible problem of interpretation. This is because the restriction inherent in those words does not, grammatically, apply to the earlier reference to servants and agents. In other words, it might be argued that a carrier is liable for his servants or agents, so long as they are acting within the course of their employment, even when not being used for the performance of the contract of carriage. So, for example, if a servant or agent of the carrier performing some task unrelated to the carriage of a consignment of goods negligently causes a fire at the carrier's premises which damages the goods while awaiting transit, the situation would be covered by Article 3. The result would be different, however, if the same damage resulted from a fire started by a contractor (not being an agent) present on the carrier's premises for some purpose unconnected with the carriage: such a person would not be a "person of whose service (the carrier) makes use for the performance of the carriage" within Article 3.

3.7 The difficulty of such an approach is most apparent in relation to the use of the word "agents". If this word is unqualified in any way, then Article 3 could operate to make the carrier liable for persons carrying out functions on his behalf but quite unrelated to the carriage. It would seem unlikely that this was intended, so an approach which would avoid such a wide basis of liability would be to treat the qualifying words "of whose services he makes use for the performance of the carriage" as applying equally to agents. The difficulty of adopting such an approach (apart from that of grammatical construction which has already been noted) would be that it would be necessary to apply the same qualification to the carrier's liability for his servants, which would represent a narrower basis of liability than normal English common law notions of vicarious liability. This problem, however, is apparent rather than real. The carrier would not be liable under Article 3 for acts of his servants which are unconnected with the carriage, but would remain liable under the appropriate national law for any such acts under the normal rules of vicarious liability.[5]

---

3. See Kahn-Freund, *Law of Carriage by Inland Transport*, 4th edn., London, Stevens, Chapter 22. An alternative basis of liability applicable to the use of sub-contractors as well as servants or agents is the non-delegable duty that the carrier has deriving from his position as bailee or quasi-bailee. See *Morris* v. *C.W. Martin & Sons Ltd.* [1966] 1 Q.B. 716, *Metaalhandel* v. *Ardfields Transport Ltd.* [1988] 1 Lloyd's Rep. 197, and see generally Palmer, *Bailment*, 2nd edn., London, Sweet & Maxwell.

4. In *Sun Alliance & London Insurance PLC* v. *PBC International* [1994] 1 Lloyd's Rep. 395, Hirst, L.J., considered that, in construing the words of an insurance policy covering CMR liability, a person who turns up impersonating the contractual carrier is not a person whose services are made use of by the carrier. By contrast, a person to whom the carrier entrusted the goods under a contract of carriage would be such a person notwithstanding that his intention is to steal the goods. This distinction would also seem to be relevant to the CMR Convention. Cf., further, *Harrison & Crossfield Ltd.* v. *L.N.W. Railway* [1917] 2 K.B. 755.

5. Article 28(1) would not operate to protect the carrier in any such claim, since it is not a claim arising out of carriage under the Convention. Nor would Article 28(2) protect the servant or agent responsible, for in such circumstances he is not a person for whom the carrier is liable under the terms of Article 3. On Article 28 generally, see *infra*, Chapter 9, paras 9.65–9.71.

3.8 The better view would therefore seem to be that the words "agents", "servants" and "other persons" are all equally qualified by the words "of whose services he makes use for the performance of the carriage". In so far as such an interpretation might conflict with traditional English rules of construction, this would seem an appropriate occasion for the "new approach" now judicially recognised in relation to the construction of international conventions.[6] Further, having regard to the French text,[7] there does not appear to be any equivalent grammatical difficulty: being arranged rather differently, the qualifying words would appear to refer equally to servants and agents as to other persons.

3.9 This difference of arrangement stems mainly from the fact that the equivalent of the words "agents and servants" is, in the French text, the single word "*préposés*". It would appear that the English and French versions of the text in this respect have slightly different meanings, since the French word "*préposés*" does not appear to include independent contractors,[8] who could come within the word "agent" in the English text. However, so long as the view put forward above as to the application of the qualifying words is adopted,[9] it would not seem that this difference can give rise to any conflict in practice, since an independent contractor would in any event come within the wider expression "any other person whose services [the carrier] makes use . . . ". For the same reason the traditional English distinction between servants and independent contractors for the purposes of determining the liability of the employer will be of no significance in relation to Article 3 in those cases where the contractor cannot be regarded as the carrier's agent.

3.10 Having dealt at some length with this possible problem of interpretation in relation to this aspect of Article 3, it must be stressed that it is only likely to be of relevance in rare cases, since in the nature of things damage, loss or delay in relation to the carriage of a consignment of goods is most likely to be caused in the performance of the carriage, and will thus in any event be within Article 3. In such cases, therefore, the carrier will be liable for the acts or omissions of any intermediary he employs for the performance of the carriage.

3.11 It would also seem clear that the carrier will also be liable for any acts and omissions attributable to that intermediary on the basis of the intermediary's own vicarious liability. On the other hand, Article 3 does not render any person for whom the carrier is responsible personally liable to the sender. Any such liability must be sought in the relevant national law.[10] Where the carrier employs a successive carrier he is responsible for the latter under Article 3, whether or not Article 34 applies.[11] However, it would seem that where there is a chain of successive carriers, the original carrier will be responsible under Article 3 only for that carrier with whom he has a contractual relationship, since he is responsible under Article 3 for the acts and omissions only of persons "of whose service

---

6. See *supra*, paras 0.5–0.13.

7. As to whether it is permissible to have regard to the French text, see *supra*, paras 0.14–0.17.

8. See Report of the Brussels Conference of 1967 (concerned with the amendment of the Hague Rules) at p. 293: "*l'expression anglaise* 'servant or agent (not being an independent contractor)' *peut être traduite en français par un seul mot*: '*Préposé*'."

9. This is indeed a strong argument for the adoption of that view since the contrary interpretation would mean that a different result might be produced under the different texts in circumstances where damage, etc., is caused by an independent contractor, acting as an agent for the purposes of English law, but not being used for the performance of the carriage.

10. OGH., 25.9.68 (1973) 8 E.T.L. 309. BGH., 9.2.79 (1980) 15 E.T.L. 84. But note that any person so sued may be able to rely on Article 28(2), *infra*, Chapter 9, paras 9.65–9.71.

11. App. Paris, 4.7.84 (1985) B.T. 158.

he makes use". If each successive carrier appoints a sub-contractor without reference back there will be no continuity of liability, whereas if the original carrier is responsible for the appointment of each successive carrier, he will remain liable throughout. This question will, however, often be academic as Article 34 provides that where carriage governed by a single contract is performed by successive road carriers, each of them shall be responsible for the performance of the whole operation.[12] In other words, where Article 34 applies, the cargo interests can proceed against the carrier with whom they originally contracted, as the first carrier, even where damage has been caused by a later successive carrier with whom the first carrier had no direct contractual relations.

Nature of the liability

3.12 It is important to realise that Article 3 does not impose any kind of absolute liability on the carrier. It does not, for example, put him under an "*obligation de resultat*".[13] It does not in any way derogate from the general provisions regarding the liability of the carrier contained in Articles 17–29, so does not extend the carrier's liability beyond that laid down in those Articles. Article 3 only means that, for the purposes of those Articles, the acts or omissions of the persons specified in Article 3 are to be treated as those of the carrier. Thus, by way of example, under Article 17(2) the carrier is relieved of liability where damage has been caused "through circumstances which the carrier could not avoid and the consequences of which he was unable to prevent". Because of Article 3, it is not open to the carrier to rely on this defence simply because damage has been caused through the fault of his sub-contractor. But if the carrier can show that *his sub-contractor* could not avoid those circumstances or prevent those consequences, then he will be able to rely on Article 17(2) notwithstanding Article 3. However, in this context it should be noted that where damage is caused by the wilful misconduct or default of the carrier's agents or servants or of any other persons of whose services he makes use for the performance of the carriage, then the carrier will lose the protection of the provisions in Chapter IV which exclude or limit his liability or shift the burden of proof.[14] The agents, servants, etc., will likewise lose this protection.[15]

Scope of employment

3.13 Article 3 provides that the carrier will be liable for the acts and omissions of his servants, agents and other persons, where any act or omission of the latter has been performed while acting within the scope of their employment,[16] so where an act or omission has occurred outside the scope of such employment the CMR carrier will not be liable for it under the Convention. In such circumstances the only remedy of the claimant will be directly against the party responsible in delict or in tort. Presumably in deciding whether an act or omission is within the scope of employment or not the criteria to be

12. On Article 34 generally, see *infra*, Chapter 11.
13. OGH., 25.9.68 (1973) 8 E.T.L. 309.
14. Cf. App. Paris, 6.4.81 (1981) B.T. 567.
15. See Article 29(2), *infra*, Chapter 9, para. 9.72.
16. Thus in OGH., 22.11.77 (1979) I U.L.C. 284, where the carrier's driver was discovered crossing the border with persons not permitted to leave the country, with the result that the vehicle and goods were confiscated by the authorities, the carrier was held liable under Article 3. See also BGH., 27.6.85 (1986) 21 E.T.L. 102.

employed will be those applicable at common law, in respect of which in this country a considerable body of case law exists,[17] and which the Convention does nothing to clarify.

3.14 Finally, it is questionable whether Article 3 includes a person from whom the first or successive carrier has hired a road vehicle, or his agents and servants. This doubt arises because Article 17 contains a special provision whereby a carrier cannot avoid liability for loss, damage or delay to goods on the grounds that it has resulted from the wrongful act or omission of a "person from whom he may have hired the vehicle or of the agents or servants of the latter".[18] However, liability for such a party will be covered either by Article 3 or by Article 17(3), or by both provisions, so that the question is purely an academic one.

---

17. See Kahn-Freund, op. cit., Chapter 22; *Winfield and Jolowicz on Tort*, 15th edn., p. 703 ff. The restriction to acts connected with the carriage must, however, be kept in mind. See, e.g., a case involving the carriage of goods by sea, *Leesh River Tea Co. Ltd.* v. *British India Steam Navigation Co. Ltd., The Chyebassa* [1966] 2 Lloyd's Rep. 193. The fact that the acts of the servant or sub-contractor depart drastically from the anticipated performance should not negate the carrier's responsibility for them if they are the acts of the person entrusted by him with the performance, see *F. Morris* v. *C.W. Martin & Sons Ltd.* [1966] 1 Q.B. 716, notwithstanding that such acts are effective to destroy the protection of the haulier under his goods in transit policy, as in *SCA (Freight)* v. *Gibson* [1974] 2 Lloyd's Rep. 533.

18. Article 17(3), *infra*, para. 6.53 ff.

CHAPTER 4

Documentation

The consignment note

4.1 As with other international conventions concerning the carriage of goods, CMR provides a system for the documentation of the carriage. The transit document in relation to CMR is the consignment note, and Articles 4 to 9 of the Convention deal with its form, content and effect.

4.2 Article 4 provides as follows:—

"The contract of carriage shall be confirmed by the making out of a consignment note. The absence, irregularity or loss of the consignment note shall not affect the existence or the validity of the contract of carriage which shall remain subject to the provisions of this Convention."

The need for a consignment note

4.3 Any contract of carriage which is entered into subject to CMR must be confirmed by the making out of a consignment note, which is *"prima facie* evidence of the making of the contract of carriage, the conditions of the contract and the receipt of the goods by the carrier".[1] In *Gefco U.K. Ltd.* v. *Mason*[2] it was argued that because it was not possible to make out a consignment note in relation to an "umbrella contract" for multiple movements, it followed that the Convention could not apply to the umbrella contract, even though it clearly applied to the individual movements. It was held by the Court of Appeal that it would be inconsistent with Article 4 to hold that contracts to which the Convention applied should be limited to those contracts for which a consignment note could be made out contemporaneously. The Convention accordingly was held to apply to the umbrella contract.

4.4 The consignment note is not intended to be a document of title, but merely a method of evidencing the contract, and unlike the bill of lading in relation to carriage by sea it is not intended to facilitate the transfer of title or payment for the goods. Although the absence, irregularity or loss of the consignment note will not affect the validity of the contract of carriage, which will still remain subject to the Convention,[3] the existence of a consignment note is essential in certain circumstances to permit the operation of the Convention.[4] First, for a right of disposal under Article 12 to be exercised the first copy

---

1. Article 9(1), *infra*, para. 4.24 ff.
2. [1998] 2 Lloyd's Rep. 585.
3. Article 4. But see the Italian case at fn. 52.
4. Provisions of national law may make it advantageous for the carrier to ensure that he is in possession of a CMR consignment note, e.g. in France under the *"loi Gayssot"*, see (1998) B.T. 811.

69

of the consignment note must be produced to the carrier.[5] Secondly, the sender can only declare a value for goods or make a declaration of a special interest in delivery by entering the particulars in the consignment note.[6] It has also been held in a German decision that if there is no consignment note, no reservation can be endorsed pursuant to Article 9(2).[7] The existence of a consignment note will be very relevant for the purposes of the operation of the successive carriage provision of the Convention.[8]

4.5 Further, practical problems may arise as regards the question of proof of the existence and terms of a contract if no consignment note is issued,[9] since if a party wishes to rely upon a contract of carriage he will normally be required to prove both its existence and its terms.[10] For example, in a French decision[11] it was held that where a carrier was in possession of a TIR card belonging to another carrier this did not render the former agent of the latter. The sender's argument to this effect was rejected and he therefore failed to prove that the second carrier was party to the contract of carriage. A TIR card is simply an administrative document permitting a vehicle to cross an international frontier, rather than evidence of any contractual arrangements. Similarly, if the contractual relationship has been entered into subject to the general conditions of a forwarder over a period of some years, if there is no consignment note, this may affect the sender's ability to prove that the contract is subject to CMR.[12]

4.6 Generally, however, the absence of a note covering the complete transit will not affect the situation by virtue of the provisions of Article 4. Thus, in the case of a combined transport operation from England to France, the fact that there was no consignment note covering the whole transit did not prevent CMR from applying throughout,[13] and where a contract was concluded by telex and no consignment note issued, a Belgian court held that the party organising the carriage was a CMR carrier and not a forwarder.[14] Similarly, where no consignment note was issued and a carrier was claiming payment of freight, he was able to prove the conclusion of the contract of carriage by the other party's acceptance of the invoices, even though they were later protested, and by the evidence of his accounts clerk who established the existence of a business relationship between the parties.[15] Again, where no consignment note was issued a French court held that it was possible to prove that a forwarder had acted as sender for the purposes of the Convention by the fact that other goods carried on behalf of a third party on the same journey named the

5. Article 12(5)(a), *infra*, para. 5.2 ff.

6. See Articles 24 and 26. As for the need for a consignment note in relation to successive carriage, see *infra*, Chapter 11, paras 11.21–11.48.

7. BGH., 9.2.79 (1980) 15 E.T.L. 215.

8. See *infra*, Chapter 11. In *Gefco U.K. Ltd.* v. *Mason, supra*, the Court of Appeal regarded the consignment note as primarily relevant for the purposes of the right of stoppage in transit and to impose liability on successive carriers under Articles 34 and 35—see Kennedy L.J. at p. 590.

9. If a consignment note is issued subsequent to formation of the contract this will not affect either the validity or the existence of the contract or the application of the Convention. App. Rouen, 16.6.72 (1972) 7 E.T.L. 1040.

10. For example, by production of the order form and the sender's instructions: App. Paris, 27.6.79 (1979) B.T. 440, cited by Mercadal, Theunis, p. 32 n. 5.

11. Paris, 4.12.70 (1970) B.T. 407.

12. Comm. Antwerp, 28.6.71 (1971) J.P.A. 162—here the client knew the forwarder had no vehicles of his own. For a full discussion of the status of the parties, see *supra*, Chapter 1, paras 1.11–1.54.

13. Cass., 17.2.70 (1970) B.T. 158.

14. Hof. Brussel, 19.10.72 (1973) 8 E.T.L. 503. Cf. Supremo Tribunal Da Justiça (Portugal) 17.11.94 (1997) U.L.R. 200, where the absence of a CMR consignment note, but rather the issue of a forwarding agent's certificate of receipt, did not prevent the application of CMR to the "effective" carrier.

15. RB. Mechelen (Malines), 14.2.73 (1973) 8 E.T.L. 630.

forwarder as sender.[16] In a further French case where the consignment note was not before the court because it had been stolen along with the vehicle and its load, it was held that the contract was sufficiently proved by a Customs document which mentioned the names of the sender, the consignee and the principal carrier as well as the weight and nature of the goods.[17]

4.7 The carrier cannot therefore rely on the absence of a consignment note to avoid liability for loss or damage to goods. In the absence of a consignment note he will be assumed to have received the goods in good condition and will be liable if he fails to deliver them in the same state.[18] Equally, even in the absence of a consignment note the carrier will be liable for damage to goods where he has failed to carry out special instructions given to him regarding them.[19]

4.8 As regards the question of the irregularity of a consignment note it is clear that the courts are prepared to look behind the form to establish the true relationship existing between the parties, although in *Aqualon (UK) Ltd.* v. *Vallana Shipping Corporation*[20] Mance, J., rejected the suggestion that Article 9(1) does not make the consignment note *prima facie* evidence of the identity of the parties. In a Belgian decision it was held that a forwarder who gave instructions to the carrier but whose name did not appear on the consignment note was a party to the contract.[21] Similarly, where the consignment note only gave the name of the forwarder as sender instead of that of the shipper of the goods, the former was held to be the agent of the shipper, thus making the latter sender.[22] The fact that the consignment note refers to the contractual carrier as *expéditeur* where he employs a sub-contractor does not alter his status as carrier, the unnamed shipper being treated as sender.[23] It will be apparent therefore that the provisions of the consignment note do not permit the contractual carrier to claim to be a forwarder (*commissionnaire-expéditeur*) when from other factors it is clear that he contracted as a carrier. So, too, where the consignment note indicated that the destination was Antwerp, it was nonetheless held in the light of other evidence supplied that the terms of the contract provided for carriage to Berchem (a near suburb).[24] Similarly, the fact that a waybill in a case of international road carriage had been drawn up to include the phrase "air transport" did not prevent the application of CMR to the carriage.[25]

4.9 It is clear therefore that those courts which have had to consider the applicability of CMR where there is a defective or non-existent consignment note have been able to adopt a robust line, and it will be possible to show the existence of the contract, the terms of the contract, and the parties to the contract either without a consignment note being in existence or despite the existence of a consignment note which does not accurately reflect

---

16. Comm. Paris, 12.10.71 (1972) 1 Gaz. Pal. 132, (1973) I U.L.C. 268.
17. App. Nancy, 29.11.94 (1995) B.T. 17.
18. App. Limoges, 2.6.67 (1967) B.T. 273. Cf. BGH., 9.2.79 (1980) 15 E.T.L. 215, *infra* para. 4.25.
19. GH. s' Hertogenbosch, 13.1.70 (1971) 6 E.T.L. 817. But see the German decision where the date for delivery was agreed by telex but not specified in the consignment note—the judge said, though very much in passing, that this did not bind the parties for the purposes of CMR. OLG. Stuttgart, 24.1.67, U 57/66 (1968) N.J.W. 1054 No. 10. See the discussion of the need for a consignment note in relation to the delay provisions of CMR, *infra*, paras 7.1–7.2.
20. [1994] 1 Lloyd's Rep. 669.
21. Comm. Antwerp, 20.4.72 (1972) J.C.B. 466.
22. Comm. Bruxelles, 12.4.72 (1972) 7 E.T.L. 1046.
23. Comm. Bruxelles, 24.1.73 (1973) J.C.B. 201.
24. RB. Antwerpen, 3.4.77 (1977) 12 E.T.L. 411, cf. BGH, 17.4.97, *infra*, para. 4.12 fn. 36.
25. Hof. Brussel, 16.11.77 (1980) 20 E.T.L. 319.

the position. However, it is important not to underestimate the problems which can arise. From an evidential point of view, it is clearly desirable that a consignment note, accurately reflecting the terms as agreed, be drawn up so that disputes can be kept to a minimum. This is particularly so in the case of a defective consignment note in the light of the presumption created by Article 9(1).[26] Wherever there is a deficiency in the documentation it must be supplied by (possibly disputed) evidence or by the application of canons of construction which will be governed by domestic law.[27]

Form and contents of the consignment note

4.10 The Convention makes no provision as to the form the consignment note should take, and in the absence of any national regulation on the matter the carrier is free to use any form that he wishes, so that the application of CMR to the contract is quite independent of the form of the consignment note. However, the IRU has drafted a form of consignment note intended to satisfy the requirements of the Convention, a copy of which appears in its amended form in Appendix C.[28] The note includes a copy for the sender (red), the carrier (green) and consignee (blue), together with a fourth copy (black) which is unsigned and intended for national administrative purposes.

4.11 Although no format is specified, the Convention does make provision in Articles 5 and 6 for the information which should be included, and in Article 7 allocates responsibility for the inclusion and accuracy of those details. Taking those Articles in turn, Article 5 provides as follows:—

> "1. The consignment note shall be made out in three original copies signed by the sender and by the carrier. These signatures may be printed or replaced by the stamps of the sender and the carrier if the law of the country in which the consignment note has been made out so permits.[29] The first copy shall be handed to the sender, the second shall accompany the goods and the third shall be retained by the carrier.
>
> 2. When the goods which are to be carried have to be loaded in different vehicles, or are of different kinds or are divided into different lots, the sender or the carrier shall have the right to require a separate consignment note to be made out for each vehicle used, or for each kind or lot of goods."

4.12 The Convention does not state which of the parties must issue the consignment note. It therefore would not seem to matter which party does draft it, and in practice either may do so.[30] In any event, as both parties are required to sign the three copies the consignment note must in theory go through both their hands.[31] The Convention appears to assume that the carrier will be responsible for issuing the consignment note, as it states

---

26. *Infra*, para. 4.24 ff.

27. Further discussion of the need for a consignment note in relation to particular provisions of the Convention is included in the discussion of those provisions: see *infra*, paras 7.1–7.2, 7.30, 8.8, 9.50 fn. 151 and 11.21 ff.

28. There is also a version of a CMR consignment note for U.K. use approved by the FTA, the RHA and SITPRO. A copy of this version also appears in Appendix C.

29. E.g. *Goodman* v. *Eban* [1954] 1 Q.B. 550. The typed entry of the name "F" in the consignment note was not a sufficient substitute for a signature in BGH., 17.4.97 (1998) 33 E.T.L. 427, (1998) TranspR 21, (1998) U.L.R. 206.

30. The carrier may supply printed consignment notes which are filled in by the sender, e.g. as in Supreme Court, Denmark, 28.4.89 (1989) 24 E.T.L. 345.

31. Although the Convention does not mention it, presumably the signature of a forwarder as the sender's agent would be acceptable. This may, however, again call in question the status of the forwarder. See *supra*, Chapter 1, para. 1.13 ff.

that the first copy is to be handed to the sender, the second one to accompany the goods, and the third one to be retained by the carrier.[32] In practice even if the sender drafts the consignment note, and then signs it, the carrier is the party who must take over the goods and would normally not sign the note until he has done so. He would therefore return the first copy to the sender and retain the second and third copies himself. The matter would seem to be only of theoretical importance. Nevertheless the absence of a signature will render the consignment note irregular and mean that it cannot technically perform the evidential role granted to it by Article 9(1).[33] In *City Vintages Ltd.* v. *SCAC Transport International*[34] the absence of the driver's signature on the consignment note rendered it *prima facie* no evidence of the receipt of the goods. Similarly, several French cases have treated the fact that a person is named in the consignment note as the sender to be of little significance in the absence of their signature.[35] Nevertheless, in the *City Vintages* case, Steyn, J., appeared to consider the defect cured by the fact that the driver entered the trailer and seal numbers on the note, thereby adopting it as an accurate record and seemingly restoring its capacity to be *prima facie* evidence. Furthermore, in coming to his view that the consignment note did reflect the full quantity of the goods, he took into account the fact that the primary task of the driver during the loading operation would have been to have checked that he had received all the goods listed on it. It may take little, therefore, for the context in which the consignment note is issued for it to acquire an evidential value despite a defect in form and, even in the absence of it having full evidential value, the statements on it may yet have a value as part of the total evidence available. Whether the consignment note is or is not effective, its value depends ultimately on all the evidence and all the circumstances.[36]

32. Article 5(1). However, in App. Grenoble, 13.3.80 (1981) B.T. 306 the court stated that it is normally incumbent on the sender to establish the consignment note unless he has taken care to expressly place this responsibility on to the carrier. In consequence, in the absence of a consignment note, he had the burden of proving the obligations which he alleged were imposed on the carrier. It may be better to suggest that the duty lies on both parties who take the eventual risk of whatever is normally incumbent on them to prove as claimant or as defendant.

33. See BGH., 16.10.86 (1987) 22 E.T.L. 56 and BGH., 17.4.97, *infra*, fn. 36.

34. Queen's Bench Division, 1.12.87, unreported except on LEXIS.

35. App. Paris, 6.1.71 (1971) B.T. 40, App. Douai, 11.3.82 (1982) B.T. 199, App. Paris, 11.6.74 (1974) B.T. 319, cf. App. Paris, 8.11.95 (1996) B.T. 132; see Mercadal, Theunis, p. 39, Clarke, p. 63. Similarly, the fact that the driver signs the consignment note naming his employer as carrier is of little weight in showing that some other party was not a carrier: *Electronska Industrija Oour TVA* v. *Transped Oour Kintinentalna* [1986] 1 Lloyd's Rep. 49, *supra*, para. 1.29. More value as an indication of status may derive, however, from proof of the authority of the driver to make the indication, as in *Texas Instruments Ltd.* v. *Nason (Europe) Ltd.* [1991] 1 Lloyd's Rep. 146, or where the consignment note provides proof of the understanding of the person making the indication, as in *Aqualon (U.K.) Ltd.* v. *Vallana Shipping Corporation* [1994] 1 Lloyd's Rep. 669, *supra* para. 1.40 fn. 92.

36. See per Tudor Evans, J., in *Texas Instruments Ltd.* v. *Nason (Europe) Ltd.* [1991] 1 Lloyd's Rep. 146, 150. Cf. Mercadal, Theunis, who cites (p. 46 n. 9) Hof. Amsterdam, 4.1.85 (1985) S. & S. No. 123, where the court considered that whilst a consignment note which does not comply with Articles 5 and 6 does not carry the same evidential weight " . . . in each individual case it should be considered individually which weight concerning evidence should be attached to such document as may have been made out and which of the parties should have the onus of proof as to the correctness or incorrectness of details in that document as to the carried goods". In BGH., 17.4.97 (1998) 33 E.T.L. 427, (1998) TranspR 21, (1998) U.L.R. 206, a consignment note not signed by the carrier was held not to provide *prima facie* evidence of the extent of the carriage, proof of which was provided by other evidence which contradicted the consignment note. The court indicated, however, that the consignment note could have whatever evidential value is attached to such a document under applicable procedural law. Cf. HR., 23.1.87 (1987) N.J. No. 410 (see Krings, p. 156) where a consignment note made out by the driver in the course of the journey, in respect of a sealed vehicle, was held to have no probative force against the carrier.

4.13 The main difficulty in practice is that few hauliers have any systematic documentation and in many cases no documentation accompanies the goods at all. In particular, in
the case of successive carriage, it is not uncommon for the first carrier to require a
signature from the onward carrier to the effect that he has received the goods without a
consignment note having been handed over.[37] Only where a particular country has a
mandatory system of documentation with some form of administrative sanctions available
is there any guarantee that a consignment note will be issued. CMR does not assist either
of the parties if the other refuses to co-operate in the issue of the consignment note.[38]
Further, the Convention makes no provision as to when the consignment note must be
issued.[39] When it is done will therefore depend upon the parties' convenience and local
practice,[40] although if it is not to hand when the carrier takes over the goods some other
form of receipt may be necessary as evidence thereof.[41]

4.14 Article 5(2) provides that where the goods which are to be carried have to be
loaded in several vehicles, or are of different kinds, or else are divided into different lots,
either the sender or the carrier can require that a separate consignment note is to be made
out for each vehicle, kind or lot of goods respectively. This provision merely formalises
what may be done in any case for commercial convenience.[42] It was relied on in *Gefco
U.K. Ltd.* v. *Mason*[43] as confirmatory of the finding that the Convention was applicable to
an "umbrella contract" for multiple movements. Again there would seem to be no way in
which the provision can be enforced, and no penalty for non-performance as such, unless
one is provided under national law as part of an administrative requirement.

4.15 The detailed information which should be included in the consignment note is set
out in Article 6 and Article 7 concerns the responsibility for the inclusion of those
particulars. Those Articles provide as follows:—

"*Article 6*
1. The consignment note shall contain the following particulars:
    (a) the date of the consignment note and the place which is made out;
    (b) the name and address of the sender;
    (c) the name and address of the carrier;
    (d) the place and the date of taking over the goods and the place designated for deliv-
    ery;

37. For successive carriage generally, see *infra*, Chapter 11.

38. Loewe, para. 69, suggests that failure to co-operate in issuing a consignment note would justify the other
party in repudiating the contract and, if permitted under the national law, claiming damages for breach of
contract. However, this would seem difficult to support in view of the provisions of Article 4 (cf. Article III, Rule
3 of the Hague-Visby Rules: the carrier "shall ... issue a bill of lading"). Mercadal, Theunis, p. 32 n. 7,
suggests that the absence of a consignment note may be used against the carrier as an example of fraud or gross
negligence, citing App. Paris, 9.7.80 (1980) B.T. 449. See however, Clarke, p. 63 n. 12.

39. Cf. Hague-Visby Rules, Article III, Rule 3: "on demand of the shipper".

40. For example, in *M. Bardiger Ltd.* v. *Halberg Spedition APS*, Queen's Bench Division, 26.10.90,
unreported except on LEXIS, the consignment note was drawn up only after the groupage consignment had
reached the operator's terminal for onward consignment as part of a larger load. It will commonly be drawn up
only after the contract of carriage itself has been made (in some cases even after an accident has occurred, as
in App. Rouen, 16.6.72 (1972) 7 E.T.L. 1040), which affects the weight to be attached to it, per Hobhouse, J.,
Electronska Industrija Oour TVA v. *Transped Oour Kintinentalna* [1986] 1 Lloyd's Rep. 49, 51. Cf. Tudor
Evans, J., in *Texas Instruments Ltd.* v. *Nason (Europe) Ltd.* [1991] 1 Lloyd's Rep. 146, 150.

41. Cf. in particular Article 9(2), *infra*, para. 4.24 ff. If there is neither a consignment note nor any other form
of receipt available at this stage, the carrier will find it difficult to endorse any reservations under that provision.
See too the case cited at fn. 7.

42. But as to the problems which might arise in relation to the exercise of right of disposal under Article 12
where separate consignment notes are issued, see *infra*, para. 5.6 ff.

43. [1998] 2 Lloyd's Rep. 585. See *supra*, para. 1.7.

    (e) the name and address of the consignee;

    (f) the description in common use of the nature of the goods and the method of packing, and, in the case of dangerous goods, their generally recognised description;

    (g) the number of packages and their special marks and numbers;

    (h) the gross weight of the goods or their quantity otherwise expressed;

    (i) charges relating to the carriage (carriage charges, supplementary charges, Customs duties and other charges incurred from the making of the contract to the time of delivery);

    (j) the requisite instructions for Customs and other formalities;

    (k) a statement that the carriage is subject, notwithstanding any clause to the contrary, to the provisions of this Convention.

  2. Where applicable, the consignment note shall also contain the following particulars:

    (a) a statement that transhipment is not allowed;

    (b) the charges which the sender undertakes to pay;

    (c) the amount of 'cash on delivery' charges;

    (d) a declaration of the value of the goods and the amount representing special interest in delivery;

    (e) the sender's instructions to the carrier regarding insurance of the goods;

    (f) the agreed time-limit within which the carriage is to be carried out;

    (g) a list of the documents handed to the carrier.

  3. The parties may enter in the consignment note any other particulars which they may deem useful.

Article 7

  1. The sender shall be responsible for all expenses loss and damage sustained by the carrier by reason of the inaccuracy or inadequacy of:

    (a) the particulars specified in article 6, paragraph 1, (b), (d), (e), (f), (g), (h), and (j);

    (b) the particulars specified in article 6, paragraph 2;

    (c) any other particulars or instructions given by him to enable the consignment note to be made out or for the purpose of their being entered therein.

  2. If, at the request of the sender, the carrier enters in the consignment note the particulars referred to in paragraph 1 of this article, he shall be deemed, unless the contrary is proved, to have done so on behalf of the sender.

  3. If the consignment note does not contain the statement specified in article 6, paragraph 1(k), the carrier shall be liable for all expenses, loss and damage sustained through such omission by the person entitled to dispose of the goods."

4.16 Article 6(1) first requires the date and place at which the note was made to be inserted, but since, as has been seen, it is largely immaterial whether the note is issued before or after the goods are taken over by the carrier it is not of much importance. Article 6(1)(b), (c) and (e) require the name and address of the sender, carrier and consignee to be inserted. As discussed already this does not prevent the courts from treating other parties as holding these positions as the note is only *prima facie* evidence of the terms of the contract.[44] Article 6(1)(d) requires the place and date of taking over of the goods and that designated for delivery to be inserted. Both facts are relevant from the viewpoint of proving the applicability of CMR, and the place designated for delivery is relevant to the application of Articles 12 to 15 and to the question of jurisdiction.[45]

4.17 Article 6(1)(f) requires the description and nature of goods and the method of packing, and in the case of dangerous goods, their generally recognised description. In

---

44. See *supra*, para. 4.8 ff and Article 9(1).
45. See *infra*, Chapter 10, para. 10.21 ff.

practice this will require careful adherence to the provisions of ADR as to the classification and description of goods laid down therein.[46] If the shipment is to be carried subject to a combined transport shipment, it would seem that the requirements for the carriage of dangerous goods by the other modes of transport should also be satisfied.

4.18 Article 6(1)(g) and (h) require the number of packages, their marks, weight or quantity to be inserted. Any inaccuracy as regards the weight may result in the overloading of the vehicle. Article 6(1)(i) requires the insertion of charges relating to carriage, Customs duties and other charges incurred from the contract until the time of the arrival of the goods at their destination. Clearly this requirement can rarely by completely satisfied as often it is not clear what supplementary charges have been incurred until after delivery of the goods.[47] Article 6(1)(j) requires the requisite instructions for Customs and other formalities. In relation to this requirement it should be noted that Article 11(1) also requires the sender to attach the documents necessary for Customs or other formalities to the consignment note or to place them at the disposal of the carrier. If the carrier sustains any expense, loss or damage as a result of the inaccuracy in inadequacy of the particulars specified in Article 6(1)(b), (d), (e), (f), (g), (h), and (j), then the sender will be responsible.[48] Accordingly, where a carrier has incurred extra expense he can recover it from the sender in circumstances where, for example, the latter has inserted in the consignment note a weight less than the actual weight of the goods. As will be seen, the effect of Article 7 and Article 11(2) is to relieve the carrier of the obligation to inquire into the accuracy of such information.[49] Furthermore, if at the sender's request the carrier enters any of the above particulars in the consignment note he will be deemed to have done so on behalf of the sender unless the contrary is proved.[50] There appears to be no case law on this provision as yet.

4.19 Article 6(1)(k) requires the insertion of a clause paramount which states that the carriage is subject to the provisions of CMR notwithstanding any clause to the contrary. It would appear that this provision is intended to overcome the problem that arises where the provisions of the Convention are to be applied in a dispute before the courts of a State which is not a party to it.[51] The intention is to give the provisions of the Convention contractual effect between the parties if the courts in question do not under private international law accept its application. To date there is no case law on the point, so it remains to be seen whether in the event of the insertion of a clause to the contrary the courts of a non-contracting State would apply CMR or not. It must be emphasised that the clause paramount is only intended to prevent a party from evading the operation of CMR in such circumstances and does not otherwise affect the status of the parties or the

---

46. See Articles 7(1)(a), 7(2), and 22. Dangerous goods and the requirements of ADR are discussed generally *infra*, Chapter 8.

47. See the discussion of Article 23(4) *infra*, Chapter 9, para. 9.21 ff.

48. Article 7(1). The carrier must, however, ensure that the requisite instructions, as required by Article 6(1)(j) are carried out. A carrier had no recourse to the sender when the consignee's employees affixed forged customs stamps to the transport documents leading to costs and delays on the return journey. For this purpose, the carrier was held responsible for their actions under Article 3, OLG. Munich, 26.11.97 (1998) TranspR 254, see further Herber/Piper Article 13, Rdn. 8, cf. Comm. Créteil, 26.5.98 (1999) B.T. 419.

49. Colmar, 16.6.72 (1972) B.T. 320, cf. however, Comm. Perpignan, 7.4.98 (1998) B.T. 308, where the court does not appear to have taken this point, but considered that the carrier had taken such steps of verification as he could before sealing the trailer which had already been loaded.

50. Article 7(2).

51. As to jurisdiction, see *infra*, Chapter 10, para. 10.21 ff.

applicability of the Convention. It has been held in one Italian decision that the Convention does not apply unless the consignment note contains the clause paramount required by Article 6(1)(k).[52] Not only would this appear to conflict with the apparently unambiguous effect of Article 4, it is also entirely out of line with the balance of authority on the need for, and contents of, a consignment note, and it would seem that it is extremely unlikely that it will be followed elsewhere. Apart from anything else, if followed it would permit voluntary derogation from the Convention, contrary to Article 41, by the simple expedient of omitting the clause paramount from the consignment note.

4.20 If the consignment note does not comply with this provision,[53] Article 7(3) provides that the carrier will be liable for all expenses, loss and damage suffered through the omission by the person entitled to dispose of the goods.[54] Here it should be noted that the standard United Kingdom CMR policy specifically excludes the insurer from liability for any failure on the part of the carrier to comply with the provisions of Article 7(3).[55]

4.21 Article 6(2) lists certain particulars which the consignment note shall contain where applicable. Such particulars must obviously relate to the provisions agreed to by the sender and the carrier at or before the time when the consignment note is signed. The particulars are as follows:—

(a) "A statement that transhipment is not allowed." The meaning of transhipment is not defined by the Convention, but it would seem that there are two possibilities intended to be covered by this provision, first, transhipment on to another mode of transport, and second, transhipment on to another vehicle. As far as the former is concerned, this has already been discussed in relation to the multi-modal provisions of Article 2, and it was there indicated[56] that it would seem that even in the absence of a stipulation against transhipment, the unauthorised use of another mode of transport or unloading of the goods for the purposes of transport by another mode would amount to a breach of contract by the carrier, and may further mean that the carrier is deprived of his CMR defences and limits of liability by virtue of the rules relating to deviation. If such is the case, then a stipulation that transhipment is not allowed will make no significant difference to the situation in such circumstances other than to place the contractual intentions of the parties beyond doubt.

So far as transhipment in the second context is concerned,[57] it has already been explained that unloading the goods from the vehicle other than in the context of multimodal transport does not affect the application of the Convention.[58] It

---

52. Cass. (It.), 26.11.80 (1981) I U.L.C. 271, cf. Cass. (It.), 8.3.93 (1984) Il Diritto Maritimo 554. Pesce, Theunis, p. 7 n. 29, suggests that a different view appears from a later decision, citing Cass. (It.), 20.2.84, No. 1227, unpublished; similarly, App. Florence, 2.2.81 (1981) Rivista di diritto civile, 455, cited by Libouton, (1982) J.T. 693, 700, para. 29.

53. Nothing is indicated as to whether there may be liability for other omissions from or inaccuracies in the consignment note in respect of details to be inserted by the carrier, cf. Comm. Marseille, 7.5.99 (1999) B.T. 480.

54. As to the right of disposal, see Article 12. Article 7(3) might provide a remedy where the sender shows that, not being aware of the application of CMR, he failed to make a declaration of value in order to overcome the limit of liability, see (1996) B.T. 502.

55. See Lloyd's Goods in Transit (CMR) Policy, cl. 3(2), Appendix D.

56. *Supra*, para. 2.11 ff.

57. As to proof, see App. Grenoble, 13.3.80 (1981) B.T. 306.

58. *Supra*, para. 2.13.

would further appear that in the absence of a contractual stipulation that tranship-ment is not allowed, no breach of the Convention will be involved, and even where there is such a stipulation, in the absence of resulting damage, it would not appear that the transhipment itself is actionable. However, where it is actionable by virtue of a stipulation under Article 6(2)(a) coupled with resulting damage, it may again be legitimate to have regard to the principles regarding deviation, since again there may be sufficient alteration of the risks involved in the carriage.[59]

(b) "The charges which the sender undertakes to pay." Other charges will pre-sumably be payable by the consignee. If the particulars are inaccurate and the carrier fails to recover from the consignee presumably under Article 7(1)(b) the sender will be liable.

(c) "The amount of 'cash on delivery' charges." Under Article 21 if the carrier delivers the goods to the consignee without collecting the charges due under the contract of carriage he will be liable to the sender for the amount due. Article 21 is discussed in detail in Chapter 7.

(d) "A declaration of the value of the goods and the amount representing special interest in delivery." Article 24 provides for the sender to declare the value of the goods in the consignment note which will then replace the limit laid down in Article 23(3) against payment of a surcharge. Article 26 provides for the sender to declare a special interest in delivery in the consignment note against payment of a surcharge. These provisions are discussed in Chapter 9.

(e) "The sender's instructions to the carrier regarding insurance of the goods." No specific mention is made in the Convention of the question of the failure of the carrier to effect insurance when instructed to do so, so this will depend upon the provisions of the national law in question.

(f) "The agreed time-limit within which the carriage is to be carried out." Article 19 provides that delay in delivery occurs when the goods have not been delivered in the agreed time, for which the carrier will be liable under Article 17. Both these provisions are considered in Chapter 7.

(g) "A list of the documents handed to the carrier." Article 11 provides that the sender must either attach documents necessary for Customs and other formalities to the consignment note or place them at the carrier's disposal. The sender will be liable to the carrier for any damage caused by the absence, inadequacy or irregularity thereof, unless the carrier is at fault.[60]

4.22 If the carrier sustains any expenses, loss or damages by reason of the inaccuracy or inadequacy of the particulars specified in Article 6(2) then the sender will be responsi-ble for it,[61] without any limit on that liability. The parties are free to enter any other particulars in the consignment note which they may deem useful.[62] However, the sender will again be responsible for the inaccuracy or inadequacy of any other particulars or instructions given by him to enable the consignment note to be made out or for the purpose

---

59. See *supra*, para. 2.16. See also RB. Leuven, 23.10.90 (1992) 27 E.T.L. 272.
60. On Article 11 generally, see *infra*, para. 4.37 ff.
61. Article 7(1)(b).
62. Article 6(3).

of their being entered therein.[63] Furthermore, if at the sender's request the carrier enters any of the above particulars (those referred to in Article 7(1)(b) and (c)) in the consignment note he will be deemed to have done so on behalf of the sender unless the contrary is proved.[64]

4.23 Article 6(3) permits other particulars to be included as desired by the parties. Such entries may be made not only at the time of issue of the consignment note. In the case of a change of instructions under Article 12(1)[65] the fresh instructions are required to have been entered on the first copy of the consignment note when they are given to the carrier. In *Harrison & Sons Ltd.* v. *R.T. Steward Transport Ltd.*[66] the load shifted shortly after the journey began. After the driver had reported the problem and re-secured the trailer a warning was entered on the consignment note which Gatehouse, J., inferred was added under the provisions of Article 6(3).

Carrier's reservations in the consignment note

4.24 Articles 8 and 9 provide as follows:—

"*Article 8*
1. On taking over the goods, the carrier shall check:
   (a) the accuracy of the statements in the consignment note as to the number of packages and their marks and numbers, and
   (b) the apparent condition of the goods and their packaging.
2. Where the carrier has no reasonable means of checking the accuracy of the statements referred to in paragraph 1(a) of this article, he shall enter his reservations in the consignment note together with the grounds on which they are based. He shall likewise specify the grounds for any reservations which he makes with regard to the apparent condition of the goods and their packaging. Such reservations shall not bind the sender unless he has expressly agreed to be bound by them in the consignment note.
3. The sender shall be entitled to require the carrier to check the gross weight of the goods or their quantity otherwise expressed. He may also require the contents of the packages to be checked. The carrier shall be entitled to claim the cost of such checking. The result of the checks shall be entered in the consignment note.

Article 9
1. The consignment note shall be *prima facie* evidence of the making of the contract of carriage, the conditions of the contract and the receipt of the goods by the carrier.
2. If the consignment note contains no specific reservations by the carrier, it shall be presumed, unless the contrary is proved, that the goods and their packaging appeared to be in good condition when the carrier took them over and that the number of packages, their marks and numbers corresponded with the statements in the consignment note."

4.25 Article 8 makes it obligatory for the carrier to do two things on taking over the goods. First he must check the accuracy of the statements in the consignment note as to the number of packages and their marks and numbers.[67] However, if he has no reasonable means of checking the accuracy of these statements, he must enter his reservations in the

63. Article 7(1)(c).
64. Article 7(2).
65. *Infra*, para. 5.2 ff.
66. (1993) 28 E.T.L. 747.
67. CMR provides the extent of the carrier's obligation to check. The claimant's position cannot be improved or additional compensation claimed by reference to some contractual or other legal obligation to check, AG. Munich, 27.6.96 (1997) TranspR 341.

consignment note together with the grounds on which they are based. It will be a question of fact in each case whether the means of checking are reasonable or not,[68] but if neither the opportunity nor the facilities are available the carrier cannot be expected to effect such a check. If on checking them he finds a discrepancy he must enter a reservation in the consignment note, although he is not required to enter any grounds for it. The consignment note is *prima facie* evidence of the matters set out in Article 9(2),[69] and as such it reverses the usual burden of proof. However, it has been held in a German decision that this reversal is dependent on the existence of the consignment note.[70] A further German decision has held that nevertheless, exceptionally, the burden can be on the carrier.[71]

4.26 Secondly, the carrier is required to check the "apparent condition"[72] of the goods and their "packaging" on taking them over.[73] In other words, a reasonable external inspection is required.[74] If the carrier discovers any defects he must enter his reservations in the consignment note together with the grounds on which they are based. A Belgian court has decided that Article 8(1) only imposes the duty of checking the "apparent state" of the goods and the packing, it does not impose a duty of checking the packing itself.[75] In other words, the carrier cannot be blamed for not making any reservations where the packing appears externally satisfactory. However, where goods are accepted by the carrier without reservation, and therefore in apparent good order, but when delivered are found to be damaged in such a way that the damage, if present at the time of collection, would have been apparent,[76] there is a presumption of fact that the damage occurred in transit.[77]

4.27 A particular problem arises where goods are carried in a sealed trailer which is pulled by several carriers. If the goods are damaged the last carrier is likely to find himself

68. In Trib. inst. Colmar, 9.1.96 (1996) B.T. 503, the fact that Dutch Customs rules prohibited the breaking of a seal of a container after the passage through Customs did not assist the carrier in the absence of proof that verification of the contents of the container under the control of Customs, after breaking the seal, was impossible.

69. The consignment note is *prima facie* evidence also of the identity of the parties, per Mance, J., *Aqualon (UK) Ltd.* v. *Vallana Shipping Corporation* [1994] 1 Lloyd's Rep. 669, cf. OLG. Düsseldorf, 14.12.95 (1996) TranspR 155, Jung, p. 152, and the extent of the carriage, see App. Paris, 23.10.96 (1997) B.T. 104. In the absence of a consignment note, proof of the receipt of the goods may also take the form of a telex from the carrier in which he confirms the number of packages received and the nature of the goods: Mercadal, Theunis p. 37 citing Antwerp, 15.12.82 (1983–84) J.P.A. 57.

70. BGH., 9.2.79 (1980) 15 E.T.L. 215, cf. cases noted *supra* para. 4.12 fns 33 and 36.

71. This was the case where the carrier was instructed to collect a specified quantity of goods which had arrived at a port from the Far East. The carrier collected a container, in which, according to the bill of lading, the goods had been consolidated. The carrier made no check to determine the contents of the container or to determine whether it could be opened during transportation. In the absence of a consignment note, national law could be applied to require the carrier to check or inform the sender if a check was impossible: OLG. Hamm, 30.3.98 (1998) TranspR 463.

72. For the meaning of the phrase in relation to carriage by sea, see Carver, *Carriage by Sea*, 13th edn., para. 114.

73. Article 8(1)(b).

74. In *Compania Naviera Vascongada* v. *Churchill* [1906] 1 K.B. 237, 245, in respect of carriage by sea the standard expected of the carrier was said to be that of an unskilled person. From *Dent* v. *Glen Line* (1940) 67 Ll.L.Rep. 72 the carrier by sea is expected to have basic knowledge of the kind of goods normally carried. Similarly, in respect of CMR carriers, OGH., 25.9.68 (1973) 8 E.T.L. 309, 315, see Clarke, p. 69, n. 56.

75. RB. Antwerpen, 13.10.72 (1973) 8 E.T.L. 330. Cf. *Silver* v. *Ocean SS. Co.* [1930] 1 K.B. 416.

76. No presumption was applied when the goods were handled entirely by the sender without any involvement of the carrier and consisted of a liquid pumped into the carrier's road tanker through an opening in the top, RB. Amsterdam, 29.4.87 (1989) S. & S. No. 269, (1990) 18 E.L.D. Nos 182 and 183, cf. Hof. Amsterdam, 2.11.89 (1990) S. & S. No. 298, (1990) 19 E.L.D. No. 255.

77. RB. Amsterdam, 29.11.78 (1980) 8 E.L.D. 103, Comm. Charleroi, 26.6.77 (1977) 12 E.T.L. 776.

liable even if he enters a reservation to the effect that it was impossible to check the state of the goods.[78] In the case of refrigerated goods a German court has held that the carrier must not only check the apparent state of the goods but also their temperature inside the packing with a thermometer.[79] Such an interpretation makes sense in relation to the carriage of refrigerated goods given the technical problems involved, since otherwise the consignee will have difficulty in claiming against the sender where the temperature of the goods is too high when taken over by the carrier. In the absence of reservations, therefore, he will then be left with a claim against the carrier.[80] A French court has held, in the case of a shipment of frozen cherries from Italy to France which thawed, that as the goods were packed in 10 kilo cartons, for goods of this nature a check on their apparent condition also implied a check on their temperature, and that in the absence of reservations in the consignment note the carrier cannot claim in such circumstances that the goods were insufficiently refrigerated when he took them over.[81] Where the carrier is required to state his grounds for making a reservation he will be required to state what defects he has discovered and the method of checking employed.[82] Where a full form CMR consignment note is utilised, the carrier should enter his reservation in the first copy of the consignment note before handing it to the sender.[83] This is because any entry made in the other copies of the consignment note after the first copy has been handed over to the sender will not affect the sender's position as there will be no means of ascertaining whether the goods were deficient or not at the time the carrier took them over.

4.28 If the carrier fails to comply with Article 8 the sanction is that laid down by Article 9(2),[84] which provides that if the consignment note contains no specific reservations by the carrier, then it will be presumed, unless the contrary is proved,[85] that the goods and their packaging appeared to be in good condition when the carrier took them over, and that the number of packages, their marks and numbers correspond with the statements in the consignment note. It will therefore be presumed that any damage to the goods discovered at their destination will have occurred during transit, that is, between the time when the goods were taken over by the carrier and that of delivery.[86] For example, where the carrier was given special instructions by the sender for a consignment of galvanised wire coils to be carried in a specially equipped vehicle, but instead an open vehicle with a defective sheet was used, in the absence of any reservations the carrier was held to be liable when

78. On successive carriage generally, see *infra*, Chapter 11.

79. OLG. Karlsruhe, 18.10.67 1 U 227/66, D.V. No. 150, 13.

80. Cf. Loewe, para. 97. Cf. *Ulster-Swift Ltd.* v. *Taunton Meat Haulage Ltd.* [1977] 1 Lloyd's Rep. 346, discussed at para. 6.114 ff. *infra*.

81. App. Paris, 30.5.73 (1973) B.T. 304.

82. Cf. however, OLG. Düsseldorf, 24.9.92 (1993) TranspR 54. Merely for the carrier to affix the phrase "said to contain" to the indications of quantity may be an insufficient indication of the grounds for making the reservation (RB. Antwerpen, 8.5.73 (1973) J.P.A. 245, cited by Clarke, p. 73, n. 90), unless the course of dealing between the parties indicates that no further description is needed, as in RB. The Hague, 14.1.81 (1981) S. & S. No. 65, cited by Ruitinga, Theunis, p. 49. Cf. RB. Antwerpen, 26.6.85 (1985) 20 E.T.L. 582. Similarly, "with reservation", RB. Rotterdam, 15.4.88 (1990) S. & S. No. 60, (1990) 18 E.L.D. No. 787.

83. Box 18 on the IRU form of consignment note is provided for this purpose. In Comm. Nanterre, 22.5.98 (1998) B.T. 904, the absence of the reservation in the sender's copy meant that it had no effect on the sender's claim.

84. See further *infra*, para. 6.69 fn. 264.

85. In App. Rouen, 11.2.93 (1993) B.T. 583, the court would not accept the fact that the containers were delivered with their seals intact as sufficient contrary proof. According to the court, experience suggests that inviolable seals do not exist, cf. Trib. inst. Colmar, 9.1.96 (1996) B.T. 503. Similarly in Germany, LG. Bochum, 28.6.95 (1996) TranspR 336, OLG. Frankfurt, 11.2.94 (1994) OLG-Rp. 76, see Jung, p. 152.

86. Comm. Bruxelles, 26.10.72 (1973) 8 E.T.L. 516, and cf. cases cited at fn. 77 *supra*.

the goods became oxidised.[87] In other words, there is a rebuttable presumption that the damage occurred in transit which transfers the burden of proof on to the carrier. Obviously careful documentary and administrative procedures are necessary if the carrier is not to find himself in the position of having to rebut this presumption. Further, as will be seen in the next section, he may not be able to obtain compensation from the sender for any resultant damage or expenses for which the sender would otherwise be liable under the provisions of Article 10. Clearly, the carrier is particularly dependent on his driver in this respect, since in the vast majority of cases it will only be the driver who has the opportunity of entering any relevant reservations. In this connection, the IRU has approved a check list for the guidance of drivers aimed at avoiding disputes as to the condition of the goods at the commencement of the carriage and carriers will be well advised to draw the attention of their drivers to this check list.[88]

4.29 Article 8(2) further provides that where the carrier has entered reservations the sender will not be bound by them unless he has expressly agreed to be bound by them in the consignment note. In this connection Loewe is of the view that the sender must have actually agreed the reservation, rather than merely have signed the consignment note containing the reservation[89] and this conclusion would appear to be correct for three reasons.[90] First, on the wording of the Convention, Article 8(2) itself makes it clear that the sender must have "expressly agreed to be bound" by a reservation in the consignment note. Secondly, in terms of general principle, given the serious consequences for the sender in that it will thereafter be well-nigh impossible to dispute the admission, it would seem right that nothing short of express agreement should suffice. Thirdly, as Loewe points out, the sender is in any event required to sign the consignment note by Article 5(1). In the absence of such express agreement, therefore, the reservation will not be binding on the sender, but equally, so far as the carrier is concerned, the presumption against him in Article 9(2) will not apply: the reference to reservations in the latter provision makes no requirement that they be expressly agreed.[91]

4.30 Article 8(3) makes special provision whereby the sender is entitled to require the carrier to check the gross weight or quantity of the goods, together with the contents of the packages, the carrier being entitled to claim the cost of such checking. The results are to be entered in the consignment note. Irrespective of whether a check is made or not no presumption arises under Article 9, which does not refer to the provisions of Article 8(3) at all.[92] This might either be a hiatus in the Convention or else, as seems probable, unnecessary in practice and therefore intentionally excluded.[93] Presumably in the event of

---

87. App. Paris, 10.12.71 (1972) B.T. 35, (1973) I U.L.C. 274.

88. The check list appears in a leaflet published by the FTA, the RHA and SITPRO: Notes on the completion of CMR Consignment Notes 1981.

89. Loewe, para. 104.

90. See RB. The Hague, 6.6.78, docket number 77/3939 (unpublished) cited by Ruitinga, Theunis, p. 52, n. 37.

91. See also Loewe, para. 103, Ruitinga, Theunis, p. 53, App. Aix-en-Provence, 9.12.80 (1981) B.T. 143 and OLG. Düsseldorf, 24.9.92 (1993) TranspR 54, 55.

92. It has been suggested obiter in a German decision (LG. Offenburg, 21.1.69 (1971) 6 E.T.L. 283) that weights entered in the consignment note cannot be relied on as against the carrier unless there has been a check made under Article 8(3) and the result entered in the consignment note. However, the point does appear to be partly one of the burden of proof in German domestic law, and the same result will not necessarily follow if the provision falls to be considered in other jurisdictions, Cf. also, however, Cass. 18.1.2000 (2000) B.T. 74 where the Court took the view that whilst CMR does not impose a duty to weigh the goods, the carrier is generally obliged under Article 9 to verify that statements in the consignment note (including those as to weight) conform with the reality.

93. Rodière, para. 22, treats it as an omission.

disagreement on the matter between the sender and the carrier evidence could be brought as to the true facts by either party.[94]

4.31 The question also arises as to the responsibilities of the carrier where the sender is responsible for loading or stowing the goods. If the loading or stowage is clearly defective the carrier is obliged to request the sender to rectify the matter, or refuse to commence transit until it is done, or else to enter the necessary reservations in the consignment note.[95] A number of Continental decisions support the view that although the Convention does not require the carrier to check loading or stowage, the carrier is implicitly bound to do so (at least so as to ensure the stability and safety of his vehicle),[96] and if he fails to do so he cannot plead faulty loading or stowage if the fault is apparent and no reservations are made.[97] Such decisions as there are to the contrary, whereby a failure to enter reservations in respect of defective loading or stowage did not prejudice the carrier's position, have been where the defect was not apparent.[98] It would therefore appear that in the event of defective loading or stowage where the defect should have been apparent to the driver, a failure to enter appropriate reservations will give rise to a presumption against the carrier. This will, of course, materially affect the position under Article 17(4)(c) whereby the carrier is freed from liability in respect of risks through loading or stowage by the sender. Thus, in a Danish decision, it was held that the carrier was not relieved of liability under Article 17(4)(c) where it should have been obvious to the carrier that there was insufficient stowage.[99]

Defective packing of goods

4.32 Article 10 provides as follows:—

"The sender shall be liable to the carrier for damage to persons, equipment[100] or other goods, and for any expenses due to defective packing[101] of the goods, unless the defect was apparent or known to the carrier at the time when he took over the goods and he made no reservations concerning it."

94. Cf. Loewe, para. 95.
95. App. Paris, 16.5.69 (1969) 4 E.T.L. 896, (1969) B.T. 190. Comm. Corbeil-Essones, 18.4.69 (1969) 4 E.T.L. 988.
96. App. Paris, 31.3.77 (1977) B.T. 315.
97. App. Colmar, 10.7.70 (1970) B.T. 358, (1971) U.L.C. 137. App. Paris, 12.6.70 (1970) B.T. 228. Civ. Charleroi, 1.10.68 (1969) J.P.A. 129. Comm. Verviers, 18.5.68 (1968) 3 E.T.L. 1240, (1969) U.L.C. 159.
98. Paris, 21.1.70 (1970) B.T 100. App. Paris, 3.11.70 (1971) 6 E.T.L. 264, (1971) U.L.C. 144. LG. Köln, 7.12.65 45 0 142/64 (1967) D.V. No. 155. Cf. App. Bruxelles, 17.6.85 (1987) R.D.C.B. 48.
99. Cour Supreme, 2.3.79 (1980) I U.L.C. 277. A full discussion of the position in relation to Article 17(4)(c) appears *infra*, para. 6.72 ff.
100. OGH., 2.4.82 (1984) TranspR 151, cited by Clarke, p. 314, where it was held that the cost of replacing mudguards could not be recovered under Article 10. Mudguards are not equipment. Liability might still be possible by virtue of the application of national law.
101. *Quaere* whether any distinction is intended between packing and stowage in this respect. In Article 17, both the English and the French texts distinguish between the two: cf. Article 17(4)(b) and (c). In Hof. Antwerpen, 13.12.89 (1990) 25 E.T.L. 319 damage to a vehicle caused by the shifting of bundles of steel pipes was not subject to the application of CMR. Rather, the issue had to be determined by the application of national law. The court based its decision on Article 17 and did not consider Article 10, cf. App. Bruxelles, 17.6.85, *supra* fn. 98, where the court distinguished Article 10. The series of operations prior to carriage may, however, combine to cause the damage. In App. Lyon, 21.2.91 (1992) B.T. 166, a gas bottle escaped from its container and caused a road accident. The container had not been properly secured and had been loaded by the sender with its doors facing the carriageway. The sender was held to be liable under Article 10. Similarly, where the defect in packaging was such that the goods were improperly stowed by the sender, and the carrier, being unaware of the defect, was unable thereby to exercise effective supervision of the stowage, App. Liège, 27.10.92 (1993) J.T. 538, Putzeys, "Le droit des transports en Belgique (II) 1993–1995" 1996 U.L.R. 557, 565.

4.33 The first point to note in relation to this provision is that it provides only that the sender shall be liable to *the carrier* in respect of the matters referred to. It does not therefore create a right of action against the sender by third parties who have suffered loss or damage but only means that insofar as the carrier has incurred liability to third parties as a result of defective packing, he is entitled to be indemnified in respect of such liability by the sender. Any claim which the third party may have directly against the sender will depend on the national law.

4.34 Secondly, Article 10 refers to expenses resulting from defective packing. Clearly it matters not whether such expenses result either from preventative measures taken by the carrier to avoid damage or from measures taken to rectify the situation once damage has occurred. It should also be noted that the potential liability of the sender under Article 10 is unlimited. Further, so far as damage to the goods themselves is concerned, Article 17(4)(b) relieves the carrier from liability for loss or damage to the goods due to their defective packing.

4.35 The sender will be able to avoid liability under Article 10 only where the defect was apparent or was known to the carrier at the time when he took over the goods and where the carrier made no reservations concerning the matter. For example, in a Belgian decision the sender was held to be liable for damage which resulted to a carrier's trailer owing to defective packing and the consequent spillage of a load of window glass. Here the driver of the vehicle had made written reservations to the effect that two sheets of glass were damaged and had struck out the provision on the note that "*disposition du chargement et arrimage sous ma responsabilité*".[102] Similarly, in a Dutch decision, a sender employed a carrier to tow a trailer which did not reveal any apparent defects. He was held liable to the carrier for damage to the tractor which occurred when the trailer overturned, on the grounds that the carrier could establish either that the trailer was not suitable for towing or that the interior of the trailer had been defectively loaded. In such a case the defect is neither known nor apparent and the carrier is not required to make any reservations.[103]

4.36 Where, however, the party who actually hands over the goods to the carrier is not the sender of the goods he will not be a party to the contract and will therefore not be liable as a sender under Article 10 for damage caused to the carrier's vehicle due to his defective packing. He may, however, be liable to the carrier under the national law for his negligent actions in delict or tort.[104] On the wording of Article 10, the sender may still be liable to the carrier in such circumstances.

Customs and other formalities

4.37 Article 11 provides as follows:—

> "1. For the purposes of the Customs or other formalities which have to be completed before delivery of the goods, the sender shall attach the necessary documents to the consignment

---

102. Comm. Bruxelles, 12.4.72 (1972) 7 E.T.L. 1046. This clause does not appear in the IRU model of CMR consignment note.

103. RB. Amsterdam, 15.11.72 (1973) S. & S. No. 53. Cf. App. Bruxelles, 25.3.88 (1992) R.D.C.B. 780 (see Putzeys, *Le Droit Des Transports en Belgique* 1991–1992, (1992) I U.L.R. 190, 199). See also App. Liège, 27.10.92, *supra*, para. 4.32 fn. 101.

104. App. Liège, 18.12.67 (1969) 4 E.T.L. 965.

note or place them at the disposal of the carrier and shall furnish him with all the information which he requires.

2. The carrier shall not be under any duty to enquire into either the accuracy or the adequacy of such documents and information. The sender shall be liable to the carrier for any damage caused by the absence, inadequacy or irregularity of such documents and information, except in the case of some wrongful act or neglect on the part of the carrier.

3. The liability of the carrier for the consequences arising from the loss or incorrect use of the documents specified in and accompanying the consignment note or deposited with the carrier shall be that of an agent, provided that the compensation payable by the carrier shall not exceed that payable in the event of loss of the goods."

4.38 Article 11 provides that where documents are required for the purpose of Customs or other formalities which need to be completed prior to delivery, the sender is required to attach all necessary documents to the consignment note, or else to place them at the disposal of the carrier. He must also furnish the carrier with all information required for the purpose.[105] The sender is therefore required to supply the carrier with all necessary certificates and other papers relating to exchange control requirements, import and export permits, health and veterinary requirements together with any other documentation which the authorities of any State across which the shipment is to be carried may require.[106] In practice such documentation will commonly be organised by a forwarder on behalf of the shipper of the goods, even though the latter may be treated as the sender for the purposes of CMR and will therefore be liable under CMR to the carrier, if appropriate seeking recourse under his forwarding contract or national law.

4.39 The carrier has no duty to inquire into either the accuracy or adequacy of such documents and information as the sender provides,[107] and can assume that they are adequate for the transit in question.[108] Furthermore, the sender[109] will be liable to the carrier for any damage caused by the absence, inadequacy or irregularity of the documentation and information. For example, in a French case a vehicle was held up at the Franco-German frontier because the German Customs authorities required documentation in German, instead of in French as provided by the sender. It took a number of days for the translation to arrive at the crossing point during which period the vehicle was

---

105. Article 11 provides an indemnity to the carrier only in respect of Customs and other formalities which have to be completed before the delivery of the goods. It does not require the sender to make enquiries and then to inform the carrier of the fact that the consignees' premises were closed: Cass., 26.2.91 (1992) I U.L.R. 340.

106. Thus in App. Paris, 7.12.81 (1982) B.T. 73 where delay at Customs was due to the failure of the sender to supply a certificate of origin, the carrier was held not liable for the delay, in an action against him by the consignee, both on the basis of Article 11 and under Article 17(2), as amounting to "circumstances which the carrier could not avoid and the consequences of which he was unable to prevent". See also App. Lyon, 7.5.81 (1981) B.T. 410. The statement made in Comm. Créteil, 26.5.98 (1999) B.T. 418 that, since the carrier had notice that the goods had been sold on CIP terms, the sender's responsibility ceased on providing the necessary documents to effect export, appears incorrect in light of the wording of Article 11(1). The case was, however, concerned with the failure of the carrier to ensure that customs formalities were effected, having trusted to the good faith of an apparent representative of the consignee.

107. At least for this purpose. Whilst the court in OLG. Düsseldorf, 23.12.96 (1997) TranspR 422 saw no duty to check for the purpose of Article 11, it did find that the carrier should have checked that he had the standard papers for the load and that, had this been done, the delay could have been avoided. In consequence, the carrier was liable up to the limit in Article 23(5). See further, *infra* para. 4.40 fn. 113.

108. See, e.g., OLG Munich, 23.6.95 (1996) OLG-Rp. München 18 (Jung, p. 157).

109. In App. Paris, 7.12.81 (1982) B.T. 73 (see *supra*, fn. 106) it was also held that the consignee was not liable to the carrier under Article 11(2), even though the consignee was entitled to claim against the carrier under Article 13(1).

immobilised. The court held that the carrier was not liable for extra expenses incurred by the sender as a result of the delay. The purported arrangement whereby the carrier supplied the necessary documentation required for Customs transit was null and void under Article 41 as it derogated from Article 11 in as much as this provides that the carrier need not inquire into the adequacy of the documents and that conversely the sender is liable for all resultant loss and damage suffered by the carrier.[110]

4.40 Article 11(2) concludes by providing that the sender will not be liable to the carrier where the damage has been caused by some wrongful act, neglect or default on the part of the carrier. Given that the carrier is neither obliged to provide nor to check the relevant documentation, and that any attempt to impose such obligations on him would be invalid under Article 41, the only scope for the operation of this provision would appear to be where the carrier has either misused or mislaid the documentation or information with which he has been provided.[111] It is therefore at first sight curious that Article 11(3) goes on to provide for the carrier's liability in respect of the loss or incorrect use of the documents.[112] However, the explanation for this apparent duplication would seem to be that the concluding words of Article 11(2) are concerned exclusively with the carrier's claim against the sender arising from inadequate documentation; Article 11(3), on the other hand, is concerned with the liability of the carrier to the sender.[113]

4.41 The statement that the liability of the carrier under Article 11(3) is that of an "agent" is not very clear. In the French text the word "*commissionnaire*" is used. One authority considers that this in fact means a "*commissionnaire de transport*", that is, a freight forwarder.[114] Be that as it may, whether or not the carrier is to be treated as an "agent" or as a "forwarding agent" in relation to the English version of the Convention will not affect the situation, since in either case the carrier will be liable to the sender for any negligence. Since the carrier has obligations first to the sender as the party with a right of disposal of the goods, and later to the consignee when he obtains that right,[115] it follows that the consignee can then also claim against the carrier under Article 11(3). It should finally be noted that whereas the sender's liability under Article 11(2) is unlimited, the

---

110. App. Paris, 27.1.71 (1971) B.T. 115, (1971) U.L.C. 151. The fact that the German Customs usually accepted French language documents was regarded as immaterial. The carrier had supplied a list of the necessary documents without drawing attention to the Regulation which required a translation. See also a case concerning the CIM Convention, AG. Karlsruhe, 11.8.89 (1990) I U.L.R. 470, where the railway was held not liable for any wrong information it might have given to the consignor, since it is the latter's duty to ascertain which documents are necessary for the completion of Customs formalities. Cf. OLG. Düsseldorf, 12.2.81 (1982) VersR 302, cited by Clarke, p. 77, whereby the road carrier raised a missing Customs document himself and was held liable when it proved to be defective. Where, however, the carrier agrees to effect Customs formalities and fails to do so, he can be liable for the fine imposed by the Customs authorities notwithstanding that he gave the task to a Customs agent who failed to carry it out: App. Paris, 23.9.81 (1981) B.T. 538.

111. The carrier may, however, have misled the sender as to the route or method of transport, so that the wrong documents are supplied, cf. App. Paris, 15.2.82 (1982) B.T. 141.

112. "Use", for the court in OLG. Düsseldorf, 23.12.96 (1997) TranspR 422 is to be understood as the temporarily successive making use of the documents supplied by or on behalf of the sender. It involves no obligation to check the completeness of the documents supplied and the court doubted whether the procurement of the documents by the carrier can be understood as "use".

113. As to which the burden of proof is on the sender, proof which the sender failed to provide in OLG. Düsseldorf, 23.12.96, *supra*, fn. 112. In this case, however, the carrier failed to satisfy the burden of proof on him to establish a defence to the delay produced by the absence of Customs documentation, being unable to prove neither the fault of the sender nor that the loss was unavoidable.

114. Loewe, para. 116.

115. Article 12, *infra*, Chapter 5, para. 5.2 ff.

compensation payable by the carrier under Article 11(3) cannot exceed that payable in the event of loss of the goods.[116]

---

116. See Article 23, *infra*, Chapter 9.

CHAPTER 5

The Performance of the Carriage

5.1 This chapter concerns the position as between the three parties concerned, namely the consignor, the carrier and the consignee, in the period during which the goods are being carried and then on the arrival of the goods. The sort of questions which might arise are such as: who has the right to dispose of the goods while they remain in transit? What is the position of the carrier where those rights of disposal are exercised? Who is entitled to require delivery of the goods? What is the position where performance of the carriage becomes impossible? The Convention deals with issues such as these in Articles 12 to 17.

The right of disposal

5.2 Article 12 provides as follows:—

"1. The sender has the right to dispose of the goods, in particular by asking the carrier to stop the goods in transit, to change the place at which delivery is to take place or to deliver the goods to a consignee other than the consignee indicated in the consignment note.

2. This right shall cease to exist when the second copy of the consignment note is handed to the consignee or when the consignee exercises his right under article 13, paragraph 1; from that time onwards the carrier shall obey the orders of the consignee.

3. The consignee shall, however, have the right of disposal from the time when the consignment note is drawn up, if the sender makes an entry to that effect in the consignment note.

4. If in exercising his right of disposal the consignee has ordered the delivery of the goods to another person, that other person shall not be entitled to name other consignees.

5. The exercise of the right of disposal shall be subject to the following conditions:

(a) that the sender or, in the case referred to in paragraph 3 of this article, the consignee who wishes to exercise the right produces the first copy of the consignment note on which the new instructions to the carrier have been entered and indemnifies the carrier against all expenses, loss and damage involved in carrying out such instructions;

(b) that the carrying out of such instructions is possible at the time when the instructions reach the person who is to carry them out and does not either interfere with the normal working of the carrier's undertaking or prejudice the senders or consignees of other consignments;

(c) that the instructions do not result in a division of the consignment.

6. When, by reason of the provisions of paragraph 5(b) of this article, the carrier cannot carry out the instructions which he receives, he shall immediately notify the person who gave him such instructions.

7. A carrier who has not carried out the instructions given under the conditions provided for in this article, or who has carried them out without requiring the first copy of the consignment note to be produced, shall be liable to the person entitled to make a claim for any loss or damage caused thereby."

5.3 It will first be noted that the rights conferred on the sender under Article 12 are distinct from and wider than the equivalent rights under the Sale of Goods Act 1979,[1] namely the unpaid vendor's lien and the right of stoppage in transit. Those rights are restricted to circumstances where the sender is an unpaid seller or where the consignee is an insolvent buyer[2]; there are no equivalent restrictions on the rights of disposal under Article 12.

5.4 Turning to the substance of the rights conferred by Article 12, it will be observed that the basic scheme is that the carrier is, in the first instance, treated in effect as the agent of the sender,[3] so that the sender retains the right to dispose of the goods until, in most cases, the delivery of the consignment note to the consignee, which will normally coincide with delivery of the goods.[4] This right will, however, necessarily be overridden where the sender chooses to confer an immediate right of disposal on the consignee under Article 12(3). It should be borne in mind, however, that the right of disposal conferred by Article 12 does not necessarily reflect the ownership of the goods, nor does the first copy (the sender's copy) of the consignment note constitute a document of title; it is a document of proof only, and in this context will prove the right of the consignee to dispose of the goods, although only if the appropriate instructions were entered at the time the consignment note was drawn up. One commentator has detected a hiatus in the scheme of CMR here, and suggests that where the consignee is in possession of the first copy of a consignment note but without any right of disposal having been vested in him under Article 12(3) by means of an appropriate note, the sender has lost his right of disposal while the consignee has not yet acquired his right.[5] As against this, however, it might be argued that the delivery of the first copy to the consignee may on its true construction amount to an assignment of the right of disposal, but in any event it would seem that this theoretical difficulty has caused few, if any, problems in practice. If there is in truth a hiatus then the goods can only be disposed of once the second copy of the consignment note, which will have accompanied the goods, has been handed to the consignee.[6]

Restrictions on the right of disposal

5.5 The right of disposal is not unfettered, but is qualified in various respects by CMR. First, there is the procedural requirement for production of the first copy of the consign-

---

1. Distinguish also a variation of the contract agreed by *both* parties, OLG. Hamburg, 7.4.94 (1994) TranspR 444 (Clarke, p. 90, n. 94).
2. See sections 38 to 46, Sale of Goods Act 1979.
3. The sender himself may also be an agent. In *Kala Ltd.* v. *International Freight Services (U.K.) Ltd.*, Queen's Bench Division, 7.6.88, unreported except on LEXIS, the instructions of the forwarder contemplated the forwarder naming himself as sender on the CMR consignment note. The principal had thereby given the forwarder the right to control transit, a right which was said to be confirmed by CMR. This right enabled the forwarder to establish his lien. It does not appear that privity of contract had been established between the principal and the CMR carrier. Should this be the case the principal may well have the right to intervene as true sender, but not otherwise; see further *infra*.
4. If the goods are lost or delayed, then the consignee becomes entitled to claim directly against the carrier under Article 13(1), *infra*. At the time the consignee becomes entitled, the sender loses his right of disposal: Article 12(2). If, however, the consignment note is irregular by virtue of not being signed by the sender and carrier, the German courts consider that the consignee does not "acquire" (cf. Clarke, p. 95) the right of disposal, in accordance with Article 12(2), simply by obtaining the second copy of the consignment note: BGH., 15.10.98 (1999) TranspR 102, 103.
5. Loewe, para. 119.
6. Article 12(2). For the position where the goods are lost or delayed, see fn. 4.

ment note endorsed with the amended instructions.[7] Secondly, there is the requirement that the person exercising the right must indemnify the carrier against any expenses, loss or damage,[8] and the carrier can also refuse instructions which are impracticable, or which would interfere with the normal working of his undertaking, or which would prejudice the senders or consignees of other consignments.[9] The Convention offers no guidance as to when a refusal for these reasons will be justified, but in relation to interference with normal working, since any alteration to instructions will cause a degree of interference, if only through the revision of paperwork, it seems sensible to follow Loewe[10] when he suggests that the provision applies only where "the carrying out of the instructions will seriously interfere with the normal activities of the carrier's enterprise". Finally, the right can only be exercised if the instructions do not result in a division of the consignment.[11] This provision is of particular relevance to consolidated groupage shipments, and in this context prevents the freight forwarder who has organised the shipment pressing the carrier into service as a *de facto* distributor, but it is not necessarily restricted to this category of shipment.

5.6 Loewe is of the view that the carrier is not entitled to rely on Article 12(5)(c) where separate consignment notes have been issued for goods in different vehicles, or for goods of different kinds, or for goods in different lots.[12] There would seem no difficulty about this in relation to goods in different vehicles, but as regards the other two suggested categories this would mean that the sender can in effect render the carrier a *de facto* distributor by the device of issuing separate consignment notes for different parts of a consolidated shipment, originally naming the same consignee but then issuing separate instructions under Article 12(1) in respect of each. To an extent there is no problem about such an interpretation since in such circumstances the carrier will have his claim for expenses under Article 12(5)(a), but nevertheless there are objections to this view. First, such an approach is to interpret the word "consignment" in Article 12(5)(c) as "goods covered by one consignment note", which is not what the Convention says, and secondly, such an interpretation could involve the carrier in a fundamentally different operation from what he originally envisaged. It would seem, therefore, that the better view is that where the carrier is carrying what can properly be regarded as one consignment,[13] he ought to be entitled to refuse to accept instructions relating to different parts of the consignment, even if they are covered by separate consignment notes. If, however, he chooses to comply, he will be entitled to his expenses under Article 12(5)(a).

---

7. Article 12(5)(a). In a case, however, where an instruction to suspend the carriage was given by the sender before the consignment note had been made out, the German Supreme Court held the carrier to be liable for having ignored the instruction and sent the goods on to the consignee: BGH., 27.1.82 (1985) 20 E.T.L. 349, (1982) II U.L.R. 168.

8. Article 12(5)(a). Unfortunately, CMR does not specify what is meant by an indemnity in this context, so for instance, it is not clear whether, and if so in what circumstances, a carrier can insist on security for payment or payment in advance, or whether he can exercise a lien on the goods under any relevant national law in the event of non-payment.

9. Article 12(5)(b).

10. Loewe, para. 122.

11. Article 12(5)(c).

12. Loewe, para. 123.

13. So if the goods are on separate vehicles and covered by separate consignment notes, it would seem that it would be legitimate to regard these as separate consignments, so as to accord with Loewe's view in this respect.

5.7 It must be conceded that in most respects this question will be academic, since in such circumstances the carrier would be able, in appropriate cases, to rely on Article 12(5)(b), as amounting to interference with the normal working of his undertaking. There is, however, one significant difference between sub-paras (b) and (c), in that Article 12(6) provides that where the carrier wishes to rely on the former, he must immediately notify the person who has given him the revised instructions that he cannot carry them out. The consequence of failure to do so is that the carrier will be liable, without limit as to quantum, for any loss or damage caused thereby.[14] There is no corresponding obligations or liability in relation to sub-para. (c), so if the carrier has not given the necessary notification under sub-para. (b), the ability to rely on sub-para. (c) may be crucial to him.

5.8 Article 12(7) provides that where the carrier has failed to carry out amended instructions,[15] or where he has carried them out but without first requiring production of the first copy of the consignment note, he is liable to the person entitled to make a claim[16] for any loss or damage caused,[17] there being no limit on such liability.[18] The second part of this provision, namely acting on a variation without requiring production of the first copy of the consignment note, would, on the face of it, permit a sender who suffers loss as a result of the execution by the carrier of instructions which the sender has himself given to rely on the absence of compliance with the formal requirements of CMR to sustain a claim. It is submitted that the applicable principles of domestic law[19] will prove to be available to assist the carrier to defeat any such claim, in particular since it is clear that the purpose of the provision is to cover the situation where the carrier agreed to amended instructions without a sight of the first copy of the consignment note and it transpires that the person giving the instructions had no right to do so.

5.9 The attachment of conditions to the right of disposal can also give rise to a further problem, namely the absence of specific rules to cover a dispute as to whether a refusal by a carrier to accept amended instructions which is grounded in Article 12(5)(b) is or is not justified,[20] or whether, even if it can be established that a refusal was not justified, any person who has suffered loss or damage thereby is entitled to compensation, and if so, on what basis. Again, it would seem that the parties will have to have recourse to domestic law, on the footing that an unwarranted interference with the right of disposal will be a

---

14. Article 12(7), discussed in the next paragraph.

15. Unless, of course, the carrier is entitled to refuse under Article 12(5).

16. E.g. an original consignee whose right to obtain delivery has been frustrated by an irregular disposition by the sender, Lamy, paras 478(d) and 1535, Jur. 3, citing Cass., 19.7.76 (1976) B.T. 364.

17. In OLG. Munich, 23.4.93 (1993) TranspR 348, the carrier obeyed the instruction of the consignee, who had not yet obtained the right of disposal, to deliver the goods at a different place than that specified in the consignment note. The carrier was held not liable to the sender for the damage sustained by the goods during on-carriage by a different carrier to the original destination. The goods had been delivered to the intended consignee and the sender could establish his claim only if he could show that he gave or would have given fresh instructions to change the consignee or to have the goods returned to him. However, where the carrier obeys an irregular instruction of the consignee to deliver to a third party, the carrier is liable to the sender where the third party fails to pay the price of the goods, Lamy, paras. 478(d) and 1535, Jur. 4 citing Cass., 24.4.90, Lamyline.

18. Loewe, para. 124. Liability was limited, however, in BGH., 27.1.82 (1985) 20 E.T.L. 349, 353. Cf. further Clarke, p. 96.

19. E.g. under English law the sender would be estopped from denying the validity of his instructions.

20. The onus of proving that the instructions comply with Article 12(5)(b) is on the person seeking to exercise the right to disposal, see Clarke, p. 91, citing BGH., 27.1.82 (1985) 20 E.T.L. 349, 353. Cf. Maccarone, Theunis, p. 73, n. 43.

breach of contract. Finally, it should be stressed that, since there is no obligation on the carrier to notify a person giving amended instructions that they are being refused on the grounds set out in Article 12(5)(a) or (c), the CMR régime appears to assume familiarity with these provisions on the part of such persons.

Successive disposals

5.10 The sender, while he retains the right to dispose of the goods, is free to nominate an alternative consignee,[21] as is the consignee if he has acquired the right to dispose of the goods under Article 12(3). However, a consignee nominated by the original consignee under that provision does not have a similar right,[22] but there is no similar express restriction on the sender or on the consignee who is exercising his right under Article 12(3); presumably therefore either the sender or the original consignee, as the case may be, can nominate further alternative consignees so long as he retains the right to dispose of the goods and the carrier does not invoke any of the restrictions in Article 12(5). The obligation under Article 12(5)(a) to indemnify the carrier will, of course, apply equally to any such further disposals.

5.11 The German Supreme Court has held that if the carrier, faced with the refusal of the consignee to accept the goods, avails himself of the right under Article 16(2) to unload the goods[23] and to deposit them with a third party, the carriage comes to an end and the right of the sender to dispose of the goods by asking the carrier to deliver them to another consignee also ceases to exist.[24]

Disposal of goods on delivery

5.12 Article 13 provides as follows:—

> "1. After arrival of the goods at the place designated for delivery, the consignee shall be entitled to require the carrier to deliver to him, against a receipt, the second copy of the consignment note and the goods. If the loss of the goods is established or if the goods have not arrived after the expiry of the period provided for in article 19, the consignee shall be entitled to enforce in his own name against the carrier any rights arising from the contract of carriage.
> 2. The consignee who avails himself of the rights granted to him under paragraph 1 of this article shall pay the charges shown to be due on the consignment note, but in the event of dispute on this matter the carrier shall not be required to deliver the goods unless security has been furnished by the consignee".

5.13 Article 13 provides that after the goods have arrived at the place designated for delivery which is specified in the consignment note[25] the consignee is entitled to obtain

---

21. Article 12(1). The fact that the sender may thereby be in breach of an underlying contract of sale should not affect his rights under CMR. See, however, Cass., 29.10.90 (1991) B.T. 189 and the comment there made in respect of the case. Cf. further OLG. Düsseldorf, 16.6.92 (1993) TranspR 17, noted *infra*, fn. 26.

22. Article 12(4). But in a Dutch case it was held that, where a consignee who had the right to dispose of the goods nominated an alternative consignee, who subsequently indicated that the goods were to be delivered to the original consignee via a freight forwarder, this instruction did not infringe Article 12(4) as in effect there had been no change in the consignee. Arrond. Amsterdam, 16.2.66 (1966) S. & S. No. 69.

23. See *infra*, para. 5.37 ff.

24. BGH., 5.2.87 (1988) II U.L.R. 708, (1987) 22 E.T.L. 574.

25. See Article 6(1)(d).

both the second copy of the consignment note and the goods themselves.[26] It should be noted that the consignee is not himself required to hand over to the carrier the first copy of the consignment note although he is required to provide the carrier with a receipt, the form of which is not specified by the Convention. Since under current transport practice it is unlikely that the consignee will ever see the first copy of the consignment note[27] or, if he does, it is likely to arrive by post after the actual arrival of the goods, it seems sensible that the consignee should not be required to hand over to the carrier the first, or indeed any, copy of the consignment note.

5.14 Article 13(1) also lays down the procedure to be followed in the event of the loss or non-arrival[28] of the goods. First, it may be established that the goods are lost, or alternatively that there has been a delay in delivery in that they have not arrived in the period laid down by Article 19. That period is either the agreed time limit or else, in the absence of such limit, the time it would be reasonable to allow to a diligent carrier having regard to the circumstances of the case. If either of these circumstances can be established, the consignee will be entitled to enforce in his own name against the carrier any rights arising from the contract of carriage.

5.15 Consequently, if the goods have been lost the consignee may be able to rely upon Article 17(1) whereby the carrier will be liable for the total or partial loss of the goods when it has occurred between the time of taking over and the time of delivery as well as for any delay in delivery. If the consignee cannot prove that the goods have in fact been lost he may be able to rely upon the provisions of Article 20 whereby failure to deliver within 30 days following the expiry of the agreed time limit or, in the absence of such a limit, within 60 days from the time when the carrier has taken over the goods is to be treated as conclusive evidence of loss, in which case the consignee is entitled to claim the amount lost.[29]

5.16 It will be noted that under Article 13(1) it is the consignee who is entitled to require the carrier to deliver the goods and the second copy of the consignment note to him. The consignee will normally be named and identified in the particulars of the consignment note as provided by Article 6. Not being a negotiable instrument, a consignment note will never be issued to bearer or to order. However, it should be remembered that the original consignee will cease to be so entitled where the sender has amended his instructions as regards delivery, naming a new consignee or instructing the carrier to return the goods to him under Article 12(1). Alternatively, the consignee himself may have

---

26. This right is qualified by any conditions which the sender has imposed on delivery and contained in a consignment note or instructions to the carrier, e.g. that delivery is only to be effected against a banker's order for a specified amount. In OLG. Düsseldorf, 16.6.92 (1993) TranspR 17, the consignment note indicated that delivery was to be against a banker's order. The consignee tendered an order for a lesser amount, delivery was refused and instructions given to return the goods. The consignee claimed that the agreement was for the lesser amount but no change in the sender's instructions had been received by the carrier. The carrier was not liable to the consignee, who could not claim that the goods were lost when they had been returned in accordance with a valid instruction. As the court said: "if the sender makes a reservation to the carrier involving instructions that he is not allowed to reserve to himself in his relationship with the consignee, then he and not the carrier is liable to the consignee."

27. The only circumstances in which the consignee is likely to have the consignment note is where the right to dispose of the goods has been transferred to the consignee under Article 12(3) and he holds the first copy of the consignment note for the purposes of giving amended instructions.

28. It would seem that the position will be the same where the goods have been damaged: see *infra*, para. 5.24.

29. For Articles 19 and 20 generally, see *infra*, Chapter 7, paras 7.1–7.29.

exercised his right of disposal under Article 12(3) and named a third party as consignee.[30]

5.17 If the consignee of goods wishes[31] to avail himself of the rights granted to him under Article 13(1) he must pay the charges[32] which are shown to be due on the consignment note[33] and in the event of a dispute as to the amount due the carrier need not deliver the goods to the consignee unless security has been furnished by the latter. This provision is potentially unsatisfactory, in so far as it appears to place an absolute obligation on the consignee to furnish security, regardless of the existence of a legitimate claim by the consignee against the carrier. This problem will not arise in the event of a total loss of the goods since there is obviously then no question of any delivery, but may arise in the case of delay, damage or partial loss.[34]

The right to enforce the contract

5.18 The object of Articles 12 and 13 is to establish who at any particular time has the right to dispose of the goods comprised in a particular contract for the international carriage of goods by road. The object does not seem to have been fully achieved in that courts in different jurisdictions have given conflicting decisions as to the interpretation of the two Articles, and also in that the Articles do not provide an exhaustive code of rules and have to be supplemented by national law. As a minimum, the courts are required to recognise the right given to the consignee by Article 13(1) to enforce the contract in his own name.[35] The German courts, as a matter of interpretation of CMR rather than the application of national law, have linked this with the right of disposal, so that the

30. N.B. also that, while national law may provide for the assignment of the consignee's rights in varying ways, it is submitted that the obligation of the carrier as to the person to whom he must make delivery will be regulated by CMR alone. (See Rodière, para. 39.)

31. This has a conditional sense, yet Article 13(2) is not expressed in conditional terms and might be construed as imposing independent obligations subject to the right of security on delivery granted to the carrier. Thus the right of suit may not be dependent on payment, a view supported by App. Venezia, 31.10.74 (1975) 10 E.T.L. 242. Clarke takes the conditional line, see p. 134.

32. A recent change in French law, whereby Article 101 of the *Code de Commerce* has been amended (by the *loi Gayssot* of 6.2.1998) to provide that both sender and consignee are guarantors of the carriage charges, may have clouded the position in France since even in international carriage the consignee are seen as based on this law rather than Article 13(1), see Comm. Nanterre, 17.9.99, (1999) B.T. 710.

33. Controversy surrounds the question whether the consignee is required to pay carriage charges which do not appear on a consignment note. If the purpose of Article 13(2) is for the consignee to be informed of what he is obliged to pay before taking delivery, then the absence of a charge on the consignment note will render him under no obligation to pay it, cf. App. Paris, 19.11.81 (1982) B.T. 62. (cf., in Germany, OLG. Hamm, 12.11.73 (1974) N.J.W. 1056, but note the possibility of incorporation of tariffs by reference, BGH., 25.4.91 (1991) TranspR 312. See further, Herber/Piper, Article 13, Rdn. 22, 26, cf. OLG. Munich, 9.10.92 (1993 TranspR 75)). If, however, the purpose is seen as confined to that of proof only, and the consignee is considered to become a party to the contract by the demand he makes for delivery, then he thereby becomes liable to pay the carriage charges notwithstanding its absence from the consignment note, see Libouton, (1982) J.T. 693, 703 para. 46, citing Comm. Tongres, 22.4.76 (1976) J.C.B. 365. See also Maccarone, Theunis, 62, 69. A distinction may be possible between freight charges and other disbursements made in respect of the goods, e.g. import duties, for which the carrier claims a refund, cf. AG. Tempelhof, 18.6.98 (1998) TranspR 403, see also OLG. Munich, 9.4.97 (1997) TranspR 368. Not all the obligations under the contract are imposed on the consignee, however, since CMR expressly imposes some obligations solely on the sender, e.g., Article 10 *supra*, para. 4.32 ff.

34. Article 23 sets out the entitlement of the consignee to damages. In particular, in the event of partial loss, a proportion of the transport charges, etc., must be refunded, while in the case of delay the measure of damages is tied to the amount of the transport charges.

35. There is no merit in the proposition that the Convention does not cover entitlement to sue: Hof. 's Hertogenbosch, 20.6.88 (1989) S. & S. No. 261, (1990) 18 E.L.D. No. 187.

consignee becomes entitled to sue in his own name for damage to the goods from the time he acquires the right of disposal of the goods.[36] The right to delivery carries with it the right to receive the goods in a complete and undamaged condition, so that the consignee's right of action is not limited to loss and delay which is expressly referred to in Article 13.[37] Once acquired, on the arrival of the goods at the place designated for delivery, the consignee does not lose his rights by refusing to accept delivery of damaged goods unless it is made clear to the carrier that the consignee intends also to reject the right to claim compensation.[38] Logically, one might expect that the application of the right of disposal would mean that, as between sender and consignee, only one of them would have the right to sue at any one time.[39] However, this does not appear to be the case, the German courts being concerned rather to determine the time when the consignee's rights are established.[40] Consequently, both sender and consignee can possess joint rights against the carrier.[41] This is especially true where the goods are lost, since the right of the consignee under the second sentence of Article 13(1) is not dependent on his having had the right of disposal transferred to him.[42] The possibility of concurrent rights is also accepted in France,[43] Belgium,[44] Holland,[45] and Austria.[46] Clearly there is recognition that the rights of the consignee are not exclusive and do not exhaust the possibility of a claim by another party by virtue of their position as sender.

5.19 The possibility of there being more than one person with a right of action requires a mechanism to protect the carrier from having to pay twice in respect of the same claim. CMR itself does not provide this,[47] so that the answer must be provided by the national

---

36. BGH., 6.7.79 (1980) 15 E.T.L. 863. See Haak, pp. 257 et seq., who considers this case a move away from previous cases of the Supreme Court applying national law, e.g. BGH., 21.12.73 (1975) 10 E.T.L. 92.

37. BGH., 6.7.79 (1980) 15 E.T.L. 863, 869, BGH., 15.10.98 (1999) TranspR 102, 104, Herber/Piper, Art. 13 Rdn. 5. A right of action in respect of damage is consistently assumed in Continental case law. See, e.g., OLG. Karlsruhe, 24.5.67 (1967) U.L.C. 289, RB. Antwerpen, 7.12.73 (1976) 11 E.T.L. 295, App. Bruxelles, 30.10.75 (1976) 11 E.T.L. 238. See also OGH., 17.2.82, cited by Clarke, p. 131.

38. BGH., 15.10.98 (1999) TranspR 102 (1999) U.L.R. 1036. The consignee may, however, be limited in his claim if he refuses the whole consignment on the spurious basis that it is difficult to separate the damaged parts from those which are sound: OLG. Hamm, 6.2.97 (1998) TranspR 34.

39. Cf. Clarke, p. 135. See also OLG. Karlsruhe, 24.5.67 (1967) U.L.R. 289.

40. Haak, p. 263, says that BGH., 6.7.79 (*supra*, fn. 37) did not wish to adopt the exclusivity aspect of the doctrine of disposal (contrary to the view of the distinguished German writer Helm. See his *Frachtrecht*, Art. 17 CMR, Anm. 30). The decision appears contradictory since it proceeds largely on the basis of various parts of CMR which expressly give rights to the party with the right of disposal. Nevertheless the court deals, at the end of the judgment, with the possibility of double claims. The emphasis by the court on provisions of CMR granting particular rights of action to the party with the right of disposal are based on the German translation which differs from the authentic versions; see Haak, p. 264 and Clarke, p. 140.

41. BGH., 24.10.91 (1992) 27 E.T.L. 839, 844. BGH., 6.5.81 (1982) 17 E.T.L. 313, 317.

42. BGH., 28.4.88 (1989) 24 E.T.L. 89, 93, BGH., 15.10.98 (1999) TranspR 102, 103.

43. Lamy, para. 555.

44. Hof. Brussel, 5.12.68 (1969) 4 E.T.L. 958. App. Bruxelles, 30.10.75 (1976) 11 E.T.L. 238. RB. Antwerpen, 7.12.73 (1976) 11 E.T.L. 295. Hof. Brussel, 16.11.77 (1980) 15 E.T.L. 319, affirmed Cass., 13.6.80 (1980) 15 E.T.L. 851. Hof. Brussel, 7.2.92 (1993) 28 E.T.L. 286.

45. Arrond. Den Haag, 18.2.69 (1969) S. & S. No. 58. Arrond. s' Hertogenbosch, 11.12.64 (1967) S. & S. No. 5.

46. OLG. Linz, 27.11.89 (1990) I U.L.R. 442.

47. Unless it is interpreted as affording rights of action only to the person with the right of disposal, which can be one person only, subject perhaps to agency principles. See *infra*, para. 5.20.

law.[48] One solution is to release the carrier once he has paid one of the parties.[49] Another is to require the claimant to prove a sufficient interest to sue.[50]

5.20 Where the contract of carriage has been made by a forwarder, the possibility of intervention by the principal is recognised in some states. In France, for example, the principal of a *commissionnaire de transport*[51] can intervene in relation to the contract of carriage and be identified as the real sender or real consignee.[52] In the absence of such intervention the forwarder, as the party actually contracting, may also sue.[53] The courts may require the forwarder to prove that he has an interest in doing so, which can include

48. Cf. BGH., 24.10.91 (1992) 27 E.T.L. 839, 844 in relation to the effect of claims made by individual creditors.

49. BGH., 6.7.79 (1980) 15 E.T.L. 863, 869 where the court stated "there is no justification for the concern raised by the defendant that the sender could possibly claim against them a second time. If the receiver has, with entitlement, asserted claims for compensation for damage to the goods then the carrier, should they be sued by the sender, can object to the sender that they have fulfilled their obligation to pay compensation associated with the contract and are therefore free", see also BGH., 15.10.98 (1999) TranspR 102, 105. Cf. Hof. Antwerpen, 13.10.86 (1987) 22 E.T.L. 443 where the court stated that the right to bring an action founded on CMR constitutes an interest such as required by the Judicial Code. The German courts may, however, still require some proof of interest: see BGH., 6.5.81 (1982) 17 E.T.L. 313, 317, and OLG. Munich, 5.7.89 (1990) I U.L.R. 438, *infra*, fn. 54.

50. In France, the right to claim must be exercised by the party who has effectively suffered the prejudice: Lamy, para. 396. A sender who had been indemnified by his insurer had no title to sue in App. Versailles, 7.3.96 (1996) B.T. 633. This is not completely dependent on the actual division of risk or property made, e.g., by an underlying contract of sale. Thus it is not strictly necessary, e.g., for the sender to prove that he is at risk under the contract of sale if he shows that he has in fact taken the loss upon himself: Lamy, para. 393 and see e.g. App. Montpellier, 1.12.87 (1988) B.T. 13, cf. App. Paris, 29.9.99 (1999) B.T. 747. For Holland, see Haak, p. 268 n. 86 where the varying positions taken by the courts are traced. Alternatively, the court may derive the interest from the right of disposal: see App. Bruxelles, 30.10.75 (1976) 11 E.T.L. 238, 243 where the court said that the right of the sender derives from the fact that he has contracted with the carrier and has the right to dispose of the goods until this right is extinguished in accordance with Article 12(2). Rights derived from the right of disposal could have special significance for courts which consider it possible for a party to have the status of sender even though not the party who originally contracted with the carrier, as seems to have occurred in HR., 7.12.73 (1974) 9 E.T.L. 724, *infra*, fn. 57. A different approach would regard the question of whether there was sufficient interest in the claimant as *res inter alios actae* unless there were multiple claims against the carrier, in which case the division of risk, e.g., in an underlying contract of sale, determines who has the better interest. See Hof. Brussel, 16.11.77 (1980) 15 E.T.L. 319, affirmed Cass., 13.6.80 (1980) 15 E.T.L. 851 where, this not being a case where the carrier was faced with double claims, the consignee buyer, being a party with an interest in the carriage, did not have to prove ownership or risk. Proof that the seller had promised to replace the goods did not defeat his right of action.

51. See *supra*, para. 1.50 ff.

52. Article 101 *Code de Commerce*, Lamy, para. 555. See also Cass., 23.9.71 (1972) B.T. 113. It seems possible, however, that even apart from Article 101, the French courts accept that CMR, being less formalistic than other transport régimes (e.g., that relating to carriage by air), enables the court to look behind the appearances of the consignment note to find, e.g., the real consignee. See Cass., 22.2.94 (1994) B.T. 263, and also Cass., 12.5.80 (1980) B.T. 370 where the principal's known agent was named as consignee and the court treated the principal as *destinataire réel* (Libouton, (1982) J.T. 693, 704 para. 54), and also see *infra*, para. 5.21. Similarly in Germany, procedural rules are available to enable the principal to intervene on a contract to which his agent is a party, e.g. as receiving forwarder. In BGH., 6.5.81 (1982) 17 E.T.L. 313 goods had been ordered by the plaintiff which were despatched by the seller under a groupage contract with the defendant. The consignment note named a receiving forwarder as consignee, although ultimately the goods were delivered to the plaintiff. The forwarding company was the agent of the plaintiff and had been named by him for incorporation into the consignment note. Although the lower court considered that under freight law the forwarder was technically the consignee, it took the view that the forwarder had tacitly agreed to the plaintiff exercising the claim. The forwarder had acquired only a formal legal right. The party truly affected was the plaintiff and was entitled to sue. This was upheld by the Supreme Court. The finding that the forwarder was the primary consignee was not contested. The plaintiff, having a legal interest to sue, could be authorised by the forwarder to take over the action. This need not be express, but can be implied from the circumstances.

53. B.G.H., 20.4.89 (1989) II U.L.R. 806. See also RB. Antwerpen, 16.5.97 (1999) 34 E.T.L. 119.

the fact that he has a liability towards his principal,[54] or that the agent is suing on behalf of his principal.[55]

5.21 Further difficulties arise from uncertainty as to the importance of the consignment note in this respect. Sometimes the courts assume that only the persons named in the consignment note are properly to be regarded as the CMR sender or consignee.[56] This appears to have been the assumption of the Supreme Court of the Netherlands which decided[57] that, while CMR was drafted on the basis that the CMR contract would be made between the named carrier and the named sender, and therefore makes no provision as to liability when this is not the case, the person who has actually contracted with the carrier can maintain an action despite the provisions of Articles 12 and 13. Any such action must of course be under national law. Alternatively, application of Articles 4 and 9 may enable the court to reach the view that persons other than those named can be contracting parties,[58] within the scope of CMR. This can be just as true in identifying the consignee as well as the original contracting party.[59] Nevertheless, the consignee is not necessarily the person who receives the goods[60]; rather, it is necessary for the person whom the sender intends to have the right to receive the goods to be identified to the carrier,[61] and proof of this will normally come from the consignment note,[62] unless a change of instruction has been made by the sender in accordance with Article 12. In a recent French case[63] goods were consigned under a CMR consignment note. Metpas sold goods to Tisselva, who were importing them. A consignment note designated Metpas as sender and a bank as the consignee. Tisselva was indicated as a notify party along with Ste. Mory, who were agents of the sender and to whose address the goods were to be delivered. Ste. Mory was to re-consign the goods to the ultimate purchaser. Despite being indicated as the party to receive the goods, Ste. Mory could not be considered as the real consignee (*destinataire réel*). As agents of the sender, they were indicated in the part of the note referring to

---

54. OLG. Munich, 5.7.89 (1990) I U.L.R. 438, where the court made the point that the same is true of a carrier seeking to sue on the contract he has made, see also for Austria OGH., 26.11.96 (1997) TranspR 281. In France, the *commissionnaire de transport* must either show that he has indemnified his principal or that he is under an obligation to do so: Mercadal, para. 253, see also Lamy, vol. 2, para. 61 and see Cass., 13.6.89 (1989) B.T. 527, App. Reims, 22.5.96 (1997) B.T. 447.

55. HR., 11.3.77 (1977) N.J. 521, cited by Haak, p. 271. In Belgium, the courts may only need to be satisfied that the principal of the forwarder is not opposed to actions by him: see Libouton, *Auxiliaires*, para. 54.

56. This is similar to the strict position taken in respect of CIM, see Yates, para. 4.1.2.54.3.

57. HR., 7.12.73 (1974) 9 E.T.L. 724. The facts were that the contract was made by the Dutch importer, but the name of the Belgian supplier was shown as the sender, and that of the importer's customer as consignee. See the criticism of Haak, p. 254. Cf. BGH., 6.5.81, *supra*, fn. 52.

58. E.g. App. Aix-en-Provence, 25.11.82 (1983) B.T. 591, where the seller was named as sender but was not a party to the contract of carriage, the goods being at the risk and under the control of the consignee buyer whose agent had assisted at the loading and had signed the consignment note, which was also signed by the driver in the absence of any representative of the seller. Cf. Comm. Antwerp, 20.4.72 (1972) J.C.B. 466, involving a claim by a *commissionnaire-expéditeur* who gave the instructions to the carrier but whose name did not appear in the consignment note.

59. Cass., 22.2.94 (1994) B.T. 263. See also BGH., 15.0.98 (1999) TranspR 102, 103.

60. App. Aix-en-Provence, 28.5.91 (1993) B.T. 204, where the company in whose premises the consignee, named as such, was resident and which was charged only with receiving the goods for the account of the latter had no right of action against the carrier. The fact that the company's servant stamped the consignee's box in the consignment note did not have the effect of transferring that status to them.

61. See, e.g., the non-CMR case App. Aix-en-Provence, 19.6.80 (1981) B.T. 103.

62. In Hof. Antwerpen, 23.2.93 (1993) 28 E.T.L. 934 the consignee was identified from the Customs transit document attached to the consignment note. See also App. Paris, 2.12.81 (1982) B.T. 73 and Cass. (Belg.), 18.6.90 (1993) B.T. 665.

63. App. Paris, 7.4.94 (1994) B.T. 410. Cf. generally Cashmore, "Who are the Consignors and Consignees for the Purposes of a Contract of Carriage?", (1990) J.B.L. 377.

"instructions of the sender". Thus to the reader of the consignment the real consignee was not identifiable. Consequently, the insurer which had indemnified Ste. Mory had no right of subrogation through Ste. Mory in its position as agent of the sender, not even if it were the agent of Tisselva, the owner of the goods, since that company was neither the sender nor the consignee, the real consignee being the ultimate purchaser.[64]

5.22 These matters have received limited treatment in England to date. In *Texas Instruments Ltd.* v. *Nason (Europe) Ltd.*[65] goods were consigned under a CMR consignment note to a buyer to whom property and risk passed at the time of the carriage. Tudor Evans, J., held that the seller, by whose order the goods were consigned, acted as agent for the buyer who was thereby the carrier's co-contractant with title to sue.[66] Thus he applied the traditional approach in English law of presuming agency in a seller when the intention under the contract of sale is to pass the property to the buyer prior to or at the time of delivery to the carrier.[67] However, if he were wrong in this view, he considered that the buyer had an unqualified right to sue by reason of Article 13 of CMR.[68]

5.23 In conclusion, it is suggested that so long as the sender and the consignee can be identified, CMR affords both parties the right to sue, and to obtain substantial damages, with the result that the carrier is released by payment to the first to present the claim. Regardless of how conflicting claims are to be resolved or the rights *inter se* adjusted,[69] the relationship between sender and consignee is best regarded as *res inter alios actae* to the carrier[70] except to the extent that the relationship involves considerations of agency.[71]

64. Cf. Cass., 12.5.80 (1980) B.T. 370 where the named consignee was known to be the agent of the true consignee.

65. [1991] 1 Lloyd's Rep. 146.

66. Conversely, the terms of sale are unlikely to affect the position of the seller, as apparent sender, in respect of an action by the carrier against him, cf. App. Paris, 8.11.95 (1996) B.T. 132.

67. *Dawes* v. *Peck* (1799) 8 T.R. 330. See generally Glass & Cashmore, para. 1.04 et seq., and Cashmore, *Contract*. Contrast *E.W. Taylor* v. *Bell* [1968] 2 Lloyd's Rep. 63, 70. It is said, furthermore, that the carrier is entitled to assume that the consignee is the owner; see Clarke, pp. 428–429 and Glass & Cashmore, para. 1.12. This assumption should not, however, affect the title to sue at common law, but is a protection for the carrier. In *Aqualon (UK) Ltd.* v. *Vallana Shipping Corporation* [1994] 1 Lloyd's Rep. 669, Mance, J., suggested that although there was slender evidence of the precise relationship between the seller and the buyer, named respectively as sender and consignee in the consignment note, it may be realistic to assume that the sale contemplated CMR carriage and on this assumption the combination of Article 12 of CMR and section 19 of the Sale of Goods Act 1979 (treating this Act as applicable in the absence of evidence that foreign law either governs or differs) may mean that property in the goods did not pass on collection by the carrier. Mance, J., referred to the unreported decision of Webster, J., in *Courtaulds PLC* v. *Detrafor*, Queen's Bench Division, 26.1.90, unreported, which counsel contended supported the view that the named sender could claim for damages even if it contracted as agent for its buyer consignee. In *Aqualon*, Mance, J., preferred not to decide the point. On the facts before him in the *Courtaulds* case, Webster, J., considered that the sender had contracted both on his own behalf and on behalf of the consignee, he being concerned, in essence, with the issue of whether the consignee had title to sue.

68. In *Courtaulds PLC* v. *Detrafor*, Queen's Bench Division, 26.1.90, unreported, Webster, J., said: "In my judgment rights under the Convention are conferred upon the sender as such, in this case the first plaintiff, and at least in the case of loss or non-arrival of the goods, on the consignee as such." See also RB. Antwerpen, 16.5.97 (1999) 34 E.T.L. 119.

69. E.g. on principles of agency where this is the relationship between sender and consignee or presumed agency in the case of immediate seller and buyer. See generally *Bowstead on Agency*, 16th edn., p. 429 et seq.

70. Cf. *The Sanix Ace* [1987] 1 Lloyd's Rep. 465.

71. Thus the terms of sale between seller and buyer, to the extent that they assist in founding an actual or presumed agency, can be relevant to determine who are the original parties to the contract of carriage. This will be of importance to a buyer who is not named as consignee but has had goods directed to a sub-buyer who may not be considered his agent for the purposes of making him the real consignee. If it is not correct to suggest that

Damage to the goods

5.24 Article 13 only expressly confers a right of action on the consignee in cases where the goods have been lost or delayed, no mention being made of the position in relation to damage. However, since under Articles 12(2) and 13(1) the right of disposal passes to the consignee on delivery, it would seem to follow that the consignee will equally have a right of action against the carrier from that time in respect of any damage that may have occurred, and such seems to have been consistently assumed in the Continental case law.[72] As to whether or not such right of action is exclusive to the consignee, the same problem arises as discussed in the previous paragraph.

Impossibility

Impossibility in relation to the carriage

5.25 Article 14 provides as follows:—

> "1. If for any reason it is or becomes impossible to carry out the contract in accordance with the terms laid down in the consignment note before the goods reach the place designated for delivery, the carrier shall ask for instructions from the person entitled to dispose of the goods in accordance with the provisions of article 12.
> 2. Nevertheless, if circumstances are such as to allow the carriage to be carried out under conditions differing from those laid down in the consignment note and if the carrier has been unable to obtain instructions in reasonable time from the person entitled to dispose of the goods in accordance with the provisions of article 12, he shall take such steps as seem to him to be in the best interests of the person entitled to dispose of the goods."

5.26 Article 14(1) is expressed to apply in all circumstances, there being no explicit limitations of causes of impossibility. In particular, the consignment note may specify in some detail the way in which the contract is to be performed[73] and it would seem that Article 14(1) will apply if it becomes impossible to comply with any of these requirements. On the other hand, the Article applies only to impossibility[74]; it will therefore not apply where it has merely become onerous, difficult or expensive to perform. The rights and liabilities of the parties in such circumstances would be governed by the domestic law.[75] The carrier must ensure that he seeks instructions from the person currently entitled to dispose of the goods,[76] and failure to do so will render him liable to an action at the

once title to sue is established under CMR it follows that a person with such title can recover substantial damages, then the further issue of what is sufficient to establish a right to substantial damages will arise. This issue would have to be determined on the basis of the principles expressed in *The Albazero* [1977] A.C. 774. Note that the Contracts (Rights of Third Parties) Act 1999 does not apply to CMR cases: section 6(5)(b) (except for the enforcement of exemption or limitation clauses).

72. See, e.g., OLG. Karlsruhe, 24.5.67 (1967) U.L.C. 289. RB. Antwerpen, 7.12.73 (1976) 11 E.T.L. 295. App. Bruxelles, 30.10.75 (1976) 11 E.T.L. 238. See further *supra*, para. 5.18 fn. 37.

73. Cf. Article 6(2) and 6(3). This can include the identity of the party to be contacted by the carrier in case of difficulty. So in *Harrison & Sons* v. *R.T. Steward Transport Ltd.* (1993) 28 E.T.L. 747 the consignment note stated "In case of problems please revert to Somasi", Somasi being either a freight forwarder or a sub-contracting carrier involved in the movement. There seems to be no reason why a sender cannot identify another party to receive notification of difficulty.

74. Not, e.g., a strike at the Customs post at the entry to the Mont Blanc tunnel which results in a period of waiting, unless this will affect the time limit. See Lamy, para. 1537, Jur. 3. Cf. App. Paris, 27.5.80 (1980) B.T. 435, App. Orléans, 12.11.96 (1997) B.T. 159, (1997) U.L.R. 628.

75. Cf. Loewe, para. 128. Under English law, the mere fact that the contractual obligations of the carrier have become more onerous than anticipated will not affect the position: see *Tsakiroglou & Co. Ltd.* v. *Noblee & Thorl G.m.b.H.* [1961] 1 Lloyd's Rep. 329, [1962] A.C. 93.

76. Cf. Article 12(2) and 12(3).

hands of the person properly so entitled.[77] Furthermore the carrier will be liable if, instead of responding to the instructions, he engages in a discussion as to the costs involved and so delays delivery even beyond an extended period granted by the fresh instructions.[78]

5.27 Where the carrier fails to obtain instructions, but acts on his own responsibility, he will be liable for any loss, whether or not he acted reasonably: the provisions of CMR operate to deprive him of any defence of implied agency or the like.[79] It seems that CMR does not itself provide for the case where loss results from the fulfilment of instructions given on the basis of incomplete or misleading information from the carrier, but it is submitted that in such a case the carrier's liability will depend upon whether or not he has been negligent.

5.28 Where the carrier cannot obtain instructions "in reasonable time"[80] he may, if it is possible, take such steps as appear to be in the best interests of the person entitled to dispose of the goods to carry out the carriage in some alternative way. While it seems clear that if the carrier continues to carry the goods in a different manner he will remain liable as CMR carrier,[81] it is by no means so clear that he will remain so if he arranges for onward carriage by a third party. It may be that the third party will become a successive carrier under the original CMR contract, in which case the original carrier will remain liable for the onward carriage under Article 34.[82] It is possible, however, that the original carrier will be treated as a forwarder in respect of the further carriage. In support of such a view, it might be argued that the duty imposed by Article 14(2) to act "in the best interests of the person entitled" does perhaps suggest the character of a freight forwarder.[83] If this latter view be correct, the onward carriage will be governed by a new contract; it will depend upon the circumstances whether this in its turn will also be a CMR contract.

5.29 Article 14 must also be read in conjunction with Article 16(2) which permits the carrier, in circumstances where carriage has become impossible within the meaning of Article 14(1), to unload the goods for the account of the person entitled to dispose of them, whereupon the carriage is deemed to be at an end.[84] It is important to keep in mind that both Article 16(2) and Article 14(2) qualify and amplify Article 14(1) which must therefore be read subject to them. It may be that further performance has become impossible within the meaning of Article 14(1), and the carrier would prefer to exercise his option under Article 16(2). It would seem that he may do so instead of seeking instructions. There is no problem where onward carriage has become impossible; where, however, the carriage could be completed in some alternative manner, if the carrier has sought instructions but has not received them, Article 14 imposes on him a duty to act in the best interests of the person entitled to dispose of the goods. He can only unload and store the goods if that is in the best interests of the person entitled. In a case where the carrier wishes to unload and store, but believes that this might not be in the best interests

---

77. Arrond. Arnhem, 28.1.71 (1973) S. & S. No. 82, (1977) II U.L.C. 207.
78. App. Paris, 22.6.82 (1982) B.T. 432.
79. App. Liège, 18.12.67 (1969) 4 E.T.L. 965.
80. What is a reasonable time is, of course, a question of fact to be decided on the circumstances of each case.
81. Subject to the provisions of Article 2. See *supra*, Chapter 2.
82. On successive carriage generally, see *infra*, Chapter 11.
83. Cf. Clarke, p. 145 and Glass & Cashmore, para. 3.34.
84. For full text and discussion of Article 16(2), see *infra*, para. 5.37 ff.

of the person entitled, he will therefore be well advised to refrain from seeking instructions. While this clearly frustrates the primary intention of Article 14(1) it appears to follow from the true construction of the three provisions referred to.

5.30 Where, in pursuance of his responsibilities under Article 14(2), the carrier has elected to continue to carry the goods, and during the course of that further carriage difficulties arise, it is arguable that the carrier cannot then avail himself of Article 16(2). This is because the new difficulty does not make it "impossible to carry out the contract in accordance with the terms laid down in the consignment note", since *ex hypothesi* carriage is not being performed pursuant to the consignment note and therefore this is not the "[case] referred to in Article 14, paragraph 1 ... ". Accordingly, it would seem to follow that Article 16(2) cannot apply, and the carrier can again only unload and store the goods if this can be said to be in the best interests of the person entitled to dispose of them.

Impossibility in relation to delivery of the goods

5.31 Article 15 provides as follows:—

> "1. Where circumstances prevent delivery of the goods after their arrival at the place designated for delivery, the carrier shall ask the sender for his instructions. If the consignee refuses the goods the sender shall be entitled to dispose of them without being obliged to produce the first copy of the consignment note.
> 2. Even if he has refused the goods, the consignee may nevertheless require delivery so long as the carrier has not received instructions to the contrary from the sender.
> 3. When circumstances preventing delivery of the goods arise after the consignee, in exercise of his rights under article 12, paragraph 3, has given an order for the goods to be delivered to another person, paragraphs 1 and 2 of this article shall apply as if the consignee were the sender and that other person were the consignee."

5.32 Article 15(1) makes provision for the situation that arises where circumstances prevent the delivery of the goods after their arrival at the place designated for delivery. The provision makes a distinction between two situations. First, the inability to deliver may result from extraneous causes. Secondly, the consignee may refuse to accept delivery of the goods. Although the carrier must ask the sender for his instructions in either case,[85] where the consignee refuses to accept delivery of the goods the sender may dispose of them without being required to produce the first copy of the consignment note. This rule is a practical necessity, as the sender may no longer have possession of the first copy of the note, and some action on his part is essential if the consignee will not accept the goods. Presumably in all other cases the sender will be required to produce the first copy of the consignment note. It is not clear what will happen where the sender cannot produce the consignment note because he has sent it to the consignee, as the Convention makes no provision for this eventuality. One authority has suggested that this can be overcome by

---

85. Failure to do so renders the carrier liable for resulting loss or damage, e.g. due to the fact that the goods have become unsaleable (App. Lyon, 24.3.83 (1984) B.T. 166) or have depreciated while languishing in the carrier's warehouse (App. Montpellier, 8.1.87 (1987) B.T. 589) or vehicle (Cass., 23.5.77 (1977) B.T. 388, (1978) I U.L.R. 381), or because the sender has sent further goods to the consignee in ignorance of his previous refusal (Cass., 13.12.82 (1983) B.T. 64, upholding App. Paris, 15.2.80). See generally Lamy, para. 481.

reference to domestic law and the other provisions in the Convention.[86] Probably the only possibility is for the carrier to unload the goods under the provisions of Article 16(2) or to sell them if perishable under Article 16(3).[87]

5.33 However, even if the consignee has refused to accept the goods he can still change his mind and require delivery provided the carrier has not actually received instructions to the contrary from the sender.[88]

5.34 If the circumstances preventing delivery of the goods arise after the consignee has given instructions for the goods to be delivered to a third party where he has been given the right of disposal of the goods under Article 12(3), Article 16(1) and (2) will apply as if the consignee were the sender and the third party were the consignee.

Expenses

5.35 Article 16(1) provides as follows:—

> "1. The carrier shall be entitled to recover the cost of his request for instructions and any expenses entailed in carrying out such instructions, unless such expenses were caused by the wrongful act or neglect of the carrier."

5.36 Clearly, the provisions of Articles 14 and 15 may well involve the carrier in additional expenditure. Accordingly, Article 16(1) provides that the carrier will be entitled to recover the cost of his request for instructions under Articles 14 and 15 together with any expenses entailed in carrying them out,[89] unless such expenses were caused by "the wrongful act or neglect of the carrier". The question therefore arises as to the exact scope of this proviso, and the French text is perhaps more precise as it uses the phrase "la conséquence de *sa faute*". It would appear that this proviso is intended to cover not only fault of the carrier which occurs in relation to the request for instructions or to the carrying out of such instructions, but that it also applies to the events which have themselves caused the circumstances which preclude the performance of the contract of carriage in accordance with the terms laid down in the consignment note or have prevented delivery of the goods on arrival. Further, where such expenses have been caused by the fault of a servant or agent, it would seem that Article 3 will apply in the normal way. Since under that provision any fault of such persons is to be treated as being that of the carrier,[90] any such fault will operate to deprive the carrier of his claim for expenses. The position would be the same where the circumstances have arisen due to the fault of a successive carrier.[91] Apart from the expenses payable under Article 16(1), it is possible for the parties to agree that further sums be paid to the carrier such as a demurrage charge for delay in completion

---

86. Rodière, para. 49. Ultimately a party must have access to the goods to which he is entitled, and a duty to the person with the immediate right to possession would seem to be available under English law even if the right of disposal cannot be exercised by the sender, if he can show that the consignee does not intend to exercise the right granted by Article 15(2). A distinction between a mere right to obtain the goods and the more extensive rights attached to a right of disposal is suggested by the decision in BGH., 5.2.87 (1987) 22 E.T.L. 574.

87. See *infra*, para. 5.43 ff.

88. Article 15(2).

89. The reference to "cost" would seem to suggest that the carrier cannot include an element of profit in the amount claimed, see, e.g. OLG. Munich, 12.4.91 (1991) TranspR 298, (1992) VersR 724 (Clarke, p. 110, n. 45). Contrast, however, OLG. Cologne, 26.8.94 (1995) TranspR 68, where such an element was included, although without explanation, see Jung, p. 153.

90. For Article 3 generally, see *supra*, Chapter 3.

91. For successive carriage, see *infra*, Chapter 11.

of the carriage, for example because of delays at Customs points, beyond a specified period.[92]

Unloading and storing the goods

5.37 Article 16(2) provides as follows:—

> "2. In the cases referred to in article 14, paragraph 1, and in article 15, the carrier may immediately unload the goods for account of the person entitled to dispose of them and thereupon the carriage shall be deemed to be at an end. The carrier shall then hold the goods on behalf of the person so entitled. He may however entrust them to a third party, and in that case he shall not be under any liability except for the exercise of reasonable care in the choice of such third party. The charges due under the consignment note and all other expenses shall remain chargeable against the goods."[93]

5.38 Article 16(2) provides for an entirely different response on the part of the carrier where he is prevented from performing the contract of carriage in accordance with the consignment note under Article 14(1) or from delivering the goods under Article 15. Provision is made whereby he may immediately unload the goods for account of the party entitled to dispose of the goods, in which event the carriage shall be deemed to be at an end. As has already been mentioned, while the provision does not preclude a request for instructions under Articles 14(1) or 15 coupled with action under Article 16(2), it is submitted that there may be grounds for refraining from submitting a request in the circumstances described above.[94]

5.39 Article 16(2) refers to the goods being unloaded "for the account of the person entitled to dispose of them"; it would seem therefore that where the carrier is exercising his rights under this provision, all the costs of the unloading will be payable by that person, regardless of which party would have been responsible for unloading the goods had they reached their destination. On the other hand, the carriage is not deemed to be at an end until after the unloading, so any damage caused during the unloading would be the carrier's responsibility. Thereafter, since completion of the unloading is in effect equated to due delivery, any continuing liability of a carrier who elects to store the goods himself will be regulated by domestic law. This may well raise interesting questions as to the terms on which the goods are being stored. First, it may be necessary to establish which national law is applicable.[95] Secondly, the carrier may wish to avail himself of clauses of his own terms of business which exclude or limit the liabilities imposed on him by the general law. This may not be a problem if he and the sender are already in contractual relations governed by those terms, but otherwise there may be some difficulty in establishing that they have indeed been incorporated into the contract for storage, if indeed there is a contract, rather than a non-contractual bailment.

---

92. See, e.g., *Kala Ltd.* v. *International Freight Services (UK) Ltd.*, Queen's Bench Division, 7.6.88, unreported except on LEXIS. See also, App. Donai, 10.2.2000 (2000) B.T. 210.

93. The fact that the carrier is unable to complete delivery, having reached the destination, should not deprive him of his right to the carriage charges, as in a non-CMR case, App. Paris, 21.12.82 (1983) B.T. 233, where the carrier was required to deposit the goods with the Customs authorities after the consignee failed to take delivery.

94. See *supra*, para. 5.29.

95. Loewe, para. 139, suggests, reasonably enough, that one should ignore the "shadow" of CMR and apply the normal rules of private international law.

5.40 Article 16(2) also permits the carrier to "entrust [the goods] to a third party"[96] in which case the former will not be "under any liability except for the exercise of reasonable care in the choice of such third party". In other words, if the carrier deposits the goods with a warehouseman, he will only be liable if he fails to exercise reasonable care in selecting him. Where, however, goods are later found to be missing the carrier must prove that he did actually warehouse them.[97] Whether this duty is more or less than would be applicable if the contract was subject to common law will depend upon the form the relationship takes. If the carrier is considered as entering into a contract with the warehouseman as agent for the interested party, as is provided in some general conditions of contract, his liability would be approximately the same. If, however, the carrier remains bailee of the goods and enters into a contract of sub-bailment with the warehouseman, he will remain liable as bailee unless he has contractually restricted his liability. Whether the cargo interests will be bound by the terms of contract employed by the warehouseman in the latter case will also depend upon the form the relationship takes.[98]

5.41 Article 16(2) also provides that where the carrier unloads the goods and either holds them himself on behalf of the cargo interests, or else deposits them with a warehouseman, the charges which are due under the terms of the consignment note together with all other expenses will remain chargeable against the goods. The Convention does not make provision for the exercise of any rights of lien over goods carried, so the exercise of any such right will remain subject to national law. Presumably in most cases the courts seised of the matter will apply such rights as exist under their domestic law.[99]

5.42 If the carrier, faced with the refusal of the consignee to accept the goods, avails himself of his rights according to Article 16(2) to unload the goods and to deposit them with a third party, the carriage comes to an end, and the right of the sender to dispose of the goods by asking the carrier to deliver them to another consignee also ceases to exist.[100]

Sale of goods by the carrier

5.43 Article 16(3)–(5) provides as follows:—

> "3. The carrier may sell the goods, without awaiting instructions from the person entitled to dispose of them, if the goods are perishable or their condition warrants such a course, or when the storage expenses would be out of proportion to the value of the goods. He may also proceed to the sale of the goods in other cases if after the expiry of a reasonable period he has not received from the person entitled to dispose of the goods instructions to the contrary which he may reasonably be required to carry out.
>
> 4. If the goods have been sold pursuant to this article, the proceeds of sale, after deduction of the expenses chargeable against the goods, shall be placed at the disposal of the person entitled to dispose of the goods. If these charges exceed the proceeds of sale, the carrier shall be entitled to the difference.

---

96. The carrier may have no choice, as in App. Paris, 21.12.82 (1983) B.T. 233, *supra*, fn. 93.
97. KG. Rotterdam, 24.5.66 (1966) 1 E.T.L. 729.
98. On the question of sub-bailment see Hill, "Warehousing and the Sub-Contractor", [1977] LMCLQ 373.
99. Loewe, para. 140, considers that conflict of laws rules should be applied, but it is rare in practice for a court to apply rights *in rem* other than those of their domestic law owing to the problem of enforcing them. For the position regarding carriers' liens in this country, see *Halsbury's Laws*, 4th edn., Vol. 5, paras. 447–449.
100. BGH., 5.2.87 (1988) II U.L.R. 708 and (1987) 22 E.T.L. 574.

5. The procedure in the case of sale shall be determined by the law or custom of the place where the goods are situated."

5.44 Provision is thus made for the sale of goods by the carrier in certain circumstances. First, if the goods are of a perishable nature or their condition warrants such a course the carrier is free to sell them without waiting for instructions from the person entitled to dispose of them. Secondly, where storage expenses would be out of proportion to the value of the goods the carrier is again free to sell them without waiting for instructions from the person entitled to dispose of them. Presumably, however, in both cases if the carrier has in fact already received instructions from the latter he would be bound to follow them, and could not sell the goods if instructed to the contrary. Thirdly, the carrier may sell the goods in other cases, that is, where the goods are neither perishable nor of small value, if after the expiry of a reasonable time he had not received from the person entitled to dispose of the goods instructions to the contrary which he may reasonably be required to carry out.

5.45 As to the reasonableness of any such instructions, there is no explicit cross-reference here to Article 12(5)(b),[101] although the two clauses do deal with the same stage of the performance of the contract. In theory, the carrier may seek to say that although the instructions he has received are admittedly not within Article 12(5)(b), they are not such as he may "reasonably be required to carry out" under Article 16(3). While each case will have to be considered on its own facts, difficulties are unlikely to arise in practice since it will normally be open to the carrier to argue that any instructions which he considers it would not be reasonable to require him to carry out would interfere with the normal working of his undertaking so as to also be within Article 12(5)(b). It will be recalled, however, that where the carrier wishes to refuse instructions under the latter provision, he is required by Article 12(6) to notify immediately the person exercising the right of disposal. Article 16(3) imposes no equivalent requirement, but if the carrier is at all uncertain as to whether he is entitled to refuse instructions under Article 16(3), and might thus be forced to rely on Article 12(5)(b), it is clearly in his interests to protect himself by giving such notification.

5.46 Article 16 is silent as to whether, at the time of sale, the goods must have been unloaded as provided for by Article 16(2) and are being held either by the carrier himself or by a third party.[102] It may well be that in some cases (e.g. perishable goods) there will be no unloading. But, where the goods have not been unloaded and the sale involves further carriage, this will of course be subject to a fresh contract of carriage with the buyer.

5.47 Where the goods have been sold pursuant to Article 16 the proceeds of sale are to be placed at the disposal of the person entitled to the goods subject to the deduction of expenses chargeable to the goods. If the latter exceed the proceeds of the sale, the carrier is entitled to claim the difference.[103] The procedure in the case of such a sale will be governed by the law or trade custom of the place where the goods are situated.[104]

---

101. See *supra*, para. 5.2 ff.

102. Loewe, para. 142, is of the opinion that Article 16(3) presupposes that carriage has ended pursuant to Article 16(2). He also is of the view that before the carrier can sell the goods pursuant to the second sentence of Article 16(3), it is implicit that he should seek first the instructions of the person entitled to dispose of the goods.

103. Article 16(4).

104. Article 16(5).

CHAPTER 6

Liability of the Carrier

6.1 Articles 17 and 18 of the Convention provide as follows:— *of loss, damage.*

contributed to the cause of loss, damage

"Article 17

1. The carrier shall be liable for the total or partial loss of the goods and for damage thereto   *general*
   occurring between the time when he takes over the goods and the time of delivery, as well
   as for any delay in delivery.
2. The carrier shall however be relieved of liability if the loss, damage or delay was caused
   by the wrongful act or neglect of the claimant, by the instructions of the claimant given   *exception*
   otherwise than as the result of a wrongful act or neglect on the part of the carrier, by
   inherent vice of the goods or through circumstances which the carrier could not avoid and
   the consequences of which he was unable to prevent.
3. The carrier shall not be relieved of liability by reason of the defective condition of the
   vehicle used by him in order to perform the carriage, or by reason of the wrongful act or   *vehicle*
   neglect of the person from whom he may have hired the vehicle or of the agents or servants
   of the latter.
4. Subject to article 18, paragraphs 2 to 5, the carrier shall be relieved of liability when the
   loss or damage arises from the special risks inherent in one or more of the following
   circumstances:

   vehicle (a) use of open unsheeted vehicles, when their use has been expressly agreed and
       specified in the consignment note;

   package (b) the lack of, or defective condition of packing in the case of goods which, by their   *exceptions*
       nature, are liable to wastage or to be damaged when not packed or when not properly
       packed;

   sender (c) handling, loading, stowage or unloading of the goods by the sender, the consignee or
       persons acting on behalf of the sender or the consignee;

   nature (d) the nature of certain kinds of goods which particularly exposes them to total or partial
       loss or to damage, especially through breakage, rust, decay, desiccation, leakage,   *care*
       normal wastage, or the action of moth or vermin; *etc.*

   Label (e) insufficiency or inadequacy of marks or numbers on the packages;

   living animal (f) the carriage of livestock.   *Livestock*

5. Where under this article the carrier is not under any liability in respect of some of the
   factors causing the loss, damage or delay, he shall only be liable to the extent that those
   factors for which he is liable under this article have contributed to the loss, damage or
   delay.

Article 18

1. The burden of proving that loss, damage or delay was due to one of the causes specified
   in article 17, paragraph 2, shall rest upon the carrier.
2. When the carrier establishes that in the circumstances of the case, the loss or damage could
   be attributed to one or more of the special risks referred to in article 17, paragraph 4, it
   shall be presumed that it was so caused. The claimant shall however be entitled to prove
   that the loss or damage was not, in fact, attributable either wholly or partly to one of these
   risks.

107

3. This presumption shall not apply in the circumstances set out in article 17, paragraph 4(a), if there has been an abnormal shortage, or a loss of any package.

4. If the carriage is performed in vehicles specially equipped to protect the goods from the effects of heat, cold, variations in temperature or the humidity of the air, the carrier shall not be entitled to claim the benefit of article 17, paragraph 4(d) unless he proves that all steps incumbent on him in the circumstances with respect to the choice, maintenance and use of such equipment were taken and that he complied with any special instructions issued to him.

5. The carrier shall not be entitled to claim the benefit of article 17, paragraph 4(f), unless he proves that all steps normally incumbent on him in the circumstances were taken and that he complied with any special instructions issued to him."

General

6.2 In these Articles the CMR provisions as to liability of the carrier for loss, damage or delay are to be found. The question therefore arises as to what rules apply if the circumstances in question do not involve loss, damage or delay to the goods. In the case of *Shell Chemicals UK Ltd.* v. *P. & O. Roadtanks Ltd.*,[1] which concerned contamination to a refinery resulting from delivery of the wrong goods, Saville, J., held that although the circumstances were not within Article 17, it did not follow that the carrier was not under any liability and held that the carrier was fully liable at common law without the benefit of any of the limits of liability contained in the Convention.[2] However, in *Sidhu* v. *British Airways*,[3] a case under the Warsaw Convention it was held by the House of Lords that although that Convention did not purport to deal with all matters relating to contracts of international carriage by air, in areas with which it does deal, including the liability of the carrier, the Convention was intended to be exclusive of any resort to domestic law. In the light of that case, the decision in the *Shell Chemicals* case should be regarded with some caution, although in the only case which has directly raised this issue, *Gefco (U.K.) Ltd.* v. *John Mason*,[4] the court was able to avoid treating it as wrongly decided.[5]

6.3 Turning to the provisions of Article 17, *prima facie*, where loss, damage or delay has occurred in relation to goods carried subject to the Convention, the carrier is liable for such loss, damage or delay.[6] In certain circumstances he may avoid that liability,[7] but in the detailed provisions regulating those circumstances[8] we are brought firmly up against a central problem of conventions such as CMR. This is that the Convention is intended to be applied by the national courts with widely disparate judicial and procedural approaches. In this instance we are dealing with questions both of the burden of proof, and of rebuttable presumptions of fact. It is therefore not surprising that the courts of different countries have adopted variant approaches to the interpretation of these Articles. Some consideration of these different approaches is necessary.

6.4 The first approach is that whereby the carrier is absolutely liable for any accident which occurs to the goods unless he can prove that the act or default is due to certain defined events which relieve him of liability. This system is followed principally in

---

1. [1993] 1 Lloyd's Rep. 114.
2. See too LG. Bremen, 6.5.65 (1966) 1 E.T.L. 691.
3. [1997] 2 Lloyd's Rep. 76.
4. Q.B.D. (Birmingham District Registry), 21.2.2000 (unreported).
5. See the discussion *supra*, paras 0.27–0.35.
6. Article 17(1).
7. Articles 17(2) and 17(4).
8. Article 18.

France, together with those countries whose systems are based upon that of France.[9] It is also applicable in certain other countries, including Austria, Czechoslovakia and Hungary, in respect of loss or damage to the goods. The carrier will be relieved of liability where he can prove one of the matters set out in Article 17(2). The last of these, i.e., "circumstances which the carrier could not avoid and the consequences of which he was unable to prevent", appears to cover two distinct situations usually designated *force majeure* and *cas fortuit*. The former indicates something imposed upon the carrier by a third party with either *de facto* or *de jure* coercive power. This would include act of State, governmental requisition or the effects of strikes or other industrial action. However, it will also cover an accidental occurrence which is foreign to the normal course of events in the environment in which it occurs, an event which is neither within the power of man nor his judgment to foresee or prevent, such as an accident of nature. *Cas fortuit*, on the other hand, is an accidental occurrence due to chance, which is extraneous to the carrier's normal activities, and which it is not possible to foresee or prevent, but which occurs during and in the course of performance of the contract, as for example, a fire in a warehouse. In France and Belgium, however, the two terms *force majeure* and *cas fortuit* are commonly accepted as synonymous, used as alternative terms for the same situations, covering accidents caused by external events which could not be foreseen, such as floods, storms, lightning, act of State, etc. There is, however, a dispute among the jurists as to whether the two terms are truly synonymous.[10] At common law the liability of the common carrier is similar as he is considered to be an insurer of goods subject to certain excepted perils.[11]

6.5 The second approach is to look at the carrier's liability as primarily founded on negligence. It is, however, clear that in many countries[12] the liability is more strict, in that the burden of proof is put on the carrier to disprove negligence. Thus in this context it will be for the carrier to show that the loss, damage or delay arose through circumstances which a careful carrier could not have foreseen or avoided. Under English law, as noted above, a common carrier is an insurer of the goods except in respect of "excepted perils". Apart from this, the common and private carrier have a dual liability, in contract and in the tort of negligence. In contract liability is determined by a failure to achieve the stipulated result, in tort by whether there has been a failure to use reasonable care.[13]

6.6 As regards liability of the carrier under CMR, it seems fairly clear that he is subject to an *obligation de résultat*.[14] However, it has been held by the German Supreme Court[15]

---

9. I.e., Belgium, Holland, Luxembourg, Italy, Spain, Portugal and Romania.

10. Rodière & Mercadal, "Droit des transports terrestres et aériens", 5th edn., 1990, Dalloz, Paris, para. 224, et seq. Also Rodière, "Droit des transports", t.2, Nos. 862–867.

11. See Kahn-Freund, *The Law of Carriage by Inland Transport*, 4th edn., Stevens, 1965. See also Palmer, *Bailment*, 2nd edn., Sweet & Maxwell, 1991, Chapter 15.

12. E.g., Germany, Switzerland, Scandinavia, Austria, Hungary and the Czech Republic apply the rule to cases of delay only.

13. For detailed discussion of what amounts to unavoidable circumstances under Article 17(2), see *infra*, paras 6.28–6.52.

14. Libouton, op. cit., para. 35. See also Comm. Antwerpen, 27.9.71 (1971) J.P.A. 169, Comm. Liège, 13.12.77 (1980) I U.L.R. 270, Comm. Antwerpen, 10.10.80 (1982) 17 E.T.L. 64 and Comm. Paris, 9.6.83 (1983) B.T. 457.

15. BGH., 21.12.66 (1969) 4 E.T.L. 888, (1967) U.L.C. 283. See also BGH., 28.2.75 (1975) 10 E.T.L. 516. This is expressed as *"Gefahrdungshaftung"* as opposed to *"Verschuldenshaftung"* with a reversed burden of proof. The latter concept finds support in Austria, see OGH., 10.7.91 (1992) 27 E.T.L. 833, 837, OGH., 21.2.96 (1996) TranspR 422, 423, which lend support to the view of Helm, *Frachtrecht*, Article 17 CMR Anm. 4. It has been suggested that, given the emphasis on the burden of proof and the level of duty imposed on the carrier (see

that liability under CMR is based upon risk, that is, *responsabilité objective*.[16] Alternatively, it has been suggested that liability under CMR is a *responsabilité subjective*, that is a liability based upon a presumption of negligence which it is open to the carrier to disprove.[17] In practice it is fairly clear that given the fine juridical distinctions drawn in the various legal systems as to the basis of the carrier's liability, it will be far from easy to obtain a consensus as to the basis of liability under CMR, however desirable this might be.[18] Looked at from the empirical viewpoint of the common law it seems unlikely that the problem is likely to affect the interpretation of CMR to any noticeable degree. The starting point for an English judge is the words used in the Convention, to which he will endeavour to give their natural meaning[19] rather than the more theoretical analysis often favoured on the Continent.

Period of carrier's liability

6.7 Under Article 17(1) the carrier is liable for loss or damage to the goods which occurs "between the time when he takes over the goods and the time of delivery . . . ".[20] The carrier's liability does not therefore only begin with the carriage proper. It commences from the moment when the goods have been handed over, either to the carrier himself or to his agent who is entrusted with their receipt.

6.8 To date there does not appear to have been much discussion of what will constitute a "taking over" of the goods by the carrier, and it would seem that each situation will turn upon its own particular facts.[21] One possibility is that the "taking over" of the goods occurs at the moment when they pass from the control of the sender to that of the carrier, irrespective of when the carriage actually begins.[22] It is suggested, however, that "control" in such circumstances must be control for the purposes of carriage and not for other purposes such as warehousing or packing prior to the commencement of carriage. Only when such ancillary operations are completed will the carrier be deemed to take over the goods for the purpose of Convention.[23] Until that moment his liability will be governed

---

infra), there is little difference in practice between the two concepts, see Haak, p. 125. See further, Clarke, p. 215.

16. Loewe treats it as such: para. 148.

17. Libouton, op. cit., para. 35. More recent statements of the same writer take note, however, of those cases which express the liability as an obligation of result or similar concept, see Theunis, p. 79 and (1982) J.T. 713, para. 56.

18. See Comm. Liège, 13.12.77 (1980) I U.L.R. 270, and OGH., 16.3.77 (1978) I U.L.R. 370, which stress that it is important for the court to strive for a uniform interpretation of CMR untrammelled by national concepts.

19. See, e.g., Mustill, J., in *J.J. Silber Ltd.* v. *Islander Trucking Ltd.* [1985] 2 Lloyd's Rep. 243 (*infra* at para. 6.29 ff.) at p. 246.

20. Temporarily placing the goods back into the possession and control of the sender (e.g. where the sender is a groupage operator) might be thought to temporarily suspend transit. Alternatively, it might be thought of simply as providing the context whereby any loss damage or delay is more likely to be ascribed to a defence available to the carrier, such as fault of the claimant, see Cass., 20.1.98 (1998) B.T. 87, (1999) U.L.R. 1034 (*infra*, para. 6.20).

21. In App. Paris, 6.10.95 (1995) B.T. 764 there was no proof that the carrier had "taken over" empty pallets for redelivery to the sender. Nor, in the absence of fault, was the carrier liable for failure to secure their return from the consignees, being under no duty to compel them to do so.

22. E.g. Comm. Anvers, 8.1.93 (1995) J.P.A. 332 (Putzeys "Le droit des transports en Belgique (II) 1993–1995", (1996) U.L.R. 557, 565). Cf., in carriage by air, *Swiss Bank Corp.* v. *Brink's-Mat Ltd.* [1986] Q.B. 853.

23. Cf., however, Clarke, para. 27, cf. GH. 's Gravenhage, 15.6.79 (1980) 15 E.T.L. 871. See also App. Versailles, 6.7.94 (1994) B.T. 759 where the court held that, in the absence of contrary agreement, the storage of goods prior to transport, even where this had lasted for a month, could only be viewed as accessory to the

by domestic law. Obviously if goods are handed over to a carrier at his depot for subsequent carriage an element of warehousing must result. A distinction must therefore be drawn between warehousing which is necessarily ancillary to the performance of the contract of carriage and that which results from a specific contractual agreement between the sender and the carrier for the goods to be warehoused for a period of time prior to the performance of the contract of carriage. A similar distinction must be drawn where a carrier has undertaken to pack goods on behalf of the sender prior to the performance of a contract of carriage. It remains to be seen, however, how the courts will deal with such a situation. It has been held, however, that where the phrase *"prise en charge sur camion"* (taken over on lorry) is used the carrier will be deemed to have taken over the goods at the moment at which they are loaded in the vehicle.[24] There is also Dutch authority to the effect that loading the goods on to the vehicle amounts to taking over the goods for carriage.[25]

Delivery

6.9 The Convention does not offer much guidance in deciding what will constitute "delivery" of the goods. It merely regulates certain matters related to the question of delivery. First, it regulates the rights of the consignee.[26] Secondly, it deals with the situation that arises where it is either impossible to carry out the contract in accordance with the terms of the contract, or where circumstances prevent delivery of the goods after their arrival at the place designated for delivery.[27] Thirdly, it makes provision for a right of sale of the goods by the carrier under certain circumstances.[28] Finally, delivery is relevant under the Convention for the purpose of calculating the period of limitation.[29] None of these provisions, however, contain any definition of delivery.[30]

6.10 Delivery in general terms is the handing over of possession to another party.[31] At common law the carrier is bound to deliver the goods to the right person, that is generally

---

principal operation of carriage, so that the loss of the goods during this period was subject to CMR, see further Cass., 22.4.97 (1997) B.T. 350; but cf. OGH. Wien, 15.4.93 (1993) TranspR 425. See further, in the context of a multimodal transport, OGH., 13.7.94 (1995) TranspR 21 (Mayer, p. 171).

24. App. Paris, 16.5.69 (1969) B.T. 190, (1969) 4 E.T.L. 896, (1970) U.L.C. 127.

25. GH. 's Gravenhage, 15.6.79 (1980) 15 E.T.L. 871; but cf. App. Lyon 5.5.95 (1995) B.T. 451.

26. Articles 12 and 13.

27. Articles 14 and 15.

28. Article 16.

29. See Article 32(1)(a), discussed *infra*, Chapter 10, para. 10.55 ff, in particular the further discussion of the meaning of delivery in that context at paras 10.59–10.62.

30. The view has been expressed that, in consequence, national law may be used to supply the omission, in particular the law of the place of delivery, Lamy, para. 482 (see App. Amiens, 23.11.90 (1991) B.T. 292, App. Paris, 22.10.97 (1997) B.T. 779, (1998) U.L.R. 220, App. Paris, 6.7.95 (1995) B.T. 561 where, having used Italian law to supply a definition of "delivery", the court went on to apply the French *"contrat type"* derived from the law of the place of conclusion of the contract to determine the party responsible for unloading (see, however, Trib. gde inst., Annecy 5.11.96 (1997) B.T. 278 where the court applied Article 4(4) of the Rome Convention of 19.6.1980), cf., further, Clarke, para. 37). The court may also take into account what arrangements may have been made by the parties and any custom of the trade. The court, however, is not thereby entitled to disregard the framework in which the provisions relevant to delivery are set, and must supply a meaning to the concept which is consistent with the scheme of CMR, see Hof. the Hague, 24.11.92 (1993) S. & S. No. 128, Cleton, "Digest of Netherlands case law", (1992) II U.L.R. 186, 192. The carrier is not permitted to impose a definition of delivery which has the effect of permitting him to derogate from his proper responsibilities under the Convention.

31. Cf. clauses 6 and 7, R.H.A. Conditions. Where the goods are offloaded from the road vehicle for the purpose of onward transportation by a different mode CMR responsibility must come to an end at some point notwithstanding the non-involvement of the consignee. In OGH., 19.1.94 (1994) TranspR 437, (1996) B.T. 84

the consignee[32] or his agent.[33] Particular care to establish the authority of the recipient is necessary if the carrier delivers otherwise than at the address indicated on the consignment note.[34] Under some legal systems delivery will take place when the carrier permits the consignee to unload the goods from the vehicle.[35] However, the fact that Article 17(4)(c) expressly exempts the carrier from liability for loss or damage due, *inter alia*, to " . . . unloading by the consignee" suggests that delivery has not at that stage taken place.[36] Furthermore, delivery may not occur until after unloading has taken place.[37] Where physical custody and the right to exercise authority over the goods are transferred simultaneously as soon as the vehicle reaches its destination practical problems will be few. It would, however, seem that, in the absence of a contrary stipulation, a carrier who warehouses goods on arrival at destination has not yet delivered them.[38] On the other hand, it is open to the parties to agree that the liabilities of the carrier as carrier under CMR should cease on offloading, with the warehousing regulated by a separate, non-CMR

(Mayer, p. 170) this occurred only at the point where the goods (cranes) were fixed to the railway wagons. In OLG. Düsseldorf, 26.10.94 (1996) TranspR 152, this occurred only when the goods were formally taken on to the ship; cf., however, App. Grenoble, 20.11.96 (1997) B.T. 538, *infra*, para. 6.11 fn. 43.

32. In LG. Munich, 22.1.97 (1998) TranspR 166, the carrier was not liable in circumstances where the consignee purported to fake a misdelivery.

33. See Kahn-Freund, op. cit., p. 298, et seq. See also Palmer, *Bailment*, op. cit., p. 1002 et seq. Whilst a driver might normally be expected to be entitled to go by appearances at the correct address, in one case a carrier was liable without limit (see *infra*, para. 9.72) when his driver, with poor English, arrived at complex premises in England, followed an instruction to discharge on the pavement and obtained an illegible signature on the consignment note whereas on previous occasions the name of the receiver had always been indicated: App. Paris, 6.5.99 (1999) B.T. 495.

34. OLG. Düsseldorf, 20.3.97 (1997) TranspR 425, OLG. Hamburg, 30.11.95 (1996) TranspR 280 (Clarke, p. 283, Jung, p. 157), Comm. Paris, 3.2.99 (1999) B.T. 443, App. Paris 2.4.98 (1998) B.T. 399, cf., however, Maritime and Commercial Court of Copenhagen, 16.9.96 (1998) 33 E.T.L. 70.

35. In Belgium, delivery is said to be a juridical act distinct from the unloading of the goods, Putzeys, para. 502. The French courts view delivery as the taking of effective possession by the consignee, Lamy, para. 482. Common to the approach in both countries is a requirement of conscious and deliberate acceptance of the goods by the consignee. Thus in App. Amiens, 23.11.90 (1991) B.T. 292, despite the appeal of the carrier for the application of Belgian law, he could not claim that there had been delivery while the machine remained on his vehicle, in the absence of proof that the consignee or his servant had entered the vehicle to verify its condition. The carrier could not therefore escape the presumption of responsibility under CMR where the machine fell while being shifted about by the driver within the vehicle.
As to whether the carrier could nevertheless obtain the benefit of Article 17(4)(c) by showing that the driver was acting for the consignee, see *infra*, paras 6.75–6.77.

36. Loewe, para. 149 (cf., Haak, p. 183 n. 338). This is well illustrated by a Dutch decision where delivery of the goods was to take place in the open loading bay of the consignee and the vehicle was parked there over the weekend. Rain damage sustained during that period was held to be within the period of transit for the purposes of CMR: Hof. Arnhem, 6.12.78 (1979) S. & S. No. 353, (1980) 8 E.L.D. 29 (criticised by Haak, p. 184). Cf., however, Cass., 22.11.88 (1988) II U.L.R. 745 where, on arrival of the lorry, the consignee signed the consignment note without making any reservations (see Article 30(1), *infra*, para. 10.2 ff.) before proceeding to unload the machine. The court accepted that the lower court could decide that, having signed the consignment note and proceeded, at his premises, to unload the machine, the consignee had taken delivery so that damage sustained during the process of unloading occurred after delivery for the purposes of Article 32 (see *infra*, para. 10.51) of CMR. See also App. Paris, 6.7.95 (1995) B.T. 561. The carrier must, however, enable the consignee to take control without obstacle e.g. by not placing the vehicle in such a position as to make it difficult to unload, LG. Hamburg, 26.10.94 (1995) TranspR 293.

37. Hof. Brussel, 24.1.69 (1969) 4 E.T.L. 937 and Hof. Brussel, 17.6.71 (1972) 7 E.T.L. 595.

38. See Hof. 's Hertogenbosch, 21.12.65 (1966) 1 E.T.L. 698. GH. 's Gravenhage, 10.5.78 (1978) 13 E.T.L. 607. See also App. Paris, 31.1.84 (1984) B.T. 543, Supremo Tribunal da Justiça (Portugal), 17.11.94 (1997) U.L.R. 200. Cf., however, unloading and warehousing the goods with the intention of taking advantage of Article 16(2) (*supra*, para. 5.37 ff.) with the effect that liability under Article 17 thereby comes to an end. BGH., 5.2.87 (1987) 22 E.T.L. 574 suggests that the contract of carriage is to be regarded as terminated, cf. OGH Wien, 15.4.93 (1993) TranspR 425.

contract under the relevant national law.[39] This is not a breach of the non-derogation provisions of CMR,[40] since the suggested clause does not seek to evade the liability of the carrier under CMR, but to define the period during which the carrier is subject to CMR. However, the German courts have twice held that delivery occurs on the transfer of control of or the right to exercise authority over the goods from the carrier to the consignee.[41]

6.11 Delivery will only take place if the goods are accepted by the consignee.[42] The mere arrival of the vehicle at its destination will not constitute delivery.[43] Equally, delivery requires more than the unilateral act of the carrier purporting to relinquish his control over the goods.[44] Thus the mere act of sending a notice of arrival does not amount to delivery,[45] nor does the mere fact of the driver attaching the hose to the vehicle for the purposes of discharging its contents.[46] If the consignee's office instructs the driver to proceed to the

39. HR., 20.4.79 (1981) 9 E.L.D. 317. However, such an agreement was not found in the notice sent to the consignee, after arrival of the goods at the Customs depot designated as the delivery point, that warehousing charges would become payable after a certain time.

40. Article 41(1), *infra*, Chapter 12.

41. OLG. Zweibrücken, 23.9.66 I.U. 40/67 (1967) N.J.W. 717. LG. Frankfurt am Main, 14.5.65 3/30, 228/64 cited by Willenburg, (1968) N.J.W. 1022. See also HR. 24.3.95, *infra*, fn. 51.

42. So where the perishable goods had been refused by the consignee and the carrier left them on the vehicle for several days thereafter, he was liable for their deterioration since transit had not ended and the carrier should have taken the steps indicated by Articles 15(1) and 16(3): Cass., 23.5.77 (1978) 1 U.L.R. 381. Cf. OLG. Hamburg, 31.3.94, *infra*, para. 6.16 fn. 67.

43. Arrond. Rotterdam, 15.1.71 (1971) 6 E.T.L. 417. And see *Moto Vespa S.A.* v. *MAT (Brittania Express) Ltd.* [1979] 1 Lloyd's Rep. 175 discussed *infra*, paras 10.60–10.62. See also Hof. Ghent, 20.11.75 (1976) 11 E.T.L. 231, App. Paris, 15.6.84 (1984) B.T. 545, App. Paris, 22.10.97 (1997) B.T. 779, Hof. Antwerpen, 13.2.85 (1986) 21 E.T.L. 183. Cf. App. Grenoble, 20.11.96 (1997) B.T. 538, where delivery was considered to have occurred when, on arrival at the port, documentation was issued assigning a place to the trailer to await embarkation notwithstanding the lack of a receipt issued by the port agents.

44. Delivery is not achieved by leaving the goods on the public highway near the consignee's premises, App. Cambrai, 12.9.78 (1978) B.T. 445, (cited by Clarke, para. 37, n. 1) nor by leaving the goods at the Customs depot to await inspection after a Customs agent of the consignee, who was not qualified to inspect the goods, had noted on the consignment note "we certify only the arrival and unloading of the lorry, counting of the load as received by Customs must await definitive inspection", Cass., 24.11.87 (1988) B.T. 42. See also Comm. Antwerp, 4.6.74 (1976) J.C.B. 66 (see Libouton, Theunis, 79, 80) and OLG. Hamburg, 14.5.96 (1997) TranspR 101 (Clarke, p. 117, n. 92), cf. App. Paris, 15.5.79 (1979) B.T. 304, cited by Libouton, (1982) J.T. 713, para. 58. Where a carrier had offloaded a machine in the consignee's yard, and the latter refused to sign a discharge for it, there was no delivery for the purposes of Article 105 of the French *Code de Commerce*, Cass., 20.5.86 (1986) B.T. 446. Unhitching a semi-trailer and handing over the documentation did not amount to delivery in App. Aix-en-Provence, 15.11.95 (1996) B.T. 133. See also App. Paris, 22.10.97 (1997) B.T. 779, where no complaint could be made of delay in unloading the vehicle by the consignee who had not been told of its arrival. However, leaving the lorry in the consignee's enclosed yard with the consignee's permission was delivery in Trib. gde. inst. Annecy, 5.11.96 (1997) B.T. 278, contrast, however, Hof. The Hague, 24.11.92 (1993) S. & S. 128 (see Cleton, "Digest of Netherlands case law", (1992) II U.L.R. 186, 192.

45. HR., 20.4.79 (1981) 9 E.L.T. 317, GH. 's Gravenhage, 10.5.78 (1978) 13 E.T.L. 607.

46. RB. Antwerpen, 29.3.77 (1977) 12 E.T.L. 293, see Libouton, Theunis, 80. Cf. GH. 's Gravenhage, 27.3.83 (1984) S. & S. No. 68, where the court proceeded on the basis that delivery had occurred where the driver pumped the liquid into a tank which was already full, see Haak, p. 184 and p. 339 (but cf., RB. Antwerpen, 19.11.91 (1992) 26 E.T.L. 127, see Clarke, p. 118), Clarke, para. 37 n. 3. The basis of liability was the failure of the carrier to seek instructions, presumably prior to delivery. This case may confuse unloading with delivery. Arguably, it may support a view that the consignee has held out a particular place as being the place of delivery, as in where the carrier is instructed to deliver to a trade fair, see OGH., 11.12.86 (1987) Stra GuV 18, see Clarke, para. 37c, cf. further LG. Cologne, 24.9.93 (1994) TranspR 114, where the consignee gave the driver the key to the silo and left it up to him to find it. The driver was no longer acting for the carrier when the goods were put into the wrong silo. The fact that the consignee has not been placed in the position of being able to inspect the goods and thus make the necessary reservations provided for by Article 30 (see *infra*) is more likely to mean that the carrier will not be considered to have effected delivery, cf. App. Paris, 15.6.84 (1984) B.T. 545, unless he has clearly foregone the opportunity.

former's warehouse delivery will only take place at the latter point.[47] Delivery may still have been effected even if the consignee only unloads part of the load.[48] It should, however, be noted that the court treated this case as one under Article 16(2). The facts were that a container duly arrived at its destination shortly before Christmas. The driver was due to return with the empty container forthwith. On learning that the container would not be unloaded until after Christmas the sender and the carrier agreed that the driver should return without the container. Another driver later collected the ostensibly empty trailer. It was later found to be still partly loaded and further delay occurred in returning the balance of the load to the consignee. Presumably the court found that "circumstances [had arisen to] prevent delivery of the goods after their arrival at the place fixed for delivery".[49] These circumstances arose after unloading had started. Clearly delivery had not then been effected, or Article 15(1) could not have applied. The court must, therefore, have considered that the agreement to leave the part empty container amounted to "unloading the goods", thus terminating the carriage under Article 16(2). This seems very questionable, but by making this finding the court did not have to go on to consider whether the carrier was still potentially liable. It is submitted that the court should either have held that arrival at destination was delivery in this case or gone on to consider Article 17.[50] In a more recent Dutch decision a consignment of chocolate had arrived at the consignee's premises and, after part had been unloaded, the consignee asked the carrier to take the remaining chocolate back to his depot. This was with a view to the carrier using heating mats in the vehicle to warm the chocolate which could then be more easily removed. The court held that this subsequent transport was outside CMR.[51]

6.12 It is not sufficient merely for delivery to be made to an authorised person. It must also be made at the place laid down in the consignment note.[52] However, the carrier will not be liable if the consignee takes delivery and gives a complete discharge on the consignment note, but later complains that the goods were not delivered at the exact point specified in the consignment note.[53] Conversely, delivery is not complete unless made to an authorised person.[54] In *Eastern Kayam Carpets Ltd.* v. *Eastern United Freight Ltd.*,[55]

---

47. OLG. Düsseldorf, 1.2.68 U. 187/67, quoted by Willenburg (1968) N.J.W. 1022. See also OLG. Nürnberg, 21.12.89 (1991) TranspR 99 (cited by Clarke, para. 37a, n. 9); App. Paris, 14.12.77 (1978) B.T. 289, cited by Libouton, (1982) J.T. 693, 702 para. 45, who contrasts RB. Dordrecht, 7.2.79 (1980) S. & S. No. 11, where on-carriage to a different consignee after arrival at the premises of the original consignee was held to be a fresh contract of carriage, subject to internal law (similarly, where a fresh internal consignment note is issued after a brief period of storage of the goods with the carrier before despatch to a changed destination, App. Versailles, 23.11.95 (1996) B.T. 257, 3.2.97 (1997) B.T. 491); cf., however, Cass., 6.6.95 (1995) B.T. 474. Also, where final delivery is delayed because delivery cannot be effected at the time of arrival (OGH., 16.3.77 (1978) I U.L.R. 370, App. Paris, 4.3.85 (1985) B.T. 396, see Clarke, para. 37a), or because the damaged goods were initially rejected by the consignee, who later decided to have them (Hof. 's Hertogenbosch, 21.12.65 (1966) 1 E.T.L. 698), delivery occurs at the later time. See further BGH., 29.11.84 (1985) 20 E.T.L. 448.

48. Arrond. Rotterdam, 15.1.71 (1971) 6 E.T.L. 417.

49. Article 15(1).

50. What went wrong was that the consignee's agent did not unload fully. The carrier should therefore have been able to rely on Article 17(4)(c).

51. HR., 24.3.95 (1995) S. & S. No. 74, (1996) N.J. No. 317, Krings p. 768.

52. Unless varied under Article 12(1).

53. Comm. Antwerp., 28.6.71 (1971) J.P.A. 162, confirmed Hof. Brussel, 19.10.72 (1973) 8 E.T.L. 503, see also OLG. Zweibrücken, 23.9.66 (1966) N.J.W. 1717, OLG. Hamm, 11.3.76 (1976) N.J.W. 2077, see Clarke, para. 37, n. 97 and 96, and OLG. Munich, 23.4.93 (1993) TranspR 348.

54. Cass., 12.12.89 (1990) B.T. 283, OLG. Munich, 27.3.81 (1982) VersR 264 (Clarke, para. 35, n. 62, para. 37, n. 80).

55. Queen's Bench Division, 6.12.83, unreported except on LEXIS.

the plaintiff's claim against the carrier for delivering the goods to the consignee buyer in breach of a cash against documents arrangement[56] was upheld on the basis of national law rather than CMR.[57] Hirst, J., considered that the earliest moment at which the loss occurred was immediately after delivery, i.e. when the goods first got into the hands of the buyers. Nevertheless, the sender is entitled to impose conditions on delivery as part of the consideration by which the carrier obtains custody of the goods,[58] so that an alternative approach would be that delivery to a consignee in breach of the original contract of carriage is to be viewed as delivery to an unauthorised person and therefore not delivery at all.[59]

6.13 The burden of proving the fact of delivery will rest on the carrier,[60] who is accordingly well advised to ensure that he both obtains the receipt provided for by Article 13(1) and that it clearly identifies the consignee who is to be bound by it. In a French case[61] the carrier produced a signed consignment note, but the consignee denied that he had signed it. In the absence of a stamp identifying the signature on the consignment note,[62] the court upheld the decision that proof of delivery had not been established. The court accepted that the same means of proof under national law are also applicable to CMR cases. The burden of proving that the event which caused the loss, damage or delay occurred between the taking over the goods and their delivery rests on the claimant.[63]

Avoidance of liability by the carrier under Article 17(2)

6.14 Article 17(2) specifies four sets of circumstances in which the carrier will be relieved of liability. Each will be considered in turn.

"wrongful act or neglect of the claimant"

6.15 Where the loss, damage or delay is caused by the wrongful act or neglect of the claimant the carrier will be relieved of liability. The claimant may be either the sender or

---

56. This not being a case within Article 21, see *infra*, para. 7.30 ff.
57. But see the discussion *supra*, para. 0.25 ff.
58. OLG. Düsseldorf, 16.6.92 (1993) TranspR 17.
59. This is not the case where the carrier is in breach of Article 21, since that Article itself contemplates delivery in those circumstances and provides the remedy. Where the carrier delivers in breach of an instruction given subsequent to the beginning of the transit, or delivers under an instruction without delivery of the first copy of the consignment note, Article 12(7), see *supra*, para. 5.8, provides the appropriate remedy.
60. See e.g. OLG. Hamm, 2.12.91 (1992) TranspR 79.
61. Cass., 15.5.84 (1984) B.T. 526, cf. Cass., 5.4.95 (1995) B.T. 660, and Cass., 5.7.88 (1988) B.T. 530 where evidence that the carrier's driver had telephoned from the consignee's office on the day of delivery as indicated in the transport document enabled the carrier to prove delivery.
62. Obtaining an undated and illegible receipt on the consignment note and not obtaining the company stamp of the consignee was gross negligence (see *infra*, para. 9.81) in OLG. Munich, 28.1.98 (1998) TranspR 256, cf. App. Paris, 6.5.99 (1999) B.T. 495. In several cases carriers have been found liable where faked receipts, company stamps or Customs stamps have been used, see, e.g., App. Paris, 2.4.98 (1998) B.T. 399, LG. Hamburg, 14.5.96 (1998) TranspR 164, LG. Hamburg, 23.1.96 (1998) TranspR 117, OLG. Hamburg, 30.11.95 (1996) TranspR 280 (Jung, p. 157). Particular difficulty has arisen in respect of delivery in Russia (see Tilche, (1998) B.T. 392) and the carrier's liability is likely to be engaged if the driver delivers at a place other than the expected or indicated point of delivery. Cf. the Maritime and Commercial Court of Copenhagen, 16.9.96 (1998) 33 E.T.L. 70 where the fraud was perpetrated at the premises indicated in the consignment note as those of the consignee. The carrier was not liable notwithstanding that a different place was indicated as the "notified delivery address" in the consignment note.
63. BGH., 8.6.88 (1988) 23 E.T.L. 705, (1988) II U.L.R. 717, in relation to damage. The English headnote in the U.L.R. report implies that it is sufficient that the event which causes the damage occurs between these times rather than the damage itself. Cf. OLG. Hamm, 11.3.76 (1976) N.J.W. 2077, discussed by Clarke, para. 33d(i).

the consignee, he may therefore be the actual seller or buyer of the goods or else a freight forwarder.[64] It would appear from the wording of the provision that it is not sufficient for the loss, etc., to have been caused by the claimant, but that it must result from a wrongful act or neglect on his part. However, where the loss, etc., has resulted from instructions given to the carrier by the claimant it would not appear that the act or neglect must be wrongful.[65]

6.16 It may occur that although the sender has caused the loss, etc., by his wrongful act or neglect the consignee is in fact the claimant and vice versa. Continental commentators are of the opinion that although Article 17(2) states that the loss, etc., must be caused by the wrongful act or neglect of the claimant, this will not preclude the carrier from relying on the defence in such circumstances.[66] This view does, however, seem to fly in the face of the plain words of the Article, and in effect treats the words "the claimant" as meaning "a possible claimant".[67] While it is undoubtedly hard that the carrier should face this risk, that is no ground for rewriting the Article, and in practice the carrier may well be able to rely either on the last limb of Article 17(2) or on Article 17(4)(b), (c) or (e). An alternative possibility would be that while the carrier may be liable under CMR, he can seek an indemnity under national law from the person actually responsible.

6.17 The wrongful act or neglect of the claimant may occur before,[68] during or after the actual transit. However, the courts will tend to examine the causal nexus carefully. Thus where in the course of carriage a carrier unloads goods of which the consignee cannot take delivery, the fact that the goods fell to the ground during unloading does not result from the instructions of the claimant for the purposes of Article 17(2), as the fall of the goods is not a necessary consequence of the instructions.[69] Similarly, in one case the lorry carrying the goods was delayed at a Customs car park because of defects in the documents supplied by the sender, and was then stolen while left unattended. It was held that the loss of the goods was not sufficiently related to the fault of the sender to excuse the carrier.[70] In another case, after a road accident a large cylinder intended for a paper factory was extracted from its packing by the claimant's employees and rolled along the ground to clear it off the roadway and reloaded onto another vehicle. Here the court rejected the

64. See Chapter 1 as to the status of the forwarder in relation to the Convention, and Chapter 5, paras 5.18–5.24 on the right to enforce the contract.

65. Loewe, para. 151. As to the position where the instructions resulted from the carrier's wrongful act or neglect see Article 16(1) *supra*, Chapter 5, paras 5.35–5.36.

66. See Rodière, para. 71, Loewe, para. 151. See also App. Paris, 2.12.81 (1982) B.T. 73, where the carrier escaped liability to the consignee for delay in transit at the Customs post caused by the failure of the sender to attach a certificate of origin.

67. A refusal of the goods by the consignee would not, at any rate, be attributable to the sender as a wrongful act since the carrier must deal with this under the requirements of Articles 15 and 16 (*supra*, Chapter 5, paras 5.31–5.47), OLG. Hamburg, 31.3.94 (1995) TranspR 245 (Clarke, p. 261, Jung, p. 153).

68. As where by delaying the loading of the vehicle the sender causes the anticipated ferry to be missed, OLG. Düsseldorf, 27.2.97 (1998) TranspR 194, or by failing to complete fully the declaration of the goods in the consignment note or TIR carnet, OLG. Munich, 23.6.95 (1996) OLG-Rp. München 18, LG. Hamburg, 10.11.95 (1996) TranspR 338 (Jung, p. 157).

69. RB. Amsterdam, 12.4.72 (1972) S. & S. No. 102. Similarly, where the carrier is instructed to deliver part of the consignment, he must secure the stability of the vehicle, if necessary, before continuing with the journey, see Cass., 10.6.86 (1986) B.T. 608, App. Paris, 10.10.88 (1989) B.T. 408.

70. Cass. (Belg.), 12.12.80 (1981) 16 E.T.L. 250, see also Cass. (Belg.), 3.1.97 (1997) Pas. I No. 5 (Krings, p. 160), (1997–1998) R.W. 256, (1998) 33 E.T.L. 825. Cf. the decision in RB. Dordrecht where the court accepted the defence, in a claim arising from a theft, that the carrier had to wait an additional five days in the port of Reggio di Calabria as a result of bad organisation by the sender: reported by Humphreys and De Peuter, "Highway Robbery in Europe, Theft Under CMR" (1992) 27 E.T.L. 735.

argument that the damage was partly due to the act of the claimant. The additional damage in moving the cylinder was a direct result of the road accident.[71]

6.18 In such a situation the court must consider whether the chain of causality between the accident and the loss was broken by the fault of the claimant.[72] Where, for example, a consignment of beer was lost through being carried in a contaminated tank the carrier claimed that under Article 17(2) the brewery as sender was at fault because, after refusing to use the tank on the grounds that it gave out bad odours, it accepted it the following day after cleaning it before use. The court rejected this argument on the grounds that the customary washing of the tank by the brewery before filling did not imply that the brewery assumed responsibility for ensuring that the tank was suitable for the transport of beer. The carrier must ensure that his vehicle is in a good state.[73] The customary washing of the tank by the brewery did not constitute an acceptance of the vehicle.[74] Similarly, where a consignment of orange juice loaded in a semi-trailer tank was lost due to the tank not being properly cleaned, the carrier could not plead that the sender's knowledge of the fact constituted negligence under Article 17(2), if the sender only ordered the tank to be filled on the strength of a written undertaking by the driver that the bad smell (of vinegar) would not damage the juice.[75]

6.19 In such cases it would seem that much will depend on the respective specialist knowledge of the parties. So, for example, a shipper who instructs a carrier to load a liquid which he knows requires special conditions without checking that those conditions exist may well be said to cause any resultant loss and bring the case within this limb of Article 17(2). On the other hand, a carrier should know that wine carried in winter, while capable of withstanding a modicum of cold weather, is susceptible to damage if subjected to extreme cold. Since such conditions are not rare in Northern Europe, the carrier should not need to be told about the necessity of providing protection. He should either provide it, or refuse to carry the goods if he cannot.[76] Equally, slight overloading of a vehicle at the sender's instigation would rarely constitute a wrongful act or neglect so as to excuse liability for a burst tyre resulting in the vehicle being set alight.[77] In one case the sender had instructed the carrier to set the thermostat on the refrigeration equipment at 2°C for the purposes of the carriage of a consignment of nectarines, which then arrived damaged by frost. However, since the carrier was well aware of the conditions necessary for the carriage of such goods and since the sender would not know of the internal temperature of the vehicle, the carrier was held to be unable to claim that the instructions of the sender were the sole cause and had to accept a share of the responsibility.[78] In another case goods were destined for delivery to a yard near which an incinerator was situated. Despite the

---

71. Comm. Tournai, 21.11.72 (1972) J.P.A. 446. See, however, Libouton, Theunis, p. 81.

72. An alternative is to apportion liability under Article 17(5), see *infra*, para. 6.107 ff. and see Cass., 19.4.82 (1983) 18 E.T.L. 13, *infra*, para. 6.19 fn. 78.

73. Article 17(3).

74. Comm. Bruxelles, 14.2.69 (1972) J.C.B. 548, and on appeal App. Bruxelles, 15.4.71 (1972) J.C.B. 552.

75. App. Colmar, 11.1.72 (1972) B.T. 90, (1973) I U.L.C. 276.

76. App. Nimes, 18.5.88 (1988) B.T. 472. In some cases, the courts may accept the possibility of making a reservation, see further *infra*, para. 6.69 ff., in relation to defective packing. In *Cicatiello* v. *Anglo European Shipping Services Ltd.* [1994] 1 Lloyd's Rep. 678, 686, counsel suggested a duty on both parties to assess the risks involved in transporting the load. However, in that case, the suggestion was rejected on the facts.

77. Comm. Bruxelles, 3.10.70 (1970) J.P.A. 487.

78. Cass., 19.4.82 (1982) B.T. 309, (1983) 18 E.T.L. 13. Cf. App. Aix-en-Provence, 7.9.95 (1995) B.T. 833.

poor labelling of the goods the carrier, on notice of their nature and value, was liable for consigning the goods to the flames.[79]

6.20 Probably the main areas of operation in respect of which the claimant may be guilty of a wrongful act or neglect resulting in loss, etc., to the goods will be those performed by the sender, such as defective packing of the goods, faulty handling, loading, stowage and, so far as the consignee is concerned, unloading. However, these matters are covered specifically, *inter alia*, by the provisions of Article 17(4). Otherwise the principal occasions when the carrier is likely to invoke the protection of this provision in Article 17(2) are where the claimant has been negligent in supplying or failing to supply relevant information,[80] or failing to supply proper documentation,[81] or failure to give instructions as to the appropriate precautions in respect of a particular risk or to take necessary precautions other than those envisaged within Article 17(4).[82] A further example is provided by a case where the sender was a groupage operator and the carrier, having collected a first load of goods, left the vehicle with the sender so that other goods could be loaded into it. The subsequent disappearance of the goods was ascribed to the fault of the claimant with consequent relief from liability of the carrier.[83]

6.21 The burden of proving that the loss, damage or delay was due to the wrongful act or neglect of the claimant will rest upon the carrier.[84]

"the instructions of the claimant given otherwise than as the result of a wrongful act or neglect on the part of the carrier"

6.22 Where the loss, damage or delay is caused by the instructions of the claimant,[85] which have been given otherwise than as a result of the wrongful act or neglect of the carrier, then the latter will be relieved of liability under Article 17(2). Thus in a decision of the French Cour de Cassation a carrier was instructed by the sender to carry 20 tonnes of apples from France to the United Kingdom. The shipment was loaded on a single vehicle. On arrival in the United Kingdom the vehicle was stopped because the load was

---

79. Cass., 14.10.97 (1997) B.T. 800 (no appeal was made to Article 17(4)(e), see *infra*, para. 6.104). Nor could the carrier rely on the limits of liability on the French interpretation of Article 29 (see, *infra*, para. 9.80). The comment to the case considers the decision to be severe and hardly conducive to the duty of the sender to contribute to the good execution of the transport by avoiding all ambiguity.

80. E.g. as to the unusual height in the centre of gravity of the goods, Cass., 5.7.77 (1977) B.T. 402, or failing to inform the carrier of the particular sensitivity of the goods to damage by frost, App. Paris, 25.3.69 (1969) B.T. 174, see Lamy, para. 1545, Jur. 6. In App. Liège, 8.3.96 (1996) B.T. 411 the carrier was exonerated because the sender made no objection to the defective state of the cover used by the carrier. Since there was a risk that the carrier might not appreciate the need for a particular type of cover, special vigilance was required by the sender.

81. E.g. for Customs purposes, see App. Paris, 27.1.71 (1971) B.T. 115, (1971) U.L.C. 151, App. Paris, 2.12.81 (1982) B.T. 73.

82. See the suggestion of counsel in *Cicatiello v. Anglo European Shipping Services Ltd.* [1994] 1 Lloyd's Rep. 678, 686, referred to above in fn. 76, which was rejected because the judge considered that there was no failure on the part of either party to take the precautions necessary to protect the goods from theft by armed robbery, see *infra*, para. 6.48.

83. Cass., 20.1.98 (1998) B.T. 87, (1999) U.L.R. 1034.

84. Article 18(1). Cf. the position under Article 17(4).

85. E.g. where instructions as to the temperature of the goods leads inevitably to damage, cf. Cass., 19.4.82 (1982) B.T. 309, (1983) 18 E.T.L. 13, *supra*, fn. 78. Cf., however, the difficulty in showing a causal relation between an instruction to take a particular ferry and a robbery occurring after the vehicle had arrived, inevitably, outside the working hours of the recipient, BGH., 8.10.98 (1999) TranspR 59. As to whether the carrier is permitted to rely on the faulty instructions of a party other than the claimant himself, see the discussion *supra*, para. 6.16.

in excess of domestic road traffic regulations. The excess was loaded onto another vehicle with resultant delay in delivery. The carrier claimed for his charges and the sender counterclaimed for the amount of carriage charges as damages for the delay. The court found that the carrier had informed the sender of the United Kingdom regulations, but the latter had written to the carrier leaving it to him as to the actual quantity loaded. The request for transport by the sender was therefore held not to be an "instruction" for the purposes of Article 17(2) and the carrier was therefore liable for the delay.[86]

6.23 In another French appellate decision the carrier was able to obtain the benefit of this provision. The facts were that the driver, on reaching his destination, placed the vehicle in a position for unloading to take place by means of a mobile bridge. The consignee then instructed him to move his vehicle again. In the course of this operation a cable fixed to the mobile bridge, which the driver could not see, knocked part of the consignment off the vehicle. It was held, applying Article 17(2), that the instruction given by the consignee to move the vehicle was the cause of the accident, and did not result from the wrongful act or neglect of the carrier, who was therefore not liable.[87]

6.24 Where the instructions are unclear or imprecise, the carrier can be liable if he fails to get clarification and thereby fails to carry out the instructions as intended by the sender. Thus in one case the instruction to the carrier stated that insurance was to be paid by the consignee. It was unclear whether this meant that the carrier or the sender was to obtain insurance, but it was held that the carrier was liable since he knew the value of the goods and the special need to insure the goods for the particular transit. If he had doubts about the meaning of the order, he should have obtained clarification.[88] An instruction to do the impossible, however, can in some circumstances relieve the carrier. This was the case where a forwarder, having had several days to engage a carrier to deliver by a certain date left insufficient time before doing so. As a professional he should have realised the impossibility of the carrier complying with the time limit he had given.[89] Liability was shared, however, where the carrier's driver must have known that the stability of the load was such that it could not tolerate the instruction given by the consignee.[90]

"inherent vice of the goods"

6.25 Where the loss, damage or delay is caused by inherent vice of the goods, again the carrier will be relieved of liability. A similar rule exists in the case of all modes of transport.[91] Inherent vice may be defined as some defect or unfitness which is latent in the goods themselves which may result in the damage or destruction of the goods while carried.[92] In other words, there is something inherent in the goods which renders them unfit to withstand the ordinary incidents of carriage. The carrier must, however, exercise the normal degree of care which by contract he is required to do in relation to the goods

---

86. Cass., 5.7.76 (1976) B.T. 377, (1977) II U.L.C. 204, (1979) 7 E.L.D. 442.
87. App. Paris, 25.11.77 (1978) B.T. 66.
88. Comm. Paris, 30.6.81 (1981) B.T. 448.
89. Comm. Avignon, 12.5.95 (1995) B.T. 795, cf. Cass., 22.10.96 (1996) B.T. 758.
90. LG. Hamburg, 26.10.94 (1995) TranspR 293.
91. In relation to carriage by sea, see *Scrutton on Charterparties*, 18th edn., Article 107.
92. Frozen vegetables which had passed their "sell-by" date were not necessarily inherently defective (thus explaining their state of deterioration on arrival) since it was known that they could be expected to remain useable for industrial purposes for several months, App. Paris, 19.6.97 (1997) B.T. 643.

in question. Two factors must therefore be considered. First, the nature of the goods themselves, and secondly, the normal methods of operation by which the carrier carries the goods.[93] A defect in goods may create no problems as long as the goods are stationary, but may cause damage when the goods are moved, even though the carrier employs all reasonable care in handling them.[94]

6.26 Special provision is made, however, in Article 17(4)(d) in respect of "the *nature* of certain kinds of goods which particularly exposes them to total or partial loss or damage". A distinction must therefore be drawn between the actual *nature* of the goods and some vice which is not of the nature of the goods.[95] If a problem exists in all goods of the same nature this will not constitute inherent vice. For the latter to exist it will be necessary for the goods in question to possess some exceptional defect which is not normally found in goods of the same type when subjected to carriage.[96]

6.27 The distinction between loss through inherent vice and that arising out of the nature of the goods is of importance as regards the burden of proof laid down by the Convention. Where the loss is caused by inherent vice, Article 18(1) provides that the burden of proof of how the loss occurred rests on the carrier.[97] If, on the other hand, the loss, etc., has resulted from the nature of the goods as laid down in Article 17(4)(d), then the carrier may obtain the benefit of Article 18(2) whereby, under certain circumstances, the burden of proof may be shifted on to the claimant. Where, however, goods are to be carried in vehicles subject to regulated temperatures the carrier must also fulfil the requirements of Article 18(4) regarding equipment, maintenance, etc., if he wishes to obtain the benefit of Article 17(4)(d). No such requirement exists, however, if he only purports to rely on Article 17(2). The question is of importance in relation to the carriage of refrigerated meat and similar commodities. In *Ulster-Swift Ltd.* v. *Taunton Meat Haulage Ltd.*[98] the Court of Appeal rejected both the plea of inherent vice under Article 17(2) and the carrier's plea that the damage was due to the nature of the goods under Article 17(4)(d) in respect of refrigerated meat which arrived damaged by overheating.[99] In doing so, the court upheld the finding of Donaldson, J., that the carrier had not proved inherent vice. In the context of refrigerated carriage, the deterioration of the meat could not, of itself, be so described. It would, however, cover a situation in which the meat was shipped at a temperature such that deterioration was inevitable, despite the taking of all reasonable and proper precautions by its carrier, but this had not been proved.[100] In a French case, however, the carrier managed to escape liability under Article 17(2) where

---

93. See Rodière, para. 74.

94. See Kahn-Freund, op. cit., p. 250.

95. See Libouton, Theunis, p. 81 and Chao, (1987) B.T. 393. The distinction is easily missed, see App. Agen, 26.11.85 (1986) B.T. 235 where both characteristics were present. See further *infra*, fn. 102.

96. E.g. all sheet metal must be protected from rain. If it is not, the damage is not due to inherent vice, but to bad packing or sheeting. Comm. Bruxelles, 4.2.72 (1972) 7 E.T.L. 573, 583, (1972) J.C.B. 648.

97. It is not sufficient for the carrier to bring forth a theory that inherent vice is a possible cause. Adequate proof of this must be given, see App. Nimes, 29.10.80 (1981) B.T. 256, (1981) I U.L.R. 267, and App. Paris, 10.11.81 (1982) B.T. 183, Cf. App. Besançon, 27.10.99 (2000) B.T. 94. See further the discussion *infra*, para. 6.103.

98. [1977] 1 Lloyd's Rep. 346, discussed *infra*, para. 6.114 ff. Cf. Cass., 8.12.98 (1999) B.T. 352.

99. See Hill, "The Interpretation of CMR in the English Courts", (1977) E.T.L. 212, 214, and Glass, "The Divided Heart of CMR", (1979) 14 E.T.L. 687. See also App. Paris, 30.5.73 (1973) B.T. 304 where a shipment of frozen cherries from Italy to France arrived unfrozen. *Inter alia* the carrier sought the benefit of Article 17(2) because the goods were subject to inherent vice—this was rejected by the court. See also *infra*, para. 6.95 ff.

100. [1975] 2 Lloyd's Rep. 502, 506.

the evidence established that the meat had been insufficiently cooled prior to transport.[101] It seems to be particularly difficult for the court clearly to maintain the distinction in cases of refrigerated transport[102] or to recognise it at all. Thus in *Centrocoop Export–Import S.A. v. Brit European Transport Ltd.*,[103] Bingham, J., found that the cause of the damage was that the meat was insufficiently chilled before loading yet concluded that the loss thereby arose from a special risk inherent in the nature of the goods which particularly exposed them to damage through decay.[104] Where, however, synthetic fibres are destroyed by fire while in transit, if they are subject by their nature to spontaneous ignition, then the carrier can take advantage of Article 17(4)(d) and Article 18(2), but if the ignition only results from the faulty processing of the fibre, such as through insufficient rinsing, cooling or drying, then, as these faults are not due to the nature of the synthetic fibres, they will constitute inherent vice in the goods and the carrier will be subject to Articles 17(2) and 18(1).[105] It has also been held by a French court that where fatty acids carried in transport tanks suffer a chemical reaction which was set up by the ferrous metal of which the tanks are commonly constructed this will constitute an inherent vice in the goods, under Article 17(2). This will relieve the carrier of liability unless he has been requested to take necessary precautions in the matter.[106]

"circumstances which the carrier could not avoid and the consequences of which he was unable to prevent"

6.28 This expression is somewhat unspecific and even loose to the mind of a common lawyer. The history of the provision is that it was taken verbatim from the 1952 text of CIM.[107] It must therefore be assumed that the intention of the draftsman was to achieve the same result under each régime. There is, however, a difficulty in that the 1952 text of CIM was a restatement or consolidation of an earlier treaty in which the corresponding provision excluded the liability of the carrier in any case of *force majeure*. It is thus argued that, as the revision was not intended to affect the sense, and as the two provisions are to be construed in the same sense, then this part of Article 17(2) should be read as a *force majeure* clause. The question whether there is a binding rule of interpretation requiring this result is a complex one, a full discussion of which is outside the scope of this work. The better view appears to be that, not only can the provision be approached *de novo*, but that its ambit may be wider than that of *force majeure*, given in particular that *force*

---

101. App. Aix-en-Provence, 12.5.87 (1987) B.T. 400. See also App. Paris, 14.2.91 (1991) B.T. 289.
102. See Arrond. Arnhem, 28.1.71 (first instance) and Hof. Arnhem, 10.4.73 (appeal) (1973) S. & S. No. 82, (1977) II U.L.R. 207, criticised by Libouton, (1974) J.T. 511, who in Theunis, at p. 93, cites Comm. Liège, 23.2.76 (1977) J.C.B. 441 where it was pointed out that the rules of proof set out in Article 18(2) concern solely the damage that occurs during carriage and may not be invoked to establish the allegedly defective state of the goods before they were taken over. The problem is not confined to refrigerated transport. In a French case concerning the carriage of new potatoes which arrived in a rotted state the court appeared to confuse the two defences, which was not surprising since the goods both had a sensitive nature and were suffering from an inherent vice: App. Agen, 26.11.85 (1986) B.T. 235. Cf., also, Trib. gde. inst. Sarreguemines, 3.2.98 (1998) B.T. 801, App. Orléans, 29.1.98 (1999) B.T. 383, but see, however, the discussion *infra*, para. 6.103.
103. [1984] 2 Lloyd's Rep. 618. Cf. App. Orléans, 29.1.98, *supra*, fn. 102.
104. See further, however, the discussion *supra*, text to fn. 100.
105. Comm. Bruxelles, 3.10.70 (1970) J.P.A. 487, 490.
106. App. Paris, 28.10.69 (1970) B.T. 7.
107. Article 27(2) CIM. (See now Article 36(2) of the CIM Rules 1980.)

majeure is in any event a concept that is not interpreted uniformly in all jurisdictions.[108] Thus while in some jurisdictions *force majeure* includes only matters external to the business operations of the carrier, Article 17(2) does not, in terms, exclude these. Indeed Article 17(3) is only necessary for this reason.[109] In addition, it has been held that this provision does not import a requirement of unforeseeability which is a part of some national definitions of *force majeure*.[110] For example, in Germany, where the definition of *force majeure* is "an extraordinary event, external to an enterprise, which is unforeseeable, cannot be averted even by using the greatest possible care, and which does not have to be taken account of and allowed for by the party concerned by reason of its frequency", there is a separate category of "inevitable occurrence" which may have its origin within the enterprise of the party concerned. Article 17(2) is held to cover both of these categories of occurrence.[111]

6.29 English authority confirms that CMR does not reproduce a concept of *force majeure*. The traditional starting point for the English judge has been to consider the words of the Convention in their natural and ordinary meaning.[112] The most extensive analysis to date is that of Mustill, J., in *J.J. Silber* v. *Islander Trucking Ltd.* in the context of a claim relating to goods seized by armed robbers[113] and it is likely that this will now be treated as the starting point for future discussion.

6.30 Several possible interpretations were canvassed before him. He rejected first a view that the words mean that the carrier is liable as an insurer for all loss or damage occurring in transit, save only in those cases specifically referred to in the first part of Article 17(2) and in Article 17(4).[114] In other words, that view treats the words as operating as an extra criterion to be satisfied if the carrier is to rely on the more specific exceptions. This view is manifestly in conflict with the wording of Article 17(2). The second view states that the carrier is liable in any case where, if the carrier had done something different from what he actually did, the loss would not have occurred. This too could be rejected insofar as the word "avoid" is given a passive sense. The mere fact that

---

108. Cf. Rodière, para. 75 and Loewe, para. 154, and see Comm. Liège, 13.12.77 (1980) I U.L.C. 270 and OLG. Düsseldorf, 27.3.80 (1981) I U.L.C. 256.

109. See BGH., 28.2.75 (1975) N.J.W. 1597, (1975) II U.L.R. 370, OGH., 16.3.77, (1978) I U.L.R. 370.

110. In France, Cass., 27.1.81 (1981) B.T. 219 allowed an appeal of a decision which had held that there was no defence under Article 17(2) on the basis that the event was foreseeable. In this case, the driver had arrived at the destination at a time when he could not make delivery and he waited in the locked vehicle. Shortly afterwards, robbers forced an entry, overpowered the driver and stole the consignment, cf. Cass., 14.5.91 (1992) 27 E.T.L. 134. Cf. App. Lyon, 7.7.83 (1983) B.T. 589 where a carrier unsuccessfully pleaded *force majeure* in respect of goods destroyed by a fire which had started in neighbouring premises, since the carrier knew that inflammable products were stored at those premises, cf. Trib. gde. inst. Metz, 19.8.97 (1997) B.T. 683. For the French concept of *force majeure*, foreseeability is judged as at the time of the conclusion of the contract, see Mercadal, para. 227 and Lamy, para. 307 (the more recent trend is to lessen the requirement of unforseeability at least in respect of armed robbery, e.g., Cass. 1.10.97 (1997) B.T. 708, (1998) 33 E.T.L. 47. See also, a CMR case where the appeal court rejected the application of *force majeure* by the lower court: App. Douai, 18.5.2000 (2000) B.T. 486). Cf. further Comm. Carpentras, 19.2.93 (1994) B.T. 636 where in respect of a CMR case the court took into account the foreseeability of the risk at the time of the making of the contract of carriage, cf., Cass., 4.2.97 (1997) B.T. 183 (see Tilche, (1997) B.T. 272), see further *infra*, para. 6.34.

111. OLG. München, 16.1.74 (1974) 9 E.T.L. 615 where Article 17(2) was held applicable in the case of a loss in a road accident caused partly by adverse weather conditions. See also BGH., 21.12.66 (1969) 4 E.T.L. 888, BGH., 28.2.75 (1975) II U.L.R. 370, (1975) N.J.W. 1597.

112. *Michael Galley Footwear Ltd.* v. *Dominic Iaboni* [1985] 2 Lloyd's Rep. 251, 254, cf. *Thermo Engineers Ltd.* v. *Ferrymasters Ltd.* [1981] 1 Lloyd's Rep. 200.

113. [1985] 2 Lloyd's Rep. 243; the facts are discussed further, *infra*, para. 6.47.

114. Mustill, J., in fact referred here to Article 17(3). However, since that provision does not provide any defence, he must presumably have been intending to refer to the defences provided by Article 17(4).

the carrier "could have" avoided the loss by the chance of starting the journey later, on this view, would make the loss avoidable and would deprive the words almost entirely of content. Rather "the exception contemplates that the carrier could not have done anything, or abstained from doing anything, with the intention and with the effect of preventing a loss of the type which actually occurred".[115]

6.31 However, even this solution must be qualified further to avoid absurdity, there being many perils which a carrier could avoid if endowed with unlimited foresight and resources. An armed robbery can be prevented if the carrier employs an armoured vehicle surrounded by scores of armed guards. The parties to the Convention can scarcely have intended to force the carrier to such extreme measures in order to avoid liability for events not of his own making.[116] Some qualification is clearly therefore necessary. However, the learned judge rejected a qualification based on *force majeure* or Act of God for two reasons. First, whatever exactly these expressions may involve, they are concerned with extraneous events, whereas the words of the exception require that attention is focused on the acts or omissions of the carrier. Secondly, if the draftsman had intended to refer to the concept of *force majeure*, he could have said so explicitly. The judge found force in the suggestion that the draftsman had consciously abstained from employing terms of art, because in some legal systems a particular expression might not be known, and in others its significance might be unclear. A similar reason persuaded him also to reject the contention that the standard set is one of reasonable care.

6.32 However, in order to eliminate the absurdity of an almost absolute liability, the judge was of the view that it must be possible for a carrier to respond that a suggested precaution is too much to expect of him, so that a degree of practicality must be relevant. However, as soon as this is done, it will be seen that it reintroduces an element of reasonableness,[117] but not with a view to setting a standard which can be equated with the common law duty of care, but rather with a view to excluding precautions which are wholly unreasonable.[118] The learned judge concluded therefore that Article 17(2) sets a

---

115. At p. 246, cf. the *Michael Galley Footwear* case, *supra*, fn. 112, at p. 252. Where there is a failure in some other respect which provides the occasion whereby the event can occur and result in damage, the French courts have found the carrier to be liable under a principle of *force majeure*. In App. Paris, 8.3.91 (1991) B.T. 292, the carrier was liable when his driver stopped at a service station to make his door more secure and was attacked by robbers. The stop enabled the robbery to take place, and had the vehicle been properly maintained the stop would not have been necessary. The stop was not therefore unforeseeable. Similar reasoning can be found in respect of CMR cases. In Comm. Liège, 13.12.77 (1980) I U.L.R. 270, the carrier started the journey at a much later time than envisaged under the contract, and as a result the lorry had to be parked overnight. While not a direct cause of the loss, it required the carrier to be extra careful in protecting the goods from theft.

116. See also Comm. Liège, 13.12.77, (*supra*, fn. 115) where the court suggested that the carrier does not have to prove the absolute impossibility of avoiding the event, since this would oblige the carrier to perform heroics in defence of the goods. See also LG. Bremen, 8.4.98 (1998) TranspR 469. Failure to take the precautions necessary to avoid being placed in circumstances requiring such heroics naturally displaces the defence, Hof. Antwerpen, 7.11.95 (1998) 33 E.T.L. 114.

117. Mustill, J., at p. 247, found it natural to find the reference to a feasible duty of care qualified by the word "reasonably" in the expression utilised by the court in an Austrian decision, "*Vernunfigerweise Zumutbarer Sorgfalt*", to which he had been referred, OGH., 29.6.83 (1984) 19 E.T.L. 526, (1986) II U.L.R. 602. Similar expressions, varying in nuance (see Clarke, p. 275), are to be found in other German and Austrian decisions: BGH., 21.12.66 (1969) 4 E.T.L. 888, OLG. Düsseldorf, 27.3.80 (1983) 18 E.T.L. 89, OLG. Hamm, 6.12.93 (1994) TranspR 62, OGH., 10.7.91 (1992) 27 E.T.L. 833. In BGH., 5.6.81 (1982) 17 E.T.L. 301, the court referred to the especially conscientious carrier taking the greatest care that could be asked of him.

118. Similarly, the Supreme Court of Finland, 18.5.84 (1986) II U.L.R. 634 held that the provision implies a stricter duty of care as regards carriage. Comm. Liège, 13.12.77 (1980) I U.L.R. 270, 273 referred to the diligence necessary to achieve the contractual obligation (i.e. *obligation de résultat*, see *supra*, para. 6.6), cf. RB. Turnhout, 30.6.97 (1998) 33 E.T.L. 139, where reference was to all precautions demanded from a conscientious

standard which is somewhere between a requirement to take every conceivable precaution, however extreme, within the limits of the law,[119] and on the other hand a duty to do no more than act reasonably in accordance with prudent current practice. In order to express that position, Mustill, J., thought that the words "could not avoid" should be treated as including the rider "even with the utmost care". In deciding whether this criterion is satisfied in the individual case, the court will need to look at all the circumstances which can, in an appropriate case, include the financial practicability of the suggested precautions.[120]

6.33 In *Michael Galley Footwear Ltd.* v. *Dominic Iaboni*,[121] Hodgson, J., rejected as irrelevant the criteria normally considered in relation to negligence, such as the magnitude and likelihood of the risk, the gravity of the consequences and the cost and practicality of overcoming the risk. After *Silber*, it seems that these factors can be drawn into the necessary consideration of the circumstances of the particular case but, as Mustill, J., indicated, they require a different balance to be struck from the "negligence equation" rejected by Hodgson, J., Mustill, J., in *Silber* expressed an alternative formulation of the test by indicating that the court must ask itself whether there were precautions which could have been taken, subject to the constraints that such precautions were not extravagant beyond the bounds of common sense,[122] illegal or obviously useless.[123]

6.34 While "unforeseeability" is not a technical requirement of the defence,[124] the degree to which a particular risk is foreseeable forms part of the circumstances which can affect the view taken by a court as to the precautions to be expected of the carrier.[125] The

carrier. The Spanish Supreme Court, 20.12.85 (1986) II U.L.R. 630 referred to the degree of diligence that should be observed according to the purport of the obligation and which corresponds to the circumstances of those involved and the time and place. Given the result (see *infra*, para. 000) the court would seem to have applied a standard of due diligence rather than the stricter duty. The Dutch Supreme Court, 17.4.98 (1998) N.J. No. 602, (1999) 34 E.T.L. 82 stated the test as whether the carrier took all measures in the given circumstances that could reasonably be expected of a careful carrier (see, however, the result, *infra*, fn. 199.

119. "Utmost care" did not therefore require the driver to breach daily rest requirements laid down by EC regulations, see p. 249, and see also the *Michael Galley Footwear* case. Cf. however HR., 17.4.98 (1998) N.J. No. 602, (1999) 34 E.T.L. 82 where the court thought it reasonable to expect the driver to continue a further 50–60 km in Italy to find a protected parking area even if this was in technical breach of rules concerning driving time. Consequently the armed robbery was not unavoidable when he parked at night beside the closed gates of an illuminated industrial park. See also, Cleton, "International transport Law Conventions: a digest of recent decisions by courts in the Netherlands", (1999) U.L.R. 997, 1002.

120. See, e.g., Comm. Paris, 9.6.83 (1983) B.T. 457 (curiously also involving Islander Trucking) where the court held that the cost of garage parking for one night to obviate the risk of robbery was not unreasonable in view of the total cost of this kind of transport. Express recognition of this factor can be found in Germany in KG. Berlin, 11.1.95 (1995) TranspR 342 (Jung, p. 156), cf. LG. Bremen, 8.4.98 (1998) TranspR 469, OLG. Zweibrücken, 17.12.96 (1997) TranspR 369.

121. [1985] 2 Lloyd's Rep. 251.

122. A precaution which effectively prohibits night driving in Italy in all circumstances was thought to be too extreme in BGH., 13.11.97 (1998) TranspR 250, (1998) 33 E.T.L. 829, cf., however, RB Gent, 11.2.97 (1998) 33 E.T.L. 121, Hof. Antwerpen, 7.11.95 (1998) 33 E.T.L. 114.

123. In *M. Bardiger Ltd.* v. *Halberg Spedition APS*, 26.10.90, unreported except on LEXIS, Evans, J., expressed a preference for this latter formulation of the test since it made no reference to the concept of "care". In *Cicatiello* v. *Anglo European Shipping Services Ltd.* [1994] 1 Lloyd's Rep. 678, 682, Judge Marr-Johnson adopted the former analysis of Mustill, J.

124. App. Paris, 6.5.98 (1998) B.T. 724.

125. Cf. App. Luxembourg, 11.11.98 (1999) U.L.R. 1038, *infra.*, para 6.47 fn. 194, Comm. Bobigny, 20.1.2000 (2000) B.T. 192. Cf. however, App. Douai, 18.5.2000 (2000) B.T. 486. The Spanish Supreme Court, 20.12.85 (1986) II U.L.R. 630 considered that a degree of foresight greater than that to be expected of a prudent person in normal circumstances is not to be expected. Foresight involves both an estimation of the risk and an estimation of the precautions necessary to overcome it. The *Silber* case would seem to suggest a level of clairvoyance greater than that to be expected from the prudent carrier.

risk, for example, of theft or armed robbery in Italy has achieved such notoriety[126] that carriers can expect to be treated as being aware of it,[127] and therefore can expect to be held liable except in the most exceptional circumstances.[128] In *Bardiger*,[129] Evans, J., however, doubted that Article 17(2) requires the carrier to go outside the confines of the contract,[130] so that a carrier of goods properly sent by groupage transport is not required to transport them in a single vehicle so as to obviate a slight risk of theft. The confines of the contract would in any event provide part of the circumstances in respect of which financial practicability is to be considered. Thus in *Cicatiello* v. *Anglo European Shipping Services Ltd.*[131] financial impracticability was the basis for finding that utmost care did not require the provision of two drivers.

6.35 Further analysis of the relevant case law enables the meaning of the provision as it has been interpreted by the various courts to be understood. The main problem areas can be classified as follows: those relating to the condition of the vehicle itself and the associated issue of the burden of proof, traffic conditions, natural phenomena and theft or robbery of the load. Other circumstances have also been considered.[132] It should first be appreciated, however, that it will not be enough for the carrier to demonstrate "unavoidable circumstances", but that he will also have to show the causal connection between those circumstances and the loss, damage or delay.[133] It should be further noted that both the circumstances and the consequences must be unavoidable. In one French decision,[134] the court held that a carrier, whose driver was unable to avoid a collision with a large object falling from the preceding vehicle but who could have reduced the consequences

---

126. An average of one vehicle stolen every hour in 1991, see Humphreys and De Peuter, op. cit., at p. 742. More recently the risks associated with delivering in Russia may require particular diligence, App. Paris, 2.4.98 (1998) B.T. 399, OLG. Düsseldorf, 20.3.97 (1997) TranspR 425, LG. Munich, 22.1.97 (1998) TranspR 166, OLG. Hamburg, 30.11.95 (1996) TranspR 380 (Jung, p. 157), cf., however, Maritime and Commercial Court of Copenhagen, 16.9.96 (1998) 33 E.T.L. 70.

127. The Supreme Court of Finland, 18.5.84 (1986) II U.L.R. 634 said that the information on this coming from insurance sources was such that the carrier must have been aware of the risk. In Hof. Leeuwarden, 23.10.96 (1998) N.J. No. 58 (upheld HR., 17.4.98 (1998) N.J. No. 602, (1999) 34 E.T.L. 82) it was indicated that before January 1991 it could be assumed to be known that the risk of theft in Italy was particularly great. See also numerous French decisions taking into account the notoriety of Italy: Cass., 14.5.91 (1992) 27 E.T.L. 134 (in this and other cases the complaint that the lower court took the foreseeability of the risk into account received little countenance), Cass., 20.5.90 (1990) B.T. 778, Cass., 18.3.86 (1986) II U.L.R. 639, App. Aix-en-Provence, 17.10.78 (1978) B.T. 585, App. Limoges, 1.3.83 (1983) B.T. 330, Comm. Paris, 9.6.83 (1983) B.T. 457. In Cass., 19.1.93 (1993) B.T. 102, the court referred in this context to the provisions of the carrier's insurance, which laid down precise rules of security in respect of the stopping places to be utilised by the vehicle.

128. "The inference is, therefore, that national courts will take a less than lenient approach in their interpretation of what constitutes an 'unavoidable circumstance' where the loss has occurred in Italy": Humphreys & De Peuter, op. cit., at p. 740. See, however, *Cicatiello* v. *Anglo European Shipping Services Ltd.* [1994] 1 Lloyd's Rep. 678, discussed *infra*, para. 6.48. See further RB. Turnhout, 30.6.97 (1998) 33 E.T.L. 139 and Laurussen, "Diefstal in Italië onder Artikel 17 Lid 2 CMR", (1998) 33 E.T.L. 121. Cf. Comm. Anves, 9.4.95 (1995) J.P.A. 26 with Comm. Anvers, 8.1.93 (1995) J.P.A. 332 (Putzeys, "Le droit des transports en Belgique (II) 1993–1995", (1996) U.L.R. 557, 566).

129. *Supra*, fn. 123.

130. Cf. LG. Bremen, 8.4.98 (1998) TranspR 469.

131. *Supra*, fn. 128, at pp. 682–683.

132. It should be noted, however, that routine commercial circumstances are not covered by Article 17(2). In a French case (App. Paris, 3.11.77 (1978) B.T. 14) where yarn was damaged by water penetration during a 10 day delay caused by the failure of a third party to make Customs documents available it was held that the carrier could not rely on this part of Article 17(2) to avoid liability.

133. E.g. App. Paris, 30.9.87 (1988) B.T. 59.

134. Cass., 22.7.86 (1986) B.T. 591.

by braking, was thereby partially responsible for the damage. The court therefore applied Article 17(5)[135] to reduce the damages payable.

Condition of the vehicle and the burden of proof

6.36 It is important to understand at the outset that Article 17(2) is primarily concerned with matters other than the defective condition of the vehicle, which is made the subject of separate provision in Article 17(3).[136] However, a certain amount of Continental case law has been concerned with the situation arising where the vehicle suffers a burst tyre. These cases do not fall within Article 17(3) since they do not involve the defective condition of the vehicle, but rather concern vehicles in good working order which malfunction for some extraneous and unavoidable reason. The issue which they involve is the nature of the proof required of the carrier in order to be relieved of liability, bearing in mind that, under Article 18(1),[137] the burden of proof is on the carrier.

6.37 In such cases, as a Continental jurist has pointed out,[138] there are two approaches to the matter. Under the first, the courts have required the carrier to prove the exact cause of the burst, which of itself is not an "unavoidable circumstance". It is not sufficient for the carrier to establish that he has taken all possible steps to avoid the burst.[139] Merely because a burst tyre is in good condition at the start of a journey is not proof of "unavoidable circumstances" as it does not exclude the possibility that the damage was caused by the negligence of the driver.[140] The carrier must also accept responsibility for the condition of his own equipment,[141] and even where fire results from the bursting of a new tyre, the carrier will not be able to rely on the defence of "unavoidable circumstances" since a manufacturing defect in a tyre, even a new one and which would not constitute a ground for relief, could not be ruled out in circumstances where the specific cause remains uncertain.[142]

6.38 The second approach is for the courts to allow the carrier to prove that he has exercised due diligence as regards the selection and maintenance of his tyres.[143] In a Dutch decision it was held that the carrier is not required to prove the existence of any particular fact or circumstances relieving him of liability. He can rely on a number of established facts which in the absence of any other explanation would permit a reasonable supposition that the damage in question might be attributed to "unavoidable circumstances", the consequences of which he was unable to prevent. He will, however, be required to prove that the external factor could neither have been noticed nor prevented by

---

135. See *infra*, para. 6.107 ff.

136. *Walek & Co.* v. *Chapman and Ball (International) Ltd.* [1980] 2 Lloyd's Rep. 279, discussed *infra*, para. 6.54. The breakdown of the vehicle was held not to be an unavoidable circumstance in App. Rennes, 19.3.87 (1988) B.T. 105.

137. *Infra*, para. 6.112.

138. Libouton, (1973) 8 E.T.L. 2, para. 42.

139. Arrond. Amsterdam, 28.10.64 (1966) 1 E.T.L. 718, (1966) U.L.C. 96. Fire caused by the bursting of two almost new tyres, side-by-side, through friction.

140. Hof. Brussel, 17.6.71 (1971) 6 E.T.L. 825.

141. RB. Antwerpen, 9.4.69 (1969) 4 E.T.L. 1028.

142. Comm. Bruxelles, 3.10.70 (1970) J.P.A. 487. See also Libouton, Theunis, p. 83 and also OGH., 10.7.91 (1992) 27 E.T.L. 833.

143. Libouton, (1973) 8 E.T.L. 2, para. 42.

his employees while exercising normal care.[144] So where reduction of tyre pressure was caused by a sharp object causing the heating or burning of a tyre where the tyres were in good condition and had been checked both before and during carriage the carrier was permitted to rely on the defence in Article 17(2).[145] In a German decision it has been held that where a tyre catches fire the carrier will be liable unless he can prove that the fire was not caused by the defective condition of the vehicle in question.[146]

6.39 The difference between the two approaches towards the burden of proof can be summarised as follows: while in each case this is on the carrier, under the first approach the carrier must in effect exclude all but his own explanation,[147] while under the second he must merely establish due diligence and a likelihood of his explanation being correct. The latter approach was in effect adopted in *Sidney G. Jones Ltd.* v. *Martin Bencher Ltd.*,[148] where the defendant carrier argued that the cause of the accident was a tyre blow-out. Having reviewed the evidence, Popplewell, J., held that the accident was in fact caused by the driver falling asleep. However, he clearly indicated, albeit obiter, that since it was accepted by the parties that the vehicle was properly maintained, if the accident was because of a blow-out, then this of itself would be sufficient to provide a defence under Article 17(2). Apart from these cases involving burst or ignited tyres, a carrier has generally found it difficult to establish sufficient evidence of due care to be relieved of liability in the absence of evidence of the precise circumstances.[149] On the other hand, there may be proof of the immediate circumstances surrounding the event which caused the loss,[150] so that the focus is then on the precautions that were taken or which might have been taken to avoid it.[151] In *Silber* v. *Islander Trucking*,[152] there was little proof of the care

144. GH. Amsterdam, 21.10.65 (1966) 1 E.T.L. 305, (1966) U.L.C. 109. The ignition was caused by two tyres bursting simultaneously. Clearly a carrier could not be relieved of liability when the driver, who had been told by another driver that smoke was escaping from one of his wheels, carelessly left the lorry to search for spare tyres, when the first signs of the fire must have been perceptible, Libouton, Theunis, p. 84 citing Comm. Ghent, 27.5.69 (1972) J.T. 398.

145. Arrond. Rotterdam, 20.4.65 (1966) 1 E.T.L. 137. Cf. Arrond. Rotterdam, 21.1.69 (1969) 4 E.T.L. 998 and on appeal GH. The Hague, 13.2.70 (1970) S. & S. No. 47, where testing with a gauge before commencement of carriage, together with a visual check at each stop, was considered adequate, although on appeal the carrier was required to furnish full technical information regarding the tyre in question.

146. OLG. Düsseldorf, 18.11.71 (1973) II U.L.C. 428. See also OLG. Düsseldorf, 27.3.80 (1981) I U.L.C. 256: carrier protected under Article 17(2) where he shows the loss, etc., occurred despite the exercise of due care.

147. In Brussels, 31.12.79 (1980) J.T. 388, cited by Libouton, Theunis, p. 84, the carrier was required to prove that a burst tyre was due neither to a manufacturing defect, nor to a mistake, nor to excessive speed, nor to any wrongful act of the driver, and in addition that the accident could not have been avoided after the tyre burst.

148. [1986] 1 Lloyd's Rep. 54.

149. See BGH., 5.6.81 (1982) 17 E.T.L. 301. RB. Alkmaar, 5.6.67 (1967) 2 E.T.L. 1013. In one case (reported by Libouton, Theunis, p. 85: Antwerp, 4.4.84 (1985–86) R.W. 184) the carrier could not escape liability where the cause of fire in a consignment was unknown. It was not sufficient to prove that the tyres did not catch fire and that the floor of the lorry suffered only superficial fire damage, as these circumstances did not exclude the possibility that it originated in another part of the vehicle, nor did the fact that the fire broke out in the consignment necessarily prove that there was an inherent defect in the goods or lead to the conclusion that unavoidable circumstances were involved. Similarly, in Comm. Liège, 13.12.77 (1980) I U.L.R. 270, the court said that the carrier must establish the circumstances *in concreto* that he has been unable to avoid.

150. Proof of this must be sufficient: Libouton, Theunis, p. 91, citing Comm. Tongres, 6.10.83 (1984) RDC 693, where in a case of robbery with violence the carrier did not furnish absolutely conclusive proof of the alleged circumstances, when he produced only a Police Statement containing the evidence of the driver, whom the court nevertheless refused to examine under oath. See further Hof. Antwerpen, 10.10.95 (1998) 33 E.T.L. 108, 6.2.96 (1998) E.T.L. 118, and the cases noted *infra*, para. 6.40 fn. 162.

151. See e.g. OLG. Hamburg, 7.12.95 (1996) TranspR 283 (Jung, p. 157) where the carrier showed that he had taken all imaginable precautions.

152. *Silber* v. *Islander Trucking* [1985] 2 Lloyd's Rep. 243.

taken by the defendants to avoid the robbery, the immediate circumstances of which were known. Mustill, J.,[153] suggested that while it was tempting to consider that the defendants had thereby failed to prove the exercise of any care, far less the utmost care, and then go on to concentrate on the issue of causation, this was not the best approach. He was of the view that the court must first do the best it can, and ask itself whether there were any precautions which could have been taken, and only if that is the case, consider whether, if taken, they would have prevented the loss.[154] In respect of the burden of proof, it was not for the carrier to formulate a list of all the steps which could conceivably have been thought appropriate, and then to methodically demonstrate, one by one, either that they were not called for even by the requirement of the utmost care, or that, if they had been taken, it would have made no difference to the outcome. Instead, Mustill, J., was of the view that it is for the plaintiff to suggest (although not to prove) what the carrier ought to have done, and it is then for the carrier to rebut, if he can, the specific complaints thus put forward. Although this suggests a merely speculative role for the plaintiff, he did go on to remark, in respect of a particular suggestion,[155] that he thought something more solid was required of them than merely putting forward the suggestion with nothing to back it up.[156] Continental courts seem generally to be more rigorous in requiring exact proof by the defendant that no measure available could have been taken by him to avoid the loss.[157]

Traffic conditions

6.40 Traffic conditions as such will rarely assist the carrier under the proviso to Article 17(2), since the need to take evasive action is a foreseeable risk in modern traffic conditions.[158] Equally, the need to brake suddenly will generally not constitute "unavoid-

153. See the discussion at pp. 247–249, cf., however, *Cicatiello* v. *Anglo European Shipping Services Ltd.* [1994] 1 Lloyd's Rep. 678, 682.

154. It may be possible to determine with some certainty that particular precautions would have been useless in the circumstances (see, e.g., the *Cicatiello* case, at p. 684, see also LG. Darmstadt, 28.2.96 (Jung, p. 157), LG. Bremen, 8.4.98 (1998) TranspR 469, OLG. Zweibrücken, 17.12.96 (1997) TranspR 369, OGH., 19.1.94 (1994) TranspR 282 (Mayer, p. 173)) but equally it can be difficult to predict with certainty whether a different result would have been achieved by a precaution which would have altered the circumstances. In a French decision, the carrier argued unsuccessfully that, given the brutality of the armed robbers, the suggested precaution of making the stop at a service station would have been useless: Cass., 5.1.88 (1988) B.T. 103. A carrier had more success in Germany where the court accepted that, given the nature of the goods and the criminal energy required to commit the armed robbery against a moving vehicle at night, spending the night at a guarded parking place in Italy involved no reduction of risk: BGH., 13.11.97 (1998) TranspR 250, (1998) 33 E.T.L. 829.

155. That the carrier should have employed a convoy of vehicles.

156. If some evidence is to be required from the plaintiff, this is perhaps better viewed as involving an evidential burden placed on the plaintiff where the defendant has done enough on his evidence to establish a *prima facie* defence. Notably the only evidence in respect of nearby security parks (see *infra*, para. 6.47) was supplied by the plaintiffs (see also in this respect *Cicatiello*, *supra*, fn. 153, at p. 685). Cf., however, Evans, J., in *M. Bardiger* v. *Halberg Spedition APS*, 26.10.90, unreported except on LEXIS, who considered the burden of proving the impracticability of suggested precautions to be on the defendant.

157. In France, in cases involving robbery or theft, the starting point for the court often involves criticism of the measures in fact taken by the carrier in view of the risk, so that the carrier will need to show that no other measure could have been taken by him to avoid it; see Comm. Paris, 9.6.83 (1983) B.T. 457, App. Poitiers, 7.2.83 (1983) B.T. 455, (1984) II U.L.R. 409, App. Limoges, 1.3.83 (1983) B.T. 330, Cass., 14.5.91 (1992) 27 E.T.L. 124. Cf., however, App. Riom, 15.9.99 (1999) B.T. 746 (although the decision mainly relates to the issue of whether there was *faute lourde* (see *infra*, para. 9.80) which involves reversal of the burden of proof). For Germany, see BGH., 8.10.98 (1999) TranspR 59, OLG. Oldenburg, 25.9.96 (1997) TranspR 229. See also Danish Supreme Court, 24.6.97 (1998) 33 E.T.L. 52.

158. Comm. Namur, 22.7.65 (1966) 1 E.T.L. 133, (1965) U.L.C. 355.

able circumstances",[159] unless quite exceptional and unforeseeable circumstances arise.[160] It has also been held in a French decision that a detour and consequent delay necessitated by the state of the road did not constitute "unavoidable circumstances".[161] However, if the carrier can show that he took proper precautions in this respect, then he will be protected. Thus in a Dutch decision it was held that where the driver had made inquiries and been told that the Gotthard tunnel was open, the carrier was able to rely on Article 17(2) in respect of the resulting delay when it turned out to be closed.[162] However, the fact that the driver falls asleep at the wheel is clearly not an "unavoidable circumstance",[163] nor is a collision with a parked vehicle at night where the driver, in near perfect driving conditions, mistook its rear lights for those of a moving vehicle.[164]

6.41 Where a collision or other form of traffic accident results entirely from the fault of a third party, the carrier will be able to rely on Article 17(2) only if it is possible to show that the circumstances could not have been avoided even with the exercise of the greatest care.[165] Thus where an articulated vehicle was correctly parked on the side of the road and properly lit, but was hit at night by another vehicle, it was held by a French court that as the carrier was not at fault himself and could not avoid the act of the third party he was not liable.[166] It should be noted, however, that where the cargo interest wishes to bring an action directly against the third party who is at fault, the latter will not be able to invoke the protection of CMR or any of the stipulations in the original contract of carriage to which he is not a party.[167]

---

159. RB. Antwerpen, 30.12.70 (1970) J.P.A. 123. Paris, 21.1.70 (1970) B.T. 100. Paris, 27.2.70 (1970) B.T. 110. Similarly in Comm. Brussels, 9.1.78 (1977–78) J.P.A. 278 (see Libouton, Theunis, p. 83) the fact that a car stopped suddenly in front of a lorry in a built-up area was not an unavoidable circumstance and did not exonerate a carrier who swerved and lost control of his vehicle because the road was muddy. By reducing his speed and keeping a safe distance from the vehicle in front, the accident could have been avoided. On the other hand, in Hof. Antwerpen, 13.10.86 (1987) 22 E.T.L. 443 the carrier was held not to be liable where the driver of a vehicle travelling in the second lane of a motorway applied his brakes without warning and changed lanes and thus caused the resulting collision, cf. OLG. Munich, 16.1.74 (1974) 9 E.T.L. 615, RB. Antwerpen, 31.10.97 (1998) 33 E.T.L. 835, 30.1.98 (1998) 33 E.T.L. 842.

160. Vredegerecht le Kanton, Antwerpen, 26.10.71 (1972) 7 E.T.L. 1058. In RB. Rotterdam, 19.6.70 (1973) S. & S. No. 73, where a machine was dislodged from a vehicle which braked at an unexpected road subsidence which could not be avoided owing to oncoming traffic, it was held the carrier may be able to avoid liability if he can prove that neither the subsidence nor the consequences of driving into the subsidence could be avoided. Cf. Cass., 22.7.86 (1986) B.T. 491, *supra*, para. 6.35 fn. 134.

161. App. Paris, 18.10.73 (1973) B.T. 488.

162. RB. Amsterdam, 20.4.77 (1978) S. & S. No. 194, (1978) 6 E.L.D. 585. In Comm. Liège, 27.6.85 (1985) 20 E.T.L. 572, the carrier was held liable when he failed to take into account the restrictions on lorry traffic in Germany during the weekend when collecting the goods late on Friday afternoon. Proof by the carrier of such restrictions is necessary if reliance is to be placed on them. The evidence solely of the driver was insufficient in App. Paris, 29.9.99 (1999) B.T. 747, cf. Comm. Bruxelles, 31.5.96 (1996) 31 E.T.L. 586 where there was evidence of other drivers and police reports in support.

163. GH. Dordrecht, 18.5.66 (1968) 3 E.T.L. 416, (1968) U.L.C. 155.

164. Hof. 's Gravenhage, 3.6.76 (1977) S. & S. No. 3.

165. BGH., 28.2.75 (1975) 10 E.T.L. 516. See also the cases referred to in fn. 159.

166. App. Paris, 13.12.76 (1977) B.T. 37. (But see fn. 186 *infra*.) See also RB. Turnhout, 29.3.76 (1976) 11 E.T.L. 775 where the carrier was held not liable when the vehicle hit an unusually low bridge in the absence of any warning signs. Cf. domestic French decisions concerning the defence of *force majeure* where the carrier has likewise been exonerated in circumstances where the free height of a bridge has been incorrectly marked (App. Bordeaux, 8.11.78 (1978) B.T. 568), or where the carrier has been misled as to the free height by an official (Comm. Marseille, 23.10.87 (1988) B.T. 43). In general, however, the carrier has not been exonerated where there has been no such error, App. Paris, 15.10.92 (1992) B.T. 724.

167. See note on "La responsabilité du transporteur en cas de collision par la faute d'un tiers", (1977) B.T. 31.

Climatic conditions

6.42 Climatic conditions have also been put forward as ground for reliance on Article 17(2), although again with little success. It is clear that a heavy summer storm or a sudden summer cloudburst in Northern Europe are not "circumstances which the carrier could not avoid and the consequences of which he was unable to prevent".[168] In general where violent and exceptional rains which occur during transit damage the load the carrier will be liable.[169] Similarly, it appears that the effect of high winds upon a vehicle and its load is a contingency for which the carrier must make provision.[170] Nor is the occurrence of frost in winter likely to exonerate a carrier under this defence.[171] A CMR carrier has, however, had rather more success where the goods have been damaged in the course of a storm occurring while the vehicle is being carried by ferry.[172] A carrier also had success where climatic and traffic conditions combined, in that, having been forced to stop because of fog, his vehicle was hit from behind by another vehicle travelling at speed.[173]

Theft and robbery

6.43 Theft from a lorry by a third party is a situation which is difficult for a carrier to justify as an unavoidable circumstance, the consequences of which he is unable to prevent. Generally the prudent carrier will be able to prevent theft if he takes adequate precautions.[174] So where a carrier left an unguarded lorry parked at the frontier during the night he was liable for the theft of its load of cobalt.[175] In a French decision the carrier left a valuable shipment of hides in an unprotected vehicle, simply sheeted over, in Marseille docks for a period of $3\frac{1}{2}$ hours at night, having arrived after the Customs office was closed.

---

168. App. Bruxelles, 25.5.72 (1972) J.P.A. 219—a violent storm between Antwerp and Bâle. Comm. Courtrai, 25.5.72 (1972) J.P.A. 192, (1973) II U.L.C. 438—35 mm rainfall in a cloudburst on the Franco-Belgian frontier. But cf. the case noted *supra*, fn. 111. In App. Bordeaux, 13.7.82 (1983) B.T. 542, a snow storm in Turkey in December was not considered to be an unavoidable circumstance. Furthermore, on parking on an incline the carrier had not used the legs of the semi-trailer to further guarantee its immobility and so prevent it from falling into the ravine once the snow had melted.

169. Arrond. Amsterdam, 14.3.64 (1966) 1 E.T.L. 738.

170. Libouton, op. cit., para. 39.

171. In App. Amiens, 28.10.92 (1992) B.T. 766 the court considered frost in winter on Mont Blanc to be usual, and that the carrier had failed to prove an exceptionally low temperature. The carrier failed to protect the goods when faced with a delay which exacerbated the risk. Even intense cold is something that can normally be protected against and is foreseeable in winter: App. Paris, 30.9.87 (1988) B.T. 59, App. Montpellier, 28.2.85 (1985) B.T. 600. In such cases a successful defence is more likely to be founded on defective packing or sensitivity of the goods, as in the App. Montpellier case, but not in the App. Paris case. Cf. App. Nimes, 18.5.88 (1988) B.T. 472.

172. Comm. Paris, 11.5.83 (1990) 25 E.T.L. 217. RB. Ghent 19.6.90 (1991) 26 E.T.L. 377.

173. Comm. Bruxelles, 31.5.96 (1996) 31 E.T.L. 586.

174. Thus where a driver called to collect a load of parcels and after loading the goods re-entered the premises on being informed that he had forgotten some goods, and the vehicle was stolen by an individual who took advantage of his absence, it was held that the loss was not unavoidable under Article 17(2). App. Paris, 14.6.77 (1977) B.T. 354.

175. BGH., 21.12.66 (1969) 4 E.T.L. 888, (1967) U.L.C. 283. See also App. Bruxelles, 30.10.75 (1976) 11 E.T.L. 238, OGH., 29.6.83 (1984) 19 E.T.L. 526 (the vehicle being left in a parking lot near Milan which was frequented by other people but not under surveillance, see Humphreys & De Peuter, op. cit., at p. 739) and Supreme Court of Finland, 18.5.84 (1986) II U.L.R. 634, where the carrier was liable, having left the vehicle in an unguarded restaurant car park with no anti-theft device, in view of the high risk of theft in the area concerned. Cf. further Hof. Antwerpen, 30.5.79 (1981) II U.L.R. 236. Similar considerations arise where arson is committed against a vehicle left at night at the end of the week on a public road, Cass., 6.1.98 (1998) B.T. 39.

The court considered that the prudent procedure would have been to garage the vehicle for the period in question, and the carrier could not rely on Article 17(2).[176]

6.44 In another French decision the carrier who left his vehicle with a load of valuable goods on a public highway in Italy, which was then stolen despite the fact that the vehicle was locked with its anti-theft device activated, was held not to have done enough to prove that the loss was unavoidable.[177] Similarly where, even though the driver unavoidably had to leave the vehicle for a short period, it was possible for thieves to quickly discover the lorry's contents and neutralise the anti-theft device.[178] In such cases, the defendant in France may well be more concerned to avoid unlimited liability rather than to establish a defence.[179] Thus even where the carrier put his vehicle in a closed garage which was locked with a padlock and which was owned by another transport company, the carrier did not seek to raise a defence under Article 17(2) but contested rather the possibility of unlimited liability.[180] In another case where the carrier placed his locked vehicle in his own closed and locked premises, he did not escape liability when the court considered that, in view of the risk, he should have taken further measures such as the use of a guard or anti-theft device.[181]

6.45 On the other hand, the Supreme Court of Spain exonerated a carrier who, while he went for a meal, had locked the cab door of his lorry and parked it in a built-up area close to a police station, despite the fact that this took place in Italy.[182] In a recent Belgian

---

176. App. Aix-en-Provence, 11.3.69 (1969) B.T. 389, (1970) U.L.C. 122. Cf. in a Paris decision, the vehicle was left parked outside the driver's hotel overnight and the load was stolen. The court rejected the argument that the vehicle was too large to fit in a guarded park. The carrier was therefore liable as the theft was not unavoidable: App. Paris, 15.2.77 (1978) B.T. 53. Similarly in a Belgian decision a shipment of magnetic tape was parked for the night by the driver in Germany close to his home. A fire occurred during the night which an investigation later proved was criminal in origin (by unknown persons). It was held that, applying Article 17(2), the fire was an avoidable circumstance, as if the vehicle had been placed in a guarded park it would not have occurred. In any case as the vehicle was close to his employer's premises he could have parked it there: App. Bruxelles 12.12.77 (1978) B.T. 39, (1978) I U.L.R. 376 (see also BGH., 5.6.81 (1982) 17 E.T.L. 301). Cf. Arrond. Amsterdam, 13.12.72 (1973) S. & S. No. 40 where in similar circumstances but in the absence of negligence on the part of the driver the carrier was not liable for the fire. It should be noted that both the French and Belgian decisions follow the same pattern inasmuch as they all hold that the criminal act in question could have been avoided had the carrier garaged his vehicle either in a warehouse or in a guarded park. See too Cass. (Belg.), 12.12.80 (1981) 16 E.T.L. 250 where the carrier was held liable for theft from a locked lorry parked in an official Customs car park.
177. App. Toulouse, 16.3.81 (1981) B.T. 318.
178. App. Paris, 20.2.96 (1996) B.T. 658.
179. Under the French application of Article 29, see *infra*, para. 9.80. See e.g. App. Paris, 22.10.97 (1997) B.T. 779, (1998) U.L.R. 200, Comm. Paris, 3.5.99 (1999) B.T. 782, App. Paris 3.7.96 (1997) B.T. 65. Some cases are less strict, e.g. Comm Bourg-en-Bresse, 27.2.98 (1998) B.T. 358, App. Paris, 24.4.97 (1997) B.T. 468, App. Paris, 6.5.98 (1998) B.T. 724, App. Paris, 12.10.99 (1999) B.T. 802, App. Riom, 15.9.99 (1999) B.T. 746. As is illustrated by this last case, the French courts occasionally appear to respond to both *faute lourde* and unavoidable circumstances in the same breath, perhaps conflating the differing relevant criteria and burdens.
180. Successfully: Comm. Grenoble, 8.3.82 (1982) B.T. 298, cf. Cass., 13.1.81 (1981) 16 E.T.L. 686 where the fact that the carrier left an unguarded lorry, known to be loaded with valuable goods, overnight in the Paris streets, locked but not fitted with an anti-theft system, resulted in his unlimited liability for the loss of the goods.
181. App. Limoges, 1.3.83 (1983) B.T. 331. Brunat wondered whether this went a little too far, (1983) B.T. 327, cf. Comm. Liège, 13.12.77 (1980) I U.L.R. 271 to similar effect, and note also Arrond. Amsterdam, 12.9.79 (1980) S. & S. No. 34, where even parking in a guarded car park over the weekend because of the need to repair the vehicle was ineffective to protect the carrier in circumstances where the surveillance at the car park was insufficient and the vehicle contained goods attractive to thieves; see Libouton, (1982) J.T. 715.
182. 20.12.85 (1986) II U.L.R. 630. Cf. Audiencia De Barcelona, 28.9.95 (1996) U.L.R. 596, where the court found that the loss could have been avoided through more thorough scheduling of the time of arrival at the consignee's premises in Milan, which were closed, and greater care other than parking the lorry in the street (Cf. further, App. Luxembourg, 11.11.98 (1999) U.L.R. 1038 *infra*, para. 6.47 fn. 194).

decision, a carrier was held not to be liable for the theft in Italy of a vehicle and its load where he parked the vehicle between several other vehicles in a guarded parking area, locked it on both sides and activated an anti-theft device, and where he was only absent from the vehicle for a short period for the purpose of visiting the toilet.[183]

6.46 Article 17(2) fell to be considered by an English court in the context of the theft of the consignment in the case of *Michael Galley Footwear Ltd.* v. *Dominic Iaboni*.[184] In that case a consignment of shoes was stolen while two drivers, who had been driving in convoy, were eating in a nearby restaurant. Even though the stolen lorry was fitted with an adequate alarm and Hodgson, J., was prepared to accept that the nearest guarded park was farther away than the driver could legally drive under the relevant regulations, since the theft could have been avoided by one driver remaining with the lorries while the other ate, the carrier was liable.[185]

6.47 It will be noted that Hodgson, J.'s approach to this case closely resembles that adopted in the Continental case law, and it will only be in rare cases that the carrier will be able to rely on theft as "unavoidable circumstances" within Article 17(2).[186] In the course of his judgment, Hodgson, J., did indicate that had the driver remained with the lorry but been overpowered by armed robbers then the carrier would have been excused liability.[187] However, these circumstances did not excuse the carrier in *J.J. Silber Ltd.* v. *Islander Trucking Ltd.*,[188] where the driver pulled up for a rest at a well-lit tollgate area in Italy and was overwhelmed by armed robbers. Mustill, J., held that utmost care required the carrier to employ two drivers which would have enabled the break required by the EC regulations relating to driving hours to have been taken at a nearby lorry park, as to which there was some evidence adduced by the plaintiffs by Civil Evidence Act Notices.[189] Given the level of risk involved in the journey, the decision compares with Continental decisions where a similar degree of risk has been present,[190] and which require the carrier

---

183. RB. Tongeren, 27.5.92 (1992) 27 E.T.L. 853.

184. [1982] 2 All E.R. 200. The decision in this case confirmed the observations of Neill, J., as to the effect of Article 17(2) in *Thermo Engineers Ltd.* v. *Ferrymasters Ltd.* [1981] 1 Lloyd's Rep. 200.

185. See also Hof. Antwerpen, 30.5.79 (1981) II U.L.R. 236. In *M. Bardiger* v. *Halberg Spedition APS*, 26.10.90, unreported except on LEXIS, Evans, J., held the theft from the lorry not to be unavoidable where the carrier, being unable to find the consignee's premises, parked the lorry and left it whereas he could have driven to a public telephone box or to a police station.

186. See Loewe, para. 153. The logic of the theft cases would suggest that leaving a vehicle on the road involves not just a risk of theft, but also a risk of damage, which can be avoided by off-street parking. It must therefore be questioned whether the French case referred to above at fn. 166 was correctly decided. It might also be noted that Italian national law requires the presence of two drivers whenever the journey exceeds 450 km (Act of 15.6.59, No. 393). The presence of an additional driver or assistant will often be a precaution which might enable the carrier to avoid the loss, cf. *J.J. Silber* v. *Islander Trucking*, *infra*, fn. 188.

187. Similarly, it has been noted that German decisions reflect a distinction between simple theft as avoidable (e.g. even parking in a supervised lorry park will not provide a defence if, where the goods are valuable, the carrier should have made sure that unauthorised persons would not be permitted to remove the trailer with a replacement tractor unit: OLG. Oldenburg, 25.9.96 (1997) TranspR 229) and armed robbery or violent assault as more likely to be assessed as unavoidable (Jung, p. 156)—provided the court is satisfied that utmost care has been taken, cf. BGH., 8.10.98 (1999) TranspR 59, (1999) 34 E.T.L. 371.

188. [1985] 2 Lloyd's Rep. 243.

189. The evidence in respect of one of the two parks indicated in the notices suggested a level of security which, while it might not have withstood a determined gang, would have made things better and could not therefore be ignored by a person exercising the utmost due care, see p. 250. Without evidence, one cannot determine how determined the particular robbers might have been (see the comment in fn. 154 *supra*). Nevertheless, the assessment of the degree of protection that the lorry park was likely to provide was clearly less speculative than the suggested risk of hi-jack involved in making the detour from the motorway to reach it.

190. The level of risk was recognised in the *Silber* case at p. 249. In OGH., 29.6.83 (1984) 19 E.T.L. 526 where theft rather than armed robbery was involved, the risk in respect of a valuable load was such that resort

to ensure that the vehicle stops only at places which are secure,[191] unless there are exceptional circumstances.[192] In a recent Belgian decision a carrier could not rely on a violent attack as unavoidable when, on a journey in Italy and carrying a high risk load, he stopped on the road at a spot with poor surveillance.[193] The duty of utmost care naturally requires the carrier to plan the journey in the light of known risks, inform himself as to appropriate stopping places and instruct the driver accordingly.[194] On the other hand, the carrier has been excused where the occupants of a car flagged down the driver who, even

to a second driver constituted a prudent and reasonable measure given that uninterrupted surveillance of a vehicle could not otherwise be guaranteed in view of the EC regulations as to rest periods. Cf. Humphreys and De Peuter, op. cit., p. 739, who cite a first instance decision in Holland which stopped short of imposing a requirement that, for carriage to Italy, two people should always be on board the vehicle: Arrond. Amsterdam, 11.5.79 (1980) S. & S. No. 85. See also Hof. Leeuwardern, 23.10.96 (1998) N.J. No. 58 (upheld HR., 17.4.98 (1998) N.J. No. 602 (1999) 34 E.T.L. 82. See further App. Caen, 15.11.83 (1984) B.T. 131, (this being the *renvoi* from the decision of Cass., 27.1.81 (1981) B.T. 219, see *supra*, fn. 110) where the driver, arriving at the consignee's premises in Italy late in the afternoon and unable to effect delivery, parked the lorry across the road, locked the doors from the inside before settling down for the night, in the course of which armed robbers succeeded in effecting entry. The court considered that, in view of the fact that the large dimensions of the lorry did not make it particularly attractive to thieves, that parking elsewhere would probably have made no difference and that there was no secure park for large lorries nearby, that the driver had taken the normal precautions that prudence dictates to ensure continuous surveillance, the carrier was not liable. It did note also, however, that the robbery took place in 1976, before the considerable rise in their number since that time, and that there was no reason for the driver to suppose that the risk in transport to Italy was any greater than in France at the time. Contrast where the carrier, having been a victim previously, should have been well aware of the risk, Cass., 2.2.99 (1999) B.T. 175.

191. Parking to rest for the night in places of doubtful security rarely excuses the carrier: in App. Poitiers, 7.2.83 (1983) B.T. 455 the fact that the driver stopped in order to sleep overnight and parked the vehicle in open country was considered imprudent and did not excuse the carrier for the armed robbery in the absence of proof of the impossibility of reaching a more secure parking place. Similarly, parking near a service station, Comm. Paris, 9.6.83 (1983) B.T. 457, where the court pointed to the need for the carrier to provide for stops in secure places. See also Cass., 20.5.90 (1990) B.T. 778, Comm. Paris, 10.11.93 (1994) B.T. 320, Comm. Paris, 12.2.99 (1999) B.T. 213. Cf. Trib. gde. inst. Metz, 18.5.82 (1983) B.T. 347 where the carrier was excused when the driver parked by the auto-route between Milan and Turin, but not in App. Poitiers, 4.11.87 (1988) B.T. 426 where the driver parked on a road with limited traffic on it so that would-be robbers would be unlikely to be discouraged by other vehicles. Equally, parking in a motorway rest area, Cass., 14.5.91 (1992) 27 E.T.L. 124 where the court criticised the carrier for not giving the driver precise instructions with regard to safe resting places and stops at protected points, cf. App. Paris, 24.4.97 (1997) B.T. 468, see also, App. Montpellier, 18.11.99 (2000) B.T. 351. When driving in Italy, even a stop to respond to a call of nature will not provide a defence when the driver or his passenger should suspend the urge until it can be satisfied in more secure surroundings, Cass., 5.1.88 (1988) B.T. 103, Cass., 19.1.93 (1993) B.T. 102. In the latter case, the court took account of the terms of the carrier's insurance policy with regard to utilising secure stopping places. Many French insurance policies contain precise conditions in respect of security, see e.g. (1983) B.T. 334, and particular conditions for Italy, see (1983) B.T. 450. The French carrier does not lack advice, see Fourcade (1982) B.T. 398, nor information about secure stopping places in Italy, see (1983) B.T. 451, 534 and (1997) B.T. 356. In this last respect, see, however, the decision in *Cicatiello* v. *Anglo European Shipping Services Ltd.*, *infra*, fn. 6.48, where it was accepted that there is virtually no such thing as a secure overnight lorry park in Italy (p. 680) and that no reliable guide to them exists (p. 685), cf. Danish Supreme Court, 24.6.97, *infra* fn. 194. Goods in transit policies in this country will now also very often contain conditions requiring the use of anti-theft devices and secure overnight parking, see Croner's *Road Transport Operation*.

192. As in Comm. Bourg-en-Bresse, 27.2.98 (1998) B.T. 358 where the driver had been forced to wait some hours before loading was completed and, faced with nighttime traffic restrictions and fatigue, parked near an occupied toll booth where he was subjected to an armed robbery for which the carrier was excused liability.

193. RB. Turnhout, 30.6.97 (1998) 33 E.T.L. 139, cf. RB. Gent, 11.2.97 (1998) 33 E.T.L. 121, see further, Laurussen. "Diefstal in Italië onder Artikel 17 Lid 2 CMR", (1998) 33 E.T.L. 39.

194. Danish Supreme Court, 24.6.97 (1998) 33 E.T.L. 52 (the duty is reflected also in German cases concerned with whether the carrier had been grossly negligent (for purposes of Article 29, see *infra*, para. 9.81): BGH., 17.4.97 (1998) TranspR 25, (1998) 33 E.T.L. 60, (1998) U.L.R. 210, BGH., 28.5.98 (1998) TranspR 454, (1999) 34 E.T.L. 109. A carrier may also be required to show that it issued the driver with instructions regarding late arrival since it must organise the carriage with due regard to the possibility of delay, App. Luxembourg, 11.11.98 (1999) U.L.R. 1038. Cf. however, App. Douai, 18.5.2000 (2000) B.T. 486.

if he had realised the danger of stopping, could not have escaped from his attackers by adopting any other course since they were in possession of a fast vehicle and a weapon against which he could not have defended himself[195]; where the carrier was intercepted by robbers convincingly disguised as policemen[196]; where the carrier arrived at the destination during lunchtime and, being unable to effect delivery, decided to carry on to premises 100 km further on and, on stopping on the motorway because of a puncture, was overwhelmed by armed robbers.[197]

6.48 In *Cicatiello* v. *Anglo European Shipping Services Ltd.*[198] a carrier who suffered an armed robbery whilst sleeping in his cab parked at a service station was held to have exercised the utmost care. Although there were accessible lorry parks, the evidence was that they did not provide the necessary security in any event and the judge was of the view that greater risks would have been involved in leaving the motorway for the purposes of reaching such a park.[199] Unsurprisingly, in the light of the *Silber* case the plaintiffs alleged that two drivers should have been provided. However, the court accepted the defendants' evidence to the effect that the provision of a second driver would almost have doubled the cost of the freight and that the plaintiffs would not have been prepared to bear that extra cost. In any event, the court did not feel that the provision of an extra driver would have avoided the theft. The same was true of the various safety devices which the plaintiffs alleged should have been fitted, namely an intruder alarm with panic button, mortice locks for the cab lock, a steering lock, and some kind of vehicle immobilisation device, such as a cut-out for the fuel or electrics. The court carefully reviewed each of the security devices which it was alleged should have been fitted, and held that, in the circumstances of the case, the duty of the utmost care did not require the fitting of that device. In any event, the court was also of the view that none of these various devices, even if fitted and used, could or would have prevented the robbery. Referring in particular to the fact that the load consisted of pickled pelts, which would only have been of interest to an industrial organisation which was capable of processing them, the court noted that there had been no previously known instance of such a load having been stolen. Accordingly, quite apart from the violence with which the theft was effected, it was held that the load has been specifically targeted and stolen to order. In those circumstance none of the precautions referred to would have prevented the theft.

6.49 In the case of *Longmill Corporation PLC* v. *Andrea Merzario Ltd.*,[200] once again a consignment to Italy was involved. The vehicle was parked in a lay-by outside the Customs compound, which was closed by the time the driver got there. The driver alleged that he left the cab unattended as a result of being threatened by three men who pulled up in a car beside his cab and told him to go and join other drivers in a bar in the vicinity, but this allegation was not accepted by the court. The driver remained in the bar for the rest of the evening. When he returned, the tractor unit, trailer and the load had been stolen.

---

195. Cass., 21.6.88 (1988) B.T. 437, (1988) 23 E.T.L. 711, cf. OGH., 19.1.94 (1995) TranspR 65, BGH., 13.11.97 (1998) TranspR 250, (1998) 33 E.T.L. 829, App. Riom, 15.9.99 (1999) B.T. 746.

196. App. Rouen, 30.5.84 (1984) B.T. 598. See also, Comm. Nanterre, 3.4.98 (1998) B.T. 399, LG. Bremen, 8.4.98 (1998) TranspR 469, Comm. Epernay, 8.12.98 (1999) B.T. 212.

197. Cass., 10.12.91 (1992) B.T. 176. Where the driver was (justifiably) having difficulty in finding the consignee's premises and, on stepping down from the vehicle to consult a map, was attacked and robbed of the goods, the carrier was exonerated: Comm. Paris, 28.1.98 (1998) B.T. 178.

198. [1994] 1 Lloyd's Rep. 678.

199. Cf. *Silber* v. *Islander Trucking, supra,* fn. 188, at p. 250.

200. C.L.C.C.(B.L.), 6.12.95, unreported.

The court considered various precautions which could have been taken. It was held that a prudent carrier would have arrived at the destination in time to be admitted to the compound or at least to get to a viable alternative such as a motorway service station or a designated lorry park. No evidence was tendered that more secure parking could not have been found. It was also held that a prudent carrier would not have left a thief-attractive consignment unattended for some two or three hours, in particular given the notorious risks of the Milan area. Accordingly, the carrier was not protected by Article 17(2) of the Convention. However, although it was accepted that the vehicle should have been fitted with an alarm or a fuel or ignition cut-out, it was also accepted that such precautions may have been of little benefit even if they had been present.

6.50 Attempts by the carrier to rely on Article 17(2) also failed in the cases of *Levco Toys* v. *Rolenco*[201] and in *Ken Bell (International)* v. *Eric Bacon Transport Services Ltd.*[202] In the former case the consignment was stolen in a hi-jack of a vehicle which was parked overnight at a service area on a motorway in Italy. It was argued that the driver had reached the end of his permitted hours, there were no alternative secure parking areas and the driver had parked in close proximity to other vehicles, in particular another vehicle operated by the same carriers with which it was travelling in convoy. However, it was held that by better planning the driver could have reached the plaintiffs' premises that evening and either unloaded the goods immediately or parked at the plaintiffs' premises, which were relatively secure. Accordingly, the carrier was held liable.

6.51 In *Ken Bell (International)* v. *Eric Bacon Transport Services Ltd.*,[203] the driver left the motorway in Italy to travel by another route, which, although a main road, was not of motorway standard. He stopped at a service station for the night at the end of his driving hours in a well-lit area among eight or nine other units and was the subject of an armed hi-jack. It was held that he would have been safer on the motorway and that accordingly the carrier was liable. The carrier tried to argue that the vehicle had been specifically targeted, so that nothing they could have done would have made any difference to the outcome. However, the *Cicatiello* case, where, as has been seen,[204] the consignment consisted of pickled pelts and the judge was of the view that the consignment had been specifically targeted, was distinguished in the present case where the consignment consisted of frozen goods. The court was of the view that there was no adequate evidence that the load had been targeted, despite the fact that the thieves would have needed access to refrigerated storage for the stolen consignment. It was enough for the thieves to have known that it was a loaded refrigerated vehicle heading south from England, without more.

Other circumstances

6.52 Other circumstances, such as strikes, or the actions of officials into whose hands the goods must pass, may provide a defence. Thus, for example, where a dockers' strike at Marseille delayed for three days the loading on board of a vehicle containing a load of meat for Algeria, the carrier was excused the delay.[205] However, it may be difficult for the

---

201. C.L.C.C.(B.L.), 22.5.97, unreported.
202. C.L.C.C.(B.L.), 22.7.98, unreported.
203. *Supra.*
204. See *supra.* para. 6.48.
205. App. Paris, 1.7.86, unpublished, cited by Lamy, para. 1547, Jur. 4, cf., RB. The Hague, 12.7.89 (1990) 545, No. 147, (1990) 18 E.L.D. No. 788.

carrier to show that he could not have avoided the effects of the strike by taking a different route. Thus where a strike of Customs staff occurred at one office, this did not excuse the carrier who could have put the goods through another office not affected by the strike.[206] Similarly, a delay or detour caused by a strike will not excuse a carrier who can be criticised for failing to protect the goods from the consequences.[207] Where fragile goods, which had been well packed, had been opened up for Customs inspection and then not properly re-packed and re-loaded, the carrier, while not liable for the damage caused directly by the rough handling of Customs staff, was held to be responsible for the fact that this damage was aggravated by his failure to remedy the result of their actions before proceeding to delivery, and was therefore liable for a proportion of the damage.[208] In another French case, a carrier has been held unable to rely on unavoidable circumstances where he refused to continue the carriage in view of the tension in the Gulf, in circumstances where those risks were known at the time of making the contract.[209]

Defective condition of the vehicle under Article 17(3)

6.53 Article 17(3) is intended to prevent the carrier from avoiding liability for loss, damage or delay to goods where it has resulted from the defective condition of the vehicle used for the carriage. (Where the defective condition of the brakes caused a fire and, in order to extinguish it, firemen caused damage to the goods by rough handling in removing them from the vehicle; the damage was held to have resulted from the defective condition in App. Paris, 10.3.2000 (2000) B.T. 211.) It is in some respects a counterbalance to the provision in Article 17(2) whereby the carrier can avoid liability where the loss, etc., resulted from unavoidable circumstances. The carrier cannot obtain the benefit of Article 17(2) by pleading that the "defective condition" of the vehicle constituted circumstances which he could not avoid and the consequence of which he was unable to prevent. The carrier is responsible for the maintenance of his vehicle under all circumstances.[210]

6.54 Thus, in *Walek & Co.* v. *Chapman and Ball (International) Ltd.*[211] where a consignment of yarn was damaged by rain as a result of holes in the tilt cover of the trailer on which it was being carried, it was held that the defective state of the tilt should have been apparent on careful examination. In response to the carrier's argument that this amounted to "circumstances which the carrier could not avoid and the consequences of which he was unable to prevent", under Article 17(2) (the trailer and cover having been supplied by the principal carrier, the claimant in these proceedings), Mocatta, J., indicated that, in his view, that provision applied only to matters other than the defective condition of the vehicle, and held that the carrier was liable under Article 17(3).

---

206. App. Paris, 27.5.80 (1980) B.T. 435, see also App. Amiens, 28.10.92 (1992) B.T. 766 where there was a lack of proof as to when the strike of Customs staff had begun, whether it was sudden or not and whether there was any possibility of the carrier changing his itinerary. Contrast App. Orléans, 12.11.96 (1997) B.T. 159, (1997) U.L.R. 628, where the strike was general and it was reasonable for the carrier to think that it would be over by the time he reached the Customs post so that its prolongation was an unavoidable circumstance.
207. Cass., 9.10.74 (1974) B.T. 491 (Lamy, para. 1547, Jur. 4), App. Amiens, 28.10.92 (1992) B.T. 766.
208. App. Paris, 7.4.86 (1987) B.T. 9; see the criticism of the apportionment in the comment to the case.
209. Comm. Carpentras, 19.2.93 (1994) B.T. 636, see Tilche (1994) B.T. 633.
210. E.g. if the gearbox breaks down, the carrier will be liable: OLG. Zweibrücken, 23.9.66, 1 U. 40/67 (1967) N.J.W. 1717. A design defect, which makes it impossible to see a bridge height restriction due to the height of the driver's cabin is within Article 17(3): App. Colmar, 26.1.2000 (2000) B.T. 154.
211. [1980] 2 Lloyd's Rep. 279. See Glass, "CMR and hire trailers—a tilt too far?", [1981] LMCLQ 384 and cf. Cass., 17.6.97 (1997) B.T. 596.

6.55 A distinction must be drawn between the defective condition of the vehicle itself and any defects in special equipment on the vehicle which is intended to protect the goods from the effects of heat, cold, variations in temperature or the humidity of the air.[212] Separate provision is made for such equipment in Article 18(4) whereby the carrier may prove that he has taken all steps incumbent upon him in the circumstances as regards the choice, maintenance and use of such equipment, and that he has complied with any special instructions issued to him regarding the carriage of such goods. If he provides satisfactory proof to this effect he can avoid liability for the special risks laid down in Article 17(4)(d) regarding (*inter alia*) the carriage of perishable goods.[213]

6.56 As has already been noted in the discussion of Article 17(2), the defective condition of a road vehicle must be distinguished from the situation which arises where a vehicle which is in good working order malfunctions because of some extraneous and unavoidable circumstances. In the latter situation, the carrier may still rely upon the provisions of Article 17(2). Here the problem of the burst tyre, discussed above, is pertinent.[214]

6.57 A road vehicle may be in a defective condition for the carriage of goods generally or else it may be defective for the carriage of certain categories of goods. In the latter case, it has been suggested that the carrier will only be prevented from avoiding liability for loss, etc., of the goods where he is either aware of the fact that the vehicle is not suitable for the goods in question or else he should be aware of the fact.[215] Such an interpretation may be difficult to support on the wording of the text, which clearly envisages strict liability under Article 17(3).[216] A similar interpretation of Article 17(3), however, is that it is concerned with defects independent of the contracted carriage.[217] On this view, suitability is a matter for the express or implied terms of contract,[218] or, as suggested

212. But in Arrond. Rotterdam, 27.4.71 (1971) S. & S. No. 73, (1971) 6 E.T.L. 830, the temperature registered on the outside of a refrigerated vehicle indicated the temperature of the air being drawn into the system, not that of the air after cooling. It was held that this was a fault for which the carrier was liable under Article 17(3). The court considered that Article 18(4), which deals in greater detail with, *inter alia*, the use of refrigerated vehicles, was not applicable. In other words, Article 17(3) was viewed as imposing a wider duty. See also RB. Brussel 27.2.87 (1987) 22 E.T.L. 582 where a defect in the refrigeration unit, being an essential part of a refrigerated lorry, was held to be a defect in the vehicle itself. Cf. Arrond. Rotterdam, 11.1.79 (1980) S. & S. No. 45, Arrond. Rotterdam, 14.3.96 (1997) S. & S. No. 21, see further *infra*, para. 6.97 ff.

213. Related to this is the view that Article 17(3) does not qualify Article 17(4), placed as it is at a later point and qualified, rather, by Article 18(4), OLG. Hamburg, 27.10.88 (1989) VersR 719, 720, see Clarke p. 288, Herber/Piper, Art. 17, Rdn. 79, Thume, Art. 17, Rdn. 112.

214. See *supra*, paras 6.36–6.39 and RB. Antwerpen, 9.4.69 (1969) 4 E.T.L. 1028, Comm. Bruxelles, 3.10.70 (1970) J.P.A. 487. But in OLG. Düsseldorf, 18.11.71 (1973) 8 E.T.L. 510, (1973) II U.L.C. 428 it was held that there could be no presumption that a fire was caused by a third party where it was impossible to rule out a defect in the vehicle.

215. Loewe, para. 157.

216. See Cass., 22.2.94 (1994) B.T. 263, where the carrier used a large ill-equipped semi-trailer to carry works of art when a smaller properly equipped vehicle would have been more suitable. The carrier was held liable on the basis of Article 17(3). Cf. App. Toulouse, 1.12.92 (1993) B.T. 782. In App. Paris, 18.12.92 (1993) B.T. 52, the carrier was unable to escape liability under Article 17(3) for inadequate sheeting of the load by claiming that the height of the goods meant that they could not be fully covered. Sheeting of the goods is the carrier's responsibility. He should have either obtained an agreement to carry in this way or entered a reservation.

217. Haak, p. 152, Lamy, para. 505.

218. See, e.g., OLG. Nürnberg, 14.6.65 (1971) 6 E.T.L. 247, as discussed by Clarke, p. 82, where it was held that a carrier, having been instructed to carry the goods at a certain temperature, must supply equipment capable of maintaining that temperature, cf. App. Venezia, 31.10.74 (1975) 10 E.T.L. 242 where the carrier was not liable under Article 17(3) for the supply of an unrefrigerated vehicle to carry bananas since the claimant had not ordered a specially equipped vehicle.

above, dependent on the extent to which the carrier is aware of the unsuitability of his vehicle. If the goods are of a category widely carried or if the carrier specialises in the carriage of the category in question, such knowledge will presumably be imputed to him.[219] This approach is not necessarily incompatible with a view that Article 17(3) applies to such cases. The point is that once the requisite characteristics of the vehicle have been determined through the contract it is not open to the carrier to show that he could not have avoided the failure of the vehicle to comply with them, even if he could not have known that it would fail. In an Austrian decision the facts were that the carrier cleaned the vehicle in the usual way but this proved insufficient to clear it of residue from the previous transport with consequent contamination of the goods. The court applied Article 17(3) to hold the carrier liable, deeming it irrelevant that he did not know that the usual cleaning method was inadequate.[220] Here, again, it should be noted that the area where this question is most likely to arise, that is, in respect of the carriage of refrigerated goods, is subject to a separate provision in Article 17(4), discussed in detail below.[221]

6.58 A similar question which arises is as to whether the carrier can avoid liability by showing acceptance of an obviously defective vehicle by the sender. The viewpoint of certain Continental jurists is that Article 17(3) is a strict provision which is not subject to any derogation.[222] On this point no guidance can be obtained from the provisions of CIM as Article 36 CIM does not contain any provision analogous to Article 17(3). Moreover, certain Continental case law has tended to exonerate the railway carrier from liability where the shipper has failed to check the state of a wagon which is in bad condition.[223] It may be reasonable to suppose that Article 17(3) was introduced to exclude the application of this body of case law from the international carriage of goods by road.[224] However, it can be argued that where the sender accepts a vehicle which is unsuitable for particular goods whose particular characteristics are not and cannot reasonably be known, the sender must accept responsibility.[225] In such circumstances it is not really a question

---

219. App. Bruxelles, 15.4.71 (1972) J.C.B. 552. The carrier who, in preparing his vehicle for carriage, falls below the standard to be expected for goods in general is unlikely to succeed on an argument that the goods or their packaging rendered them particularly susceptible to the damage which occurred. In App. Paris, 2.12.81 (1982) B.T. 73, toy elephants wrapped in plastic arrived badly stained. The carrier was liable because the damage was caused by the dirty state of the floor of the lorry rather than the nature of the goods or the fragile state of the packaging.

220. OGH., 21.2.96 (1996) TranspR 422, see Jesser, "Unzulängliche Reinigung des Transportfahrzeugs als Mangel i.S. des Art. 17 Abs. 3 CMR", (1997) TranspR 98.

221. See para. 6.97 ff, cf. Arrond. Rotterdam, 14.3.96 (1997) S. & S. No. 21, where meat, not belonging to the claimant, stuck and affected the cooling mechanism of the vehicle with consequent damage to the goods and liability of the carrier on application of Article 17(3). The question has arisen in other contexts. E.g. in one decision the carrier was held unable to rely on the special risk in Article 17(4)(c) (see *infra*, para. 6.72 ff.) where the carriage involved chilled raspberries and the carrier supplied a vehicle set up for transporting meat on hooks with neither a grooved floor nor corrugated sides: OLG. Hamburg, 29.11.84 (1985) TranspR 130, see Thume "Haftungsprobleme bei CMR—Kuhltransporten" (1992) TranspR 1, 2. In OGH., 2.9.87 (1988) II U.L.R. 724 the issue of whether the sender's method of loading water sensitive goods without full protection was defective depended on whether, at the time of carriage, tarpaulins generally used by carriers were water permeable. If the sender could assume that the carrier would use only a water repellent tarpaulin, his loading would not be defective but the carrier would be liable under Article 17(3).

222. Rodière, para. 76.

223. See Bruxelles, 21.11.49 (1949) J.P.A. 467, *contra* Hof. Brussel, 5.4.73 (1973) 8 E.T.L. 528. See also Rodière, para. 76.

224. Libouton, (1974) J.T. 512. This view has been described as speculative, see Haak, p. 152.

225. Even where the characteristics were known, the sender was held to have taken on the risk of breakdown of a vehicle known to be defective in App. Grenoble, 4.12.96 (1997) B.T. 120.

of the vehicle being defective,[226] but rather that the sender has given inadequate instructions in respect of the goods.[227] Alternatively, such a case might be regarded as falling within Article 17(4)(d)[228] or as a case for apportionment within Article 17(5).[229] Where the carrier can appreciate the obvious unsuitability of the vehicle ordered by the sender he can expect to share responsibility by reason of his failure to warn.[230]

6.59 Article 17(3) also provides that the carrier shall not be relieved of liability "by reason of the wrongful act or neglect of the person from whom he may have hired the vehicle or of the agents or servants of the latter". From this it follows that the carrier cannot rely on Article 17(2) so as to avoid liability in respect of circumstances which resulted from the acts and omissions whether tortious or not of the owner of the hired vehicle or his servants or agents.[231] An unusual application of this provision arose in *Walek & Co.* v. *Chapman and Ball (International) Ltd.*,[232] where the third party carriers had hired a trailer with a holed tilt cover from the defendant principal carriers. It was held that the third party carriers were liable to the principal carriers under Article 17(3).[233]

Special risks under Article 17(4)

6.60 Article 17(4) lays down a list of particular circumstances that may affect the carriage, and provides that if loss or damage occurs due to the special risks inherent in those circumstances, the carrier will be relieved of liability, subject to him being able to satisfy the burden of proof which is laid down by Article 18(2) to (5). Although the provisions as to the burden of proof are set out at the end of this chapter,[234] the case law involving the special risks under Article 17(4) has inevitably involved consideration of the appropriate burden of proof, since the two are necessarily interdependent. The following discussion of the special risks provided for by Article 17(4) therefore incorporates, as appropriate, discussion of the burden of proof applicable to each of those circumstances.

(a) *"use of open unsheeted vehicles, when their use has been expressly agreed and specified in the consignment note"*

6.61 This provision is not likely to cause many problems in practice. Presumably for the purposes of this provision a vehicle will be "open" when there is no covering on the top

---

226. If the vehicle must always be viewed, objectively, as defective in such circumstances it must inevitably lead to a shared responsibility which can be regarded as unfair to the carrier, see Jesser, *supra*, para. 6.57, fn. 220, at p. 99.

227. In one case the carrier was not liable for frost damage to goods where the sender had failed to inform him of the sensitivity of the goods to such damage and the carrier had used a vehicle which was simply sheeted: App. Paris, 25.3.69 (1969) B.T. 174 (Lamy, paras 500 and 1554, Jur. 14). See further Arrond. Rotterdam, 25.5.73 (1977) S. & S. No. 67.

228. Cf. App. Liège, 8.3.96 (1996) B.T. 411.

229. In App. Amiens, 28.11.74 (1975) B.T. 23 (see Libouton, (1982) J.T. 716, para. 69) the sender was responsible for his negligence in loading a vehicle where the dirtiness (*malpropreté*) of the lorry was particularly apparent. See also Arrond. Dordrecht, 22.12.71 (1972) S. & S. No. 50. See further *infra*, para. 6.107 ff.

230. App. Aix-en-Provence, 7.9.95 (1995) B.T. 833.

231. Loewe, para. 158. See also Article 3, *supra*, Chapter 3.

232. [1980] 2 Lloyd's Rep. 279. For criticisms of this case see Glass, op. cit.

233. Cf. RB. Antwerpen, 20.6.75 (1975) 10 E.T.L. 540.

234. *Infra*, para. 6.111 ff.

of the vehicle at all, either permanent or temporary and removable.[235] Article 17(4)(a) will only apply, however, where the use of an open unsheeted vehicle has been expressly agreed[236] and specified in the consignment note. This again raises the question of whether specific mention in the consignment note is necessary where such use has been agreed, either in separate correspondence or telex or else orally, perhaps by telephone. On the strict wording of the provision the carrier would not be able to obtain the protection of Article 17(4)(a) unless specified in the consignment note, but reference should be made to the discussion of the need for a consignment note in Chapter 4. It would seem the better view is that there is no need for specific mention in the consignment note,[237] but obviously it does obviate difficulties of proof in demonstrating the agreed use of unsheeted vehicles. It is therefore desirable that where the use of an open unsheeted vehicle is contemplated by the parties that an entry to that effect should be made in the consignment note to protect the carrier and to avoid dispute.

6.62 Article 17(4)(a) is subject to Article 18(2), but the presumption therein in favour of the carrier is effectively restricted by Article 18(3) which provides that:

"This presumption shall not apply in the circumstances set out in article 17, paragraph 4(a), if there has been an abnormal shortage, or a loss of any package."

The position is therefore as follows. First, Article 17(4)(a) gives the carrier a limited protection where the use of open unsheeted vehicles has been expressly agreed upon by the parties and specified in the consignment note. The carrier will be relieved of liability where the loss or damage arises from one of the special risks inherent in carriage by open unsheeted vehicles. The relief from liability is that laid down in Articles 18(2) and (3). Accordingly, where the carrier can establish that the loss or damage can be attributed to the special risks inherent in carriage by open unsheeted vehicles,[238] it shall be presumed that the loss or damage was caused by such use. The claimant, however, is free to rebut this presumption by proving that the loss or damage was not attributable either wholly or partially to the use of an open unsheeted vehicle.[239] The carrier cannot in any case rely on

---

235. Several writers suggest that a vehicle which is open at the sides rather than on the top also falls within Article 17(4)(a), see Haak, p. 154 n. 172, Clarke, p. 307, cf. Rodière, p. 309. There may be an argument that a vehicle is to be regarded as open when the carrier supplies merely a chassis on which the sender's container or swap-body is to be carried, cf., under the former German Domestic Railway Law (EVO), the decision in LG. Frankfurt, 13.10.89 (1990) TranspR 196, (1990) Bulletin des Transports Internationaux Ferroviaires 116, where thieves gained access to the contents of a swap-body carried on a flat railway wagon and this was held to excuse the railway under the special risk of carriage in an open wagon. See also RB. 's Gravenhage, 24.6.87 (1988) S. & S. No. 73 where the court drew a distinction between a tilt trailer on the one hand and an open trailer with a sheeted load on it on the other hand, and held that the latter amounted to an unsheeted vehicle for the purposes of Article 17(4)(a). In other words, a distinction was drawn between a sheeted vehicle and a sheeted load, the latter falling within Article 17(4)(a).

236. This is clearly not fulfilled where the carrier has supplied a sheeted vehicle but the driver considered the sheet to be of no use and so he did not fix it in place, App. Rennes, 18.1.89, unpublished, cited by Lamy, paras 495 and 1549, Jur. 3. Nor is it fulfilled by simply stating on the consignment note "Instructions of sender: clean on truck No. x", Hof. Arnhem, 6.12.78 (1979) S. & S. No. 114.

237. See Comm. Bruxelles, 4.2.72 (1972) 7 E.T.L. 573 where although the court stressed the need for express agreement, no reference was made to the need to specify the use of unsheeted vehicles in the consignment note. See too OLG. Frankfurt, 25.10.77 (1978) 6 E.L.D. 530. The need for specification in the consignment note was, however, endorsed in OLG. Düsseldorf, 8.5.69 (1970) 5 E.T.L. 446, 467.

238. For examples under the equivalent rule in relation to CIM, see App. Paris, 19.3.82 (1982) B.T. 322 and App. Pau, 19.11.86 (1988) B.T. 104.

239. Article 18(2). In Comm. Courtrai, 25.5.72 (1972) J.P.A. 192, (1973) U.L.C. 438 paper was carried from Belgium to France in an open unsheeted vehicle. It was badly damaged by heavy rainfall when left for two days

this presumption where either, as stated above, the use of an open unsheeted vehicle was not expressly agreed upon and specified in the consignment note, or where there has been an abnormal shortage, or a loss of a package.[240]

6.63 It should be noted that merely because the carrier loses the benefit of the presumption under Article 18(3) this does not mean that he will necessarily lose the protection of Article 18(2) in relation to Article 17(4)(b) (inadequate packing) or Article 17(4)(c) (loss through loading, etc.). So, in a German decision, it was held that where grain or a similar commodity is carried in bulk at the sender's request, the carrier can still invoke the protection of Article 17(4)(b) or Article 17(4)(c) even where there is an abnormal shortage within the meaning of Article 18(3).[241]

(b) *"the lack of, or defective condition of packing in the case of goods which, by their nature, are liable to wastage or to be damaged when not packed or when not properly packed"*

6.64 The protection afforded by this provision[242] only applies if two criteria are met: first, that the goods are goods that are liable to wastage or damage if not packed or improperly packed, and secondly, that the loss or damage arises from the special risks inherent in absence of or defect in packing. In other words, the correct approach in any case is to ask:

1. Are these goods of a type which will tend to suffer wastage or loss if unpacked,[243] or not packed in a particular way?[244]
2. Is there an absence of, or defect in,[245] packaging?

at the frontier post. *Held*: the carrier was liable and could not avoid liability under Article 17(4)(a) as heavy thunderstorms are foreseeable in summer in Belgium, and the carrier is bound to protect goods carried in an open unsheeted vehicle against them. If this decision is correct, the ambit of Article 17(4)(a) is narrower than at first sight appears, since although the effects of rain and other inclement weather would seem to be "inherent" in the case of an open vehicle it appears that the carrier must nonetheless guard against them. To similar effect also Comm. Toulouse, 25.6.80 (1980) B.T. 518, (1982) J.T. 716 para. 70.

240. Article 18(3).

241. LG. Offenburg, 21.1.69 (1971) 6 E.T.L. 283. See also Libouton, (1973) 8 E.T.L. 2, para. 63.

242. Which, obviously, will not apply if it is agreed that the carrier will do the packing, see OLG. Hamburg, 10.7.97 (1998) TranspR 243.

243. Some goods, e.g. heavy machinery and motor cars, are commonly shipped unpacked. Their size and nature makes packing unnecessary. Cf. App. Poitiers, 31.3.71 (1971) B.T. 168. Other goods are commonly packed in a particular way—e.g. clothing which is individually wrapped or packeted and carried on rails within the vehicle. Cf. also OLG. Düsseldorf, 8.5.69 (1970) 5 E.T.L. 446 (motor cars); LG. Duisburg, 10.5.68 (1969) 4 E.T.L. 979 (motor cars).

244. Hof. Brussel, 19.12.68 (1969) 4 E.T.L. 953. To benefit from the exoneration under Article 17(4)(b) it is sufficient to show the "possibility" that damage "may have" resulted from the absence of packing. Arrond. Roermond, 4.10.73 (1974) 9 E.T.L. 201. A shipment was carried from Germany to The Netherlands. On arrival, it was found that a steel roll had come off the pallet on which it had been fixed by the sender and had been damaged. The carrier claimed that the roll had been badly fixed to the pallet and that he was relieved of liability under Article 17(4)(b) and (c). *Held*: that the burden of proof shifted to the claimant to prove the damage was *not* due to bad packing. The fact that the steel roll had come off the pallet on which it had been fixed by the sender, together with the facts that the rest of the load was undamaged, and that the steel roll's packing gave way in such a fashion that its support stayed on the lorry, proved that the damage to the roll was attributable to defective packing under Article 17(4)(b). Cf. App. Anvers, 8.10.86 (1987) R.D.C.B. 65.

245. E.g. Comm. Verviers, 3.3.86 (1987) R.D.C.B. 155. It is not sufficient for the carrier to prove the possibility that the packaging may have been defective, he must prove that it was defective; the mere occurrence of damage, even in the course of a number of shipments, did not prove this and it was up to the carrier to prove the qualities of normal packaging and the specific defects of the packing in issue, taking into account the nature of the goods, Comm. Ghent, 14.2.84 (1984) R.D.C. 722, Libouton, Theunis, 79, 92. See also Arrond. Maastricht,

    3. Is the loss or damage of a type causally connected with the absence of or defect in packaging?

6.65 It is not necessary to prove that the loss or damage did result from the packing fault since Article 18(2) applies here,[246] but it is necessary to prove that the goods were liable to wastage or damage in the absence of packing (or as the case may be, of correct packing).[247] The fact that goods are crated does not prove that they are of a kind that would tend to suffer wastage or damage if not so packed.[248]

6.66 The position, therefore, is that merely because goods are not packed prior to shipment they will not "by their nature" be considered to be liable to damage if their nature in fact permits them to stand up to the normal foreseeable risks of transit performed by a diligent carrier.[249] A factor to be considered in this respect is whether the commodity in question is normally carried unpacked without resultant damage.[250] Some commodities, such as cereals, cannot by reason of their nature be carried without some form of packing unless loaded into bulk carriers. If, therefore, such goods are carried in bulk at the sender's request the carrier will not be liable for loss or wastage, but will be protected by Article 17(4)(b).[251] A further factor would be whether the adopted method of packing normally enables the goods to arrive undamaged.[252] It should be noted that, as the provision is

21.2.74 (1976) S. & S. No. 88, Arrond. Amsterdam, 4.2.76 (1976) S. & S. No. 71, Libouton, (1982) J.T. 716 para. 71. The German Supreme Court has held that, while the court can draw on appearances derived from experience, evidence of normal carriage by the driver is not sufficient. Rather, the carrier must show concrete facts supporting a defect in packing, since otherwise it leaves room for the claimant to seek to prove that the damage could not have arisen from a defect in packing under Article 18(2), which he is not required to do, sole responsibility for proving that the packaging was defective resting with the carrier, BGH., 4.10.84 (1985) 20 E.T.L. 154, 158.

    246. See *infra*, paras 6.91 ff. and 6.115 ff. where cheeses were stacked seven high on pallets, whereas the general practice was not to exceed six, and this *could* have been the cause of the damage, the burden passed to the claimant to show that the damage was really due to some other cause: App. Besançon, 7.2.96 (1996) B.T. 682 and 706.

    247. Arrond. Roermond, 4.11.76 (1977) 12 E.T.L. 432, App. Reims, 22.5.96 (1997) B.T. 447.

    248. Hof. 's Hertogenbosch, 21.12.65 (1966) 1 E.T.L. 698, (1966) U.L.C. 119.

    249. OLG. Düsseldorf, 8.5.69 (1970) 5 E.T.L. 446. App. Poitiers, 31.3.71 (1971) B.T. 168. On the one hand, unpackaged steel sheets were not by their nature liable to damage because the absence of packaging could not give rise to damage if the sheets were properly handled and stowed, Comm. Antwerp, 28.3.66 (1966) 1 E.T.L. 712, whereas aluminium sheets which were not packed or fitted with protective or insulating elements were so liable to damage by scratching, Comm. Verviers, 27.9.76 (1977) J.C.B. 524, Libouton, Theunis, p. 86. A large table made of granite in relation to which the slightest imperfection such as marking by the legs of a forklift or splintering would render it useless obviously required protection by means of packing, App. Paris, 19.10.93 (1993) B.T. 792. Frozen vegetables which were not intended for sale to consumers did not need to be packed: App. Paris, 19.6.97 (1997) B.T. 643.

    250. LG. Duisburg, 10.5.68 (1969) 4 E.T.L. 979. App. Colmar, 10.7.70 (1970) B.T. 358, (1971) U.L.C. 137—shipment of two rolls of aluminium foil. Comm. Tournai, 11.12.80 (1981–82) J.P.A. 381, Libouton, Theunis, p. 86. In this case it was not customary to pack the goods since they were difficult to place in crates. In Tribunal De Première Instance, Geneva, 1.5.80 (1984) 19 E.T.L. 259, the admission by the carrier that the goods were normally carried unpacked meant that he could not rely on Article 17(4)(b). See further App. Orléans, 18.1.95 (1995) B.T. 129.

    251. LG. Offenburg, 21.1.69 (1971) 6 E.T.L. 283—shipment of grain in bulk.

    252. In 's Hertogenbosch, 5.1.79 (1980) S. & S. No. 22, Libouton, (1982) J.T. 719, para. 85, there was proof that other goods, similarly packed, arrived undamaged; as also in *Tetroc* v. *Cross-Con* [1981] 1 Lloyd's Rep. 192, 199 (discussed *infra*, para. 6.67) where Martin, J., accepted that the machines were packed in the usual way which was adequate for the intended journey (cf. RB. Antwerpen, 13.10.72 (1973) 8 E.T.L. 330); see also Hof. Brussel, 19.12.68 (1969) 4 E.T.L. 948 where the judicial experts reported that the projecting moulded-plastic elements of unpackaged machines were damaged, almost always at the same height, when the fastenings were removed, so that it could be concluded that the damage would not have occurred if they had been properly packaged, see Libouton, Theunis, pp. 86 and 95.

identical to that laid down in Article 36(3)(b) of the CIM Rules in respect of carriage by rail, case law on the latter may be relevant.[253]

6.67 The main difficulty as regards the application of this provision is whether damage or wastage is due to the absence of or defective condition of packing or whether it is due to other causes.[254] First, the packing may be adequate for the normal risks of carriage over the usual routes in question. Obviously the standard of packing required for transit over high grade Western European roads will differ from that required for shipments over badly maintained routes to the Middle East.[255] If, however, the packing proves inadequate merely because of abnormal circumstances such as a road accident resulting from the driver's negligence, the carrier will not be protected.[256] Secondly, the damage or wastage may not be due to packing at all, but instead result from the manner in which handling operations are carried out.[257] This aspect of Article 17(4) arose for consideration by an English court in *Tetroc Ltd.* v. *Cross-Con (International) Ltd.*[258] The plaintiffs had arranged for the carriage of four machines from The Netherlands to this country. The trailer containing the machinery went astray in transit, and when recovered the machines were found to be damaged, mainly through rust. The court was satisfied on the evidence

253. See Clarke in Yates, Part 4, paras 4.1.2.36.10–4.1.2.36.12.

254. Arrond. Almelo, 9.6.76 (1978) S. & S. No. 8—a shipment of yarn packed in sacks was carried from Turkey to The Netherlands, arriving wet at its destination. It was held that the carrier could not rely on the protection of Article 17(4)(b), on the grounds that goods which are carried in a covered road vehicle should not get wet, even if they have not been packed in waterproof packing (see also OLG., Cologne, 30.8.90 (1990) TranspR 425, OLG. Nuremberg, 12.4.91 (1992) TranspR 263, see Clarke, p. 312 n. 7). However, liability was shared where the carrier used a vehicle which was insufficiently watertight but where the goods were expensive and fragile carpets, unprotected from the danger of damage by droplets of water, App. Paris 12.7.78 (1979) B.T. 159, Libouton, (1982) J.T. 716, para. 71.

255. OGH., 18.12.84 (1985) Stra GüV 4.34 (Clarke, p. 310 n. 91), Arrond. Amsterdam, 25.6.75 (1976) S. & S. No. 49 and 1.6.77 (1978) S. & S. No. 74, App. Paris, 12.7.78 (1979) B.T. 159 (Haak, p. 156 n. 184). Where extreme weather conditions may be anticipated, the sender should allow for these in preparing the goods for transport, App. Montpellier, 28.2.85 (1985) B.T. 600, cf. App. Versailles, 29.2.84 (1984) B.T. 249, unless it was not unreasonable for him to expect that the carrier had the means available to protect the goods from the foreseeable risk of such conditions, App. Nimes, 18.5.88 (1988) B.T. 472; cf. further, however, App. Paris, 10.2.86 (1984) B.T. 558 (see *infra*), and App. Paris, 30.9.87 (1988) B.T. 59 where the carrier was held liable for failing to protect a consignment of fresh eggs from freezing during a delay at Customs. He could not plead unavoidable circumstances, and no reliance had been placed on Article 17(4)(d) (see *infra*, para. 6.95 ff.). A reservation sent by telex denying responsibility for the effect of cold was ineffective, since not included on the consignment note as envisaged by Article 17(4)(a). As is pointed out in the comment to the case, if frost is not an unavoidable circumstance, it is equally incumbent on the sender to ensure the protection of the goods or demand an adequate vehicle.

256. GH. Dordrecht, 18.5.66 (1968) 3 E.T.L. 416, (1968) U.L.C. 155: accident caused by the driver falling asleep. In App. Toulouse, 8.3.82 (1982) B.T. 247 the court pointed out that the usual role of packing is to protect the goods from the usual risk of transport, i.e. the inevitable shocks and shakes to which transported goods are necessarily exposed.

257. RB. Antwerpen, 28.3.66 (1966) 1 E.T.L. 712. *Held*: that a defect in packing will not relieve the carrier of liability under Article 17(4)(b) where the goods have been loaded and handled in an unsuitable manner by the carrier. In Hof. 's Hertogenbosch, 21.12.65 (1966) U.L.C. 119, where a crate overturned on the vehicle, the carrier could not complain that the goods should have been packed in a stronger crate, cf. Cass., 20.6.78 (1979) I U.L.R. 296. In Trib. Civile e Penale di Milano, 22.3.73 (1974) 9 E.T.L. 490, a shipment of dangerous goods was destroyed in transit from Milan to Brussels. The goods were loaded and stowed by the sender, and the carrier claimed relief under Article 17(4)(b). *Held:* that the carrier was not liable. In loading and stowing dangerous goods (such as matches), special precautions must be taken by the sender as regards the resistant nature of the packing and their differences in shape and weight. Thus, where there has been a fire and, in the light of the surrounding circumstances established by experts' surveys, it appears possible for this to have been caused by defective packing or an unsuitable method of stowage, this will suffice to relieve the carrier from liability, unless it is proved that there was some other cause for which he is responsible. It would seem that, while reading Article 17(4)(b) and (c) together, the court relied primarily on Article 17(4)(c) in reaching its decision.

258. [1981] 1 Lloyd's Rep. 192.

that the method of packing was adequate, and the inference was that the machines and their packing had been tampered with in the course of the transit. That inference, coupled with the long delay in delivery, was therefore regarded as the probable cause of the damage, so that this aspect of Article 17(4) did not assist the defendants.

6.68 To the extent therefore that the sender is not responsible for such handling operations the carrier will need to prove that the absence of or defective condition of the packing could have been the cause of wastage or damage.[259] However, where goods are not packed for transit they are not "by their nature" liable to damage merely because while being unloaded they have suffered damage where the act of unloading them has caused the loss.[260] It would appear that where a carrier notices in the course of carriage that goods have been defectively packed, he must take all steps necessary to avoid or limit the extent of any damage to them. If he fails to do so any resultant loss or damage may be apportioned between the cargo interest and himself.[261]

6.69 The question therefore arises as to whether the carrier is required to make reservations either at the time of loading or unloading if he wishes to prove that the packing was in fact inadequate.[262] Three possible positions appear from the cases: first, that the carrier retains the onus of proving that there was a defect, having lost the benefit that a reservation would have given him[263]; secondly, that the carrier has a duty to check the state of the packing and takes the risk of damage to defectively packed goods where the defect was apparent at the time of taking over the goods[264]; thirdly, that the carrier is obliged to point out any defect or inadequacy of packing which he can see and either refuse to carry the goods (unless he is able to provide adequate protection) or at least make a reservation.[265] Where the goods are obviously at risk there is much to be said for

259. Hof. Brussel, 19.12.68 (1969) 4 E.T.L. 948. See Article 18(2) discussed below, and cf. the Italian case noted at fn. 257.

260. RB. Dordrecht, 22.12.71 (1972) S. & S. No. 50.

261. OLG. Saarbrücken, 21.11.74 (1976) 11 E.T.L. 261. In such a case the duty of the carrier to contact the sender and obtain instructions as required by Article 14 can make the carrier wholly responsible for any further damage, since he cannot claim that it was due to the initial defect of packing, cf. App. Liège, 18.12.67 (1969) 4 E.T.L. 965, in an action under Article 10, *supra*, para. 4.32 ff. See, however, App. Liège, 27.10.92 (1993) J.T 538 (see Putzeys, *infra*, fn. 262 at p. 565) where the defect in stowage due to a defect in packing was not considered to make the carriage impossible, nor could the sender show that had he been contacted he would have instructed the carrier to act differently, so that the restowage could only be viewed as caused by the original failure of the sender.

262. There is no question of such a requirement unless the defect in packing is clear and externally visible: Hof. Gent, 28.4.95, (1995–1996) R.W. 747 (Putzeys, "Le droit des transports en Belgique (II) 1993–1995," (1996) U.L.R. 557, 566).

263. App. Liège, 18.12.67 (1969) 4 E.T.L. 965, Hof. Brussel, 5.12.68 (1969) 4 E.T.L. 958 (see Libouton, Theunis, p. 87).

264. A view taken in France: Cass., 12.10.81 (1981) B.T. 576, (1982) 17 E.T.L. 294, App. Aix-en-Provence, 9.12.80 (1981) B.T. 143 (where liability was shared, since there was also a defect in the loading performed by the sender), Comm. Toulouse, 8.3.82 (1982) B.T. 247, App. Chambéry, 24.6.97, (1997) B.T. 667; cf. where the state of the packaging appeared to be in order: App. Paris, 12.1.93 (1993) B.T. 187. This view is based on Article 8 (see *supra*, para. 4.24 ff.), a view that has been criticised since Article 8 is related to proof rather than to the carrier's duty of care, see Clarke, p. 317, Lamy, para. 464. See also Herber/Piper, Art. 8, Rdn. 1–3, Koller, Art. 8 Rdn. 1, OLG. Düsseldorf, 24.9.92 (1993) TranspR 54, ZG. Basel-Stadt, 19.4.91 (1992) TranspR 408, 409.

265. App. Aix-en-Provence, 9.12.80 (*supra*, fn. 264), Comm. Toulouse, 8.3.82 (*supra*, fn. 264) and App. Toulouse, 12.4.94 (1994) B.T. 714; in App. Paris, 10.2.84 (1984) B.T. 558 the carrier could see that the way that the goods were packaged would lead to damage. The goods were, however, packed in a way appropriate for carriage in the usual way which involved use of a specially equipped vehicle which the carrier had not supplied. The court considered that the sender did not need to give instructions about the carriage since the danger was apparent and the carrier should either have refused to carry or supplied an appropriate vehicle. The decision in App. Aix-en-Provence, 9.12.80 was criticised as going too far in suggesting a duty on the carrier to advise the

imposing a duty on the carrier to exercise some care to safeguard them or at least take steps to ensure that the risk is squarely placed on the sender.[266]

6.70 There is nothing in the text of CMR itself, but a parallel may be drawn with the provisions of Article 10. This deprives the carrier of his remedy against the sender for damage and expenses arising out of defective packing where this was apparent or known to the carrier at the time of taking over the goods and he made no reservations about it. It can be argued that this strengthens the argument that the carrier cannot rely on an apparent or known defect for the purposes of Article 17(4)(b) unless he has made a reservation.

6.71 Finally, where the carrier for any reason repacks goods in transit in circumstances where, under Article 14(1),[267] he is required to obtain further instructions from the sender but fails to do so, he will be liable for any ensuing loss.[268]

(c) *"handling, loading, stowage or unloading of the goods by the sender, the consignee or persons acting on behalf of the sender or the consignee"*[269]

6.72 The handling, etc., need not be performed by the sender or the consignee alone, but also by "persons acting on behalf of" them to come within the provisions of Article 17(4)(c). This would therefore include not only the servants of the sender or consignee, but also any independent contractor performing any of the operations in question on their behalf. For example, where specialised equipment or heavy loads are to be transported specialist operators may be employed to load, stow or discharge the cargo. One difficulty is that in practice it is not always easy to distinguish one operation from the other. It is necessary therefore to consider the meaning of each term in turn:

"Handling"   — this will include any physical moving of the goods prior to and subsequent to loading and unloading from the vehicle.

"Loading"   — this covers the lifting and placing of the goods on or in the vehicle.

sender that the goods required better protection, since the duty on the carrier is to check only the apparent state of the packaging and not its quality, see comment to the case and Nossovitch (1982) B.T. 463, 464; see also Lamy, para. 464, cf. Arrond. Maastricht, 21.2.74 (1976) S. & S. No. 88, Libouton (1982), J.T. 702, para. 39. Certainly, if there is an element of doubt, this should be resolved in favour of the carrier since the risk in the first place is that of the sender. Cf. App. Paris, 3.1.70 (1971) 6 E.T.L. 264 in respect of stowage, and see also App. Paris, 19.10.93 (1993) B.T. 792 where the defect of packing should have been obvious to the sender, but the possibility of this being obvious to the carrier does not appear to have been considered.

266. Cf. the decisions at common law: *Gould* v. *S.E. & Chatham Ry* [1920] 2 K.B. 186, *Barbour* v. *S.E. Ry* (1876) 34 L.T. 67 (under CMR the option of a shared liability can reflect the duty on both parties, see the comment to the decision in App. Paris, 30.9.87 (1988) B.T. 59). Such a view may underpin the requirement that a reservation made by the carrier must be accepted by the sender (applying Article 8(2) to the letter), Cass., 12.10.81 (1981) B.T. 576, (1982) 17 E.T.L. 294, a requirement considered to be utopian by Lamy, para. 464.

267. *Supra*, Chapter 5, paras 5.25–5.30.

268. App. Liège, 18.12.67 (1969) 4 E.T.L. 965.

269. It has been held that the operation of Customs staff, being imposed by the State, are not the actions of "persons acting on behalf of" the sender etc. for the purposes of Article 17(4)(c), App. Paris, 7.4.86 (1987) B.T. 9. On the other hand, a carrier has been held entitled to assume that a veterinary inspector will re-fix the hooks in a container after his inspection so that the goods will not be allowed to slide and be damaged at a later stage of the transit: LG. Bremen, 23.12.88 (1989) TranspR 267 (Thume, *supra*, para. 6.57 fn. 221, at p. 4). In Hof. Brussel, 23.12.71 (1972) 7 E.T.L. 865, a groupage operator was held to be acting as agent for the sender in respect of the loading of his goods. Since the groupage operator was held to be a CMR carrier (see *supra*, para. 1.13 ff.), this decision is doubtful except on the basis that the performing carrier was not a successive carrier (see *infra*, Chapter 11). Clearly, handling by a successive carrier is not done on behalf of the sender, Hof. Brussel, 7.2.92 (1993) 28 E.T.L. 286.

"Stowage"   — this consists of the fastening[270] or securing of the goods in the vehicle so as to ensure the stability and safety of both the vehicle and its load.[271]

"Unloading" — this covers the removal of the goods from the vehicle on to the ground or onto a wharf or other vehicle.

6.73 These operations must be performed with a view to the normal conditions which can be expected in any road transport. Thus the stowage of the goods must be such as will enable them to withstand evasive action taken by the driver,[272] or sudden braking.[273] In a decision of the German Supreme Court, the court noted the view that the sender should take into account the vibration that can be expected to result from normal hazards such as potholes in the road, but should not have to anticipate the possibility of the trailer turning over.[274]

6.74 Several areas of difficulty can be identified from the case law on this defence. Before turning to the relevant case law, however, it is first necessary to draw attention to the provisions of Article 18(2), which affects the burden of proof in relation to the special risks provided for by Article 17(4). Article 18(2) provides that:

"When the carrier establishes that in the circumstances of the case, the loss or damage could be attributed to one or more of the special risks referred to in article 17, paragraph 4, it shall be presumed that it was so caused. The claimant shall however be entitled to prove that the loss or damage was not, in fact, attributable either wholly or partly to one of these risks."

Clearly the combined provisions of Articles 17(4) and 18(2) result in a presumption in favour of the carrier where these operations have been performed on behalf of the sender and loss or wastage to the goods has resulted.[275] The problem is that the various national courts have produced a variety of interpretations as to when the carrier can benefit from this provision. Although much of the case law is concerned with Article 17(4)(c), it should be appreciated that the problem essentially concerns the general relationship between the special risks enumerated in Article 17(4) and the burden of proof in Article 18(2). Nevertheless, that relationship appears to have posed particular problems in respect of Article 17(4)(c). The various approaches to be found in the case law can be analysed by grouping the issues around several questions.[276]

---

270. See Comm. Antwerp, 4.3.69 (1969) 4 E.T.L. 1030 and Comm. Antwerp, 7.3.80 (1981) 16 E.T.L. 466. Loading, stowing and lashing have been said to be distinct operations: RB. Antwerpen, 7.3.80 (1981) 16 E.T.L. 466.

271. In one case stowage was seen as designed to stop the goods from moving rather than to separate certain goods from others, Hof. Brussel, 17.12.84 (1985) 20 E.T.L. 354.

272. Comm. Namur, 22.7.65 (1966) 1 E.T.L. 133, (1965) U.L.C. 355. The sender loaded, stowed and wedged 13 crates of polished glass on the carrier's vehicle. En route the driver was forced to take evading action to avoid an accident, causing the crates to fall off the vehicle. *Held*: that the carrier was not liable as, in the light of actual traffic conditions, such evading action constitutes a normal risk to be taken into account when stowing the goods.

273. Comm. Antwerp, 30.12.70 (1970) J.P.A. 123. See also OLG. Munich, 27.11.68 (1971) 6 E.T.L. 115, App. Brussels, 6.4.77 (1977) 12 E.T.L. 881.

274. BGH., 28.3.85 (1986) 21 E.T.L. 174, 177–178. In this case the driver, as an evasive measure, drove on to a soft verge which was lower than the carriageway and the trailer overturned as he was attempting to steer it back on to the carriageway.

275. In some cases the carrier has sought to make use of these provisions when himself making a claim for indemnity and not merely as a defence: Comm. Paris, 26.8.81 (1982) B.T. 142, App. Aix-en-Provence, 9.4.92 (1993) B.T. 527. See, however, App. Lyon, 7.10.76 (1979) I U.L.R. 289.

276. The first edition of this book drew on the analysis of the Continental case law adopted by a Belgian jurist: Wijffels, "Legal Interpretations of CMR: The Continental Viewpoint", (1976) 11 E.T.L. 209, referred to

WHO HAS THE DUTY TO PERFORM THESE OPERATIONS?

6.75 CMR does not clearly set out who is responsible for these operations which may be seen as a matter for the contract or relevant national law.[277] Where there is a doubt as to who in fact performed the operation, proof may be derived from the fact that responsibility is imposed on a particular party.[278] On one view, the acts of the servants of one of the parties are open to proof that they were undertaken under the responsibility of that party or of the other party. Thus in a Belgian appellate decision,[279] the carrier was able to show from the consignment note that the action of the sender in loading the goods was undertaken under his own responsibility. In another Belgian decision,[280] a consignment was loaded by the sender and then stowed by the carrier. In transit, the load shifted and fell off. The carrier was held liable for the loss on the grounds that it was due to inadequate stowage. He could not rely on Article 17(4)(c) as the absence of proper boarding on the sides of the vehicle to support such a cargo, together with the fact that it was only roped together, constituted defective stowage and the carrier was unable to prove an agreement that his actions in stowing the goods were for the account of the sender.[281] A different view suggests that CMR is to be taken literally, so that Article 17(4)(c) is to be applied even though the acts of the sender were under the contractual responsibility of the carrier.[282] There is no reason why CMR should regulate the extent of the performance

by the Court of Appeal in *Ulster-Swift Ltd.* v. *Taunton Meat Haulage Ltd.* [1977] 1 Lloyd's Rep. 346, 350. This still provides a useful starting point, but the development of the relevant case law suggests that a more compact analysis may now be possible.

277. In France the view is taken that where the contract is silent the law of the place of the conclusion of the contract is relevant to the operations of loading and stowage (Rodière, p. 231, Lamy, para. 466), see e.g. App. Douai, 9.4.98 (1998) B.T. 339 and it has been held that the law of the place of destination, as agreed in the contract, is appropriate to the operations of unloading, App. Amiens, 23.11.90 (1991) B.T. 292. In France this is usually determined by the *contrat types*, Lamy, para. 466, although note an early decision under CMR where the view was taken that loading and stowage are distinct operations and that the latter will be performed by the carrier unless there is an agreement to the contrary or special circumstances arise, taking into account the nature of the goods: App. Paris, 16.5.69 (1969) 4 E.T.L. 896, (1969) B.T. 190. In Belgium, the absence of agreement tends towards the assumption that the responsibility rests on the carrier, Vredegerecht Antwerpen, 26.10.71 (1972) 7 E.T.L. 1058 which concerned loading, cf. Civ. Charleroi, 1.10.68 (1969) U.L.C. 327, which placed the obligation on the sender if loading gear was required, see further exceptions in Libouton, Theunis, p. 88, Comm. Dimant, 31.10.72 (1972–3) Jur. Liège 285, Comm. Antwerp, 7.3.80 (1981) 16 E.T.L. 466 (stowage), Hof. Gent, 20.11.75 (1976) 11 E.T.L. 231 (unloading), see, however Hof. Gent, 19.1.99, *infra*, fn. 282. In Austria, in the absence of agreement, loading is the responsibility of the sender: OGH., 2.9.87 (1988) II U.L.R. 724. In Germany it has been held that the question of which party has the duty to provide for loading and stowage is determined by the applicable domestic law and in the case of Germany this places the duty on the sender: BGH., 24.9.87 (1988) II U.L.R. 713, cf. BGH., 28.3.85, *supra*, para. 6.73 fn. 274. Contrast the view of Pesce, "Uniform Law on International Carriage of Goods: A System of Special Rules", (1997) 32 E.T.L. 503, 506, n. 45, that no "gap" exists in CMR on this point.

278. Libouton, (1982) J.T. 718, para. 82.

279. Hof. Brussel, 17.12.84 (1985) 20 E.T.L. 354.

280. Hof. Brussel, 13.1.72 (1972) 7 E.T.L. 585, (1973) I U.L.C. 266.

281. Note also the Dutch authority, Binding Advice, 3.8.76 (1977) S. & S. No. 241, (1978) 6 E.L.D. 91 which makes clear the freedom of the parties to assume duties otherwise than might be implied by Article 17(4) (said in relation to packaging) although referring to Article 17(4)(c); cf. further RB. Antwerpen, 30.12.70 (1970) J.P.A. 123, RB. Antwerpen, 29.3.77 (1977) 12 E.T.L. 293.

282. BGH., 28.3.85 (1986) 21 E.T.L. 174 (cf., OLG. Düsseldorf, 25.3.93 (1994) TranspR 439, OLG. Koblenz, 6.10.89 (1990) R.I.W. 931, (1991) TranspR 93 (Clarke, p. 81 n. 34), (1991) 19 E.L.D. No. 660). On this view the parties are prohibited from altering the effect of Article 17(4)(c) by virtue of Article 41, see *infra*, Chapter 12 (cf. BGH., 27.10.78 (1979) VersR 417, discussed by Clarke, p. 329), cf. Hof. Amsterdam, 7.9.73 (1974) S. & S. No.16, see Libouton, (1982) J.T. 717, para. 74, and see further Libouton, Theunis, p. 89. Cf. OGH., 2.9.87 (1988) II U.L.R. 724, 726. In Civ. Charleroi, 1.10.68 (1969) U.L.C. 327 the carrier was held liable for the absence of stowage. Even if, as alleged, the stowage was due to be performed by the sender, the carrier both could and should have effected the stowage unless there was some special difficulty preventing him from doing

undertaken by the carrier, but a limit on the usual responsibility must be clearly proved.[283] Clearly where the carrier is unable to prove such an agreement, a printed clause seeking to restrict his liability will be rendered void by virtue of Article 41.[284] In one Belgian decision, the carrier could not claim that his servants were engaged in loading operations only to assist the sender where, had this been the case, it would have been more clearly indicated in the way that the charges were invoiced.[285]

6.76 Different issues arise where the carrier's employee loads the goods on the instructions of the sender. Conversely the sender may claim that his participation in these operations was under the supervision or instructions of the carrier.[286] In a German appellate decision, a shipment of pork was, on the instructions of the sender, too tightly loaded in a refrigerated vehicle which had insufficient capacity for the load. In an action by the sender against the carrier for damage to the goods, the court held that the carrier was not liable for the action of the driver as he was acting at the request of and on the instructions of the sender.[287] In Holland it has been held that the sender should see that his instructions are carried out.[288]

6.77 This interpretation is directly contradicted by a Belgian decision in which a consignment of cartons of tea was loaded and stowed by the carrier's driver on the instructions of the sender. The goods were damaged by damp owing partly to a hole in the tarpaulin sheet used to cover them and partly due to the loading which allowed the tarpaulin to remain in contact with the top layer of the cartons. In an action by the sender against the carrier, the court held that where the carrier's employees load a vehicle in a negligent fashion on the instructions of the sender, the carrier cannot avoid liability under Article 17(4)(c). The carrier should have refused to perform the carriage under these conditions.[289]

---

so. In Hof. Gent, 19.1.99 (1999) 34 E.T.L. 264 responsibility for stowage was considered to be regulated by Article 17(1) and Article 17(4)(c) which produces the principle that the party performing the stowage is responsible for it, cf. Hof. Gent, 14.9.94 (1994–1995) R.W. 44.

283. App. Paris, 24.5.91 (1991) B.T. 573: the participation of the sender in loading did not prove that there was an agreement that this operation was to be exclusively the work of the sender, cf. OGH., 25.9.68 (1973) 8 E.T.L. 309, 314, as to which see Clarke, p. 322 n. 70.

284. See Hof. Brussel, 13.1.72, *supra*, fn. 280., cf., Comm. Creteil, 8.6.94 (1994) B.T. 603.

285. Comm. Bruxelles, 26.10.72 (1973) 8 E.T.L. 516. Conversely, in Comm. Gand, 11.6.91 (1993) B.T. 666 the simple fact that the sender's employees engaged in the loading did not assist the carrier in the absence of proof that this was not done at the request and under the responsibility of the driver.

286. In Hof. Antwerpen, 1.5.78 (1978) 13 E.T.L. 601 it was held that where the sender alleged that he carried out the loading under the supervision of the driver, the sender is required to prove this. In App. Paris, 19.6.90 (1991) B.T. 13, where the loading was executed under the instructions of the driver, the consequent instability of the load combined with the failure of the driver to make the necessary adjustment to his driving was held to render the carrier entirely responsible for the ensuing damage.

287. OLG. Hamm, 19.2.73 (1974) 9 E.T.L. 753.

288. Arrond. Amsterdam, 1.6.77 (1978) S. & S. No. 74, confirmed by Hof. Amsterdam, 9.11.79 (1980) S. & S. No. 69, cf. OLG. Munich, 27.11.68 (1971) 6 E.T.L. 115 where it was held that it was incumbent on the driver to check that advice he had given to the sender's employees had been followed, cf. further, however, OGH., 8.10.84 (1985) Stra GüV 6.28, discussed by Clarke, p. 326.

289. Comm. Bruxelles, 26.10.72, *supra*, fn. 285. The fact that the sender's employees gave assistance in loading the vehicle neither avoided nor reduced the carrier's liability, since, as noted *supra*, the carrier's invoice indicated that loading was undertaken under the carrier's responsibility. Contrast a case where the failure of the sender to control the use by the carrier of a defective cover was held to exonerate the carrier, App. Liège, 8.3.96 (1996) B.T. 411.

NOTWITHSTANDING THAT THE SENDER PERFORMS THE OPERATIONS OF LOADING
ETC., IS THE CARRIER UNDER A DUTY TO TAKE STEPS WHERE THESE OPERATIONS
ARE OBVIOUSLY DEFECTIVELY PERFORMED?

6.78 The case discussed above clearly raised the question of the extent of the duty of the carrier where there is a visible risk of damage if he proceeds to carry the goods as loaded or stowed by the sender. The decision suggests a duty to refuse to follow an instruction which manifestly puts the goods themselves at risk. A similar duty has been placed on the carrier to ensure that defective loading and stowage, if discovered, does not cause damage to the goods.[290] In cases in which this issue has arisen, it is often combined with the question whether the carrier is required to make checks on the operations carried out by another party, so that the carrier is liable whenever he proceeds with the carriage where he could have observed that the operation was defective in some way.[291] However, putting this latter question to one side, the basic suggestion is that a carrier who proceeds with the carriage without doing anything in respect of a manifest defect will be liable for any resulting loss or damage.[292]

6.79 Such an interpretation gives rise to the further question of what steps are required of the carrier in such circumstances. As in the decision of the Belgian court referred to above, some decisions suggest that the carrier should refuse to carry the goods.[293] In an appellate Belgian decision, it was said to be a serious fault on the part of the carrier, having ascertained that the handling, loading or stowage by the sender was defective, to proceed to carry without taking steps to rectify the defect.[294] Some decisions accept the

---

290. BGH., 24.9.87 (1988) II U.L.R. 713. If the driver chooses to give a warning he may be required to ensure that he gives it to an appropriate person and not a "mere" employee of the sender without managerial responsibility, OLG. Munich, 28.7.95 (1996) TranspR 240, cf., however, OLG. Düsseldorf, 21.4.94 (1995) TranspR 347 (Jung, p. 158).

291. What he could have observed must depend on what he or the driver could be expected to observe. The carrier could not be expected to detect a defect in stowage unless informed of the special requirements of the goods in light of the method of packing adopted by the sender: App. Liège, 27.10.92 (1993) J.T. 538 (Putzeys, "Le droit des transports en Belgique (II)—1993–1995," (1996) U.L.R. 557, 565). See also Hof. Gent, 14.9.94 (1994–1995) R.W. 44 (Putzeys, *supra*, p. 565). In Hof. Gent, 19.1.99 (1999) 34 E.T.L. 264 the court indicated that a heavy goods driver cannot be expected to have specialist knowledge concerning packaging and stowage as against the consignor-manufacturer who knows the peculiarities of his product. See further App. Douai, 9.4.98 (1998) B.T. 339: the driver could not be expected to notice a defect in the stowage of a machine which obviously required a professionalism which the party performing the operation apparently had and where the driver had not been advised that its centre of gravity was off-centre. See also Cass. 5.7.77, *infra*, para. 6.80 fn. 299.

292. Cass., 3.5.76 (1976) B.T. 317, (1978) 13 E.T.L. 106, see further *infra*. See also Cass., 1.12.92 (1993) 28 E.T.L. 618.

293. Comm. Corbeil-Essones, 18.4.69 (1969) 4 E.T.L. 988 where the court held that, where stowage is inadequate, the driver should if necessary refuse to commence the journey. See also App. Paris, 27.2.70 (1970) B.T. 111, where the court decided that, as loading normally was the responsibility of the carrier, he must therefore ensure that it was properly carried out by the sender. On the other hand, the practical difficulty of refusing, or for the carrier's driver to clearly appreciate the defect in stowage, was recognised in App. Paris, 3.11.70 (1971) 6 E.T.L. 264, (1971) U.L.C. 144. In this case, under the terms of the consignment note, the sender undertook to load and stow and to accept full responsibility for the execution of those operations. Nevertheless, there remained the question whether the carrier should have refused to carry out the transport or made a reservation. Under the circumstances, the driver's failure to do so was not a serious fault for the purposes of Article 29 (see *infra*, para. 9.80 ff.). See also Hof. 's Hertogenbosch, 19.3.86 (1987) S. & S. No. 23, which permitted the doubtful carrier to rely on the expertise of the sender, see Clarke, p. 328 n. 15. See also App. Aix-en-Provence, 16.6.83 (1984) B.T. 600 and App. Anvers, 8.10.86 (1987) R.D.C.B. 65 and cases cited *supra*, fn. 291.

294. App. Liège, 6.5.70 (1970) 5 E.T.L. 716, (1971) U.L.C. 133, confirming Comm. Verviers, 18.5.68 (1968) 3 E.T.L. 1240, cf. RB. Gent, 1.12.98 (1999) 34 E.T.L. 392. See also OLG. Munich, 27.11.68 (1971) 6 E.T.L. 115 where the driver was at pains not to intervene in the loading, but his failure to fill a 15 cm gap between load and tailboard was not only sufficient to deprive his employers of the protection of Article 17(4)(c), but was also

alternative possibility of the carrier making a reservation.[295] The failure to make a reservation where the defect is apparent, at least in France, will result in liability on the part of the carrier,[296] despite the fact that provision for reservations under CMR are concerned with proof. Nevertheless, the requirement to make a reservation may be justified by the need to ensure that the potential risks of the defect have been fairly placed on the sender.[297] In a recent Belgian decision the court stated that it was necessary to combine the application of Articles 17(1), 17(4) and 18(2) of CMR with the general principles governing commercial contracts including that of good faith. This required that the carrier should inform the principal and warn him where he establishes that the stowage performed by the principal is defective and hazardous. This was achieved by means of a reservation on the consignment note indicating that the loading of the goods (a machine) was asymmetrical and unbalanced.[298]

6.80 Where there are no apparent defects, the carrier will not be liable for defective loading and stowage performed by the sender unless the sender can show loss or damage was due to another cause.[299] Where the defect in the loading becomes apparent during the course of the transit, the carrier will have a duty to take appropriate action to deal with it. Thus in a French appellate decision, the driver noticed that the load was shifting only 20 kilometres into the journey. Although he took some steps to consolidate the load, this did not prevent half of the load from later falling from the vehicle and being totally destroyed. Whilst acknowledging that the driver had tried to deal with the problem, the court considered that, given the fragile nature of the goods that he was carrying and the distance he had yet to drive, he should have returned to the sender's premises so that the stowage could be checked. The court apportioned the loss one-third to the carrier and two-thirds

serious enough to deprive the carrier of the right to invoke limitation of liability by virtue of Article 29, as to which see further *infra*, para. 9.72 ff.

295. App. Colmar, 10.7.70 (1970) B.T. 358, (1971) U.L.C. 137, concerning a consignment of rolls of aluminium paper: the carrier must check the wedging and stowage when performed by the sender (see *infra*), and if apparently defective, he must request that it be done satisfactorily; if it is not, enter reservations regarding the consequence of the defective work (cf. App. Liège, *supra*, fn. 294, where the application of the principle of serious fault seems to have been derived from the general law. A separate duty to make a reservation is derived from Article 8. If the driver finds it difficult to make a reservation, he should contact his employer.) See also App. Paris, 24.5.91 (1991) B.T. 573 where the carrier was unable to rely on this defence where the sender had loaded the goods too tightly and thus prevented the circulation of air due to the fact that he failed to make a reservation. Lamy, para. 499 suggests that it is not necessary for such reservations to be seen or accepted by the sender, but they must appear on a consignment note signed by him. Merely making a reservation was not enough, however, for the court in App. Lyon, 21.10.76 (1976) B.T. 534 and (1977) B.T. 110, see Libouton, (1982) J.T. 702, para. 40.

296. See, e.g., App. Paris, 7.7.94 (1994) B.T. 736. It is sometimes accepted, however, that the failure of the carrier to deal with an apparent defect does not necessarily render him *solely* liable, see *infra*, para. 6.83, and in particular App. Paris, 18.5.89 (1989) B.T. 577. Cf. however, Krings p. 773.

297. Cf. defective packing, *supra*. See further *infra*. Where the issue is really one of proof, the carrier faces the difficulty of discharging the burden on him since he is hard put to explain why a reservation was not made, see App. Toulouse, 12.4.94 (1994) B.T. 714, where the carrier claimed that the damage was due to a fault in handling the goods at the sender's premises. If the carrier knew this, he should have made a reservation.

298. Hof. Gent, 19.1.99 (1999) 34 E.T.L. 264. Note that the omission of this reservation from a second consignment note produced for commercial reasons was immaterial.

299. Cass., 5.7.77 (1977) B.T. 402 where a vehicle overturned due to defective stowage. The sender was at fault in not drawing the carrier's attention to the high centre of gravity of the load and the precautions to be taken on bends. There were no lateral stays on the load and its exceptional defects were not apparent to the driver. However, the latter had been driving too fast in a storm when the accident occurred. The loss was therefore apportioned equally between the parties. See also App. Rouen, 5.5.67 (1967) B.T. 195, App. Paris, 24.6.76 (1976) B.T. 378. See also Cass., 30.11.82 (1983) B.T. 129.

to the sender.[300] Similarly, even where the initial loading and stowage has been performed by the sender, where the carrier continues the journey after having discharged part of the load, he must ensure that the remaining goods are properly stowed and can be delivered safely. The carrier will be responsible if the load becomes unstable and damage results.[301]

IS AN OBLIGATION IMPOSED ON THE CARRIER TO CHECK THE OPERATIONS
PERFORMED BY THE OTHER PARTY?

6.81 There may be little to differentiate between the case where the carrier's employee admits or is taken to have seen a defect and one where the defect is so obvious that he ought to have seen it in the normal course of events. However, such instances may fall short of a positive requirement to carry out checks with a view to discovering a defect. In a recent Belgian decision, the court accepted the potential liability of the carrier for failing to deal with the manifest defect in the loading, although the court accepted that the carrier is not obliged to check the relevant operation.[302] The more the circumstances involve the overall safety of the carriage, the more the carrier may be expected to deal with any difficulty and to check that the load can be carried safely and in accordance with applicable regulations. A decision of the Supreme Court of Denmark concerned a situation where steel pipes were loaded and stowed by the sender's employees. The carrier was nevertheless held liable for damage to those pipes which fell through the side of the trailer, since considering the character, weight and insufficient fastening of the goods as well as the characteristics of the trailer (i.e. its relatively weak sides), it should have been obvious to the driver that there was a substantial risk that the goods would slide, causing damage both to the goods themselves and potentially to other traffic.[303]

6.82 On the other side of the line are decisions which place a positive obligation on the carrier to check the operations performed by a third party. This view has been developed particularly in France. In an early French appellate decision, a machine was loaded and stowed unpacked by the sender, but no reservations were made by the carrier. Although the driver noticed en route that stowage was defective, no measures were taken to avoid further damage. In an action by the consignee against the carrier, the court held that where the stowage is performed by a third party the carrier is bound to ensure that it has been

---

300. App. Metz, 28.10.87 (1988) B.T. 168.

301. Cass., 10.6.86 (1986) B.T. 608. The commentary to this case notes the underlying duty of care of the carrier (*"les soins d'un bon père de famille"*) and notes a domestic decision where the carrier was liable for serious fault in failing to check the load after having been forced to apply the brakes violently: App. Dijon, 13.9.84 (1985) B.T. 372.

302. Comm. Bruxelles, 30.4.92 (1993) B.T. 782. Cf. App. Liège, 2.10.85 (1987) R.D.C.B. 56.

303. 2.3.79 (1980) 15 E.T.L. 208. Cf. RB. Antwerpen, 7.3.80 (1981) 16 E.T.L. 466, a carrier is guilty of professional negligence if he carries out a haulage operation involving heavy goods between Antwerp and Barcelona without having taken the slightest interest in the way the goods were stowed. Statements by the sender's employees to the effect that no further stowage operations were necessary does not, *ipso facto*, release the carrier from his obligations. See also Hof. Antwerpen, 13.1.88 (1988) 23 E.T.L. 717, Hof. Antwerpen, 13.12.89 (1990) 25 E.T.L. 319. In App. Bruxelles, 12.3.69 (1969) 4 E.T.L. 931, (1970) U.L.C. 113, the carrier was held liable where damage resulted, not from the operations of loading etc., but from failure to ensure, in the case of a load of exceptional height, that the load did not exceed its anticipated dimensions. Previous transports had been effected by the carrier without incident, but a change in the loading method had increased the height with the result that the load struck a bridge.

properly carried out, and, if faulty, to request that it should be rectified. Having failed to do so, the carrier was liable.[304] This reasoning has been confirmed at the highest level.[305] It may be possible to see this obligation as an extension of or an extrapolation from the duty to check the apparent condition of the goods and the packaging expressed in Article 8.[306] The French decisions are regarded in this light, the decisions, in effect, bringing CMR into line with the domestic law.[307] More recently there is a hint of a different basis. This sees not only the question of which party has the obligation to load as one for the application of national law,[308] but includes also the range of obligations pertaining to these operations as matters to be referred to this law and this has been applied to place an obligation to check on the carrier.[309] There are also Belgian decisions where the duty has been based directly on Article 8.[310] Article 8, however, is concerned with proof rather than with a duty which affects substantive liability. The thrust of these decisions is that the carrier is under a liability regardless of proof.[311] If the carrier is unable to check, he may be liable if he fails to make a reservation to this effect.[312] A sounder basis, which appears from a Belgian decision, is a view that the professional obligation of the carrier includes an obligation to check the load and to deal with manifest defects thereby discovered.[313] This expression of the duty is more readily made in respect of safety where the professional expertise of the carrier is more likely to outweigh that of the sender.[314] Thus, in a case before the Supreme Court of Austria, loading and wedging had been carried out by the sender. En route some of the goods fell off and were damaged. In an action by the consignee against the carrier, the court held that the driver must check stowage at the commencement of the journey with the professional skill that should be expected of an

---

304. App. Paris, 16.5.69 (1969) 4 E.T.L. 988. See also Comm. Corbeil-Essones, 18.4.69 (1969) 4 E.T.L. 988, App. Paris, 27.2.70 (1970) B.T. 111, App. Colmar, 10.7.70 (1970) B.T. 358, (1971) U.L.C. 137, App. Douai, 11.3.82 (1982) B.T. 199, 200. Not all decisions are consistent with this, e.g. App. Paris, 22.9.86 (1987) B.T. 139.

305. Cass., 3.5.76 (1976) B.T. 317, (1978) 13 E.T.L. 106. Note the observations on this case to the effect that the decision lays down two basic principles: (1) the carrier has an obligation to check stowage when performed by a third party; (2) the carrier is liable where he has agreed to undertake carriage despite apparent defects in stowage. The latter aspect is considered *supra*. Together, they are expressed as a duty on the carrier to control the loading operations of the sender, see, generally, Lamy, para. 498. See also Cass., 1.12.92 (1993) 28 E.T.L. 618. In the earlier decision, the defect in stowage affected safety (in this context see also Cass., 26.10.83 (1984) B.T. 360 in respect of overloading) but the duty on the carrier was expressed in general terms. The defect in stowage affected only the goods and not general safety in Cass., 3.2.82 (1982) B.T. 285, (1982) I U.L.R. 308. Where the defect is particularly serious, the failure to check may be viewed as *faute lourde* (*infra*, para. 9.80 ff) as in Cass. 28.3.2000 (2000) B.T. 276.

306. See *supra*, para. 4.24 ff.

307. See comment to the decision of App. Paris, 24.5.91 (1991) B.T. 574; see also App. Pau, 16.12.87 (1988) B.T. 653, App. Aix-en-Provence, 9.12.80 (1981) B.T. 143, App. Reims, 22.5.96 (1997) B.T. 447, cf. in Belgium App. Anvers, 26.1.83 (1987) R.D.C.B. 43.

308. See *supra*, para. 6.75.

309. App. Aix-en-Provence, 7.9.95 (1995) B.T. 833. In a similar vein is Comm. Paris, 12.5.98 (1998) B.T. 688 where the court derived a duty to check that the stowage complied with regulations from the contract.

310. App. Liège, 6.5.70 (1970) 5 E.T.L. 716, (1971) U.L.C. 133, RB. Antwerpen, 10.10.80 (1982) 17 E.T.L. 64. The latter case also involved defective packing.

311. Cf., however, App. Liège, 27.10.92 (1993) J.T. 538 (Putzeys, "Le droit des transports en Belgique (II)—1993–1995", (1996) U.L.R. 557, 565).

312. I.e. *"réserves de défiance"*, see Comm. Thonon-les-Bains, 2.6.94 (1994) B.T. 639, cf., however, App. Aix-en-Provence, 20.12.79 (1980) B.T. 104 (see Libouton, (1982) J.T. 702, para. 40).

313. RB. Leuven, 5.4.88 (1988) 23 E.T.L. 493.

314. Cf. RB. Antwerpen, 7.3.80 (1981) 16 E.T.L. 466, Hof. Antwerpen, 13.12.89 (1990) 25 E.T.L. 319. See also Libouton, (1982) J.T. 701, para. 36. Cf. the cases cited *supra*, para. 6.78 fn. 291.

employee in his position. If he fails to do so, the carrier will be liable for resultant damage to goods where loading and stowage have been performed by the sender.[315]

6.83 It does not necessarily follow from the above that the carrier is rendered solely liable where there is an apparent defect, since some decisions accept the possibility of apportioning liability in accordance with Article 17(5).[316] Furthermore, there are cases where it has been held that the burden of proving whether a defect could have been seen by the carrier had a check been carried out, or should have been seen during whatever check was made, is on the claimant.[317]

6.84 The contrasting view is that to benefit from the presumption of non-liability under Article 17(4)(c) the carrier is not required to check in any way whether the sender has properly carried out loading and stowage operations. The most extreme view suggests that even safety matters are irrelevant for this purpose, since public law requirements do not affect liability under CMR.[318] Some decisions retain a limited reference to ensuring basic safety.[319] In a recent Belgian decision,[320] the court decided that there was no requirement on the carrier to check loading and stowage. The carrier was only liable if stowage was manifestly inadequate and a danger to traffic. This may not be inconsistent with a duty to make safety checks since, insofar as the duty to carry safely is derived from a public law duty, this would also require that a check be made that the load is safe. In general, however, this interpretation would appear to free the carrier from the duty of observing normal precautions in checking the condition of his load. It would seem that the correct distinction between the obligations of the sender and the carrier lies in the point that, irrespective of whether the former loads and/or stows the goods, the latter has a basic obligation as a road carrier to ensure that his vehicle, together with its load, is in a serviceable and roadworthy state given the possible contingencies of the journey ahead. Many of the decisions which have appeared to turn upon whether the sender or the carrier is liable for stowage can in fact be reconciled in this way.

---

315. OGH., 25.9.68 (1973) 8 E.T.L. 309. The carrier was held liable, the court also taking account of the fact that the driver failed to show due care in taking such a badly stowed load along a dangerous road in wintry conditions.

316. Cass., 10.10.89 (1989) B.T. 673, App. Paris, 18.5.89 (1989) B.T. 577. See also Cass., 20.12.83 (1984) B.T. 430, cf. App. Paris 24.5.91 (1991) B.T. 573, App. Aix-en-Provence, 7.9.95 (1995) B.T. 833. Cf. further the decision of Comm. Paris, 26.6.81 (1982) B.T. 142 that the negligence by the sender in respect of the stowage was so great as to relieve the carrier regardless of whether he was in breach of an obligation to control the stowage operations. In an earlier decision it was held that responsibility must be apportioned between the carrier and the sender in proportion to their respective fault, App. Lyon, 21.10.76 (1977) B.T. 110. See also App. Paris, 31.3.77 (1977) B.T. 315 where the defective stowage was at least partially apparent at the time when the carrier took the goods in charge. The court apportioned the blame equally between the sender, forwarder and carrier. See too App. Reims, 13.7.77 (1977) B.T. 406, App. Gand, 20.11.85 (1987) R.D.C.B. 60.

317. App. Metz, 28.10.87 (1988) B.T. 168, App. Aix-en-Provence, 21.10.88 (1989) B.T. 436, Cass., 22.7.86 (1986) B.T. 516, Cass. 31.1.95 (1995) Bull. IV No. 33, p. 27 (Krings, p. 774). See also Hof. Arnhem, 27.11.73 (1974) 9 E.T.L. 748 where a shipment of knitting machines was loaded and stowed by the sender. The machines were damaged in transit through shifting due to bad stowage. In an action by the consignee against the carrier, as the former could not prove that the carrier could have discovered the faulty stowage when he checked the load, the latter was held not liable. This is supported by Libouton, (1982) J.T. 702 para. 40. On the other hand, in App. Besançon, 15.12.82 (1983) B.T. 96 it was stated that to benefit from Article 17(4)(c) the carrier must establish *inter alia* that the alleged defect was not apparent on taking over the goods. In that case, the stowage or lack of stowage of the goods was perfectly visible.

318. Loewe, para. 161. BGH., 28.3.85 (1986) 21 E.T.L. 174, 180, cf. OLG. Düsseldorf, 1.7.95 (1996) TranspR 109 (Jung, p. 159). Cf. further, BGH., 27.10.78 (1979) VersR 417, discussed by Clarke, p. 329, and OLG. Düsseldorf, 1.6.95 (1996) R.I.W. 158.

319. OLG. Hamm, 19.2.73 (1974) 9 E.T.L. 753.

320. RB. Antwerpen, 6.5.93 (1993) 28 E.T.L. 768.

WHAT PROOF IS REQUIRED OF THE CARRIER BEFORE HE CAN OBTAIN THE
BENEFIT OF THE PRESUMPTION IN ARTICLE 18(2)?

6.85 This issue is clearly linked with the nature of the proof required of the claimant to defeat the presumption, and the courts have had difficulty in separating these two issues.[321] In general, the case law is consistent with the view that there must be proof that the relevant operation was in fact performed by the other party. Thus in a German case[322] concerning damage to trays of peaches, the test was put thus:

"To 'establish' in the sense of Article 18(2) CMR does not simply mean the mere assertion of facts which would allow (the court) to apply one of the exemptions from liability listed in Article 17(4) CMR. The carrier must rather prove the existence of one of these exemptions."

6.86 In Belgium it has been held that the easing of the burden of proof did not free the carrier from proving the reality of the exonerating factor.[323] However, the carrier may be able to rely on inference rather than direct proof. Thus it has been proved by presumption that the driver did not load and unload 771 sacks weighing 20 tonnes[324]; that he could not have loaded 475 wooden planks[325]; or 28 crates weighing together a total of 15,810 kg[326]; or 1,343 unpalletised 10-kilo crates in a single day.[327]

6.87 Once it has been shown that an operation was carried out by the sender, statements sometimes made by the courts suggest that the carrier will not be liable unless negligence has been proved. In a Belgian case[328] the sender loaded a consignment of cases of aluminium sheets onto the vehicle. The carrier shored the load. Some cases fell off and were damaged. Although the driver was criminally liable for using the vehicle with an insecure load, it was held that this did not preclude the carrier from relying on the presumption, and that the carrier was not liable unless it could be shown that he was guilty of negligence. This view, which clearly favours the carrier, is close to the view expressed in some cases that the carrier need only establish a possibility that handling, loading and stowage by the sender or consignee is the cause of the damage.[329] In a French decision[330] it was held that it is only necessary for the carrier to establish that damage *could* be attributed to insufficient wedging of the goods by the sender, even though it could not be *proved* that the wedging was either totally or partially the cause of the damage. A similar test was applied in an appellate decision in Brussels,[331] where the sender loaded the shipment himself. Here the court pointed out that handling, loading and stowage are each

---

321. The evidence may be such that the court is able to conclude as to the cause without reference to the presumption, as in App. Versailles, 1.6.95 (1995) B.T. 833 where the sheer weight of the constituent parts of the load made it unlikely that the method of stowage was the cause of the accident. Cf. Cass., 20.1.98 (1998) B.T. 364.

322. OLG. Munich, 27.11.68 (1971) 6 E.T.L. 115.

323. Libouton, Theunis, p. 92, citing Comm. Tournai, 11.12.80 (1981–82) J.P.A. 381 and Comm. Verviers, 27.9.76 (1977) J.C.B. 524.

324. App. Antwerp, 9.3.76 (1975–76) J.P.A. 180.

325. App. Antwerp, 27.4.83 (1983–84) R.W. col. 1814.

326. Hof. Brussel, 19.12.68 (1969) 4 E.T.L. 953.

327. Comm. Bruxelles, 20.12.83 (1984) R.D.C. 707. See further App. Gand, 27.2.92 (1993) B.T. 666.

328. RB. Antwerpen, 30.4.65 (1966) 1 E.T.L. 314.

329. See Comm. Antwerp, 28.3.66 (1966) 1 E.T.L. 712, criticised by Libouton, Theunis, p. 92.

330. Comm. Namur, 22.7.65 (1966) 1 E.T.L. 133.

331. Hof. Brussel, 19.12.68 (1969) 4 E.T.L. 953—carriage of aluminium plates from Belgium to London loaded by the sender which fell off in transit in Antwerp. *Held*: that the carrier was not liable. Cf. RB. Amsterdam, 29.4.87 (1989) S. & S. 269, (1990) 18 E.L.D. No. 183, where pumps used by the sender *could* have been contaminated. See also RB. Rotterdam, 10.6.88 (1989) S. & S. No. 272, (1990) 18 E.L.D. 184.

separate operations which can individually permit the exoneration of the carrier where performed by the sender or the consignee.[332] The ability of the court to say that the damage is possibly the result of these operations depends on the circumstances of the case. In the French case noted above, the court was able to extract the facts in relation to the stowage in reaching its decision that this could have been the cause of the loss, and it also had the view of an expert that the collapse was due to bad stowage.[333] The case law generally indicates that the facts must reveal the possibility of a causal link which should not merely be a matter of conjecture or hypothesis,[334] but on the other hand does not require the carrier to prove affirmatively that the operations were carried out defectively. In a further Belgian case,[335] where machinery which had been loaded, stowed and roped by the sender was damaged when it shifted in transit, it was held that the carrier had simply to establish the possibility of a causal relationship between the activities of the sender and the damage which occurred. It was not necessary to prove fault on the part of the sender.[336] Similarly, in another Belgian case, although it could not be proved that the coupling of a trailer which had been loaded by the sender failed as a result of a 12% overload, it was nevertheless obvious that the break could be attributable to this over-loading.[337] In a Dutch decision, where a heavy machine was loaded and stowed by the sender and shipped from this country to the Netherlands, considerable damage resulted when it slipped off the blocks on which it had been fixed by the sender. In an action by the consignee, it was held that the carrier was not liable since he could prove that while the machine was in transit it had slipped off the blocks to which the sender had fixed it, and that this could have caused the damage.[338]

6.88 Within this general category of cases a range of expressions can be found indicating the level of proof required. Thus in a Belgian case it was held that the carrier had to *prove* a *possible* causal relationship between stowage and damage.[339] On the other hand, in a French decision, the carrier failed in his attempt to establish that the damage *could* have resulted from a defect of stowage since the load had been carried over steep hills for most of the journey without becoming displaced.[340] There are several Belgian

---

332. Cf. RB. Antwerpen, 7.3.80 (1981) 16 E.T.L. 466.

333. See also OLG. Hamm, 4.11.71 (1974) 9 E.T.L. 499.

334. RB. Antwerpen, 2.9.71 (1971) J.P.A. 454. In Comm. Antwerp, 6.9.74 (1975) 10 E.T.L. 253 where five cartons, insulated from each other in the lorry, were the only ones to get wet out of a total of 258, this was held to mean that it was unlikely that the damage had occurred during transportation, but that it was possible that it was due to loading by the sender in the rain, see Libouton, Theunis, p. 94.

335. RB. Antwerpen, 26.5.71 (1971) 6 E.T.L. 547.

336. See also in Holland HR., 18.5.79 (1979) 7 E.L.D. 518, (1981) 9 E.L.D. 317 (1980) N.J. No 574, Krings p. 771.

337. Comm. Liège, 3.10.69 (1970) J.P.A. 478, confirmed on this point by App. Liège, 23.2.72, unpublished, although the carrier was held liable since the driver expressed no reservations and the sender was unaware of the vehicle's payload, see Libouton, Theunis, p. 94 (see also Cass., 26.10.83 (1984) B.T. 360). Contrast the French decision of App. Aix-en-Provence, 9.4.92 (1993) B.T. 527 where the court was not prepared to accept the vague indications of the expert that defective fixing of the semi-trailer to the tractor was the cause of the loss in view of the fact that part of the transit had been effected without incident. Cf. further App. Amiens, 6.5.81 (1982) B.T. 271 (confirmed, Cass., 26.10.83 (1984) B.T. 360 on the basis that the carrier was liable since he should have ensured that the vehicle was not overloaded) where, despite evidence that the sender had overloaded the vehicle by 15%, the failure by the carrier to give any proof of the circumstances surrounding the overturning of his vehicle meant that he had not established that the overload could have been a possible cause of the accident.

338. RB. Dordrecht, 22.12.71 (1972) S. & S. No. 50.

339. App. Bruxelles, 28.6.69 (1969) 4 E.T.L. 925, (1970) U.C.L. 115.

340. App. Besançon, 15.12.82 (1983) B.T. 96.

decisions which suggest a requirement of proof of a causal probability, but these have been subject to some criticism.[341]

6.89 Even more onerous for the carrier is an interpretation of Article 17(4)(c) which requires the carrier to actually prove that the loading or stowage was faulty, leaving the presumption to provide a benefit for the carrier only in respect of causation. This approach was taken in a German decision where a shipment of cartons of wallpaper was carried from Germany to Milan, which had been loaded by the sender in the presence of the carrier's driver. On arrival, some of the goods were found to have been damaged by damp due to the cartons pressing against the wet sides of the vehicle owing to the speed at which the vehicle had been driven. The court decided that the carrier could only avoid liability to the extent to which he could prove that the loading by the sender was faulty.[342] The court expressly dealt with the burden of proof as follows:—

"Within the scope of Article 18(2), it is for the carrier to prove faulty loading of the goods by the sender. Article 18(2) does, admittedly, refer only to an evidential burden of proof, but this is not satisfied by the carrier merely demonstrating the theoretical possibility of the damage being caused by faulty loading. He must rather establish the evidence proving faults in the loading. If this evidence is brought, it will be assumed that the damage was caused by faulty loading if the carriage was otherwise completely usual and without incident."

6.90 The same approach has been adopted in a decision of the French Cour de Cassation,[343] which is consistent with earlier French decisions.[344] In this case, the court was of the view that the carrier must clearly demonstrate that in the circumstances of the particular case the damage in question resulted from either faulty handling, loading or stowage. Consequently, where a consignment of lumps of granite from France to Germany was loaded and stowed by the sender, the carrier was held liable to the sender when the vehicle failed to hold the road well due to overloading, causing part of the load to fall off, as he could not prove that the faulty loading was the cause of the loss. More recently a French appellate decision puts the position in terms that the carrier must adduce precise facts sufficient to make it appear that the loading etc. was defective.[345]

WHAT KIND OF PROOF IS REQUIRED TO DEFEAT THE PRESUMPTION?

6.91 Article 18(2) permits the claimant to show that the loss was not attributable to the risk. It can be difficult, however, to separate the issue of what proof is sufficient to

---

341. Libouton, Theunis, p. 92, citing Civ. Charleroi, 1.10.68 (1969) J.P.A. 129, (1969) U.L.C. 327: "a distinct possibility of a causal link"; App. Liège, 20.1.71 (1971) 6 E.T.L. 541 (see *infra*, fn. 347): "probability of a causal link"; Comm. Antwerp, 16.1.74 (1975) J.C.B. 466; Comm. Bruxelles, 9.2.84 (1985) R.D.C. 303; Comm. Antwerp, 27.9.71 (1971) J.P.A. 169: "genuine causal link".

342. OLG. Düsseldorf, 13.1.73 (1973) 8 E.T.L. 620. See also BGH., 28.3.85 (1986) 21 E.T.L. 174, *infra*, fn. 350.

343. Cass., 9.10.74 (1974) B.T. 491.

344. App. Paris, 8.6.67 (cited by Wijffels, (1976) 11 E.T.L. 209, 219) where a shipment of electrical equipment was carried from Paris to Belgium, stowed by the sender and damaged in transit. The surveyor's report stated that there had probably been a collision between the upper part of the vehicle and some unidentified object. *Held*: that as the carrier could not prove that the damage resulted from faulty stowage by the sender he was liable. See also Strasbourg, 9.11.66 (1967) B.T. 39, which took the same approach.

345. *"Des éléments matériels précis"*: App. Besançon, 15.12.82 (1983) B.T. 96, see Lamy, para. 1551, Jur. 2. Cf., however, Cass., 31.1.95 (1995) B.T. 106, (1995) 30 E.T.L. 688.

establish the presumption from the issue of what proof from the claimant will defeat it.[346] On one interpretation, the carrier is required to prove a probable causal connection, so that if another possible cause of damage can be shown to exist by the claimant, it follows that the carrier will be unable to prove the existence of any such probable causal connection and will therefore not avoid liability under Article 17(4)(c). This interpretation was utilised in an appellate decision in Belgium where a roll of sheet metal was loaded and stowed by the sender on the carrier's vehicle. The roll fell off while the vehicle was taking a bend. In an action by the sender against the carrier, the latter was held liable on the grounds that the damage could have been due to another cause. As there was no proof as to the state of the vehicle after the accident, it was objectively possible that the damage could have been caused by the collapse of the floor of the lorry.[347] It should be remembered, however, that one of the objects of the presumption is to protect the carrier in precisely those circumstances where it is difficult to reconstruct the facts,[348] so that as long as a plausible risk can be shown to exist from the operations of the sender, the burden should pass to the claimant. It is noteworthy that in this case there was a clear clash between the experts on each side as to whether there had, in fact, been defective wedging of the goods, thus casting doubt on the existence of any risk from the operations performed by the sender.

6.92 In other decisions, the point has been made that the existence of other possible explanations of the damage does not defeat the presumption in favour of the carrier.[349] In a case where there was proof that the stowage by the sender followed the normal process conforming to standard practice, the French Cour de Cassation upheld the view of the lower court that the carrier had failed to establish the proof necessary to show that the damage could have resulted from a defect in the stowage.[350] On the other hand, it could be said that the claimant had thereby proved his case under the second sentence of Article 18(2). Thus the court in Amsterdam, having in earlier proceedings given the carrier the benefit of the presumption and so placed the burden on the claimant, upheld the claim in later proceedings where the claimant was able to show through witnesses to the loading that the heavier packages were loaded on the floor of the lorry and that there were no empty spaces between them.[351]

---

346. Cf. the cases cited *supra*, para. 6.90 fn. 344.

347. App. Liège, 20.1.71 (1971) 6 E.T.L. 541, (1971) U.L.C. 281. See also App. Amiens, 6.5.81, *supra,* fn. 337, where even though there was proof of overloading by the sender, the lack of proof from the carrier as to the circumstances in which the vehicle overturned meant that the carrier was unable to prove that the overloading was a possible cause.

348. See App. Bruxelles, 12.3.69 (1969) 4 E.T.L. 931, (1970) U.L.C. 113.

349. Hof. Brussel, 19.12.68 (1969) 4 E.T.L. 953, App. Paris, 27.1.70 (1970) B.T. 100, App. Paris, 23.12.75 (1976) B.T. 48.

350. Cass., 17.6.69 (1970) 5 E.T.L. 57. A natural approach where the view is that the burden is on the carrier to prove that there was a defect in the loading. The Court of Appeal in Paris upheld the complaint of the claimant that the lower court had rejected his claim for failure to prove that the stowage was satisfactory, even though the view of the expert was that bad stowage was not the cause. As the claimant argued, the burden is on the carrier to prove the possibility of an exonerating cause; there is no presumption of bad stowage: App. Paris, 25.1.82 (1982) B.T. 123. Similarly, the German Federal Supreme Court in BGH., 28.3.85 (1986) 21 E.T.L. 174 quashed the decision of the lower court which had held that the carrier had proved inadequate securing of the load because the court had paid insufficient attention to the report of the expert which contained facts which cast doubt on the conclusion he had reached.

351. Arrond. Amsterdam, 4.9.73 (1974) S. & S. No.16, Arrond. Amsterdam, 13.3.74 (1974) S. & S. No.77, cited by Libouton, (1982) J.T. 72, para. 87 and criticised by him for granting too easily the benefit of the presumption to the carrier.

6.93 While it will be natural for the claimant, in his efforts to successfully invoke the second sentence of Article 18(2), to attempt to prove negligence on the part of the carrier, it is not strictly necessary for him to do so. In a decision of the German Federal Supreme Court, the court held that if the claimant proved concrete circumstances associated with the transport of the goods and which, according to experience, have a tendency to have a damaging effect, the burden returns to the carrier to establish his defence under Article 17(1).[352] In this case, the driver had driven onto a soft verge in order to avoid an oncoming vehicle. When he was steering it back on to the carriageway, the trailer fell over. Whether or not there was sufficient proof of defective loading by the sender, this fact was sufficient to return the case to the lower court with a burden placed on the carrier to prove that the accident was unavoidable. The court also left it to the lower court to weigh the effect of defective loading, if this could be proved, for the purposes of apportionment under Article 17(5).

6.94 The possibility of apportionment will arise whenever there has clearly been blame on both sides in respect of the performance of various inter-related operations. For example, in a Dutch decision a consignment of vegetables was loaded and stowed by the sender for carriage from Germany to The Netherlands. Owing to the absence of partitions between the crates on the vehicle the load slipped and was damaged. In an action by the sender against the carrier, the court held that even where loading and stowage is carried out by the sender, the carrier is still bound to stabilise the load by using partitions to prevent it from slipping. These are standard equipment on a vehicle, the use of which is part of the normal duty of care required to render the vehicle serviceable. As faulty stowage and the absence of partitions contributed equally to the damage, loss was apportioned equally between the parties.[353] It is, of course, possible to apportion loss under any of the other approaches. So in another Dutch decision a large paperfolding machine was loaded by the sender. The load was affixed to wooden beams on the floor of the trailer, off which it slipped in transit. In an action by the sender against the carrier for damage to the goods, the court held that where the sender has loaded goods the carrier will avoid liability under Article 17(4)(c) even though the driver has not checked the load. However, as the damage was also partially due to the fact that the driver failed to adapt his speed to traffic conditions, the loss was to be apportioned equally between the parties.[354]

(d) *"the nature of certain kinds of goods which particularly exposes them to total or partial loss or to damage"*

6.95 The carrier will be relieved of liability for loss or damage where the nature of certain kinds of goods particularly exposes them to total or partial loss or damage subject to the provisions of Article 18(2) and 18(4). The provision then specifies by way of elaboration certain ways in which the loss or damage may occur—*"especially* through breakage, rust, decay, desiccation, leakage, normal wastage, or the action of moth or

---

352. BGH., 28.5.85 (1986) 21 E.T.L. 174, 181–182.
353. Arrond. Roermond, 2.1.69 (1969) 4 E.T.L. 1005.
354. Hof. 's Hertogenbosch, 21.12.65 (1966) 1 E.T.L. 684. See also the cases *supra*, fns. 295, 299. See also OGH., 21.3.77 (1978) II U.L.R. 292, cf. BGH., 20.10.83 (1985) 20 E.T.L. 160.

vermin". As the list is preceded by the word "especially" it is not intended to be exhaustive,[355] but merely to indicate the type of causes which may arise.[356]

6.96 However, the reference is not to general risks inherent in the goods but to a particular and special risk. In *W. Donald & Sons (Wholesale Meat Contractors) Ltd.* v. *Continental Freeze Ltd.*[357] fresh meat which had been carried in a refrigerated vehicle arrived frozen, which produced a loss in value of the meat. The carrier's appeal to Article 17(4)(d) was rejected, since the special risk in the carriage of fresh meat is the decomposition produced by decay and the loss or damage must *arise* from it.[358] What had happened could not be described as decay, which involved a natural process of deterioration, but the preservation of the meat, albeit as meat of lesser value. Although the meat could be described as having deteriorated, this was not the same as decay which was the special risk.[359] On the other hand, the particular sensitivity of some goods to damage by frost can be a special risk.[360]

6.97 This defence is first subject to the requirements of Article 18(2), that is, that when the carrier establishes that in the circumstances of the case the loss or damage could be attributed to one of the special risks under Article 17(4), it shall be presumed that it was so caused. However, it is also subject to Article 18(4), which provides as follows:—

> "4. If the carriage is performed in vehicles specially equipped to protect the goods from the effects of heat, cold, variations in temperature or the humidity of the air, the carrier shall not be entitled to claim the benefit of article 17, paragraph 4(d), unless he proves that all steps incumbent on him in the circumstances with respect to the choice, maintenance and use of such equipment were taken and that he complied with any special instructions issued to him."

6.98 The protection available to the carrier is therefore subject to two distinct burdens of proof dependent upon whether the goods are to be carried in normal vehicles, when Article 18(2) alone will be applicable, or in vehicles specially equipped to control the temperature, etc., of goods, when Article 18(4) will also be applicable.

6.99 The first question that must be asked therefore under Article 17(4)(d) is whether the goods are of such a nature as to be specially exposed to loss or damage. Goods falling in this category are of two distinct types, that is, those that are of a perishable nature and those that are not. Clearly a wide range of foodstuffs will fall in the first category, such

355. See App. Lyon, 7.10.76 (1979) I U.L.R. 289, where the court accepted that the list was not limitative but required more evidence that the suggested risk of fire from leaking gas lighters was a plausible explanation of the loss. The risks listed do not represent a genus, per Lord Murray, in *W. Donald & Son (Wholesale Meat Contractors) Ltd.* v. *Continental Freeze Ltd.*, *infra*, fn. 357.

356. It should be noted that whereas the English version uses the word "decay" the French text instead uses "*détérioration interne et spontanée*". This difference in terminology could create problems—see Rodière, para. 86.

357. [1984] S.L.T. 182.

358. See, however, Clarke p. 334.

359. Decay being a form of deterioration: "Had the concept of sub-para. (d) been to relieve the carrier of liability where loss arose through the inherent nature of certain kinds of goods to deteriorate, then there would have been no need to list the specific risks narrated. For all of them, and a great deal more besides, would have been covered by deterioration" per Lord Murray at p. 183.

360. App. Toulouse, 26.3.69 (1971) 6 E.T.L. 131, Cass., 19.4.82 (1982) B.T. 309. This is clearly indicated by Article 18(4). Some goods can perhaps be viewed as especially sensitive to both heat and cold. A malfunction of special equipment which enhances the risk merely subjects the goods to it in a way which might not have occurred if the equipment had functioned properly. The goods are still subject to the risk, but the dominant cause of the damage is the malfunction which gives rise to other issues such as the correct application of Article 18(4), as to which see *infra*, para. 6.101. Cf., further Hof. The Hague, 9.1.87 (1987) S. & S. No. 141 (see Clarke, p. 335 n. 60), App. Aix-en-Provence, 26.4.77, unpublished (see Lamy, para. 1555, Jur. 4 and Chao, Theunis, p. 121).

as grainstuffs, vegetables, plants, etc.[361] The second category will comprise all those non-perishable goods such as ferrous and non-ferrous metals, clothing and other man-made products. In both categories the carrier will be able to benefit from the presumption under Article 18(2). To benefit from the presumption the carrier must, however, show that the nature of the goods gives rise to a risk *in the circumstances of the case*.[362] If, however, either perishable or non-perishable goods require to be carried in vehicles specially equipped to protect the goods from the effects of heat, cold, variations in temperature, or the humidity of the air, then the carrier will not obtain the benefit of the presumption in Article 18(2) unless he can also prove as required under Article 18(4) that he has taken all steps incumbent on him regarding choice, maintenance, and use of equipment, and that he has complied with any special instructions given to him.

6.100 There will not usually be any real question as to whether perishable goods require special equipment. The practice of the industry is usually uniform. There is a growing body of case law to illustrate the problems arising, primarily under Article 18(4), in this connection. Thus it is for the carrier to ensure that the refrigeration equipment is actually achieving the required temperature,[363] and he cannot rely on the readings on an external temperature gauge, or a thermostat setting. It may in practice be impossible to check the interior of the vehicle because of Customs seals, but the carrier is at risk.[364] Where no criticism is made of an external gauge the courts, however, may accept the carrier's proof that the internal temperature was checked and maintained.[365] In the absence of a recording device, however, the carrier will need alternative evidence to show that the temperature

---

361. E.g.: new potatoes, App. Agen, 26.11.85 (1986) B.T. 235; nectarines, Cass., 19.4.82 (1982) B.T. 309. See also Hof. Brussel, 17.12.84 (1985) 20 E.T.L. 354, (1986) II U.L.R. 618 where the carrier was not liable for damage to artichokes and green beans caused by their sensitivity to the heat given off by celery next to which they had been loaded by the sender. In OLG. Hamm, 6.2.97 (1998) TranspR 348, Article 17(4)(d) would, if necessary provide a defence in respect of the deterioration of milk in cartons caused by a delay due to the unjustified refusal of the consignee to take delivery.

362. Proof that a load of cauliflowers remained on the lorry for several days during the month of August, RB. Roermond, 24.10.68 (1969) 4 E.T.L. 1012; carriage of chocolate biscuits in summer in a sheeted but unrefrigerated vehicle, Cass., 4.2.86 (1986) 21 E.T.L. 263, (1986) B.T. 197. In this case, the court held that the carrier does not have to show that this nature is the sole or indisputable cause of the damage. The fact that other causes, such as failure to carry the goods as agreed, may have been contributory does not prevent the carrier from relying on the presumption; rather it is for the claimant to make the case that the damage was not attributable wholly or partly to the nature of the goods (see further *infra*, para. 6.115 ff). The nature of the goods must, however, provide a plausible explanation of the loss, see App. Lyon, 7.10.76, *supra*, fn. 355.

363. Clearly the carrier must follow the instructions as to temperature and supply equipment capable of meeting those instructions (e.g. OLG. Hamburg, 27.10.88 (1989) TranspR 318, where the carrier's refrigeration equipment was not adequate for the size of shipment). In one case the carrier was liable when fruit was damaged because it was carried below the temperature specified by the sender who had also required the temperature to be recorded and the carrier did not have any recording device: App. Toulouse, 14.1.81 (1981) B.T. 158, see Chao, p. 122 (liability was shared since the goods had also been loaded defectively by the sender). Where the instructions include a time for delivery, the carrier who fails to comply cannot complain that unloading of the goods has thereby been delayed and exacerbated the risk presented by their nature: App. Paris, 4.3.85 (1985) B.T. 396.

364. Arrond. Rotterdam, 27.4.71 (1971) S. & S. No. 73, (1971) 6 E.T.L. 830, where it was held that the fact that the temperature registered on the outside of the vehicle indicated only the temperature of the air drawn in by the system and not the temperature within the system was a fault for which the carrier was liable. Arrond. Utrecht, 26.5.76 (1978) S. & S. No. 9. So too in OLG. Koblenz, 2.7.76 (1979) 7 E.L.D. 41 the carrier was liable where he had used a van which made it impossible to check from the outside whether the ventilators inside were working properly. In Comm. Paris, 22.9.80 (1980) B.T. 615, (1981) II U.L.R. 228 the failure to equip the vehicle with a thermometer precluded the carrier from proving that the required temperature was maintained during the carriage.

365. See Clarke, p. 343 n. 20 and cases there cited, in particular Hof. The Hague, 21.5.87 (1988) S. & S. No. 24.

was continuously maintained, such as regular checks of the thermometer and recording the results.[366] Such proof may be derived from the fact that the major part of the contents of a refrigerated container were, in fact, delivered in good condition.[367] Apart from the matter of proof of the continuous temperature maintenance, the obligation to make regular checks forms part of the carrier's duty. Thus it is not sufficient for the carrier simply to switch on the equipment, he must also check and control the temperature throughout the journey, either by checking the goods, or by manipulating the air vents.[368] Checking three times in 48 hours is not adequate.[369] Nor is parking the vehicle for two days without checking.[370] It is also for the carrier to ensure that the stowage of the goods allows for efficient air circulation[371] and that the goods are not removed from the refrigerated environment.[372] Even where the loading and stowage is performed by the sender, the carrier must check to ensure that this will not affect the operation of his equipment. Thus the carrier was liable where two of the vehicle's refrigeration units had ceased to function because during the journey the pallets had moved, breaking one ventilator and crushing another. This damage would have been avoided if the goods had been loaded in a manner which would have protected the cooling equipment when the vehicle braked.[373] This may be extended to ensuring that goods are not packed so tightly as to inhibit the circulation of air.[374] Further, it may be that, in the particular circumstances of the case, the cargo interests are entitled to expect that the refrigeration equipment is adequate, not simply to maintain an ambient temperature of the trailer, but also in consequence to cool the goods as necessary if they are loaded at a temperature greater than that required for the transport,[375] even where this requires an assessment to be made of the initial temperature

366. In OLG. Schleswig-Holstein, as reported by Chao, Theunis, p. 141.
367. Hof. Brussel, 17.12.84 (1985) 20 E.T.L. 354, (1986) II U.L.R. 618. Where condensation appeared inside rather than outside the packaging, this fact, among others, proved the cause of the damage to be the initial temperature of the goods rather than prolonged carriage at the wrong temperature: App. Orléans, 29.1.98 (1999) B.T. 383.
368. OLG. Hamburg, 22.7.82 (1983) VersR 63, Thume, p. 5.
369. OLG. Hamm, 18.10.84 (1985) TranspR 107. Failure by the driver to make any temperature check in a trip lasting 20 hours was held to be wilful misconduct within Article 29 (see *infra*, para. 9.72 ff.) in OLG. Munich, 16.1.91 (1992) TranspR 181. Consequently, the carrier was unable to rely on Article 17(5), even he if could prove that the fault of the sender contributed to the damage.
370. OLG. Hamburg, 2.5.85 (1985) TranspR 398, see Thume, *supra*, para. 6.57 fn. 221 p. 5. See also, App. Montpellier, 11.4.2000 (2000) B.T. 350.
371. *Schmith* v. *Den Jydske Fragmandshal* (1969) UfR 582, (1971) U.L.C. 301. In GH. The Hague, 15.6.79 (1980) 15 E.T.L. 871 the carrier was liable for damage caused to flowers because they were packed too tightly so that the refrigerated air did not adequately reach the flowers. The carrier's argument that the vehicle was too small for the load presented was rejected since he should have pointed this out to the sender (see Chao, Theunis, at p. 132).
372. *Erik* v. *Borge Langhoff & Skandinavisk Fjer Transport A/S* (1971) UfR 183, (1971) U.L.C. 303 and cf. *Schmith's* case *supra*.
373. OLG. Munich, 27.6.79 (1980) TranspR 95; see also App. Montpellier, 19.1.84, unpublished, see Chao, Theunis at p. 118, cf. App. Bordeaux, 11.7.86, unpublished, see Chao, Theunis, at p. 111, cf. App. Bordeaux, 11.7.86, unpublished, see Lamy, para. 1555, Jur. 4.
374. See Thume, *supra*, para. 6.57 fn. 221 pp. 5–6, cf. in France: Cass., 3.5.76 and 14.6.76 (1976) B.T. 317 and 342, App. Toulouse, 14.1.81 (1981) B.T. 158, App. Agen, 29.6.81 (1981) B.T. 433, see Chao, loc. cit. Cf. Comm. Anvers, 6.5.93 (1993) 28 E.T.L. 768 (1994) 624 and cf., further, where incompatible goods were stowed too closely together by the sender, Hof. Brussel, 17.12.84 (1985) 20 E.T.L. 354, (1986) II U.L.R. 618. Some qualification to this may be possible where the court considers that the carrier's duty extends only to what is apparent, App. Aix-en-Provence, 20.12.79 (1980) B.T. 104. This case, however, concerned the issue of whether there was serious fault for the purposes of Article 29, see Chao, Theunis, p. 119, cf. App. Aix-en-Provence, 16.6.83 (1984) B.T. 600.
375. *Ulster-Swift Ltd.* v. *Taunton Meat Haulage Ltd.* [1977] 1 Lloyd's Rep. 346 (See also, App. Montpellier, 21.3.2000 (2000) B.T. 350). There was, however, some expert evidence to the contrary. It is not necessary for

and state of the goods.[376] Furthermore, if the carrier follows an instruction which he knows or should realise will damage the goods, he is at fault in accepting those instructions and will be liable if he nevertheless carries them out.[377] However, if the sender wrongly specifies a vehicle without special equipment he is *prima facie* liable for resultant loss.[378] On the other hand, the courts in France have held the carrier to be liable where he should have realised that his refrigerated vehicle could not conserve the goods for the duration of the proposed transit,[379] and even where special equipment has not been requested the sender may be able to hold the carrier liable for not making use of such equipment as he might be expected to have.[380]

6.101 If ostensibly frozen goods are found to have unfrozen, there is an onus on the carrier to show that there is no fault in or defect of maintenance or use of the refrigeration equipment.[381] The courts in general appear to take a strict view of the carrier's obligations, so that there is likely to be liability whenever a defect in the equipment results in a failure to maintain the required temperature.[382] It has been suggested that this approach fails to

---

the sender to check whether the vehicle has been pre-refrigerated, this is a matter for the carrier in providing a suitable vehicle: OLG. Munich, 16.1.91 (1992) TranspR 181. Where, however, the sender loaded meat at a temperature in excess of regulations based on an EC Directive, liability was shared one-third to the sender and two-thirds to the carrier whose defective equipment was the chief cause of the damage: Supreme Court of Denmark, 16.10.86 (1987) II U.L.R. 734.

376. *Vvinners Dairy and Cheese Export* v. *Padborg Transit A/S* (1969) UfR 900, (1971) U.L.C. 301. Where the carrier fails to check the temperature of the goods, he will be presumed to have received them in apparent good condition: see Chao, Theunis, p. 117 and App. Nancy, 20.2.81 (1981) B.T. 330 (where the court noted that the carrier was a specialist in refrigerated carriage), App. Riom, 16.10.81 (1982) B.T. 11, App. Paris, 30.5.73 (1973) B.T. 304. See also, App. Montpellier, 11.4.2000 (2000) B.T. 350, cf., however, App. Paris, 14.2.91 (1991) B.T. 289 (see Lamy, para. 463 and 1520, Jur. 2), Trib. gde. inst. Sarreguemines, 3.2.98 (1998) B.T. 801 and App. Orléans, 29.1.98 (1999) B.T. 383: the frozen goods being in apparent good condition, there was no need for the carrier to check further. As already noted from the decision of OLG. Karlsruhe, 18.10.67 (see *supra*, para. 4.27 fn. 79) the carrier may be expected to use a thermometer to check the temperature of the goods. See, however, OLG Schleswig-Holstein, 9th civil division, 21.9.79, discussed by Chao, Theunis, pp. 134–135, where the court suggested that whilst the absence of reservations provides a presumption which weakens the position of the carrier, the court should not be too strict when it comes to providing evidence to the contrary since a carrier cannot be expected to have the same specialist knowledge as the sender when it comes to the preliminary treatment and the preparation of the goods for carriage. Similarly, the court in OLG. Hamm, 26.6.97 (1998) TranspR 301, thought that a "professional" check was not required of the carrier. (See further, *infra*, fn. 394).

377. Cass., 19.4.82 (1982) B.T. 309. As in this case, the carrier may expect to share the liability with the sender.

378. App. Aix-en-Provence, 10.11.76 (1977) B.T. 248 (cf. ZG. Basel-Stadt, 19.4.91 (1992) TranspR 408), App. Versailles, 29.2.84 (1984) B.T. 249, and see further App. Venezia, 31.10.74 (1975) 10 E.T.L. 242. Similarly, where the sender fails to give appropriate instructions, e.g. as to the proper method of stowage or the need to pre-cool the refrigerated container, Hof. Brussel, 17.12.84 (1985) 20 E.T.L. 354, (1986) II U.L.R. 618, cf. OLG. Munich, 16.1.91 (1992) TranspR 181, *supra*, fn. 375.

379. Cass., 15.2.82 (1982) B.T. 182 (see further Brunat, (1982) B.T. 174).

380. App. Nimes, 18.5.88 (1988) B.T. 472 where wine was damaged by severe frost. The sender could not be criticised for failing to protest at the vehicle supplied since he was not to know whether the lorry carried protective sheets or whether the carrier employed some system of protection. The court pointed out that if the carrier is not given an instruction as to the type of vehicle, he must either provide one which protects against cold or refuse to carry. In App. Paris, 30.9.87 (1988) B.T. 59 a carrier was liable to failing to protect a consignment of fresh eggs carried in an ordinary vehicle from the effects of severe cold. In neither of these cases was Article 17(4)(d) in issue; in the latter case, it was pointed out that the carrier had not sought to rely on it. The comment to the case suggests that a shared responsibility would have been more appropriate. Cf., further, App. Grenoble 28.4.99 (2000) B.T. 95.

381. OLG. Nürnberg, 14.6.65 (1971) 6 E.T.L. 247 (cf. OLG. Hamburg, 27.10.88 (1989) TranspR 318). App. Toulouse, 17.2.71 (1971) 6 E.T.L. 412.

382. Particularly in France: App. Paris, 4.3.85 (1985) B.T. 396, App. Riom, 16.10.81 (1982) B.T. 11, see Chao, Theunis, p. 127, cf. App. Paris, 6.11.79 (1979) B.T. 574 and Chao, Theunis, p. 128. There is a connection

take sufficient account of the distinction which must be drawn between Article 17(3) and Article 18(4).[383] On the other hand, it may be that Article 18(4) is concerned with proof rather than being intended to reflect a different standard from the strict liability laid down by Article 17(3).[384] In Belgium a court has held that the refrigeration unit is an essential part of a refrigerated vehicle, to the extent that a defect of the unit must be regarded as a defect of the vehicle itself. Consequently, the carrier could not invoke such a defect as a defence, even if it is described as unforeseeable by an expert.[385] Clearly, where a breakdown due to a defect in the vehicle causes a loss then this will fall within Article 17(3).[386]

6.102 There is greater difficulty where the goods are non-perishable in establishing that they are exposed by their nature to loss or damage. Thus it has been held that this provision does not cover risks arising from the means of packing,[387] or the risk of corrosion to ferrous metal.[388] However, when once it has been established that goods are by their nature particularly exposed to loss or damage, it would appear that the loosely phrased provisions of Article 18(4) regarding special equipment are capable of being applied to almost any category of goods. Even a vehicle sheeted for the carriage of galvanised wire has been held to be specially equipped for the purposes of Article 18(4).[389]

6.103 Finally, a distinction must be drawn between the question of inherent vice in goods and that of their actual nature. This problem has already been discussed in relation to Article 17(2). Basically, if a problem arises in relation to all goods of the same type this will not constitute inherent vice. For inherent vice, it will be necessary for the goods in

---

between this and the duty of the carrier to be vigilant in maintaining a check of the temperature, see *supra*, para. 6.100 Consequently, this may also underline the strictness that is generally taken in respect of the carrier's recording equipment, cf. Chao, Theunis, p. 128. Some Dutch decisions have been more favourable to the carrier: RB. Arnhem, 11.1.79 (1980) S. & S. No. 45, confirmed in Hof. Arnhem, 10.2.81 (1981) S. & S. No. 107, RB. Den Bosch, 17.10.75 (1977) N.J. 26 (Haak, p. 169 n. 255).

383. See Clarke, p. 292, and see also Chao, Theunis, p. 129 and Haak, p. 170. See further Helm, *Frachtrecht*, Art. 17 CMR, Anm. 18 & *GrossKommentar*, 425 HGB Anh. III Art. 17 CMR Rdn. 6 aE & Rdn. 18, Herber/Piper, Art. 17 Rdn. 81, Art. 18 Rdn. 26, Thume, Art. 17 Rdn. 119, and OLG. Hamburg, 27.10.88 (1989) VersR 719.

384. In *Ulster-Swift*, at first instance, Donaldson, J., [1975] 2 Lloyd's Rep. 502, 507, in effect required the carrier to prove the exact cause of the loss in order to show that he "took all steps incumbent . . . ". This is suggested as reflecting a strict liability, see Clarke, p. 339, cf. the *Centrocoop* case, *supra*, para. 6.27. An alternative possibility is that the carrier can be allowed the benefit of the presumption where no criticism can be made of his proof under Article 18(4), but that the claimant then has the burden of proving, under Article 18(2), that there was a breakdown for which the carrier is responsible, see Haak, p. 173.

385. RB. Brussel, 27.2.87 (1987) 22 E.T.L. 502.

386. Arrond. Amsterdam, 27.9.78 (1979) S. & S. No. 89.

387. Hof. 's Hertogenbosch, 21.12.65 (1966) 1 E.T.L. 698 (use of crate rather than packing case for machinery).

388. Comm. Bruxelles, 4.2.72 (1972) 7 E.T.L. 573, (1972) J.C.B. 648. See also *Tetroc Ltd.* v. *Cross-Con (International) Ltd.* [1981] 1 Lloyd's Rep. 192. Cf., however, OLG. Hamm, 2.11.95, *infra*, para. 6.103 fn. 395.

389. App. Paris, 10.12.71 (1972) B.T. 35, (1973) I U.L.C. 274. *Sed quaere* whether the oxidisation of galvanised wire can be considered as damage arising from the nature of the goods, even where the sender has given special instructions for carriage to be performed by a specially equipped vehicle, viz. a sheeted vehicle, and the carrier uses a cover which is defective. This is a novel extension of the application of Article 18(4) inasmuch as a simple sheeted vehicle is treated as a vehicle "specially equipped to protect goods from the effects of . . . humidity of the air". See, however, *Walek & Co.* v. *Chapman and Ball (International) Ltd.* [1980] 2 Lloyd's Rep. 279 and App. Liège, 8.3.96 (1996) B.T. 411. See further App. Montpellier, 28.2.85 (1985) B.T. 600, where a "cardboard-insulated" vehicle does not appear to have been specially equipped, although the carriage involved perishable goods (see Chao, Theunis, p. 115).

question to have some exceptional defect not normally found in goods of the same type when subject to carriage. Otherwise any defect will merely arise out of the nature of the goods. The distinction is of importance as regards the burden of proof placed upon the carrier. If the carrier can categorise the goods as being particularly exposed to damage through their own nature, under Article 17(4)(d), he will benefit from the presumption of non-liability under Article 18(2).[390] On the other hand, if loss or damage results from the inherent vice of the goods, as laid down in Article 17(2), then the carrier must himself prove that the loss or damage was due to inherent vice.[391] The distinction has appeared to be difficult to apply where special equipment is employed, particularly where it is alleged that the goods have not been properly prepared for carriage, for example as where refrigerated goods are inadequately pre-cooled.[392] The lack of adequate pre-cooling may be regarded as an inherent vice[393] with the apparent result that the burden of proof falls on the carrier to prove its existence. In Germany, however, it has proved possible to place an onus on the claimant to prove that the goods were adequately pre-cooled.[394] A further possibility is that some goods may fall within Article 17(4)(d) even where lack of adequate preparation for transport may ultimately be the cause of the damage. In another German decision, the particular vulnerability of steel rings to rust due to the effect of dampness in the air was accepted as a special risk, the burden being on the claimant to prove that the method of preparation adopted was adequate to guarantee protection.[395]

(e) *"insufficiency or inadequacy of marks or numbers on the packages"*

6.104 Under this defence the carrier is offered protection, subject to Article 18(2), if the packages carried are not properly marked by the sender or his agent where this results in loss or damage to the goods. In practice it is difficult to see under what circumstances the carrier is likely to be able to take advantage of this provision,[396] as under Article 8(1)(a), on taking over the goods, the carrier is bound to check the accuracy of the consignment note as to the number of packages and their marks and numbers, and if he has no reasonable means of checking their accuracy he must enter his reservations in the

---

390. RB. Antwerpen, 6.9.74 (1975) 10 E.T.L. 253, internal condensation inside plastic packing of a consignment of wool bobbins. The carrier only had to establish the possibility that the damage could be attributed to the actual nature of the goods. (Cf. Clarke, p. 332 n. 52, who considers that this should be regarded as a case of defective packing. If, however, the method of packing is that customarily adopted, it may be difficult to categorise the facts in this way. Rather, the goods as packed in the usual way may be regarded as carrying a special risk. Conversely, goods which carry a risk when not packed may present no risk when properly packed, e.g. glass: Hof. Gent, 20.6.86 (1986) 21 E.T.L. 371.)
391. Article 18(1).
392. See *supra*, paras 6.100 and 6.27.
393. As in the *Ulster-Swift* case, see [1975] 2 Lloyd's Rep. 502, 506.
394. OLG. Hamm, 11.6.90 (1990) TranspR 375 (cf., however, OLG. Schleswig, 30.8.78 (1979) VersR 141). As explained by Thume, *supra*. para. 6.57 fn. 221, p. 3, there is an analogy with defective packing. This suggests an onus on the plaintiff in relation also to Article 17(4)(b). Since Article 9 assists the plaintiff only in respect of the apparent condition of the goods and their packaging, an onus to prove the internal condition may yet rest on the claimant. In OLG. Hamm, the court held that the plaintiff had failed to provide the proof satisfying the pre-conditions of Article 17(1). Given, however, the requirement on the carrier to check the temperature of the goods, noted *supra*, para. 6.100, it will rarely be the case that a claimant will be considered to have failed in this proof.
395. OLG. Hamm, 2.11.95 (1996) TranspR 335.
396. Cf. Cass., 14.10.97 (1997) B.T. 800, *supra*, para. 6.19.

consignment note, together with the grounds on which they are based.[397] If the consignment note contains no specific reservations by the carrier it will be presumed, unless the contrary is proved, that the marks and numbers of the packages correspond with the statements in the consignment note.[398] Consequently, to avoid liability, the carrier must furnish proof that the marks and numbers do not correspond with the statements in the consignment note.[399] There appears to be little or no relevant case law to date in which this provision has been the subject of discussion.[400]

(f) *"the carriage of livestock"*

6.105 This provision does not have the effect of excluding the carriage of livestock from the operation of CMR.[401] It merely gives the carrier rather limited protection against claims by cargo interests.[402] Article 17(4)(f) is subject both to Article 18(2) and Article 18(5). The presumption in Article 18(2) is effectively restricted by Article 18(5), which provides as follows:—

"The carrier shall not be entitled to claim the benefit of article 17, paragraph 4(f), unless he proves that all steps normally incumbent on him in the circumstances were taken and that he complied with any special instructions issued to him."

The position therefore is this. Article 17(4)(f) merely gives the carrier a limited protection against claims by cargo interests by shifting the burden of proof, and the carrier will only be relieved of liability where the loss or damage arises from the special risks inherent in the carriage of livestock. The relief from liability is that laid down in Article 18(2) and (5). The carrier must establish three things. First, that in the circumstances of the case the loss or damage can be attributed to the special risks inherent in the carriage of livestock.[403] Secondly, that all steps which were normally incumbent on him in the circumstances have been taken.[404] Thirdly, that he has complied with any special instructions issued to him.[405] If the carrier can satisfy the court on these three points then under Article 18(2) it shall be presumed that the loss or damage was in fact caused by the special risks inherent in the carriage of livestock. The claimant is, however, free to rebut the presumption by proving that the loss or damage was not in fact wholly or partly attributable to the special risks inherent in the carriage of livestock. For example, if livestock is injured either as a result of a traffic accident for which the carrier cannot avoid liability under Article 17(2), or else as a result of the transit period being extended, the carrier may still be liable.[406] In effect

---

397. Article 8(2).
398. Article 9(2).
399. See Loewe, para. 166.
400. See Haak p. 173. See also OGH., 14.9.95 (1996) ZfRV 26 (Mayer, p. 173) and App. Lyon, 25.6.99 (2000) B.T. 312.
401. See, e.g., App. Bordeaux, 7.7.92 (1993) B.T. 84.
402. As in Cass., 17.6.97 (1997) B.T. 507, (1998) U.L.R. 200. See further, Krings, p. 775.
403. Article 18(2). In Cass. 17.6.97, *supra*, fn. 402, the sender ordered the loading of the animals despite their reluctance and the court concluded that they died due to stress.
404. Article 18(5). For countries party to the European Convention on the Protection of Animals in International Transport, 1968, requisite steps are fully indicated, see Clarke, p. 347 n. 46 and Lamy, para. 503. The carrier will need to comply also with relevant E.U. Directives and Regulations; 628/91, 29/95 and 411/98.
405. Article 18(5). The existence of such special instructions must be established, presumably by the claimant, cf. High Court of Denmark (Western Division), 10.9.96 (1997) 32 E.T.L. 230.
406. See Loewe, para. 167.

therefore the carrier is placed in a position similar to that applicable for the carriage of goods by refrigerated vehicle.

6.106 Article 18(5) does not say from whom the special instructions referred to therein should emanate, but presumably any party acting either on behalf of the cargo interests or as part of the administrative process involved in the transportation of livestock would be covered.

Apportionment of liability

6.107 Article 17(5) provides as follows:—

"Where under this article the carrier is not under any liability in respect of some of the factors causing the loss, damage or delay, he shall only be liable to the extent that those factors for which he is liable under this article have contributed to the loss, damage or delay."

6.108 The carrier's liability will therefore be restricted to the extent to which those factors for which he is liable under Article 17 have contributed to the loss, damage or delay in question.[407] The obvious problem in such circumstances is to decide which of the factors the carrier is liable for and which he is not.[408] This question is most likely to arise in relation to Article 17(4)(c) cases where the sender or his agent have been responsible for loading, stowing or wedging the goods and the carrier's driver has driven the vehicle at a speed or in a manner so as to endanger the load.[409] Examples of this have already been discussed in relation to Article 17(4)(c). For example, where the sender has loaded a trailer, which was driven at a speed excessive for the load carried, there may be contributory factors permitting an equal apportionment of liability.[410]

6.109 However, where one party is more at fault than the other apportionment may be adjusted in such proportion as the court thinks fit in the latter's favour, although it should be noted that the Convention does not lay down any detailed procedure for such apportionment, and this will no doubt depend upon the relevant national law.[411] For example, where a carrier both accepted the sender's obviously faulty stowage and wedging and knowingly rounded a double bend at an excessive speed, he was held to be two-thirds to blame when the trailer overturned and damaged the load.[412]

6.110 Another situation where the question of apportionment has arisen is in relation to damage resulting from the nature of the goods themselves as discussed in relation to

---

407. See, e.g., App. Paris, 12.7.78 (1979) B.T. 159, (1980) I U.L.C. 238 where damage was partly the result of the sender's negligence in not providing waterproof packaging, and partly the result of the carrier's negligence in not providing a sufficiently waterproof vehicle.

408. Cf. Hof. 's Gravenhage, 3.6.76 (1977) S. & S. No. 3. CMR is not concerned with the contributory negligence of other road users where a collision has taken place. It is not one of the circumstances which the carrier cannot avoid under Article 17(2). For that reason a carrier cannot rely on Article 17(5) in such a case.

409. In France the courts have divided liability between the carrier and the sender where the sender's defective stowage has caused the loss but in circumstances where the defect was apparent and the carrier is deemed to be at fault in failing to control the defective stowage, App. Paris, 18.5.89 (1989) B.T. 577 (see also Cass., 22.7.86 (1986) B.T. 516).

410. Hof. 's Hertogenbosch, 21.12.65 (1966) 1 E.T.L. 698. See also Arrond. Roermond, 2.1.69 (1969) 4 E.T.L. 1005, discussed *supra*, para. 6.94, and OLG. Saarbrücken, 21.11.74 (1976) 96 E.T.L. 261.

411. In this country, a court would presumably apply the Law Reform (Contributory Negligence) Act 1945.

412. App. Paris, 12.6.70 (1970) B.T. 228. Cf. Comm. Namur, 22.7.65 (1966) 1 E.T.L. 133, where insufficient wedging was held to be the sole cause of damage to goods—avoiding action in traffic is a normal risk which must be taken into account when stowing goods. See also App. Reims, 13.7.77 (1977) B.T. 406.

Article 17(4)(d) above.[413] For example, where perishable agricultural produce deteriorates in transit the damage may result either from the nature of the goods themselves or else from the carrier's failure to follow the sender's instructions or through his inordinate delay in transit, or through a combination of such factors.[414] In an Italian appellate decision a shipment of bananas in a rather ripe condition was loaded in the carrier's vehicle by the sender for carriage from Belgium to Italy. As the driver failed to follow the shortest route the journey was unduly lengthened and on arrival the goods were a total loss. In an action by the consignee against the carrier, the court held that liability was to be apportioned between the parties interested in the cargo (two-thirds) and the carrier (one-third) in respect of the loss, on the grounds that the sender was partially responsible for loading overripe fruit and the carrier for unduly extending the transit.[415] Such an apportionment will not be possible if the carrier has been guilty of wilful misconduct.[416]

Burden of proof

6.111 If the carrier wishes to benefit from the protection offered to him by the Convention in Articles 17(2) and 17(4) he must fulfil the requirements of Article 18 as regards the burden of proof. He must therefore bring evidence before the court as required by the various provisions of Article 18. If he fails to do so he will remain liable for the loss, damage or delay in delivery of the goods to the cargo interests. However, this does not in any way affect his rights of recovery from third parties.

6.112 Article 18(1) provides as follows:—

"The burden of proving that loss, damage or delay was due to one of the causes specified in article 17, paragraph 2, shall rest upon the carrier."

In other words, if the carrier wishes to claim that the loss, etc., resulted from the wrongful act, neglect or instructions of the claimant, the inherent vice of the goods or circumstances which the carrier was unable to avoid, he must bring proof to this effect. Unfortunately there is no internationally accepted code of rules of evidence, and the degree of proof required may differ from one jurisdiction to another. Here it should be pointed out that, CMR not being a Convention of the European Community, there is no Supreme Court whose decision will be accepted as binding in the interpretation of the law in question, a difficulty which arises in the case of most international transport conventions.[417]

6.113 The Continental courts have tended to take a strict approach with regard to the carrier bringing proof in a positive fashion.[418] This has been so in the case law discussed

---

413. In one French decision a carrier was held only one quarter to blame for accepting a consignment of cheese for transport in a vehicle which was neither refrigerated nor insulated: App. Aix-en-Provence, 10.11.76 (1977) B.T. 248 (Lamy, para. 1555, Jur. 5).

414. See Arrond. Roermond, 24.10.68 (1969) 4 E.T.L. 1012—a shipment of cauliflowers in transit in August. OLG. Zweibrücken, 23.9.66, 1 U. 40/67 (1967) N.J.W. 1717—a shipment of peaches. See also Comm. Paris, 22.9.80 (1980) B.T. 615, (1981) II U.L.R. 228—bad condition in which the goods were shipped was aggravated during carriage in a refrigerated vehicle.

415. App. Venezia, 31.10.74 (1975) 10 E.T.L. 242.

416. OLG. Munich, 16.1.91 (1992) TranspR 181, applying Article 29 (see *infra*, para. 9.72 ff.).

417. *Ulster-Swift Ltd.* v. *Taunton Meat Haulage Ltd.* [1977] 1 Lloyd's Rep. 346, 350. See, however, Article 47 of the Convention in Appendix B.

418. The German Supreme Court has accepted that *prima facie* proof is acceptable, so that German courts can take into account indicative facts and circumstances that, according to experience, typically point to a certain course of events, BGH., 4.10.84 (1985) 20 E.T.L. 154, 157.

earlier in relation to inherent vice and theft, although the courts have tended to be more flexible in the common problem of the vehicle's tyre catching fire.[419]

6.114 In the Court of Appeal decision of *Ulster-Swift Ltd.* v. *Taunton Meat Haulage Ltd.*,[420] Megaw, L.J., in considering the effect of Article 18(1), stated as follows:—

"So if the carrier wishes to seek exemption from art. 17, par. 1, general liability, on the basis (as in the present case) of art. 17, par. 2, exemption of loss caused by inherent vice of the goods, the burden of proving that the loss was caused by inherent vice is on the carrier. So it would be, also, in English law apart from the Convention. But in the present case we need not trouble about burden of proof at least at this point. For the Judge, on hearing all the evidence, has held affirmatively that, on balance of probabilities, there was not any relevant inherent vice of the goods. So it is not a question of any artificial or conventional burden of proof. Such burden of proof, when, as here, all the evidence which either party wishes to produce has been heard, usually as a practical matter is of importance only in the, fortunately, rare cases where, after considering all the evidence, the Court is unable to come to a conclusion as to the balance of probability. As Lord Reid said, in a different context in *McWilliams* v. *Sir William Arrol & Co.* [1962] 1 W.L.R. 295, at p. 307: 'But in the end, when all the evidence has been brought out, it rarely matters where the onus originally lay, the question is which way the balance of probability has come to rest'."[421]

In the light of this dictum it can therefore be assumed that this approach will continue to be the one adopted by the English courts and that Continental case law on this point will be mainly of assistance in offering parallel fact situations on which the courts can draw rather than of any fundamental juridical value.[422]

6.115 Article 18(2) then provides as follows:—

"When the carrier establishes that in the circumstances of the case, the loss or damage could be attributed to one or more of the special risks referred to in article 17, paragraph 4, it shall be presumed that it was so caused. The claimant shall however be entitled to prove that the loss or damage was not, in fact, attributable either wholly or partly to one of these risks."

6.116 The essence of this provision is that to benefit from the provisions of Article 17(4) the carrier must establish the possibility of the grounds on which he based his claim for relief from liability. If he succeeds in so doing the burden of proof is reversed and the claimant must instead prove that the loss or damage was not the result of the grounds on which the carrier claims relief.[423] If the claimant succeeds, the result is that we return to Article 17(1) and the carrier is liable. The reason for this approach to the matter would appear to be that in practice carriers would find it extremely difficult to prove positively that the loss or damage resulted from one of the matters referred to in Article 17(4), in particular since the loss or damage itself may make it impossible or more difficult to reconstruct the facts of the accident after the event so as to show that it so resulted.[424] However, as was seen above in relation to the discussion of Article 17(4)(c), the method

---

419. Libouton, (1973) 8 E.T.L. 2, para. 57.

420. [1977] 1 Lloyd's Rep. 346. See Glass, "The divided heart of CMR", (1979) 14 E.T.L. 687.

421. Ibid., pp. 351–352.

422. The provisions of CMR deal with the issue of who must prove the presence or absence of a particular fact and, if necessary, which of the parties carries the *non liquet*, but the question of the *evaluation* of proof is left to the determination of the relevant domestic law: BGH., 4.10.84 (1985) 20 E.T.L. 154, 157.

423. Cf. Arrond. Roermond, 18.11.71 (1972) 7 E.T.L. 416 where it was held that to reverse the presumption of relief from liability on the part of the carrier as provided for in Articles 17(4) and 18(2) the claimant must prove that the loss was due to the negligence of the carrier. However, what will constitute negligence and what will suffice to rebut the presumption will in each case depend upon the interpretation that the courts of the various jurisdictions put upon Articles 17 and 18.

424. See Libouton, op. cit., para. 58.

of interpretation of Article 18(2) has differed considerably from country to country. As Megaw, L.J., stated in *Ulster-Swift* v. *Taunton*:

"Unfortunately, there can be no doubt but that at least some of the provisions of those articles have presented difficult problems and have resulted in acutely diverging views in the Courts of other States which have adopted the Convention. This is illustrated by reference to a learned article by Dr. R. Wijffels, a Belgian attorney-at-law, who is the editor of a publication, European Transport Law. The article is entitled 'Legal Interpretation of CMR: the Continental Viewpoint'. It appears at p. 208 of Part I of C.M.R. 1976. The learned author discusses no less than 30 decisions of various Courts, in Belgium, France, West Germany, the Netherlands, Italy and Austria. He concludes that those decisions show no less than 12 different interpretations of the provisions of the Convention relating to the liability of a carrier in connection with art. 17, par. 4 (particularly sub-par. (c)) and art. 18, par. 2. These 12 different interpretations range through, as it were, almost the whole 360 deg. of the compass."[425]

Megaw, L.J., further pointed out that as a result there can be no one clear view as to the presumed purpose of the Convention as regards the provisions in question.[426] He did, however, specifically disapprove Loewe's suggestion[427] that a higher standard than the balance of probabilities is required of the claimant under the second sentence of Article 18(2).[428]

Special provisions

6.117 Article 18(3), 18(4) and 18(5) provides as follows:—

"3. This presumption shall not apply in the circumstances set out in article 17, paragraph 4(a), if there has been an abnormal shortage, or a loss of any package.
4. If the carriage is performed in vehicles specially equipped to protect the goods from the effects of heat, cold, variations in temperature or the humidity of the air, the carrier shall not be entitled to claim the benefit of article 17, paragraph 4(d), unless he proves that all steps incumbent on him in the circumstances with respect to the choice, maintenance and use of such equipment were taken and that he complied with any special instructions issued to him.
5. The carrier shall not be entitled to claim the benefit of article 17, paragraph 4(f), unless he proves that all steps normally incumbent on him in the circumstances were taken and that he complied with any special instructions issued to him."

The above provisions lay down special rules which restrict the carrier from enjoying the benefit of the presumption of non-liability under Article 18(2). A full discussion of these provisions will be found under the discussion of Article 17(4)(a),[429] (d)[430] and (f)[431] above.

---

425. [1977] 1 Lloyd's Rep. 346, 350.
426. Ibid.
427. Loewe, para. 170.
428. [1977] 1 Lloyd's Rep. 346, 352. Note Libouton's criticism of this disapproval: (1982) J.T. 719, para. 83.
429. See para. 6.61 ff.
430. See para. 6.95 ff.
431. See para. 6.105 ff.

CHAPTER 7

Delay, Non-delivery and Cash on Delivery

Delay

7.1 As has been seen, Article 17(1) imposes liability on the carrier for "any delay in delivery". Article 19 provides the definition of delay for the purposes of the Convention:

"Delay in delivery shall be said to occur when the goods have not been delivered within the agreed time-limit or when, failing an agreed time-limit, the actual duration of the carriage having regard to the circumstances of the case, and in particular, in the case of partial loads, the time required for making up a complete load in the normal way, exceeds the time it would be reasonable to allow a diligent carrier."

Agreed time limit

7.2 The Convention, in laying down rules as to what will constitute delay in relation to the delivery of goods, provides for two distinct situations.[1] First, if the parties agree a time limit between them for the delivery of goods Article 19 provides that failure to comply with it will constitute delay for the purposes of the Convention to which the penalties laid down in Article 23(5) will be applicable.[2] Presumably the time limit may be formulated either as delivery by a particular date or else as a period of time commencing with either the taking over of the goods by the carrier or the date of the contract.[3] Article 19 does not specify whether the agreed time limit must be inserted in the consignment note or not, although Article 6(2)(f) provides that where applicable the consignment note shall also contain the agreed time limit within which carriage is to be performed. Generally, Continental case law on the subject indicates that, provided there is adequate other evidence of the agreed time limit, insertion in a consignment note is not necessary. In a Paris Court of Appeal decision, a specified date was laid down for delivery of a shipment of apples from Le Havre to London which was not met by the carrier owing to a road accident. The court held that the time limit was binding even though it was not apparently entered in a consignment note, provided other documentation evidences the agreement.[4] In another French case, where the specified date for delivery was contained in a telex, this

---

1. It does not provide for premature delivery, liability for which must be addressed to national law: RB Rotterdam, 29.4.94 (1995) S. & S. No. 46 (Cleton, "Digest of Recent Decisions of Netherlands Courts concerning International Transport Law Conventions", (1996) U.L.R. 755, 759).
2. See *infra*, Chapter 9, para. 9.42.
3. But see *infra*, para. 7.6 for the implications of which is used.
4. App. Paris, 2.4.71 (1971) B.T. 155.

was held to be sufficient for the purposes of Article 19.[5] Although there is one German decision indicating that insertion in the consignment note is required,[6] later German authority follows the approach of the French courts.[7]

7.3 It is thought that the better view, at least as far as an action against the contracting carrier is concerned,[8] is that the time limit need not be inserted in the consignment note.[9] As has been seen,[10] Article 4 provides that the contract is only to be evidenced by the consignment note, and that "absence, irregularity, or loss of the consignment note shall not affect the existence or the validity of the contract of carriage". Clearly, if no consignment note is issued, no agreed time limit can be included; it is provided that the non-issue is not to affect the existence or the validity of the contract, and it follows that it cannot affect the validity of the agreed time limit under Article 19.[11] It would also seem to follow that if a consignment note is issued, but does not include the agreed time limit, this at most would constitute an "irregularity" within Article 4, and so equally should have no bearing on its validity. Clearly, insertion in the consignment note will facilitate proof of the agreed time limit, but any other proof of agreement should be equally acceptable.

7.4 On the other hand, Article 19 makes it clear that the delivery date must be agreed, so it is not open to the sender to stipulate unilaterally a delivery date by inserting it in the consignment note without the carrier's consent.[12] However, Article 9 provides that the consignment note is *prima facie* evidence of the terms of the contract of carriage,[13] so in such circumstances the onus would be on the carrier to show that he had not agreed. If a date appears clearly in the consignment note, this might be difficult. Thus, in another French decision, the court held that a delivery date entered in the consignment note was effective, notwithstanding the carrier's arguments that the date of delivery was not imperative and that the urgency of delivery had not been drawn to its attention.[14] It must

---

5. App. Amiens, 19.6.80 (1980) B.T. 490. See also App. Toulouse, 29.7.81 (1981) B.T. 381 (interesting because the *commissionnaire de transport* who instructed the carrier was held liable to share the damage resulting from the delay since, although he informed the carrier by telex of the time limit, he omitted to mention it in the consignment note, thus suggesting the possibility of partial responsibility of the sender) and App. Montpellier, 19.2.87 (1987) B.T. 578 where the time limit was indicated on a document entitled "instructions to transporters" and was signed by both parties.

6. OLG. Stuttgart, 24.1.67, U. 57/66 (1968) N.J.W. 1054 No. 10. This aspect of the decision was, however, very much incidental to the substance of the case, being mentioned very much as an aside in the report. This decision is not consistent with other German decisions; see, e.g., OLG. Düsseldorf, 18.1.79 (1979) VersR 356 and other cases cited by Haak, p. 195, n. 417. See also OLG. Düsseldorf, 27.2.97 (1998) TranspR 194, OLG. Cologne, 7.12.93 (1994) TranspR 197 (see further, Herber/Piper, Ant. 19 Rdn. 4, Clarke, p. 226).

7. See BGH., 30.9.93 (1994) 29 E.T.L. 97, 101, where there is a clear statement by the court that the time limit does not have to be included in the consignment note since the requirement is merely a matter of proof.

8. For the position concerning successive carriers under Article 34, see *infra*, Chapter 11, para. 11.21 ff.

9. Both Libouton, (1973) 8 E.T.L. 2, para. 65, p. 54 and Hardingham, "The delay provisions of CMR", [1979] LMCLQ 193 are of the same opinion.

10. *Supra*, Chapter 4.

11. The burden of proving the agreement rests on the claimant: App. Grenoble, 13.3.80 (1981) B.T. 306.

12. Nor by sending the carrier a fax specifying one: LG. Stuttgart, 27.9.91 (1992) TranspR 22. (Criticised by Thume, (1992) TranspR 403.) On the other hand, a carrier who advertises an express service guaranteeing delivery within 48 hours can hardly deny an agreement to provide it: App. Paris, 25.3.94 (1994) B.T. 389, cf. Comm. Bobigny, 25.2.94 (1994) B.T. 672.

13. See *supra*, para. 4.24 ff.

14. App. Bordeaux, 9.2.72 (1972) B.T. 114, (1974) I U.L.C. 341. See also commentary by Hémard, (1972) Rev. Trim. Dr. Comm. 454. On the question of delay in delivery and loss of perishable goods under French law see Brunat, (1977) B.T. 342.

be clear, however, that a time limit for the delivery of the goods has been agreed. In one French case, the goods were to be carried to London and the consignment note stated that the carrier must catch the ferry at Dieppe by a certain date and time. Due to a breakdown of the lorry, it missed the ferry and the goods arrived a day later. This was held to be a delay within Article 19.[15] Similarly, where the consignment note said "not after 16.5.78".[16]

No agreed time limit

7.5 The second possibility provided for by Article 19 is that where no time limit has been agreed upon by the parties,[17] delay will occur where the actual period of transit exceeds the time it would be reasonable to allow a diligent carrier. This will depend upon the particular circumstances of every case,[18] but it is expressly provided by Article 19 that particular regard is to be had, in the case of part loads, to the time required to make up a full load in the normal way. This would include groupage shipments by one or more forwarders. It has been held in Germany that a reasonable time is not exceeded unless there is a considerable deviation from the normal time required for carriage.[19]

7.6 In order to amount to delay, therefore, either an agreed time limit must be exceeded or *the carriage* must exceed a reasonable time. It would seem to follow, and there is Continental authority to such effect,[20] that where there is no agreed time for delivery and the carrier has totally failed to perform the carriage the situation is not caught by Article 19.[21] It also appears that the delay provisions may in some circumstances have no

---

15. App. Aix-en-Provence, 20.12.77 (1978) B.T. 245, (1979) I U.L.R. 299.

16. App. Paris, 2.12.81 (1982) B.T. 73.

17. An agreement that the goods be carried "as quickly as possible" does not amount to an agreed time limit, LG. Stuttgart, 27.9.91 (1992) TranspR 22.

18. Thus in App. Venezia, 31.10.74 (1975) 10 E.T.L. 242 the court assessed the normal period for carriage from Belgium to the Italian border at a maximum of 35 to 40 hours, having taken into account possible short breaks by the driver and time taken to complete Customs formalities. In Arrond. Utrecht, 12.3.80 (1980) S. & S. No. 127 (Libouton, (1982) J.T. 721 para. 94) the court took into account the nationality of the carrier along with the administrative complications attaching to the acquisition of permits for the particular transit. See also Comm. Brussels, 19.11.73 (1984) J.C.B. 441 (Libouton, Theunis at p. 96, cf. App. Trento, 21.5.98 (1999) 34 E.T.L. 380 where the court took account of the difficulties linked to the crossing of countries in the former Eastern European bloc and the political and administrative climate applicable there as well as the state of war in Iraq). In App. Lyon, 27.6.80 (1980) B.T. 509 there was liability for delay when the carrier took three months to carry the goods from France to Syria, a journey which normally took 15 days. In App. Rennes, 5.11.74 (1974) B.T. 514 a diligent carrier of fruit was said to require three days for a distance of 950 km, and in App. Poitiers, 3.2.76 (1976) B.T. 185 two days was considered sufficient for the distance from the Adriatic Sea to the French Vendée. See further Haak, p. 201, who also points out that goods can be carried too fast. Thus completing a journey from Moscow to Paris (2,900 km) in five days with fragile goods was considered to be a case of serious fault in Cass., 22.9.83 (1983) B.T. 566.

19. LG. Stuttgart, 27.9.91 (1992) TranspR 22. The court estimated the normal time and then held that the delay of two days beyond that time was not to be seen as exceeding the time it would be reasonable to allow a diligent carrier (cf., however, where perishable goods are involved: LG. Duisburg (1989) TranspR 268, see Koller, Art. 19 Rdn. 5). There was also a dispute as to the method of transport and the fact that the carrier chose the slower method did not avail the claimant, who was aware of the carrier's different methods and had not excluded the slower one.

20. App. Milano, 11.7.75 (1977) I U.L.C. 336.

21. Recently in Germany national law was applied where a time for loading was exceeded and the sender employed a different carrier to complete the transit, OLG. Hamm, 20.3.97 (1998) TranspR 297. The court rejected the view taken in an earlier decision that Article 17(1) applies to a delay in the loading period where this period exists with a view to the time of delivery (OLG. Hamm, 14.11.85 (1986) TranspR 77, Clarke, p. 227), cf. OLG. Düsseldorf, 23.12.96 (1998) TranspR 113. See also Koller, Art. 27 Rdn. 3.

application where the delay occurs prior to collection of the goods by the carrier.[22] Article 19 refers, in the absence of an agreed time limit, to "the duration of the carriage" as the yardstick for determining whether delay has occurred. It follows that unless the contract of carriage stipulates either a time limit running from the date of the contract or a particular date for delivery, if delay results not from an extended period of carriage but from a delay in collecting the goods, Article 19 can have no application.[23]

Defences

7.7 In cases of delay, it is specifically provided by Article 17(2)[24] that the carrier will be entitled to rely on the general defences therein contained.[25] As has been pointed out elsewhere,[26] logically, Article 17(2) can have no application where there is no agreed time limit, since the court is there concerned to establish whether delivery has taken place within a reasonable time in all the circumstances of the case, and so will have dealt with the factors in Article 17(2) at that stage. It would be contrary to all reason to say that delivery had been unreasonably delayed under Article 19, yet that Article 17(2) provided a reasonable excuse; and if there is no delay, then Article 17(2) is not brought into effect.[27] Clearly, however, Article 17(2) can be relevant to cases of delay where there is an agreed time limit,[28] since in such cases no question of reasonableness is involved in the initial definition of delay.[29]

7.8 In addition to Article 17(2), Article 17(3)[30] and (5)[31] will also be relevant for the purposes of delay. On the other hand, however, the defences in respect of special risks

---

22. A claim for compensation for delay in taking over the goods was held subject to national law and consequently not subject to Article 30(3) (*infra*, Chapter 10, para. 10.2) nor Article 23(5) (*infra*, Chapter 9, paras 9.42–9.46) in VG. Turnhout, 20.2.98 (1998) 33 E.T.L. 569.

23. Hardingham, op. cit., p. 195. As Hardingham points out, if damage to the goods were to result, then such could fall within Article 17(1) subject to it being shown that the damage occurred between the time the carrier takes over the goods and the time of delivery, which "could raise nice questions". Cf. Cass., 13.11.90 (1991) B.T. 136, *infra*, fn. 25.

24. *Supra*, Chapter 6, para. 6.1.

25. Such as the wrongful act of the claimant, as in Cass., 13.11.90 (1991) B.T. 136, where the driver had been kept waiting to begin the carriage due to the fault of the sender, with the result that it proved impossible to arrive at the Customs office before it closed.

26. Hardingham, op. cit., p. 194.

27. On the other hand, it may be that the question of what is a reasonable time permissible to a diligent carrier can be considered objectively in the light of the normal circumstances, leaving the carrier to establish abnormal circumstances in justification. This could still depend on the circumstances of the particular carriage, but the objective circumstances viewed at the time the contract is concluded rather than the particular difficulties that arise in the course of the carriage. This approach is supported by Thume in (1992) TranspR 403, where he points out that Article 19 uses the present tense, i.e. "exceeds the time it would be reasonable", rather than the past tense. On that view, the difficulties encountered in the course of the transit become matters for consideration under Article 17.

28. And the consequent burden of proof in Article 18(1), see e.g. OLG. Düsseldorf, 23.12.96 (1997) TranspR 422.

29. As in App. Paris, 2.12.81 (1982) B.T. 73, where the carrier was not liable for the delay, since a two-day wait at Customs was due to the fact that the sender had failed to attach a certificate of origin to the carriage documents, this being within his sphere of responsibility under Article 11 (*supra*, para. 4.37 ff.). A carrier should be aware of traffic circulation restrictions likely to affect his choice of route, and adjust his journey time and route accordingly, Comm. Toulouse, 7.2.80 (1980) B.T. 319, cf. LG. Stuttgart, 27.9.91 (1992) TranspR 22. Similarly, the known slowness of foreign Customs: App. Paris, 28.2.89 (1990) B.T. 100. The strike at a particular Customs post will not excuse him if it was possible for him to use a different one: App. Paris, 27.5.80 (1980) B.T. 435.

30. *Supra*, Chapter 6, paras 6.53–6.59.

31. *Supra*, Chapter 6, paras 6.107–6.110.

provided for by Article 17(4)[32] are expressed to relate to liability for "loss or damage", and although on the face of it it might be argued that these words are wide enough to cover loss or damage caused through delay, it would seem that such was not the intention since Article 17(2) and (5), as distinct from Article 17(4), refers to "loss, damage or delay". The apparent exclusion of these defences seems strange, since some of the factors there listed could equally cause delay. To an extent this will make no difference to the position of the carrier, since in considering his liability for delay, any matter relevant under Article 17(4) could equally be considered by the court either under Article 17(2), where there is an agreed time limit, or under Article 19, as bearing on the question of reasonableness, where there is no agreed time limit. On the other hand, the exclusion of delay from Article 17(4) has the effect of also depriving the carrier of the benefit of the reversal of the burden of proof in Article 18(2).[33] One can indeed only assume that it was precisely for this reason that delay was excluded from Article 17(4), it being thought appropriate in delay cases that the onus should always be on the carrier.[34] Not only is the carrier exclusively in a position to explain the delay, but also the measure of damages in delay cases is restricted to an amount not exceeding the carriage charges for the contract in question.[35]

Reservations in cases of delay

7.9 It is important to note in relation to claims for delay the requirements of Article 30(3) which provides as follows:—

"No compensation shall be payable for delay in delivery unless a reservation has been sent in writing to the carrier, within twenty-one days from the time that the goods were placed at the disposal of the consignee."

This provision has been applied so as to bar any claim for delay if such written notice is not given,[36] and such is its clearly stated effect. As distinct from claims for loss of or damage to goods, it is not immediately obvious why this provision was inserted in its present form, for the carrier's position would not be prejudiced by provision for a longer period. Further, if the claim is being brought by the consignor, he is dependent on fairly rapid communication of the delay by the consignee, which is perhaps less likely to happen in cases of delay than where loss of or damage to the goods has occurred. Furthermore, unlike Article 30(1) and (2), Article 30(3) does not specifically exclude Sundays and public holidays from the 21-day period, so it seems to follow that it was intended that they should count.[37]

7.10 Two French cases are worth noting in respect of the necessary reservation under Article 30(3). In one case the consignee had simply entered the date of the actual delivery of the goods on the consignment note, without further comment. This was insufficient to

---

32. *Supra*, Chapter 6, paras 6.60–6.106.
33. *Supra*, Chapter 6, paras 6.115–6.116.
34. By virtue of Article 18(1). *Supra*, para. 6.112.
35. Article 23(5). The only circumstances in which more can be payable is where there has been wilful misconduct under Article 29 or where the sender has declared a special interest in delivery under Article 26. The whole question of damages in relation to delay is considered *infra*, Chapter 9, paras 9.42–9.46.
36. Arrond. Roermond, 30.5.68 (1969) 4 E.T.L. 1019, App. Trento, 21.5.98 (1999) 34 E.T.L. 380. This is the case even if the carrier is guilty of wilful misconduct: BGH., 14.11.91 (1993) 28 E.T.L. 265, (1992) TranspR 135.
37. Rodière, p. 323. Hardingham, op. cit., p. 195.

comply with Article 30(3).[38] In the other case, a reservation sent prior to delivery and before the time limit for delivery had actually been exceeded was also held insufficient.[39] The court treated the words "within twenty-one days from the time that the goods were placed at the disposal of the consignee" as laying down not only the last permissible date for sending a reservation but also the earliest permissible date for so doing. On a strictly literal approach, this reasoning is no doubt correct, but it is very difficult to see any good reason why an earlier reservation should not be effective, at least where late delivery has become inevitable or likely, particularly in view of the fact that it is at precisely that stage that the cargo interest is likely to seek to expedite its delivery by making the legal implications clear. Although there is Dutch authority to the contrary,[40] in the light of the wording of Article 30(3) and of the above decision, it will be in the claimant's interests to submit a further reservation once the goods have actually arrived.

Non-delivery of goods

7.11 Article 20 provides as follows:—

> "1. The fact that goods have not been delivered within thirty days following the expiry of the agreed time-limit, or, if there is no agreed time-limit, within sixty days from the time when the carrier took over the goods, shall be conclusive evidence of the loss of the goods, and the person entitled to make a claim may thereupon treat them as lost.
> 2. The person so entitled may, on receipt of compensation for the missing goods, request in writing that he shall be notified immediately should the goods be recovered in the course of the year following the payment of compensation. He shall be given a written acknowledgement of such request.
> 3. Within the thirty days following receipt of such notification, the person entitled as aforesaid may require the goods to be delivered to him against payment of the charges shown to be due on the consignment note and also against refund of the compensation he received less any charges included therein but without prejudice to any claims to compensation for delay in delivery under article 23 and, where applicable, article 26.
> 4. In the absence of the request mentioned in paragraph 2 or of any instructions given within the period of thirty days specified in paragraph 3, or if the goods are not recovered until more than one year after the payment of compensation, the carrier shall be entitled to deal with them in accordance with the law of the place where the goods are situated."

Goods treated as lost

7.12 Where goods have not been delivered within 30 days following the expiry of the agreed time limit, or within 60 days from the time when the carrier took over the goods in the absence of an agreed time limit, this will be treated as conclusive evidence of the

---

38. App. Aix-en-Provence, 26.11.80 (1981) B.T. 185. An actual reservation entered on the consignment note will suffice however (Cass., 29.4.75 (1975) B.T. 298). The reservation is equally valid whether it is sent by the sender or by the consignee: App. Limoges, 7.11.77 (1978) I U.L.R. 383.

39. App. Amiens, 19.6.80 (1980) B.T. 490. See also App. Orléans, 20.1.82 (1982) B.T. 233. A German court was of the same view in relation to written reservation under the equivalent provision of the Warsaw Convention: OLG. Frankfurt, 15.1.80 (1980) I U.L.C. 228.

40. RB. Utrecht, 12.3.80 (1981) 9 E.L.D. 317, where there were repeated complaints by telex to the carrier during the carriage itself: it was held that the absence of a subsequent written protest under Article 30(3) did not prejudice the claim, since the carrier was clearly aware of the intention to claim.

loss of the goods,[41] and so the cargo interests are not required to furnish any further proof as to the loss.[42] The first question that arises is what will constitute "delivery" for the purposes of this provision, and reference should be made to the earlier discussion of this matter.[43]

7.13 The uncompromising wording of Article 20(1) whereby "non-delivery shall be conclusive evidence of the loss of the goods" can create considerable problems for the carrier and his insurer, as even if the goods are undamaged and their location known the carrier may still be liable under Article 20(1),[44] for example, a shipment may be sent by road from Western Europe to Eastern Europe or the Middle East, and during transit the vehicle may be involved in a road accident involving injury or damage to a third party. In such circumstances it is not uncommon for the local authorities to impound the vehicle and load pending a settlement of any third party claims or until a guarantee of settlement can be given. If for any reason the carrier's motor underwriter delays in dealing with the third party claim, then the period within which delivery must be effected under Article 20(1) may have expired, thus rendering the carrier (and his CMR underwriter) liable as for a total loss.[45] A similar result may arise where a vehicle has broken down in an inaccessible spot or otherwise come to grief, though with minimal damage to its load.

7.14 The distinction between mere delay in delivery as laid down in Article 19 and a constructive total loss which will arise when the artificial time limit under Article 20 expires is of fundamental importance for the carrier, given the different measure of damages applicable in each case. As stated earlier, under Article 23(5) the measure of damages for delay cannot exceed the amount of the carriage charges, which will generally be far less than the amount recoverable in cases of total loss. The question of damages in claims for loss is dealt with in detail in Chapter 9, but, in brief, under Article 23(1) and (2) damages are assessed on the basis of the value of the goods at the time and place of acceptance for carriage, subject to the overall limit of liability provided for by Article 23(3) unless the cargo interests have invoked the provisions of Articles 24 or 26. In addition, carriage charges, Customs duties and other charges incurred in respect of the carriage will be recoverable under Article 23(4).

7.15 Article 20 has also been considered relevant in the context of the issue of when goods are lost for the purposes of the commencement of the limitation period under Article 32.[46] This use of the provision is considered further in Chapter 10.

---

41. The claimant should not be required to so treat it, as there may be a benefit in being able to rely on the fact that the time limit in Article 32 (*infra*, Chapter 10, paras 10.51–10.80) begins to run only on later delivery, OLG. Düsseldorf, 23.11.89 (1990) TranspR 63 (see Clarke, p. 157). A claimant must, however, be clear that the goods are being treated as lost before being able to base a claim on Article 20, BGH., 15.10.98 (1999) TranspR 102, 104.

42. Unlike Article 30 (*infra*, Chapter 10, paras 10.2–10.20) Sundays and public holidays are not expressly excepted and so are included in the calculation of the respective periods of time, OLG. Düsseldorf, 25.9.97 (1999) TranspR 159, Cf. however, Koller, Art. 20 Rdn. 1, Herber/Piper, Art. 20 Rdn. 50.

43. See *supra*, para. 6.9 ff. and Rodière, para. 38.

44. Cf. LG. Hamburg, 29.7.94 (1994) TranspR 448.

45. To cover such contingencies it is advisable for an international carrier to take out some form of legal expenses insurance to enable the requisite bonds and guarantees to be made immediately available to ensure the speedy release of the vehicle, driver and load.

46. In *Worldwide Carriers v. Ardtran International Ltd.* [1983] 1 All E.R. 692, and *ICI PLC v. MAT Transport Ltd.* [1987] 1 Lloyd's Rep. 354. Cf., however, Arrond. Middlebourg, 26.6.63 (1964) U.L.C. 167 where no use was made of Article 20 when the consignee claimed a right to sue under Article 13 (*supra*, Chapter 5) in circumstances where the goods were damaged in an accident and were returned to the sender in accordance with his instructions. The redelivery to the sender took place within the time limit for delivery originally agreed in

Defences

7.16 Where Article 20 is invoked, since its effect is the presumed loss of the goods, the defences in Article 17(2),[47] (4)[48] and (5)[49] must be available to the carrier. So far as Article 17(2) is concerned, unless the carrier can either show fault on the part of the claimant or inherent vice of the goods, he is left with showing "circumstances which [he] could not avoid and the consequences of which he was unable to prevent" and the limited scope given to this phrase has already been considered.

7.17 That Article 17(4) should be available in this context is curious, since, as explained above, the authors of the Convention seem to have thought it necessary to exclude delay from its ambit.[50] By such delay being sufficiently prolonged as to constitute presumed loss, it then becomes available. Further, if the cargo interest then invokes Article 20(3)[51] by requiring delivery of the goods and claims in respect of delay only, then Article 17(4) again will cease to be applicable.[52] Presumably the reason for this difference of approach is that, as explained above, if the claimant is only claiming in respect of delay, the potential liability in most cases will be less than that in claims for loss of the goods, so justifying a more onerous burden of proof on the carrier in relation to delay than in cases of loss.

The claimant's right of election

7.18 Where the cargo interest is compensated for the loss of the goods under Article 20(1) he can request the carrier in writing[53] that he be notified immediately if the goods are recovered within a year of payment. The carrier is bound to acknowledge the request in writing,[54] although his failure to do so does not appear to have any legal consequences. It is presumably intended to avoid future disputes as to whether the carrier has received such notification, which may be of significance in relation to the provisions of Article 20(4).[55] In the event of the goods being recovered and the cargo interest notified, the latter may, within 30 days of notification, require the goods to be delivered to him, subject to him paying the charges shown to be due on the consignment note and refunding the compensation, less any charges included therein.[56]

7.19 Whether or not the cargo interest takes advantage of this provision will obviously depend on a number of factors, not least on whether or not he has any continuing need for the goods, and on whether or not he was adequately compensated by the original payment. So, for example, if the compensation was not adequate on account of the goods being

the contract of carriage. The goods were not "lost" so the consignee could not claim the benefit of Article 13.

47. *Supra*, Chapter 6, paras 6.14–6.52.
48. *Supra*, Chapter 6, paras 6.60–6.106.
49. *Supra*, Chapter 6, paras 6.107–6.110.
50. *Supra*, para. 7.8.
51. Discussed in the next paragraph.
52. It is true that such matters as are referred to in Article 17(4) must also fall within the general defences of Article 17(2), but attention has been drawn above to the knock-on effect of depriving the carrier of the benefit of Article 18(2).
53. There is no precise time stipulated within which this must be done, although the implication of the wording is that it should be at or about the time of payment of compensation.
54. Article 20(2).
55. See *infra*, para. 7.26.
56. Article 20(3).

worth more than the limit of liability provided by Article 23(3),[57] it will clearly be in his interests to require delivery of the goods. If he does so, Article 20(3) expressly provides that this will not prejudice any claim for delay under Article 23[58] or, where applicable, Article 26[59] but it must be remembered that in claims for delay, Article 30(3) requires a written reservation to be made within 21 days of the goods being placed at the disposal of the consignee.[60]

7.20 The reference in Article 20(3) to the deduction of charges from the refund of compensation means that if the compensation included items awarded or paid under Article 23(4),[61] these do not have to be repaid, unless they are payable in any event by virtue of the requirement that the charges shown to be due on the consignment note are to be paid. Again the question arises as to what the position would be where charges to which the carrier would be contractually entitled do not appear in the consignment note. For the reasons explained above,[62] it is not thought that this should affect the position, although inclusion in the consignment note would facilitate proof of the contractual arrangements.[63]

7.21 It has already been explained that the carrier can be caught by Article 20, and obliged to pay compensation in respect of the total loss of the goods in circumstances which may be very harsh. On its strict wording, the cargo interest would be entitled to refuse delivery and insist on full compensation, if it were commercially advantageous for him to do so, however marginal the failure to comply with the time limit laid down in Article 20(1). The potential inequity of such a situation has led one commentator to suggest that if the goods are recovered by the carrier prior to the claimant commencing legal action or during the course of the proceedings, then the claimant is obliged to take delivery and is limited to a claim for delay and for any legal cost incurred.[64]

7.22 The major objection to such an approach is that it does not seem that it can be reconciled with the apparently unambiguous wording of Article 20(1) that delay beyond the prescribed limits "shall be conclusive evidence of the loss of the goods, and the person entitled to make a claim may thereupon treat them as lost". Further, to adopt Loewe's view could work substantial injustice to the claimant. It would mean that he could, in effect, only treat them as lost at his peril, for if they were to re-appear at any time prior to judgment he would be restricted to his claim for delay, which in turn would be restricted to the amount of the carriage charges by virtue of Article 23(5). So, for example, if in reliance on Article 20(1) he were to have in the meantime purchased replacement goods once the time limit had expired, he might find himself with unrecoverable losses resulting from being compelled to take delivery of the original goods.

7.23 The same result is indicated by the fact that under Article 20(3) the claimant can, provided he has complied with Article 20(2), require the delivery of the goods, and as against this express right it seems unlikely that the courts would be prepared to imply any

---

57. *Infra*, Chapter 9, paras 9.9–9.14.
58. *Infra*, Chapter 9, paras 9.42–9.46 and 9.55–9.59.
59. *Infra*, Chapter 9, paras 9.50–9.59.
60. *Supra*, para. 7.9–7.10.
61. *Infra*, Chapter 9, paras 9.21–9.41.
62. *Supra*, paras 7.2–7.3, see further, para. 5.17 fn. 33.
63. In the event of any dispute it would seem that the carrier can take advantage of Article 13(2) (*supra*, para. 000 ff.) and refuse delivery unless security is furnished.
64. Loewe, para. 177.

obligation to accept delivery.[65] It must be conceded that there is much force in Loewe's point that Article 20 might operate harshly against the carrier, but for the above reasons it is not thought that his proposed solution either could or should be adopted. In a Belgian decision, the sender of the goods was held to be entitled to rely on Article 20, notwithstanding the reappearance of the goods and their presentation at the destination, the claimant having manifestly invoked the benefit of it.[66]

7.24 A preferable solution can be found in the normal principles of mitigation of damages,[67] by virtue of which the plaintiff must take all reasonable steps to keep his loss to a minimum,[68] and, by way of necessary corollary, must do nothing which unreasonably increases his loss.[69] Thus, applying these principles to the situation at issue here, if the claimant were to refuse to accept delivery of the goods after the expiry of the Article 20(1) time limit without any reasonable excuse, then he could, if appropriate, be penalised in damages if he thereby unnecessarily increased his losses. Conversely, if he had good reason for acting as he has, he would be entitled to compensation in full, notwithstanding the availability of the original goods.[70]

7.25 There is nothing in Article 20 to prohibit such an approach, but a possible problem lies with the provision of Article 23(1) and (2) which stipulate that the damages payable in cases of loss are to be fixed according to the value of the goods at the time and place they are taken over for carriage. What is not entirely clear is whether that sum must invariably be awarded in cases of total loss, or whether it represents only a basic guideline. This question is discussed below,[71] but it is thought that the latter view is to be preferred, and such would permit the court to consider issues of mitigation in the manner outlined above.

National law

7.26 Leaving aside therefore the possibility of the cargo interest being compelled to accept delivery, the question then arises as to what happens to the goods if they are recovered after the time limit in Article 20(1) has expired and the consignee does not want to take delivery. Article 20(4) covers this situation by providing that if the cargo interest

---

65. Equally, the claimant is not obliged to treat the goods as lost if they are found before he has received compensation, per Staughton, J., in *ICI PLC* v. *MAT Transport Ltd.* [1987] 1 Lloyd's Rep. 354, 360, although he considered this not to be the case where the issue involved limitation and the running of time; see *infra*, para. 10.70 and cf. OLG. Düsseldorf, 23.11.89 (1990) TranspR 63.

66. Hof. Brussel; 6.4.84 (1984) 19 E.T.L. 431, (1986) II U.L.R. 607. When the goods eventually arrived, having been mistakenly placed on a ship which took them on a world tour, they were initially accepted by the consignee's agent but then rejected since the consignee had already received a replacement from the sender. The sender was reimbursed for his loss by the first carrier who was seeking recourse from the successive carrier. The fact that the goods were eventually returned to the sender was held to be of no significance, since the goods were held to have lost all their economic value, being highly specific to the requirements of the consignee (the court, in effect, adopting a notion of constructive total loss, see *infra*, paras 10.64–10.67).

67. See generally *Chitty on Contracts*, 28th edn., para. 27–085 ff.

68. *Brace* v. *Calder* [1895] 2 Q.B. 253.

69. *Payzu Ltd.* v. *Saunders* [1919] 2 K.B. 58.

70. Similar principles are recognised in France, where after a refusal by the consignee the goods are abandoned to the carrier and a claim for total loss is made based on Article 20. To protect the carrier from abuse, it must be shown either that the goods are no longer saleable or that the consignee could no longer have put them to their normal use: App. Colmar, 1.6.90 (1991) B.T. 502.

71. Chapter 9, paras 9.3–9.8.

fails to furnish a written request under Article 20(2),[72] or to give instructions for delivery within the 30 days specified in Article 20(3), or if the goods are not recovered for more than a year after payment of compensation,[73] then the carrier is free to deal with them in any way permitted by the *lex situs*. Given that the carrier will at that stage necessarily be obliged to pay, or will already have paid, damages assessed on the basis of the value of the goods, it is difficult to see what limitation which could be placed upon his right to deal with the goods as he wishes.[74]

7.27 Loewe treats the reference to *lex situs* as referring to the provisions of the national law which may be applicable where the carrier has failed to notify the cargo interests that the goods have been recovered or where the carrier has fraudulently concealed the recovery of the goods until the year specified in Article 20(4) has elapsed, which might occur in the event of the goods being worth more than the limit of compensation provided for by Article 24(3).[75]

7.28 If this be right, and as already indicated it is difficult to envisage any other limitation, insofar as any issue of fraud is involved, the reference to the *lex situs* raises a difficulty of its own, quite apart from the general difficulty of ascertaining the provisions of the *lex situs* when proceedings are brought in a different country from that in which the goods are situated. This is because the Convention itself stipulates an extended limitation period of three years in cases of wilful misconduct,[76] and it is therefore not clear whether that extended period would apply, or whether the claimant would be dependent on establishing an extended period of limitation under the *lex situs*. In the interests of uniformity, it is submitted that the Convention period ought to apply, and that the reference to the *lex situs* in Article 20(4) should not be interpreted as overriding the extended Convention period. If, however, the claimant can show that under the *lex situs* he can bring his claim beyond the end of that three-year period, then he should be entitled to rely on that further extended period. Turning to situations where the carrier has failed to inform the cargo interest of the recovery of the goods in circumstances not amounting to wilful misconduct or equivalent default, the Convention provides no extension of the limitation period in such circumstances, so if there is an extended period under the *lex situs*, there is no equivalent difficulty in giving effect to it in Convention cases.

7.29 In this country, it is provided by the Limitation Act 1980[77] that in cases of fraud, deliberate concealment or mistake the limitation period does not start to run until the plaintiff has discovered, or could with reasonable diligence have discovered, the fraud, deliberate concealment or mistake. It would appear that the Limitation Act 1980 does not, however, apply to cases under the Convention[78] so it would be necessary to rely on the general equitable jurisdiction on which the Act was based.[79] It is doubtful whether the limitation period would be postponed if the failure to inform the cargo interest of the

---

72. Article 20(2) refers to the person so entitled having received compensation. It is possible that this person may have indicated, prior to receiving compensation, that no further interest is retained in the recovery of the goods. In such a case the court in OLG. Düsseldorf, 20.3.97 (1998) TranspR 32 held that the carrier was free to proceed to the sale of the goods before the payment of compensation and that, in consequence, the claimant had no interest in the proceeds of sale.

73. In this situation the cargo interest will therefore have lost the right to call for delivery in any event.

74. However, cf. Clarke, p. 361.

75. Loewe, para. 181.

76. Article 32(1). *Infra*, Chapter 10, para. 10.51.

77. Section 32.

78. Limitation Act 1980, section 39. But see the discussion at 10.113–10.114, *infra*.

79. For the position under the 1980 Act see *Halsbury's Laws*, 4th edn., Vol. 28, para. 916 ff.

recovery of the goods resulted from negligence, without any element of fraud or deliberate concealment.[80]

Cash on delivery

7.30 Article 21 provides as follows:—

"Should the goods have been delivered to the consignee without collection of the 'cash on delivery' charge which should have been collected by the carrier under the terms of the contract of carriage, the carrier shall be liable to the sender for compensation not exceeding the amount of such charge without prejudice to his right of action against the consignee."

Where goods are despatched on a C.O.D. basis therefore, and the carrier delivers the goods without collecting the C.O.D. charges, he will be liable to the sender for those charges. Before he will be liable on this basis, however, a provision for the collection of these charges must have been included in the terms of the contract of carriage. Again, the first issue that arises is as to whether the provision for C.O.D. must be included in the consignment note to be enforceable. Although Article 6(2)(c) states that the consignment note is to specify the amount of the C.O.D. charges, for the same reasons as discussed above in relation to the agreed time limit for the purposes of Article 19,[81] it is thought that, although such inclusion would facilitate proof of the agreed term,[82] inclusion in the consignment note is not essential.[83] The authorities in this respect referred to in relation to Article 19 can be relied on in support of this view, as can a French decision where a claim under Article 21 failed on the basis that the consignor had not given his instructions in this respect sufficiently clearly. The judge does not seem to have regarded inclusion in the consignment note as such to have been essential.[84]

7.31 The consignment note in this context is, however, important in another respect in that, as explained earlier, under Article 13[85] the consignee can on arrival of the goods require the carrier to deliver them to him on payment of the charges shown to be due on the consignment note. In this respect too, as is argued above, it is not thought that inclusion of the C.O.D. charges in the consignment note is essential, although again it will facilitate proof of the contractual arrangements between the parties.[86] But for the avoidance of disputes on arrival it is clearly in the carrier's interests to ensure inclusion of the C.O.D. charges in the consignment note, although in the event of such a dispute, the carrier

---

80. *Phillips-Higgins* v. *Harper* [1954] 1 All E.R. 116.

81. *Supra*, paras 7.2–7.3. See also the general discussion of the need for a consignment note in Chapter 4, paras 4.3–4.9.

82. In a French decision the requirement to pay was indicated solely on an invoice despatched with goods and, even if the carrier knew the contents, the invoice was not a contractual document so far as the contract of carriage was concerned and did not have the effect of imposing an obligation to collect on the carrier. The fact that the carrier had collected the money three times previously was of no moment, there being nothing in the documents to indicate that the money was to be paid to him: App. Paris, (1997) B.T. 159 (date of case not given).

83. Although again the position may be different for the purposes of claims against successive carriers. See *infra*, Chapter 11, paras 11.21–11.48.

84. App. Paris, 21.10.70 (1970) B.T. 307, (1971) U.L.C. 141. See also BGH., 10.2.82 (1983) 18 E.T.L. 32, Hof. 's Hertogenbosch, 13.1.70 (1971) 6 E.T.L. 817, OLG. Düsseldorf, 13.12.90 (1991) TranspR 91 (cited by Clarke, p. 124 n. 44), OLG. Düsseldorf, 11.11.93 (1994) TranspR 441.

85. *Supra*, para. 5.12 ff.

86. See *supra*, Chapter 5.

could in any event invoke the proviso to Article 13(2)[87] so as to enable him to require security from the consignee before delivering the goods.

Scope

7.32 Turning to the question to what items can form the subject matter of such a clause, the Convention provides no definition of a "cash on delivery" charge, but it would seem that it will include all charges which the consignee may have agreed to pay, not just the cost of the goods themselves. So, for example, the consignee might have agreed with the sender to pay the carriage charges, costs of packing, insurance, etc., and these can equally be made the subject matter of Article 21. So in *Eastern Kayam Carpets Ltd.* v. *Eastern United Freight Ltd.*[88] it was held that the provision was not confined to freight charges and the like, and that it extended to cover a stipulation requiring payment of the price of the goods to the transporter on delivery.

7.33 On the other hand, several decisions indicate that Article 21 does not extend to some analogous provisions in the contract of carriage. Thus in a Dutch decision, it was held that an instruction by the sender to the carrier that delivery was only to be effected in exchange for the original despatch note was not equivalent to a C.O.D. clause for the purposes of Article 21.[89] Similarly, an instruction to the carrier not to deliver unless the consignee could produce proof of payment has been held to be outside Article 21.[90] In another Dutch decision, however, a provision whereby a carrier was not to unload a shipment until a banker's draft had been signed was held to be within Article 21[91]; clearly such a clause does represent a much closer equivalent with the conventional C.O.D. situation. This last decision was accepted and adopted by Hirst, J., in *Eastern Kayam Carpets Ltd.* v. *Eastern United Freight Ltd.*[92] However, he held that a cash against documents transaction under which the defendant carrier was to deliver the goods only against production by the consignee of an original copy of his house bill of lading which thereby evidenced payment of the cash to a bank was outside the scope of Article 21.[93] If the stipulation is not within Article 21, then any liability of the carrier will depend on the national law.[94] However, if the first instance decision in *Gefco (U.K.) Ltd.* v. *John Mason*[95] is adopted in this context, it would appear to have the consequence that if the instruction to the carrier does not fall within Article 21, then there will be no remedy.[96]

---

87. *Supra*, para. 5.12 ff.

88. Queen's Bench Division, 6.12.83, unreported except on LEXIS. Equally it is irrelevant that the payment is intended to cover previous sales to the consignee, App. Aix-en-Provence, 6.11.81 (1982) B.T. 258. In this case the carrier had agreed to the C.O.D. request sent by telex after the carriage had begun.

89. Arrond. Amsterdam, 15.6.66 (1970) S. & S. No. 101. Cf. OLG. Düsseldorf, 21.4.94 (1995) TranspR 67, OLG Düsseldorf 19.6.86 (1988) VersR 77, see further Koller, Art. 21 Rdn. 1.

90. GH. 's Hertogenbosch, 13.1.70 (1971) 6 E.T.L. 817, (1971) S. & S. No. 10. Cf., however, Cass., 6.6.95 (1995) B.T. 474.

91. Arrond. Breda, 16.2.69 (1970) 5 E.T.L. 67, (1970) U.L.C. 298.

92. *Supra*, fn. 88.

93. Similarly, CMR does not constrain the liability of a carrier to indemnify the sender if the carrier fails in a duty to encash or return an accepted Bill of Exchange if this is what has been agreed, provided that the sender's loss is in fact due to the failure of the carrier: App. Paris, 23.6.94 (1994) B.T. 692, cf. App. Toulouse, 18.2.92 (1992) B.T. 781.

94. The carrier in the *Eastern Kayam Carpets* case was held liable on the basis of the common law. See, however, *supra*, para. 6.12.

95. Q.B.D., (Birmingham District Registry) 21.2.00 (unreported).

96. See the discussion *supra*, paras 0.27–0.35. Cf. however, Lamy, paras 485 and 1542 jur 6, Herber/Piper, Art. 21 Rdn. 7, Thume, Art. 21 Rdn. 51 ff, Koller, CMR Art. 21 Rdn. 1.

7.34 A related point is that the sender must take care to ensure that he has communicated his instructions to the carrier with sufficient clarity. Thus in a French case, it was held that the phrase *"paiement contre documents"* appearing on a copy of the goods invoice was not within Article 21, since the carrier could legitimately think that this was addressed to the consignee to remind him of the method of payment under the contract of sale.[97] Similarly in another case the phrase *"contre documents pour solde de notre facture"* (against documents in settlement of our invoice) mixed in among other instructions in no logical order was not sufficiently clear to fall within Article 21.[98]

7.35 As to whether the carrier is entitled to accept a cheque in payment of the C.O.D. charges, this is a matter for the relevant national law.[99] In this country the general common law position is that authorising an agent (which is, in effect, what the carrier is for the purpose of collecting the charges) to collect money does not *prima facie* include authority to receive payment by cheque but "it seems likely that a practice of accepting payment by cheque, whether in favour of the agent or principal, will fairly readily be recognised nowadays as being in the normal course of business, and hence within the scope of an agent's authority . . . ".[100] So in the *Eastern Kayam Carpets* case[101] Hirst, J., observed that "a sight draft is comparable to a cheque which, as numerous Order 14 cases in the Court of Appeal show, is under English law to be treated as equivalent to cash". But for the carrier to be sure that he is protected, it will be in his interests to seek specific instructions in respect of payment by cheque rather than having to rely on an implied authority, and it will, of course, always be open to the sender to stipulate the method of payment required, which would override any such implied authority.[102] Where the sender does so, the carrier must exercise reasonable care in carrying out the instruction.[103]

Measure of damages

7.36 If the carrier fails to comply with the requirement as to the collection of C.O.D. charges it is clear that the provisions of the Convention which limit the carrier's liability[104]

97. App. Paris, 19.1.78 (1978) B.T. 161. Even where the goods invoice indicated C.O.D., this was not a sufficiently clear instruction to the carrier: Comm. Pontoise, 26.9.78 (1978) B.T. 569, App. Aix-en-Provence, 2.3.79 (unpublished), see Libouton, (1982) J.T. 721, para. 96. The instruction *"notre client vous remettra lors de la livraison un chèque émis à notre nom et que vous devez nous faire parvenir"* was insufficiently precise and unequivocal in App. Anvers, 8.5.91 (1993) J.P.A. 362 (see Putzeys, "Le droit des transports en Belgique (II) 1993–1995", (1996) U.L.R. 557, 566). See further Hof. Gent, 13.9.95 (1997) 32 E.T.L. 602, Hof. 's Hertogenbosch, 16.3.88 (1989) S. & S. No. 258, (1990) 18 E.L.D. No. 186.

98. App. Paris, 21.10.70 (1970) B.T. 307, (1971) U.L.C. 141. See too App. Paris, 26.1.73 (1973) B.T. 73 and Comm. Paris, 22.7.76 (1976) B.T. 416.

99. BGH., 10.2.82 (1983) 18 E.T.L. 32, BGH., 25.10.95 (1996) 31 E.T.L. 404 (1996) TranspR 118, (1997) U.L.R. 414.

100. *Bowstead on Agency*, 16th edn., para. 3–022. Not so in Germany: BGH., 10.2.82 (1983) 18 E.T.L. 32, or in Holland: Arrond. Rotterdam, 1.10.76 (1977) S. & S. No. 23, or in Belgium: Hof. Gent 22.12.94 (1995) R.W. 262 (see Putzeys, *supra*, para. 734, fn. 96 at p. 567). But it does in France: App. Paris, 31.1.83 (1983) B.T. 183.

101. *Supra*, fn. 87.

102. In a case where local foreign exchange regulations forbade the collection of large amounts of cash it was held that the carrier was obliged to seek alternative means of payment as safe as cash. If he was unable to, he should seek instructions in accordance with Article 15 (*supra*, Chapter 5, paras 5.31–5.34): BGH., 25.10.95 (1996) 31 E.T.L. 5.31–5.34 (1996) TranspR 118, (1997) U.L.R. 414.

103. See Cass. (Belg.), 18.2.94 (1994) 29 E.T.L. 464 (see Putzeys, *supra*, para. 7.34, fn. 96, at p. 566), where the carrier was instructed to deliver against receipt of certified cheques and was liable for accepting two documents written in Greek with little resemblance to cheques. The carrier could not rely on the fact that the sender would likewise have thought these documents to be certified cheques. Cf. App. Versailles, 2.3.2000 (2000) B.T. 408.

104. *Infra*, Chapter 9.

do not apply, but the question then arises as to whether the sender is automatically entitled to the unpaid C.O.D. charges in full or whether he must prove that he has suffered loss in the normal way. The English text would appear to support the latter interpretation by use of the words "not exceeding the amount of such charge" in Article 21, which clearly contemplates an award of less than that sum.[105] On the other hand, turning to the equivalent provision in the French text, this provides that the carrier shall "*indemniser l'expéditeur à concurrence du montant du remboursement*". Although it may be open to doubt whether the word "*indemniser*" has any special significance,[106] here the words "*à concurrence du*" would seem to indicate a different meaning from the equivalent words of the English text, and that the amount of the C.O.D. charges can be recovered in any event.[107] The status of the French text in this country has been discussed in some detail above,[108] but the better view appears to be that in cases of ambiguity it is permissible to have regard to it. However, it is doubtful whether the English text is ambiguous in this respect, nor does it become so by reference to the French text; there is in effect a substantive difference between the two versions, and it would therefore not appear legitimate to have regard to the French text.

7.37 Turning to the merits of the alternative approaches, on the face of it to impose liability on the carrier to the full extent of the C.O.D. charges without requiring the sender to prove his loss could operate extremely harshly against the carrier. Article 21 expressly preserves the carrier's right against the consignee in respect of the C.O.D. charges, but does not specify whether in any claim by the carrier against the consignee the latter can plead any defences that would have been available in an action against him in the event, for example, of the goods being in some way defective. If the consignee can plead such matters by way of defence to the carrier's claim, this would leave the carrier exposed to the possibility of being unable to recover part, or indeed all, of those charges from the consignee, which seems an excessively harsh penalty to impose. If this be correct therefore, it is thought that the English text should be preferred and that the sender should be required to prove his loss, which would involve the prior disposal of any claims by the consignee against the sender.[109]

7.38 This would, however, seem an unnecessarily complex solution. The alternative possibility would be to permit the carrier full recovery against the consignee regardless of the latter's right vis-à-vis the sender.[110] Representing as it does a situation where the parties are placed in the position they would have been in had the contract been properly performed, this would seem to be the preferable approach. The carrier is left with the possibility that he might be unable to recover the C.O.D. charges, for example, by reason of the consignee's insolvency, but there is justice in that since that situation would have been avoided had the C.O.D. charges been collected at the proper time. The consignee is

---

105. Adopted in Germany, BGH., 10.10.91 (1992) 27 E.T.L. 262, (1992) TranspR 100.

106. Cf. Hardingham, op. cit., p. 197.

107. Libouton expresses the view that the carrier is liable in full: (1973) 8 E.T.L. 2, para. 67. See further the erudite discussion of the meaning of these French words by Clarke, p. 129 n. 76. See also, App. Versailles, 2.3.2000 (2000) B.T. 408.

108. See paras 0.14–0.17.

109. Cf. App. Paris, 23.11.95 (1996) B.T. 257 where the sender was awarded the full amount because he could show that failure to obtain a bank endorsement deprived him of the opportunity to obtain payment.

110. French decisions support this. App. Paris, 8.12.81 (1981) B.T. 61. App. Paris, 19.11.81 (1981) B.T. 62. The latter case concerned delivery against documents. See also in Belgium, Hof. Gand, 22.12.94 (1995) R.W. 262 (see Putzeys, *supra*, para. 7.34, fn. 96 at p. 567).

left to proceed against the sender in respect of any breaches of contract, but that, too, would have been the position had the C.O.D. charges been paid at the proper time.

7.39 For these reasons it is thought that the better approach is that contained in the French text, and that the sender should be entitled to the C.O.D. charges in full without the need to prove his loss. However, to avoid potential injustice to the carrier, he in his turn must be entitled to full recovery from the consignee, regardless of any claims the latter may have against the sender. It should be noted that Article 27 also applies in respect of Article 21, so that the sender can claim interest on the sum in accordance with this provision.[111]

7.40 As a final point, it should be noted that the Lloyd's Goods in Transit (CMR) policy[112] does not cover the carrier in respect of any failure to collect a C.O.D. charge.

---

111. BGH., 10.10.91 (1992) 27 E.T.L. 262, (1992) TranspR 100.
112. See Appendix D.

CHAPTER 8

Dangerous Goods

8.1 Special provision is made in respect of dangerous goods by Article 22 of the Convention, which provides as follows:—

"1. When the sender hands goods of a dangerous nature to the carrier, he shall inform the carrier of the exact nature of the danger and indicate, if necessary, the precautions to be taken. If this information has not been entered in the consignment note, the burden of proving, by some other means, that the carrier knew the exact nature of the danger constituted by the carriage of the said goods shall rest upon the sender or the consignee.

2. Goods of a dangerous nature which, in the circumstances referred to in paragraph 1 of this article, the carrier did not know were dangerous, may, at any time or place, be unloaded, destroyed or rendered harmless by the carrier without compensation; further, the sender shall be liable for all expenses, loss or damage arising out of their handing over for carriage or of their carriage."

Goods of a dangerous nature

8.2 The first question that arises is as to what is meant by "goods of a dangerous nature", since the Convention itself offers no definition. Clearly, however, it must refer to goods which are dangerous in themselves, rather than goods which only become dangerous, for example, as a result of the way in which they are carried. An improperly secured cargo may certainly cause danger, but unless that cargo has some inherent danger it would not be a carriage of dangerous goods within Article 22. Often, of course, such will be the nature of the goods that there is no difficulty in classifying them as dangerous, and in such cases the definition offered by Loewe may provide a workable guide: "All goods are to be considered as dangerous, if, in normal road transport, they present an immediate risk."[1] Inevitably, however, borderline cases will arise, and in such cases it would seem a logical approach to have regard to the European Agreement concerning the Carriage of Dangerous Goods by Road (ADR).[2] Any general treatment of ADR is beyond the scope of this book since it is concerned with establishing a detailed régime of documentation, packaging and method of carriage in relation to the international carriage of dangerous goods,

---

1. Loewe, para. 186
2. Signed at Geneva 1967. Cmnd. 3769 (1968). In force for the U.K. from 2.7.68. In contrast to CMR, CIM itself incorporates the International Regulations concerning the Carriage of Dangerous Goods by Rail (RID) (Article 4(d) and Annex I CIM). However, reference to ADR in CMR was clearly impossible, having regard to their respective dates, but it should be noted that both the approved forms of consignment note call for the insertion of the ADR class, number and letter, if any.

rather than with regulating the relationship between the various parties to the contract of carriage.[3]

8.3 It should not be forgotten, however, that ADR applies to every international carriage of goods by road, regardless of whether that carriage falls within CMR, so it will also cover, for example, carriage by containers.[4]

8.4 Unfortunately, ADR itself does not provide any general definition of dangerous goods, referring instead to "those substances and articles the international carriage by road of which is prohibited by, or authorised only on certain conditions by, annexes A and B".[5] Annex A then goes on to list a very large number of substances and articles classified as dangerous according to the nature of the risk they present.[6] It would seem likely that goods listed as dangerous in the Annexes to ADR will equally be regarded as dangerous for the purposes of CMR.[7] The converse, however, namely that goods not listed in ADR are not to be regarded as dangerous for the purposes of CMR, would not be accurate for *"les choses allant vite et le droit risquant de ne pouvoir jamais les rattraper"*.[8] In other words, in this technological age, it would be impossible for ADR to keep constantly abreast of all new developments, even assuming that it represents a comprehensive list of those substances already available that may be dangerous. In the event, therefore, of goods being tendered for carriage which are analogous to a type listed in ADR, or which represent a recent development of a type there listed, then it would seem that those goods can properly be regarded as dangerous for the purposes of CMR.[9]

8.5 The phrase "goods of a dangerous nature" also appears in Article IV, Rule 6 of the Hague and Hague-Visby Rules. Its meaning in that context was recently considered by the House of Lords in *The Giannis NK*,[10] which concerned the question of whether a consignment of groundnuts amounted to goods of a dangerous nature by virtue of the fact that it was contaminated by khapra beetle. The vessel was considerably delayed as a result of having been put into quarantine and then fumigated, and incurred associated expenses.

---

3. There is also the Convention on Civil Liability for Damage Caused During Carriage of Dangerous Goods by Road, Rail and Inland Navigation Vessels (CRTD) 1989, which is not yet in force. This will provide for the liability of the carrier in respect of claims, other than claims arising out of any contract for the carriage of goods, for damage caused during the carriage of dangerous goods. The carrier will, however, have a defence if he proves that the consignor failed to meet his obligations to inform him of the dangerous nature of the goods, and that neither he nor his servants or agents knew or ought to have known of their nature. Dangerous goods are defined by reference to ADR.

4. For the position of container transport in relation to CMR, see *supra*, paras 2.8–2.11.

5. Article 1(b) ADR.

6. The latest edition of ADR is that of 1995, applicable from 1.1.95. This edition is published by H.M.S.O., and includes the terms of the agreement, Annexes A and B and an editorial introduction prepared by the Department of Transport. In accordance with an additional Protocol of 30.9.57, dangerous goods carried under ADR to or from the United Kingdom must comply both with the provisions of ADR and with the United Kingdom Conditions for the Carriage of Goods by Sea until a special appendix containing provision for road/sea carriage of dangerous goods between the Continent and the United Kingdom is agreed. The very bulk and complexity of the Annexes has led one commentator to observe: *"L'ADR contient un nombre impressionant d'indications; il s'agit de les trouver, de s'y retrouver et surtout de ne pas s'y perdre. L'ADR est, elle aussi, une matière dangereuse"* (Wijffels, "Le Régime Juridique du Transport International de Marchandises Dangereuses par Route (CMR-ADR)" (1969) 4 E.T.L. 870, para. 28).

7. Support for this view can be gained from the French case cited in fn. 9.

8. L. A. and M. Drancourt, "Plaidoyer pour l'Avenir", quoted by Wijffels, op. cit., para. 13.

9. Thus in App. Lyons, 7.10.76 (1979) I U.L.R. 289, affirmed Cass., 20.6.78 (1979) I U.L.R. 296, the court was prepared to assimilate gas lighters, which are not mentioned in ADR, to gas bottles, which are. Cf., however, App. Nancy, 3.6.98 (1998) B.T. 550.

10. *Effort Shipping Co. Ltd.* v. *Linden Management S.A., The Giannis NK* [1998] 1 Lloyd's Rep. 337. For dangerous goods in relation to sea carriage generally, see Carver, *Carriage by Sea*, 13th edn., para. 691.

Further, the vessel was obliged to dump at sea another cargo of wheat which was on board, and faced a claim from the owners of that cargo.

8.6 It was held by the House of Lords that the words "goods of a dangerous nature" were to be given a broad meaning and were not confined to goods which were inflammable or explosive. It was sufficient that the goods were dangerous to other goods on the vessel, even if they would not endanger the vessel itself.[11] Further, the risk of direct physical damage to the other goods was not required. It was enough that the resulting quarantine and dumping of the other cargo was to be anticipated from the presence of the infestation.

8.7 As a result, the shippers were liable for all damages and expenses arising directly or indirectly from the shipment of the contaminated cargo. Liability was not dependent on whether the shipper knew or should have known of the dangerous nature of the cargo.

Other relevant provisions of CMR

8.8 Assuming that the goods forming the subject matter of the carriage are dangerous within the meaning of the Convention, a number of provisions of the Convention other than Article 22 are also relevant. First, it will be recalled that Article 6(1)(f)[12] provides that the consignment note must contain "the method of packing, and in the case of dangerous goods,[13] their generally recognised description", and that by virtue of Article 7(1)(a)[14] the sender is responsible for all expenses, loss and damage consequent upon the inadequacy or inaccuracy of those particulars.[15] It should not be forgotten, however, that the consignment note is only *prima facie* evidence of the terms of the contract,[16] so even in the absence of the required particulars from the consignment note, it would still be open to the sender to prove that the carrier was otherwise aware of the nature of the goods.

8.9 So far as these documentary aspects of CMR are concerned, it should be noted that in respect of goods listed in ADR a "transport document" is required, which must accompany the goods and contain, if appropriate, instructions to be carried out in the event of an accident.[17] If the sender has issued the carrier with an ADR transport document, this may well amount to compliance with Article 6(1)(f).

8.10 The other provision of CMR (over and above Article 22) which may be of major relevance in cases involving dangerous goods is Article 10 which, it will be recalled, provides that the sender is liable to the carrier for damage and expenses resulting from defective packing unless the defect was apparent or known to the carrier and no reservations were made on taking over the goods.[18] Again, ADR makes detailed provision regarding the loading, handling and stowage of dangerous goods, and compliance with

---

11. However, it has been held in Germany that the danger must be to property other than the vehicle and its load, OLG. Düseldorf, 23.1.92 (1992) TranspR 218 (see Clarke in Yates, para. 3.1.2.22.2 n. 1), cf. Comm. Paris, 31.3.92 (1992) B.T. 331 where the chemical involved was neither toxic nor classed as dangerous but was corrosive to the aluminium tank used to carry it (see *infra*, para. 8.14 fn 32).

12. *Supra*, paras 4.15–4.17.

13. This presumably means the same as "goods of a dangerous nature" in Article 22.

14. *Supra*, paras. 4.15–4.20.

15. If the carrier has entered the particulars in the consignment note at the request of the sender, he is deemed to have done so on behalf of the sender: Article 7(2) *supra*, para. 4.15 ff.

16. Article 9(1), *supra*, para. 4.24 ff.

17. Annex A, marginal 2002(3) and (4); Annex B, marginal 10 381, cf. App. Colmar, 12.12.93 (1995) B.T. 793.

18. *Supra*, para. 4.32 ff.

those requirements will again presumably be relevant, though not conclusive, as to liability under Article 10. The provisions of Articles 17(4)[19] and 18(2)[20] will also be relevant,[21] and in view of the presumption contained in the latter the sender will clearly be well advised to ensure compliance with ADR. However, compliance with ADR will clearly be irrelevant if any failure so to comply is not causally linked to the damage which has occurred.[22]

Rights and obligations under Article 22

8.11 Returning to Article 22 itself, the general requirement of paragraph (1) is that the sender notify the carrier of the "exact nature of the danger and indicate, if necessary, the precautions to be taken". It will be apparent that there is an overlap between Article 22(1) and the requirements of Article 6(1)(f) as to the contents of the consignment note, and it may well be that the information contained in the consignment note pursuant to Article 6(1)(f) is sufficient to comply with Article 22(1). However, if the information contained in the consignment note is insufficient to comply with Article 22(1), then the same paragraph goes on to provide that the burden of proving that the carrier was aware of the exact nature of the danger rests on the sender or consignee. In Germany it has been held that the sender must prove that, at the moment of delivery of dangerous goods to the driver, he drew the attention of the driver to the exact nature of the danger and also that he indicated to him the precautionary measures which should be taken. Thus, where a non-specialist driver had failed to enter the ADR class of the danger in the consignment note, the mere fact that he received an accident notice mentioning the possible dangers was insufficient to relieve the sender from liability. The sender was required to expressly draw the driver's attention to its contents.[23]

8.12 Clearly, the information which must be given to the carrier in order to comply with Article 22(1) will vary with the circumstances of every case, as will the evidence which must be brought to show that the carrier was aware of the danger, but again the provisions of ADR, in particular those as to the ADR transport document, are likely to be regarded as relevant though not conclusive. Thus, in relation to a consignment of gas lighters, it has been held in a French decision that where the carrier was in possession of a goods docket which clearly described the consignment as consisting of gas lighters, this was sufficient to comply with Article 22(1), the nature of the danger produced by such goods being regarded as common knowledge. The carrier was thus unable to invoke the provisions of Article 22(2).[24] Insofar as there was a breach of the ADR requirements,[25] this was treated

19. *Supra*, Chapter 6, paras 6.60–6.106.

20. *Supra*, para. 6.115.

21. But only in relation to a carrier's defence to a claim against him for damage to the goods, not in a claim by the carrier against the sender for loss or damage: App. Lyons, 7.10.76 (1979) I U.L.R. 289, affirmed Cass., 20.6.78 (1979) I U.L.R. 296, (1978) 13 E.T.L. 734.

22. See *infra*, cases cited in fn. 24 and fn. 26.

23. BGH., 16.10.86 (1987) 22 E.T.L. 56, (1987) II U.L.R. 708. See also Comm. Antwerp, 25.11.84 (1975–76) J.P.A. 70, Gottrau, Theunis, p. 199 (mere mention of chemical name insufficient), cf. Comm. Paris, 31.3.92 (1992) B.T. 331, where the fact that the sender and carrier had engaged in previous transports of the same goods when appropriate warnings were given was sufficient to relieve the sender on the occasion when the warning was not given; see also Cass., 16.10.90 (1990) B.T. 797, cf. App. Nancy. 3.6.98 (1998) B.T. 550.

24. App. Lyons, 7.10.76 (1979) I U.L.R. 289, affirmed Cass., 20.6.78 (1979) I U.L.R. 296, (1978) 13 E.T.L. 734.

25. Alleged in respect of the ADR requirements as to labelling of packages, transport documents, and warning plaques on vehicles.

as being of no causative effect, the accident having been caused entirely by the negligence of the carrier's driver in colliding with a bridge.[26]

8.13 There are two distinct aspects to Article 22(2). The first permits the carrier, where unaware of the dangerous nature of the goods, to unload, destroy, or render them harmless at any time and wherever the goods happen to be, without incurring any liability to the cargo interests. There is no limitation expressed as to the carrier's rights in this respect, so the question arises as to whether he must act reasonably. On the one hand, no such requirement is expressed and it might be argued that to impose any such requirement would be unreasonable, given that the carrier may be faced with an urgent decision as to what to do with the goods in circumstances where he has not been briefed as to their nature. As against this, however, such circumstances could be taken into account in deciding whether the carrier has acted reasonably.[27] Further, the requirement that the carrier be aware of the "exact nature of the goods" would mean, in the absence of any requirement of reasonableness, that the carrier could act with impunity in relation to the goods in circumstances where the sender is only technically in breach of Article 22(1) and where there is no situation of imminent danger. For these reasons, it would seem that the better view is that a requirement that the carrier act reasonably in exercising his rights under Article 22(2) is to be implied into that paragraph.[28]

8.14 The second aspect of Article 22(2) renders the sender liable "for all expenses, loss or damage arising out of [the] handing over for carriage or of [the] carriage" of the dangerous goods.[29] Liability of the sender is again dependent on the carrier being unaware of the exact nature of the danger.[30] Clearly, there may here be an overlap with Article 10[31] which renders the sender liable for "damage to persons, equipment or other goods[32] and for any expenses due to defective packing . . . ". However, at first sight it might be argued that Article 22(2) is wider than Article 10 in that the latter provision expressly refers to liability "to the carrier", which words do not appear in Article 22(2), hence suggesting that Article 22(2) extends to conferring a direct right of action against the sender on any third party injured or who has suffered damage. Such would be to ignore, however, the

26. In this respect the decision accords with normal English rules of causation. See, e.g., *McWilliams* v. *Sir William Arrol & Co.* [1962] 1 W.L.R. 295. However, in circumstances where the cause of the damage is uncertain, see *McGhee* v. *National Coal Board* [1973] 1 W.L.R. 1.

27. Viewed in this way Article 22(2) would accord with the defence of necessity, available under English law as a defence to claims in tort where the defendant has acted reasonably to avert some threatened harm. See generally *Winfield and Jolowicz on Tort*, 15th edn., pp. 877–881.

28. Loewe, para. 187, is of the same view, cf. Gottrau, Theunis, p. 201. See also *Scrutton on Charterparties*, 20th edn., p. 453, on the equivalent provision in the Hague-Visby Rules.

29. The issue of how to determine who is the sender is considered *supra*, para. 1.12 ff. Both the sender and the person who despatched the goods on his behalf were liable in App. Colmar, 12.12.93, *infra*, fn 30, the latter on the basis of fault.

30. The fact that the sender, as a buyer, has had no opportunity to see the goods is irrelevant, especially where he knows that he is buying dangerous goods. He must answer for the fault of those who acted for him: App. Colmar, 12.12.93 (1995) B.T. 793 (in respect of defects in the documents and in the loading of the goods).

31. E.g. App. Colmar, 12.12.93, *supra*, fn. 30.

32. Although not expressly mentioned therein, clearly the wording of Article 22(2) is wide enough to cover claims in respect of damage to persons, equipment or other goods; see *The Giannis NK* [1998] 1 Lloyd's Rep. 337, discussed *supra*, paras 8.5–8.7. In Belgium, it has been held that CMR only governs damage caused to the goods and not damage caused to the vehicle by them: Hof. Antwerpen, 13.12.89 (1990) 25 E.T.L. 319 (damage caused by bundles of steel pipes; neither Article 10 nor Article 22 was addressed by the court). This is clearly incorrect, at least insofar as Article 10 expressly creates liability in respect of the carrier's equipment, and many cases proceed on the basis of a potential liability for such damage, e.g., Comm. Paris, 31.3.92 (1992) B.T. 331, BGH., 16.10.86 (1987) 22 E.T.L. 56, (1987) II U.L.R. 708. Article 22 may be wider, in not being confined to equipment; see *supra*, para. 4.32 fn. 100.

overriding limitation of the Convention to "every *contract* for the carriage of goods", from which it is clear that the Convention can only operate to regulate the liability of the actual parties to the contract of carriage.[33] Article 22(2) does not, therefore, confer any rights upon an injured third party, and indeed if it is only a third party who has suffered damage, the carrier would only have a claim for nominal damages against the sender, unless he has incurred any personal liability to the third party as a result of the sender's breach of Article 22.[34] In these latter circumstances, the carrier could assign his rights against the sender to the third party,[35] and to that extent a third party could proceed against the sender under the Convention, but otherwise any claim by the third party against the sender would have to be in tort or delict under the appropriate national law.[36] As an exception to this, it should be noted that a successive carrier under Article 34 is constituted a party to the original contract of carriage by that Article, so would be able to sue the sender direct under Article 22.[37]

33. Article 1(1) *supra*, Chapter 1, paras 000–000. The only exception to this is contained in Article 28 which would not affect the present question.

34. That is unless the carrier is suing in respect of damage to goods belonging to other customers, in which case he will have the right to obtain full damages as a bailee: *The Winkfield* [1902] P. 42. Cf. further App. Bruxelles, 25.1.78 (1978) B.T. 429, see Clarke, p. 269 n. 19.

35. Carriage of Goods by Road Act 1965, section 14(2)(e).

36. Dr Hill has expressed the view elsewhere that section 14(2)(d) of the Carriage of Goods by Road Act 1965 operates to confer a direct right of action against the sender on the carrier's servants or agents. (Hill, "The carriage of dangerous goods by land", [1978] LMCLQ 74). With respect, for the reason given in the text, this would not appear to be correct. As to the effect of section 14(2), see the discussion *infra*, para. 11.15 fn. 26. Dr Hill's article contains a useful summary of the English common law position in relation to the carriage of dangerous goods.

37. On successive carriage generally, see *infra*, Chapter 11.

CHAPTER 9

Compensation

9.1 This chapter is concerned with the quantum of the carrier's liability to the cargo interests, which is dealt with by Articles 23 to 29 of the Convention. Article 23 contains the basic provisions as to the extent of that liability, but before turning to a detailed consideration of those provisions, some general explanation is required concerning Article 23(3), which contains the general limit of that liability. In its original form[1] that limit was defined in terms of the gold franc. That formula was conceived in a world financial system whereby the price of gold remained stable, as a way of establishing a uniform fixed unit of account which would not vary from country to country despite fluctuations in the exchange rates of individual currencies. This system, however, was dependent on the artificial device of declared official gold parities by members of the International Monetary Fund (IMF), but since 1971 the price of gold has fluctuated freely, producing a situation where there were in effect two gold values: the artificial official rate,[2] and the free market rate, the latter being considerably higher than the former and, moreover, constantly fluctuating.

9.2 Neither the Convention itself nor the Carriage of Goods by Road Act 1965 provided any mechanism for the conversion of the gold franc into the national currency, and it therefore became a matter of considerable controversy as to whether the official rate or the market rate was to be used in establishing the upper level of compensation under Article 23(3). As a result of these uncertainties, the Inland Transport Committee of the E.C.E. adopted a protocol to CMR[3] which has been ratified by this country, and the necessary amending legislation, contained in the Carriage by Air and Road Act 1979, is now in force.[4] That Act, in accordance with the provisions of the Protocol, substitutes a new para. (3) and adds a new para. (7) to Article 23, the general aim of which is to substitute reference to the Special Drawing Right (SDR) as defined by the IMF for the gold franc for the purposes of calculating the upper level of liability. These changes are considered in detail later in this Chapter,[5] but for present purposes it is important to note that Article 23 is set out here in its amended form. Article 23 therefore now provides as follows:—

> "1. When, under the provisions of this Convention, a carrier is liable for compensation in respect of total or partial loss of goods, such compensation shall be calculated by reference to the value of the goods at the place and time at which they were accepted for carriage.

1. Set out *infra*, para. 9.12.
2. Since 1973, the official rate has been $42.2222 per fine ounce.
3. Cmnd. 7480, H.M.S.O.
4. As from 28.12.80; see S.I. 1980, No. 1966.
5. See para. 9.9 ff.

2. The value of the goods shall be fixed accordingly to the commodity exchange price or, if there is no such price, according to the current market price or, if there is no commodity exchange price or current market price, by reference to the normal value of goods of the same kind and quality.
3. Compensation shall not however, exceed 8.33 units of account per kilogram of gross weight short.[6]
4. In addition, the carriage charges, Customs duties and other charges incurred in respect of the carriage of the goods shall be refunded in full in case of total loss and in proportion to the loss sustained in case of partial loss, but no further damages shall be payable.
5. In the case of delay, if the claimant proves that damage has resulted therefrom the carrier shall pay compensation for such damage not exceeding the carriage charges.
6. Higher compensation may only be claimed where the value of the goods or a special interest in delivery has been declared in accordance with articles 24 and 26.
7. The unit of account mentioned in this Convention is the Special Drawing Right as defined by the International Monetary Fund. The amount mentioned in paragraph 3 of this article shall be converted into the national currency of the State of the Court seised of the case on the basis of the value of that currency on the date of the judgment or the date agreed upon by the Parties."[7]

The basis of calculation

9.3 Where under the Convention a carrier is held liable for total or partial loss[8] of goods, the measure of damages payable will be calculated by reference to the value of the goods at the place and time at which they were accepted for carriage.[9] This provision is diametrically opposed to the English common law rules for assessing damages for loss or damage to goods in transit.[10] It was stated in the Court of Appeal that at common law "you look at the value at the end of the intended journey and subtract savings from that value. Under the Convention you look at the value at the beginning of the intended journey and then add to that value any extra losses suffered which properly fall within para. 4 of Article 23, or in special cases within Article 24 or Article 26".[11]

9.4 These rules, and the limitations imposed by them, cease to be relevant in circumstances where the claim is not one involving loss, damage or delay to the goods but involves a claim for loss based on the contract of carriage or some applicable rule of national law which is not restricted by CMR.[12] In *Shell Chemicals U.K. Ltd.* v. *P. & O. Roadtanks Ltd.*,[13] a tank carrier instructed to deliver chemicals to a refinery mistakenly

---

6. Carriage by Air and Road Act 1979, section 4(2)(a).

7. Ibid., section 4(2)(b).

8. Article 23(1)(2) and (4) deals with compensation for loss, but by virtue of Article 25(1) (*infra*, para. 9.47 ff.) the rules are the same in respect of claims for damage to the goods.

9. Article 23(1). In a French case concerning goods for temporary exportation the court held that their value must be fixed on the day of their departure, provisionally, from Fance: App. Chambery, 24.6.97 (1997) B.T. 667.

10. Thus loss of market, or loss of interest on capital represented by the goods, is not recoverable under Article 23, see Cass., 3.2.87 (1987) B.T. 270. See, however, Article 26, *infra*, para. 9.50 ff.

11. *James Buchanan & Co. Ltd.* v. *Babco Forwarding & Shipping (U.K.) Ltd.* [1977] 1 Lloyd's Rep. 234, 240–241, *per* Roskill, L.J. Thus, it has been held in Germany that, when determining a diminution of value in the case of damage, the price payable for the damaged goods at destination is not relevant, but regard is to be had to the value such damaged goods would have had at the place and time of taking them over, OLG. Düsseldorf, 14.7.83 (1984) TranspR 16, see Glockner, Theunis, p. 99.

12. Furthermore, the limitations on compensation provided by these rules can also be broken in cases of wilful misconduct (*infra*, para. 9.72 ff.) or where there has been failure to comply with Article 6(k) so that the carrier incurs the penalty provided for by Article 7(3), App. Aix-en-Provence, 22.3.90 (1990) B.T. 796.

13. [1993] 1 Lloyd's Rep. 114 (see Clarke, [1993] LMCLQ 156) affirmed on different grounds by the Court of Appeal (27.5.94), [1995] 1 Lloyd's Rep. 297.

delivered detergent, which was inadvertently pumped into the refinery causing loss and expense. This was recoverable without limitation by reference to the provisions of CMR.[14] In a French decision, the failure of the goods to reach their destination was classified not as loss, damage or delay, but a failure by the carrier in the execution of the contract of carriage entitling the court to award damages on the basis of the law applicable to the contract rather than CMR.[15] A difficulty which emerges from the case law is the conflicting analyses of facts adopted by the individual courts. Whereas one court may see a particular set of facts as being within the scope of the concepts of loss, damage[16] or delay,[17] another court may see similar facts as outside their scope and so properly referable to national law.[18] Thus the French decision just noted can be contrasted with a German decision where goods which had partly deteriorated were refused an entry visa to France and had to be returned to the sender who sold them to another buyer. According to the court, this amounted, in effect, to a total loss justifying the recovery of costs of return freight and other charges on the basis of Article 23(4).[19]

9.5 However, assuming the application of Article 23, the value of the goods for the purposes of Article 23(1) is determined by reference to the commodity exchange price[20] if there is such a price or, if not, the current market price,[21] or, failing this, the normal value of goods of the same kind and quality.[22]

9.6 In practice the sale price indicated on the invoice will be used as the basis of calculation[23] unless it can be argued that this is not the true value of the goods at the place where they were accepted for carriage within the meaning of the Convention.[24] Such an

---

14. See also OLG. Munich, 3.5.89 (1991) TranspR 61 (however, see the discussion of this case by Clarke at p. 246), RB. Antwerpen, 7.1.77 (1977) 7 E.T.L. 420, H.R., 15.4.94 (1994) S. & S. No. 73, (1995) N.J. No. 114, p. 465, (1994–5) U.L.R. 378, Krings, p. 778. In BGH., 14.7.93 (1993) 28 E.T.L. 917, the plaintiff had relied on the carrier's assurance that the vehicle, which was late, would arrive soon, and as a consequence made a loss of profit by not taking steps that would have maintained his sub-sale. This loss was held to be recoverable, not on the basis of the delay but on the basis of liability for the misstatement which was not based on CMR and consequently not restricted by the limitation in Article 23(5).

15. Comm. Carpentras, 19.2.93 (1994) B.T. 636. Cf. further OLG. Düsseldorf, 9.3.95, *infra*, para. 000.

16. *Infra*, para. 10.59 fn. 139.

17. See *supra*, para. 7.1.

18. Cf. App. Toulouse, 29.3.94 (1994) B.T. 736 where the court held that the limitations of Article 23 were not applicable where there was delivery without proof of payment contrary to instructions but, oddly, the court nevertheless applied Article 27 (see *infra*, para. 9.60).

19. BGH., 3.7.74 (1974) II U.L.R. 216, see Clarke, pp. 220–221, cf. Arrond. Middlebourg, 26.6.63 (1964) U.L.C. 167. Cf. BGH., 27.10.78 (1979) N.J.W. 2473, criticised by Libouton, (1982) J.T. 722, para. 98.

20. This "plainly refers to goods regularly dealt with on a recognised commodity exchange. If it be permissible to look at the French text of the Convention, the phrase '*d'après le cours en bourse*' strongly supports this view", *James Buchanan & Co. Ltd.* v. *Babco Forwarding & Shipping (U.K.) Ltd.*, *supra* fn. 11., per Roskill, L.J., at pp. 239–240. So, e.g., the London exchange price of tin was applied in RB. Amsterdam, 19.1.77 (1979) 7 E.L.D. 401.

21. In App. Paris, 9.7.80 (1980) B.T. 449 the court applied the current market price even though the real cost of manufacture of the machine destined for an exhibition was greater.

22. Article 23(2). OLG. Nürnberg, 14.6.65 (1971) 6 E.T.L. 247, cf. Cass., 11.1.94 (1994) B.T. 212. It matters not that the application of these rules enables the plaintiff to recover more than if the carriage had, in fact, been properly performed: Supreme Court of Denmark, 11.11.68 (1971) U.L.C. 305. However, stipulation of a higher value in insurance documents will not entitle recovery of the higher figure, nor will the excess be recoverable under Article 23(4) (see *infra*, para. 9.21), App. Paris, 22.4.92 (1992) B.T. 362.

23. Provided that it can be shown that the invoice in fact relates to the goods lost or damaged, see OLG. Oldenburg, 25.9.96 (1997) TranspR 229.

24. *William Tatton & Co. Ltd.* v. *Ferrymasters Ltd.* [1974] 1 Lloyd's Rep 203, cf. App. Chambery, 27.6.84 (1985) B.T. 159. See also OGH., 13.7.94 (1995) TranspR 285 (Mayer, p. 174). *Quaere*, however, whether this would apply in the event of an artificially low, or high, invoice price, for example between parent and subsidiary company. It was conceded by counsel for the defendants in the *Buchanan* case before Master Jacob (unreported) that in such circumstances the invoice price would not apply. See Cass., 8.2.82 (1983) 18 E.T.L. 43 where

argument can apply, for example, where the goods have been sold some time before the carriage has commenced and the value of the goods has increased in the meantime.[25]

9.7 The question of establishing the value of goods for the purposes of Article 23(1) was discussed at length in *James Buchanan & Co. Ltd.* v. *Babco Forwarding & Shipping (U.K.) Ltd.* both by the Court of Appeal and the House of Lords.[26] In that case the plaintiffs, who were whisky distillers, agreed to sell a shipment of whisky to Iranian buyers. The whisky was sold on f.o.b. terms. It was loaded into a container, ex-bonded warehouse in Glasgow, on a box trailer belonging to the defendants for onward carriage to a North Sea port, and ultimately by road to Teheran. The whisky was stolen while left unattended in London. The carrier's liability was not disputed, nor was it disputed that the contract was subject to the Convention, and the sole question before the court was the quantum of damage. As the whisky had been stolen in the United Kingdom the Revenue authorities claimed excise duty from the plaintiffs, which the latter paid as they were bound by law to do. This they claimed in turn from the carrier under Article 23. If the whisky had been successfully exported no such duty would have been payable. The f.o.b. Felixstowe value of the whisky was approximately £7,000 and the excise duty was almost £30,000. Although in the result the House of Lords, by a majority, decided in favour of the plaintiffs,[27] it was held that the value of whisky at the time and place at which it was accepted for carriage was to be fixed by reference to its current market price, which was held to be its current export market price, as evidenced by the plaintiff's invoices and by the evidence of the plaintiff's financial director to the effect that "the export market is a definite market and that this transaction was a normal piece of business".[28] It therefore did not include the excise duty as claimed by the plaintiffs.[29] Further, since the invoice price was f.o.b. it included an element in respect of carriage charges, which was also to be deducted for the purposes of Article 23(1) and (2).

9.8 Hardingham[30] is of the view that it was inappropriate to use the second limb of Article 23(2) (current market price) and that the third limb (normal value of goods of the

---

recovery was based on the normal sale price which would include the supplier's profit and not merely the cost of manufacture which was remitted under the arrangement between associated companies. See also App. Anvers, 9.10.85 (1988) J.P.A. No. 105 and Comm. Paris 3.12.96 (1997) B.T. 182, where the relevant invoice was that between the sender and consignee, thus including the margin of the sender over his supplier. The discount provided to a company which had a special relationship with the supplier was also disregarded for this purpose in Cass., 10.1.83 (1983) B.T. 154.

25. App. Amiens, 18.5.81 (1982) B.T. 208, Cass., 7.12.83 (1984) B.T. 538, cf. App. Paris, 30.5.84 (1985) B.T. 75 where, after being purchased in Scotland, the goods were temporarily despatched to Italy for fashioning, before being finally carried to France. The value was fixed according to the sale price and the invoice of the Italian fashioner. Clearly, the value cannot be determined by reference to the resale price to the plaintiff's ultimate buyer, App. Paris, 6.7.88 (1989) B.T. 270. In *B. Paradise Ltd.* v. *Islander Trucking Ltd.*, Queen's Bench Division, 28.1.85, unreported except on LEXIS, it was held by Webster, J., that it was proper to include as part of the value of the goods for the purposes of the Convention a 15% mark-up on their admitted value. This was stated to be on the basis of the evidence before him, but the nature of that evidence is unfortunately not apparent from the judgment.

26. [1977] 1 Lloyd's Rep. 234 (Court of Appeal); [1978] 1 Lloyd's Rep. 203 (House of Lords). See Hardingham, "Damages under CMR—the decision of the House of Lords", [1978] LMCLQ 51.

27. *Infra*, para. 9.25 ff.

28. Cf. OLG. Hamburg, 15.1.98 (1998) TranspR 290.

29. Cf. where loss of EC subsidy is occasioned by the failure to export due to the carrier's default. In France this has been held to be part of the price of the goods and thus recoverable under Article 23(1): App. Paris, 8.6.82 (1982) B.T. 564, App. Paris, 30.9.87 (1988) B.T. 59. See the contrary opinion of Thume, "Entschädigung nach Art. 23 CMR und Entgang einer Exportsubvention" (1995) TranspR 55, and cf. AG. Reutlingen, 30.4.96 (1996) TranspR 292.

30. Op. cit., p. 55.

same kind and quality) should have been used, on the basis that such goods normally purchased at that time and place would bear duty, the full amount therefore being recoverable. It seems unlikely that reliance on the third limb would on its own have made any difference to the result, since all three judges in the Court of Appeal considered the third limb and reached the same conclusion as under the second limb. However, Lord Denning, M.R., and Roskill, L.J., in the Court of Appeal, and Lord Wilberforce (with whose reasoning in this respect Lord Salmon agreed) in the House of Lords, were much concerned that if the market price or normal value were to include the excise duty, the plaintiffs would obtain a windfall in the event of the whisky having disappeared in circumstances such that the duty did not have to be paid. Hardingham argues that such a result would not follow, since it is provided only that compensation is to be calculated *by reference to* the value of the goods, not that such sum is recoverable in any event. Some support for such a view can be drawn from the comments on the wording of Article IV, Rule 5(b) of the Schedule to the Carriage of Goods by Sea Act 1971 in *Scrutton on Charterparties*.[31] It is there argued that the use of the same words indicates an intention that the value of the goods is intended only as a yardstick, thus admitting the possibility that more, or less, might be recoverable in the circumstances of a particular case. On the other hand, the use of the word "fixed" in Article 23(2) (which also appears in the same context in Rule 5(b)) perhaps suggests the contrary, but it can equally be argued that the word "fixed" in this context means no more than "assessed".

The limit of liability

9.9 As has already been explained,[32] in relation to those countries which have ratified the Protocol,[33] the position is now considerably simplified.[34] The carrier's maximum liability for the purposes of Article 23(1) and (2) is now fixed by Article 23(3) in its amended form at 8.33 units of account per kilogram of gross weight short.[35] "Unit of account" is explained by the new Article 23(7) as being the SDR as defined by the IMF. The SDR represents a notional unit of account, calculated on the basis of a "basket" of 16 currencies, which accordingly does not fluctuate in value to the same extent as any one currency within the basket. In essence, therefore, its object is to produce a unit of account free of the fluctuations of individual currencies.[36] Although doubts have been expressed as to the suitability of the SDR, based mainly on its failure adequately to take account of inflation,[37] such doubts can, it is hoped, be met by periodic reviews of the limit set. In this connection, Article 10 of the Protocol provides that any party to the Protocol can, after it has been in force for three years, request a conference for the review of the Protocol, such conference to be held if one quarter of the parties concur. A new section 8A, introduced

31. 20th edn., at p. 452.
32. *Supra*, para. 9.2.
33. Austria, Belgium, Denmark, Estonia, Finland, France, Georgia, Germany, Greece, Hungary, Iran, Ireland, Italy, Latvia, Lithuania, Luxembourg, Macedonia, Netherlands, Norway, Portugal, Romania, Spain, Sweden, Switzerland, Tunisia, Turkmenistar, United Kingdom (Guernsey, Gibraltar, Isle of Man), Uzbekistan.
34. See, however, Glockner, Theunis, p. 101 and Evans, Theunis, p. 183.
35. In France, an attempt to limit the carrier's liability by reference to a limit per consignment was held to be void under Article 41 (*infra*, Chapter 12): Cass., 17.5.83 (1983) B.T. 445
36. See generally Bristow, "Gold franc—Replacement of unit of account", [1978] LMCLQ 31.
37. See Tobolewski, "The Special Drawing Right in liability conventions: An acceptable solution?", [1979] LMCLQ 169. See further Loewe, "Le CMR a 40 ans", (1996) U.L.R. 429, 435–436.

into the 1965 Act by section 3(3) of the 1979 Act, gives power for any resulting amendment to be made by Order in Council.

9.10 Further, whatever doubts there may be as to the suitability of the new SDR limit, it does have the great advantage of producing a readily ascertainable[38] maximum figure in an area previously the subject of much controversy. Another major difficulty is also resolved by the Protocol, namely the date at which the rate is to be calculated. The new Article 23(7) removes all doubt in this respect by providing that the SDR amount is to be converted into the national currency of the court on the date of judgment[39] or the date agreed by the parties.[40]

9.11 The previous uncertainties in relation in Article 23(3) will, of course, continue in claims brought in jurisdictions which have not ratified the Protocol, and the choice of jurisdiction in relation to a CMR claim may well be influenced by having regard to which States have ratified it and which have not. For this reason, some consideration of the position under the Convention in its original form is required.

9.12 Article 23(3) in its original form provided:

> "3. Compensation shall not, however, exceed 25 francs per kilogram of gross weight[41] short. 'Franc' means the gold franc weighing 10/31 of a gramme and being of millesimal fineness 900."

As already indicated, the major controversy in relation to this provision was as to whether, in valuing the franc, regard was to be had to the official value or the market value of gold,[42] the latter producing a considerably increased possible maximum liability. This question never arose before the courts of this country, but cases in Continental jurisdictions have produced contrasting results,[43] so that Loewe was only able to describe the question as an open one.[44]

9.13 However, even for those countries which have not ratified the Protocol, some guidance as to the intention of the Convention in its original form can be deduced from the value of the SDR provided for by the Protocol. This is because the calculation based on the SDR produces a figure much closer to that based on the fixed price of gold than to that based on the market rate, and a court might well take the view that it was unlikely to have been the intention of the authors of the Protocol to reduce radically the limits of compensation, thus permitting the inference that the correct rate to apply is the fixed price rate.

9.14 Such inference is further supported by the fact that the Protocol makes provision for States which are not members of the IMF. It in effect stipulates, by way of an additional

---

38. Section 5(1) of the 1979 Act provides that the value on a particular day of one SDR shall be treated as equal to such sum in Sterling as the IMF have fixed as being equivalent to one SDR for that day, or, if no sum has been so fixed for that day, for the last previous day for which a sum has been so fixed. Section 5(2) provides that a Treasury Certificate certifying either of these facts shall be conclusive as evidence of them. The figure is calculated daily and published in the financial press.

39. See BGH., 5.6.81 (1982) 17 E.T.L. 310.

40. Then if it is to be paid in another country, Article 27(2) will apply. *Infra*, para. 9.60 ff. Unusually, by section 4(4) of the 1979 Act, the new SDR limit was given retrospective effect. Presumably, however, this has long since ceased to be of any relevance.

41. Intended to include the packing of the goods, Glockner, Theunis, p. 99, Clarke, p. 364.

42. See *supra*.

43. See, e.g., the cases cited by Tobolewski, op. cit., at p. 171 and OLG. Hamburg, 2.2.74 (1974) 9 E.T.L. 701. Also App. Bruxelles, 30.10.75 (1976) 11 E.T.L. 238, and see Haak, p. 211 *et seq*; (1982) 17 E.T.L. 274 (Scandinavian countries) and (1982) 17 E.T.L. 330 (Italy).

44. Loewe, para. 191.

provision in para. (7) and further new paras (8) and (9) to Article 23,[45] that States not members of the IMF can either make their own calculation of the value of their own currency in terms of the SDR, or, where the national law does not permit this, can declare a limit under Article 23(3) as 25 monetary units, such monetary units being defined in the same way as the gold franc in the original Article 23(3). It then provides that the conversion of that monetary unit "shall be made in such a manner as to express in the national currency of the State as far as possible the same real value for the amount of [8.33 SDRs]". As between the fixed rate and the market rate the former will, as already explained, be far closer to the new SDR limit, and so again it would seem to follow that, at least as far as the authors of the Protocol were concerned, the fixed rate is the appropriate rate under the unamended Convention. On the other hand, however, it has been held in this country in the context of the Hague Rules that the market rate is to be applied.[46]

Attempts to evade the limit

9.15 Section 3 of the Carriage of Goods by Road Act 1965 enables a court in this country, in any claim in which liability is limited by virtue of Article 23, to take account of any other proceedings which have been or are likely to be commenced, either in this country or elsewhere, to enforce that liability. The object of this provision was to ensure that there would be no duplication of compensation by the claimant commencing proceedings in a number of different countries, obtaining compensation each time limited by the Convention but unaffected by previous awards.[47] However, the vast majority of any such attempts at successive claims will be caught in any event by Article 31(2).

9.16 In *Ghandour* v. *Circle International Ltd.* at first instance it had been held that the carrier was estopped from relying on the limit of liability contained in Article 23(3) due to representations in the correspondence which had taken place after the loss in which the carriers had been seeking evidence of the value of the consignment. The Court of Appeal[48] reached the opposite conclusion. It was held that there was no unequivocal representation by the carrier that the limit was not going to be relied on. A separate argument based on misrepresentation was rejected on the basis that there was no evidence of reliance by the plaintiff.

9.17 In the same case, the first instance judge also rejected the carrier's reliance on Article 23(3) on the basis that the carrier had failed to adduce adequate evidence of the weight of the goods. This too was rejected by the Court of Appeal. The sender was responsible for providing the carrier with a statement as to the weight,[49] and indeed had done so: it was the plaintiff's figure for the weight which appeared in the freight quotation.

The calculation of the limit

9.18 For most purposes, the fact that the Convention provides for the carrier's maximum liability to be calculated with reference to the number of kilograms short overcomes

---

45. These provisions would obviously be superfluous in this country and are therefore not incorporated into the 1979 Act.
46. *The Rosa S* [1988] 2 Lloyd's Rep. 574.
47. Hansard, Vol. 715, p. 1063.
48. Court of Appeal, 2.11.99, unreported except on LEXIS.
49. Although not specified in the judgment, presumably the court was referring to Article 7 of the Convention.

the problem of *pro rata* calculations where the limit is based on larger units of weight.[50] Problems can arise, however, as to the method of calculation to be employed. Thus, for example, in a French decision a *commissionnaire de transport* arranged a consolidated shipment of several packages owned by different senders but covered by one consignment note which referred to the total number of packages and the total weight, but also supplied the carrier with both the description and the weight of each package. The court held that the limit was not to be calculated on the basis of the complete load, which was a total loss, but instead on the basis of each package according to its own weight.[51] The advantage to the carrier of such an interpretation is that the shippers of goods worth more than the limit in Article 23(3) are unable to benefit under a global sum from shippers whose goods were worth less than that amount, and such an interpretation produces an equitable result in relation to each individual consignor, avoiding windfalls to consignors of valuable goods which happen to be consolidated into consignments consisting of goods valued at less than the limitation figure.

9.19 It must be emphasised, however, that this French decision is dependent upon the fact that a groupage shipment was involved and that the forwarder's position as an intermediary enabled the court to view each shipper as being separately linked with the actual carrier.[52] On the other hand, a decision of the German Supreme Court reached the opposite conclusion so that, where goods were sent under a single contract of carriage, the carrier was not permitted to apply the limit to that half of the consignment which exceeded the limit whilst discounting it in respect of that half which fell below it. It was held that, in the interests of certainty, the weight of the whole consignment was to be brought into the calculation of the limit.[53] This approach has the advantage of rendering unnecessary any question of how a consignment should be split so as to apply the limit to different parts of it. In the case of a consolidated shipment the consignment would, in many cases, naturally divide into the goods belonging to the different shippers. If the rule were to be interpreted in a similar way in respect of consignments which did not involve the consolidation of goods belonging to different shippers, the courts would need to develop criteria to determine how the different parts of a consignment are to be separated for this purpose.

9.20 A similar issue arises in respect of the "per package" limitation of liability in the Hague Rules which has proved especially difficult to apply to the carriage of goods in containers. This issue is now resolved in English law by virtue of Article IV(5)(c) of the Hague-Visby Rules.[54] This provides that where a container, pallet or other similar article of transport has been used to consolidate goods, the number of packages or units listed in

---

50. As, e.g., arose in *F. S. Stowell Ltd.* v. *Nichols & Co. (Brighton) Ltd.* [1963] 2 Lloyd's Rep. 275 where it was held that a contractual limitation per ton was to be treated as meaning per ton or part of a ton. Presumably, were it to be significant in relation to the kilogram limit in Article 23(3), the same construction would be adopted.

51. Paris, 10.12.71 (1972) B.T. 19.

52. See also Cass. (Belg.), 3.6.76 (1976) II U.L.R. 220 in respect of the equivalent position in CIM.

53. BGH., 30.1.81 (1981) 16 E.T.L. 455. Similarly, in France, App. Paris, 15.6.84 (1984) B.T. 545, App. Paris, 25.2.87 (unpublished, see Lamy, para. 1576, Jur. 5) cf. however, where only part of the consignment is lost, Cass. 22.4.86 (1986) Bull. iv No. 66 p. 57, Krings, p. 778.

54. As given the force of law by the Carriage of Goods by Sea Act 1971. See also the consideration of similar issues in relation to the unamended Warsaw Convention in *Data Card Corp.* v. *Air Express International Corp.* [1983] 2 All E.R. 639, and in relation to the amended Warsaw Convention, *Applied Implants Technology Ltd* v. *Lufthansa Cargo AG* [2000] 1 All ER (Comm) 958.

the bill of lading as packed in such article or transport is to be deemed the number of packages or units for the purposes of calculating the liability of the carrier.[55] Adopting a similar approach, the description of the goods on the consignment note might equally be taken as a characterisation by the parties giving an indication of whether they intended the contents of the consignment to be treated individually rather than as a whole. Such an approach is consistent with the French decision referred to above where the consignment note clearly indicated that the shipment contained separate packages: "195 colis groupage: 7,837 kgs.".

Other expenses

9.21 Article 23(4) makes provision for the claimant to recover certain charges in addition to the value of the goods. Unfortunately the scope of this provision is far from clear. It provides that "carriage charges, Customs duties and other charges incurred in respect of the carriage of the goods" are recoverable,[56] and that these charges will be refunded in full in the case of total loss and in proportion to the loss sustained where only partial loss occurs.[57] The clause concludes by emphasising that no further damages will be payable.

9.22 Taking the matters specified in Article 23(4) in turn, clearly all carriage charges are recoverable.[58] This generally causes few problems where there is a total loss, but difficulties may occur where there is only a partial loss, and the question arises of dealing with those goods remaining whether damaged or not. So, for example, it has been held that where a vehicle breaks down the carrier will be liable not only for the original charges but also for the additional expenses incurred of reloading the goods and the charges for the remainder of the journey, although the Convention makes no specific provisions for such a situation.[59] Similarly, where the consignee himself makes arrangements to pick up the goods from a broken-down vehicle and deliver them, it has been held that his expenses will be chargeable to the carrier.[60] On the other hand, in *William Tatton & Co. Ltd.* v. *Ferrymasters Ltd.*[61] Browne, J., disallowed an item of claim relating to the return carriage of damaged goods. This aspect of the decision has, however, been disapproved in the

---

55. A similar rule placing emphasis on the characterisation by the parties in the shipping documents has been adopted under the developed American and Canadian case law in respect of the unamended Hague Rules as adopted in those countries. See *Binladen BSB Landscapting* v. *M.V. Nedlloyd Rotterdam* [1985] A.M.C. 2113 (U.S. 2nd Cir.), *Monica Textile Corporation* v. *S.S. Tana* [1991] A.M.C. 2378 (U.S. 2nd Cir.), *The Tindefjell* [1973] 2 Lloyd's Rep. 253 (Canadian Federal Court). See also Yates, Part 6, para. 6.2.4.3.8. and cf., however, the decision of the Court of Appeal in *The River Gurara*, [1997] 4 All E.R. 498 (see Glass, (1998) 1 *Shipping & Transport Law International* 10).

56. The French version of Article 23(4) provides: "*Sont en outre rembourses le prix du transport, les droits de douane et les autres frais encourus du transport de la marchandise.*"

57. See e.g. BGH., 14.12.88 (1989) R.I.W. 389, (1989) VersR 309 (Clarke, p. 369).

58. See *James Buchanan & Co. Ltd.* v. *Babco Forwarding & Shipping (U.K.) Ltd.* [1977] 1 Lloyd's Rep. 234, 239 where it was accepted that carriage charges would have to be refunded. See also *William Tatton & Co. Ltd.* v. *Ferrymasters Ltd.* [1974] 1 Lloyd's Rep. 203, where, on partial damage, a proportion of the carriage charges was awarded. It has also been held in a Dutch decision that where unloading costs are payable by the consignee, these are to be reduced to the extent to which the value of the goods is diminished: KG. Delft, 13.5.65 (1966) 1 E.T.L. 722.

59. Arrond. The Hague, 18.2.69 (1969) S. & S. No. 58.

60. Arrond. Roermond, 16.11.67 (1969) S. & S. No. 57.

61. [1974] 1 Lloyd's Rep. 203.

Court of Appeal,[62] and in *Thermo Engineers Ltd.* v. *Ferrymasters Ltd.*, Neill, J., indicated obiter that had he held CMR to be applicable, the claim for return freight in that case would in principle have been recoverable.[63] A similar view is taken in France[64] and applied also to the supplementary charge for onward carriage to a new consignee after justified refusal by the original consignee.[65]

9.23 The question of what amounts to Customs duties for the purposes of Article 23(4) may produce varying answers from country to country.[66] However, certain charges which may be regulated on a similar legislative basis, such as excise duties in this country, will not fall under this head,[67] nor would it seem that other forms of taxation such as VAT can be included within the definition of Customs duties.[68]

9.24 Generally, in relation to Article 23(4), it is important to bear in mind that whereas claims in respect of loss, damage or delay are subject to maximum limits under Article 23, there is no limit as regards the compensation payable under this provision. This partly explains the significance of the remaining head of liability under Article 23(4), namely "other charges incurred in respect of the carriage of the goods".

The Buchanan case

9.25 The question of the correct meaning of this provision was discussed at considerable length by both the Court of Appeal and the House of Lords in *James Buchanan & Co. Ltd.* v. *Babco Forwarding & Shipping (U.K.) Ltd.*,[69] the facts of which have been set out above.[70] In view of the importance of the issues raised, it is necessary to look at this decision in some detail.

9.26 In the result, a unanimous Court of Appeal, and the House of Lords by a bare majority, decided that the excise duty payable was recoverable in full from the carrier as being a charge incurred in respect of the carriage. There was, however, some divergence in the reasoning of the various judges. Lord Denning, M.R.,[71] was of the opinion that there was a gap in the provisions of the Convention, and set about filling that gap by reference to the "schematic and teleological" method of interpretation—otherwise referred to as the

---

62. *James Buchanan & Co. Ltd.* v. *Babco Forwarding & Shipping (U.K.) Ltd.* [1977] 1 Lloyd's Rep. 234, 238 and 241–242.

63. [1981] 1 Lloyd's Rep. 200, 207.

64. App. Paris, 13.2.85, unpublished, see Lamy, para. 1578, Jur. 4, who also cites two unpublished cases where the sender recovered the charges for return to him for repair or survey as well as of redelivery: Comm. Castres, 25.5.81, App. Versailles, 4.11.87. Cf. OLG. Munich, 5.7.89 (1990) TranspR 16, (1990) I U.L.R. 438.

65. App. Rennes, 19.3.87 (1988) B.T. 105. This was not applicable, however, to the cost of sending a replacement, App. Paris, 17.12.84, unpublished, Lamy, para. 1578, Jur. 6.

66. In App. Douai, 19.6.81 (1981) B.T. 512 the court held that "Customs duties" meant those duties normally incurred during transit and not those incurred because the goods had been stolen and presumed sold (Cf. the *Buchanan* case, *supra*, para. 9.7). Lamy, para. 1578, Jur. 7, says that this is incorrect and cf. App. Paris, 25.3.82 (1982) B.T. 434.

67. See the *Buchanan* case, *supra*, where it was common ground between the parties that the excise duty payable was not Customs duty within Article 23(4). Nor do Customs duties include import tax, OLG. Munich, 17.7.91 (1991) TranspR 427 (Clarke, p. 369, n. 98).

68. See App. Paris, 30.3.73 (1973) B.T. 195, discussed *infra*, para. 9.34.

69. [1977] 1 Lloyd's Rep. 234 (Court of Appeal); [1978] 1 Lloyd's Rep. 119 (House of Lords).

70. See *supra*, para. 9.7.

71. [1977] 1 Lloyd's Rep. 234, 235–238.

purposive approach—by having regard to the purpose behind the provision, rather than the strict literal method of interpretation. So viewed, he concluded that the sender "should also be compensated for any additional expense that he has incurred directly by reason of the loss",[72] thus including any expense rendered useless by reason of the loss, and any loss directly resulting, but excluding loss of profit and other consequential losses "not in the nature of an expense".[73] The House of Lords was unanimous in rejecting this approach.

9.27 Roskill and Lawton, L.JJ., on the other hand, regarded the text as ambiguous and felt able to derive assistance from the French text, to which they regarded it as permissible to have regard in cases of ambiguity.[74] The equivalent words in the French text are "*les autres frais encourus à l'occasion du transport*". Without stating any precise formula Roskill, L.J., expressed his agreement with the judgment of Master Jacob at first instance[75] and regarded the equivalent French words as "quite general in their nature and wide in their compass"[76] so as to include the excise duty. Lawton, L.J., did provide a formula: the English phrase is to be construed as meaning "any other expenses which the owner of the goods has to pay as a result of the carriage of the goods".[77]

9.28 In the House of Lords, Lord Wilberforce accepted the reasoning of Master Jacob at first instance and held that the words "in respect of" are wide enough to cover the way in which the goods were carried, miscarried or lost, and that they thus covered the excise duty.[78]

9.29 Viscount Dilhorne, pointing out that on a strict interpretation "charges in respect of the carriage" meant carriage charges, held that the words must be construed as meaning "in consequence of" or "arising out of" the carriage.[79] The other judge in the majority, Lord Salmon, also adopted the reasoning of Master Jacob and held that the excise duty was recoverable as being an expense incurred "in consequence of the way in which the goods were carried".[80]

9.30 As will be seen, the various judgments did not produce any clear test which can be applied in other circumstances which may fall to be considered. Two of the judges in the majority in the House of Lords used the phrase "in consequence of", and Lawton, L.J., in the Court of Appeal used the similar phrase "as a result of". Both phrases literally applied could extend the scope of the carrier's potential liability considerably beyond that under the normal common law rules of remoteness, which, generally speaking, is restricted to matters which are or ought to be in the contemplation of the parties at the time of the contract.[81] So to use an example from elsewhere,[82] if losses resulted from a penalty clause to which the claimant was subject, this would be in consequence of the carriage,

---

72. At p. 237.
73. At p. 238.
74. On the question of the legitimacy of having regard to the French text, see *supra*, paras. 0.14–0.17.
75. Unreported, but the material part of his judgment is set out in the judgment of Lord Edmund-Davies in the House of Lords at p. 130. Master Jacob's judgment is considered *infra*, para. 9.31.
76. [1977] 1 Lloyd's Rep. 234, 241.
77. At p. 244.
78. [1978] 1 Lloyd's Rep. 119, 124.
79. At p. 126.
80. At p. 127.
81. See *infra*, para. 9.55 ff.
82. Hardingham, op. cit.

and so on the face of it recoverable, although unless the carrier was aware of it, it would normally be too remote under normal English common law principles. The only judgment which effectively deals with the issue is that which at first sight seems the most expansive in its approach. Dealing with the words "but no further damages shall be recoverable", Lord Denning, M.R., was of the view that this would exclude claims for loss of profit.[83] Later, however, although not specifically referring to the same words, he refers to "consequential damage which is not in the nature of an expense" as also being excluded.[84] This would seem to raise the implication that in his judgment consequential damage which *is* in the nature of an expense can be recovered.

9.31 One is left with the question, therefore, as to whether the decision of the House of Lords leaves any restriction on which expenses are to be recoverable as "incurred in respect of the carriage". It is suggested that the answer lies with the judgment of Master Jacob,[85] approved by Roskill, L.J., in the Court of Appeal and Lords Wilberforce and Salmon in the House of Lords. He first concluded that the excise duty fell within the words of the Convention on the basis that it arose "having regard to the way in which the goods were carried". But, most importantly for present purposes, he continued:

"Mr. Buckley [Counsel for the defendant] argued that excise duty is different in different countries, and foreign carriers would not know what the domestic situation was. In my view, because it is international, persons concerned with international carriage must take the law as they find it or learn about it. If, here, excise duty is payable because the goods are lost during carriage, then that is a matter which is or ought to be within the knowledge of an international carrier."

This aspect of Master Jacob's judgment was not referred to by any of the judges in the Court of Appeal or the House of Lords, but it is submitted that it is a crucial limitation on the width of expression used by the House of Lords. In the present case, the carriers had actual or constructive knowledge that if the goods were stolen then the expense would be incurred, and so they could be held liable without offending normal rules of remoteness, and to that extent it is therefore perhaps not surprising that the various judges felt no need to deal with the point. In the event, however, of a claim arising based on an expense of which the carrier has neither actual nor constructive knowledge, for example, the penalty clause in a claimant's contract with a third party referred to above, then, even though arising in consequence of the carriage, it will not be recoverable. This would then accord with those Continental decisions which have refused to award damages in respect of consequential losses.[86]

---

83. [1977] 1 Lloyd's Rep. 234, 237.
84. At p. 238.
85. As already indicated Master Jacob's judgment is unreported, but the material part of his judgment is set out in the judgment of Lord Edmund-Davies in the House of Lords at p. 130.
86. So, e.g., claims have been rejected for: commercial losses, App. Paris, 30.5.73 (1973) B.T. 304; loss of profits, OLG. Düsseldorf, 18.11.71 (1973) 8 E.T.L. 510, (1973) II U.L.C. 428, App. Toulouse, 5.12.79 (1980) B.T. 13; loss of market, App. Paris, 30.5.84 (1985) B.T. 75, OGH., 15.2.79 (1979) II U.L.R. 229; prolonged stoppage and drop in factory production, Comm. Tournai, 21.11.72 (1972) J.P.A. 446. There are two French cases which have awarded losses on a broader basis: App. Poitiers, 31.3.71 (1971) B.T. 168, where damages were awarded for delay in getting a damaged piece of equipment into working order (criticised in the *Observations* at p. 170) and App. Paris, 9.6.67 (1969) 4 E.T.L. 911, where damages were awarded to take account of the fact that the claimant "has been put to expense, has been caused anxiety, and has had to pursue an action through to an appeal" (p. 916). Neither case is felt by Libouton to be good law: (1973) 8 E.T.L. 2, para. 70. See also Cass., 3.2.87 (1987) B.T. 270.

9.32 It would also seem that some restriction is placed on the width of what can be recovered as "other charges incurred in respect of the carriage" by the word "charges" itself, which appears in the French text as *"frais"*. Both words would seem to connote an obligation to make a payment to a third party, and would therefore not be appropriate to describe "losses", such as loss of production, which would not result from such an obligation. This is presumably what Lord Denning, M.R., had in mind when he spoke of "consequential damage which is not in the nature of an expense" as not being recoverable.[87]

9.33 An apparently paradoxical result of the House of Lords' decision in *Buchanan* v. *Babco*[88] is that it may well operate to prevent recovery by the claimant of such items as the costs of packing, insurance and quality certificates, which might properly be thought to be "charges in respect of the carriage", since they arise not "in consequence of" the carriage, but more in preparation for it.[89] This possible result was expressly referred to by Lord Dilhorne,[90] but considered by him to be of minimal significance in comparison with the much larger sums at stake in cases like that under consideration. But if it be a question of only one or other type of charge being recoverable, it would seem inherently more likely that the former was intended since some such charges will be applicable to almost every contract within the Convention.[91]

Other case law

9.34 In *Buchanan* v. *Babco*,[92] as in the *Ulster-Swift*[93] case discussed earlier,[94] attention was drawn to the desirability of a uniformity of approach by the parties to the Convention.[95] The practical difficulty of achieving such an approach is well illustrated by the fact that in the *Buchanan* case the House of Lords was faced with two Continental decisions on very similar facts which came to opposite conclusions. In a French decision,[96] goods were damaged during carriage and the consignees refused to accept delivery in France. The claimant forwarder was therefore unable to recover Value Added Tax paid when the goods were imported into France. It was held that the tax was recoverable from the carrier. On the other hand, in a Dutch decision,[97] a consignment of cigarettes was stolen from an export shipment loaded on a vehicle in Amsterdam prior to the goods leaving the country. As a result excise duty was payable by the sender and, as in the *Buchanan* case, would not

---

87. [1977] 1 Lloyd's Rep. 234, 238.
88. *Supra.*
89. The minority thought that these charges could be recovered: [1978] 1 Lloyd's Rep. 119, 132.
90. [1978] 1 Lloyd's Rep. 119, 126.
91. See further discussion of such items of expense, *infra*, para. 2.34 ff.
92. *Supra.*
93. [1977] 1 Lloyd's Rep. 346.
94. See *supra*, para. 6.11 ff.
95. See, e.g., Megaw, L.J. [1977] 1 Lloyd's Rep. 346, 348.
96. App. Paris, 30.3.73 (1973) B.T. 195 (affirmed, Cass., 28.7.75 (1975) B.T. 442). This case was not referred to by the Court of Appeal. Cf. App. Poitiers, 23.8.90 (1990) B.T. 586, Comm. Paris, 10.2.93 (1993) B.T. 332, Hof. Antwerpen, 10.11.81 (1983–84) J.P.A. 177, OGH., 25.1.90 (1990) TranspR 235. Generally, however, French courts are said to confine recoverable charges to those strictly tied to the transport, Lamy, para. 526. See e.g. Cass., 14.3.95 (1995) B.T. 248, cf. however Cass., 3.3.98 (1998) B.T. 212.
97. Arrond. Amsterdam, 30.3.77 (1978) II U.L.C. 299. This case was decided between the decisions of the Court of Appeal and House of Lords in *Buchanan*. Cf. Hof. 's Hertogenbosch, 4.2.86 (1987) S. & S. No. 25, Hof. 's Gravenhage, 14.8.86 (1987) S. & S. No. 24, cited by Clarke, p. 372, n. 24.

have been payable had the goods left the country. The court rejected the claimant's argument that the duty was recoverable under Article 23(4) in the following terms[98]:

" . . . the charges in question must in any case have been incurred in direct connexion with the carriage such as it should have been performed. [Here the duty was] more of a subsequent levy or administrative fine than an item of charges in respect of carriage."

Similarly, in another Dutch decision, it was held that penalties and legal costs incurred in relation to the T form under the Community Transit Scheme were not recoverable under Article 23(4). The case concerned the theft of part of a cargo of tin, and again the relevant charges would not have been levied if the goods had left the country.[99]

9.35 There are other examples of the lack of uniformity of approach in Continental decisions under Article 23(4). So in one German decision concerning the defective refrigeration of a consignment of fish resulting in partial loss, it was held that the claimant could recover part of the carriage charges for the outward journey, but was not entitled to a refund of expenses arising out of the return trip, storage charges, deep freeze charges or the fees payable for an expert survey.[100] Yet, in another German decision, it was held that the claimant could recover the cost of the insurance premiums, and also the costs of correspondence and telephone charges subsequent to the loss.[101] In Holland it has been held that valuation costs are recoverable, and storage charges if necessary to obtain a better price on the market,[102] but in another decision it was held that although reloading costs were recoverable, the claimant was not entitled to the travelling expenses of persons travelling to inspect the goods.[103] In France, it has been held that the cost of steps taken for the preservation of a cargo of fruit was recoverable.[104] Similarly, the costs of sending experts to survey the damage have been recovered in France.[105]

9.36 It will be seen that there are two basic categories of such expenses, those preparatory to the carriage, such as insurance,[106] packing, and certificates, and those

---

98. Quoted with approval by Lord Edmund-Davies in his dissenting judgment in *Buchanan, supra,* at p. 133. (See also, BGH., 13.2.80 (1981) I U.L.R. 260; cf., however, Supreme Court, Denmark, 4.5.87 (1994) 29 E.T.L. 360 which saw Dutch law as permitting recovery of excise duty under Article 23(4)). However, the majority favoured the French decision.

99. RB. Amsterdam, 19.1.77 (1979) 7 E.L.D. 401 and Hof. Amsterdam, 24.2.94 (1995) S. & S. No. 18 (Cleton, (1996) U.L.R. 755, 758).

100. BGH., 3.7.74 (1975) 10 E.T.L. 75, cf. OLG. Celle, 29.10.98 (1999) TranspR 106. Similarly, in respect of expert's fees, LG. Hamburg, 25.2.85 (1985) TranspR 188, cf. OLG. Munich, 5.7.89 (1990) TranspR 16. See also in Austria OLG. Wien, 23.2.89 (1990) TranspR 157, (1989) II U.L.R. 811. German opinion generally favours a narrow view, e.g Herber/Piper, Art. 23 Rdn. 27, Koller, Art. 23 Rdn. 10 and others cited in Clarke, p. 369, n. 4.

101. OLG. Düsseldorf, 18.11.71 (1973) 8 E.T.L. 510, (1973) II U.L.C. 428.

102. RB. Rotterdam, 20.2.76 (1978) 6 E.L.D. 291. Cf. Arrond. Rotterdam, 12.4.72 (1976) I U.L.C. 253: costs of experts' survey included within Article 23(1). See also KG. Delft, (1966) 1 E.T.L. 722 (unloading and storage charges recovered under Article 23(4)). In Belgium, it has been held that the cost of an expert's report which did not contribute to the sorting and preservation of the goods in sound condition could not be awarded since they are not costs necessarily incurred in the course of the carriage of the goods, Hof. Antwerpen, 23.3.83 (1983) 18 E.T.L. 518, cf. Hof. Antwerpen, 9.10.85 (1988) J.P.A. 105. Costs of destroying the goods were not recovered in RB. Antwerpen, 13.12.96 (1996) J.P.A. 365.

103. RB. Amsterdam, 12.4.72 (1972) S. & S. No. 102.

104. App. Paris, 30.5.73 (1973) B.T. 304. In App. Paris, 25.3.88 (1989) B.T. 46, the costs of testing bottles of propane gas after the accident were recovered, although as Lamy points out, at para. 1578, Jur. 19, it is hard to determine whether this was under mitigation or under Article 23(4). See also Cass., 3.3.98 (1998) B.T. 212 in respect of salvage and storage costs.

105. App. Poitiers, 31.3.71 and 23.8.90 (1971) B.T. 168, (1990) B.T. 586, (1991) II U.L.R. 324, cf. App. Aix-en-Provence, 9.12.80 (1981) B.T. 143, App. Riom, 16.10.81 (1982) B.T. 11.

106. Insurance costs were not recovered in App. Aix-en-Provence, 20.3.85 (1985) Rev. Scapel 41 (Lamy, para. 1578 Jur. 17). Contrast in Germany, *supra,* para. 9.35, fn. 101.

subsequent to the damage, such as storage and valuation charges. Loewe[107] is of the view that the latter category is recoverable so long as incurred reasonably, but that the former is outside the scope of the paragraph, being merely preparatory to the carriage. It will be recalled that Viscount Dilhorne in *Buchanan* v. *Babco*[108] indicated that, as a result of the decision in that case, preparatory expenses may not be recoverable.

9.37 Claims in respect of both types of expense were considered by Browne, J., in *William Tatton & Co. Ltd.* v. *Ferrymasters Ltd.*[109] which arose from damage to machinery being carried from England to Italy. The plaintiff's claim included items in respect of packaging, carriage charges and the costs of the return carriage, of warehousing the damaged machinery, and of visits by experts to inspect the machinery both at the place of damage and in England. The defendants were held liable for the first two items, but not for the remainder, on the basis that the relevant contract, with which the Convention was concerned, was for carriage from England to Italy, and that therefore the carriage back and subsequent costs were outside the scope of the Convention.

9.38 This case must now be treated with some caution as a result of the disapproval of it expressed in *Buchanan* v. *Babco* by Lord Denning, M.R., and Roskill, L.J.,[110] and in *Thermo Engineers Ltd.* v. *Ferrymasters Ltd.* by Neill, J.[111] It was not followed by Staughton, J., in *ICI PLC* v. *MAT Transport Ltd.*[112] who held that survey fees were recoverable under Article 23(4). He also indicated that he would in any event view the survey fees in that particular case as part of the cost of realising the value of the damaged goods, referring to an Amsterdam District Court decision to that effect. Accordingly, as he put it, "in meal or in malt", the survey fee was recoverable under the Convention. Clearly in many cases survey fees will be incurred by cargo interests and their insurers in circumstances where there is no question of any salvage sale being possible. Although Staughton, J., relied on both grounds, it seems clear from his decision that he would have reached the same decision on the basis of Article 23(4) alone.

9.39 Of course, in most cases the survey fees will be incurred by or on behalf of the cargo interests' insurers, whereas in this country the legal proceedings will be brought in the name of the actual party interested in the cargo. Unsurprisingly, it has therefore been argued that the survey fees are not properly claimable by the plaintiff on the basis that it is not an expense incurred by the plaintiff. However, this argument was rejected in the case of *M. Bardiger Ltd.* v. *Halberg Spedition APS.*[113]

9.40 However, there is a limit to the circumstances in which survey fees will be recoverable. In *Ken Bell (International)* v. *Eric Bacon Transport Services Ltd.*,[114] which concerned a vehicle hi-jack, the role of the surveyor had been to obtain evidence as to the circumstances of the hi-jack. The judge was of the view that their fees were recoverable (if at all) as costs of the action rather than as damages. He drew a distinction from fees incurred by a surveyor in supervising repairs, whose fees are part of the cost of the repairs,

107. Loewe, para. 193.
108. [1978] 1 Lloyd's Rep. 119, 126, *supra*, para. 9.33.
109. [1974] 1 Lloyd's Rep. 203.
110. [1977] 1 Lloyd's Rep. 234 and 241–242 respectively, although, as Roskill, L.J., pointed out, the court had not had the benefit of argument on the case, not having been referred to it during the hearing.
111. [1981] 1 Lloyd's Rep. 200, 207.
112. [1987] 1 Lloyd's Rep. 354.
113. Queen's Bench Division, 26.10.90, unreported except on LEXIS. The same contention was also rejected in the case of *Ken Bell (International)* v. *Eric Bacon Transport Services Ltd.*, *infra*.
114. C.L.C.C.(B.L.), 22.7.98, unreported.

and held that on this occasion the survey fees were not recoverable under Article 23(4).

9.41 Such disapproval as there has been of the decision in *William Tatton & Co. Ltd. v. Ferrymasters Ltd.*[115] has only been as to the disallowed items of claim, so it remains good authority, subject to Viscount Dilhorne's observations referred to above, for the view that packing costs are recoverable, which in turn would suggest that other costs preparatory to the carriage are equally recoverable. It has been suggested above,[116] in relation to Viscount Dilhorne's comments, that it seems likely that the Convention was intended to apply to such costs as packing as these will necessarily arise in relation to almost every carriage, and it would furthermore seem that such preparatory expenses are just as much wasted expenditure as the other matters referred to in Article 23(4). To return to the wording of Article 23(4), it refers to "other charges incurred in respect of the carriage", which would surely seem wide enough to include not only costs arising in consequence of the carriage, but equally costs arising in preparation for it.

Delay

9.42 Where delay, as defined in Article 19[117] occurs, if the claimant can prove that damage has resulted therefrom,[118] the carrier will be liable to pay compensation for such damage up to but not exceeding the amount of the carriage charges for the transit in question.[119] Clearly it is first necessary for the claimant to satisfy the requirements of Article 19 and to prove that actionable delay has occurred. The measure of damages applicable will then be limited to the carriage charges paid or payable,[120] and will be based upon the charges due for the whole transit and not merely on those due for the section of the journey in which the delay occurred.[121] Also, it has been held in France that the charges are those due for the whole contract, so that where the consignment is split between several vehicles, only some of which are delayed, it is all the carriage charges and not just those in respect of the delayed vehicles which are brought into account for the purposes of the limit.[122] It would seem that charges for this purpose will be calculated so as to exclude Customs duties and other associated charges referred to in Article 23(4).[123] It should be noted that the sum of the carriage charges is merely the upper limit of

---

115. *Supra*, fn. 109.

116. See *supra*, para. 9.33.

117. *Supra*, Chapter 7, paras 7.1–7.6.

118. In OLG. Düsseldorf, 23.12.96 (1997) TranspR 422, the court felt able to assume damage to the extent of the limit of liability by the fact that the defendant had been willing to settle for a much higher amount!

119. Article 23(5). A different basis for damages may apply where there is failure to complete the carriage. In App. Reims, 5.5.80 (1980) B.T. 299 the carrier abandoned the transport which was completed by a fresh carrier. The expense of this exercise was recovered without regard to the limit in Article 23(5), since it was not treated as a case of "delay". As explained by Libouton, (1982) J.T. 724, para. 110, the charges were incurred to avoid a loss which might otherwise have been considered to be established under Article 20, *supra*, para. 7.11 ff. Cf. *infra*, fn. 120.

120. App. Paris, 2.4.71 (1971) B.T. 155. See also Cass., 5.7.76 (1976) B.T. 377, (1977) II U.L.C. 204, (1979) 7 E.L.D. 442. In Germany, a fixed-date agreement which, under German law, would involve a right of rescission and consequent complete recovery of carriage charges was held not to have the same consequence under CMR since the sender's remedy lay in Article 17 and Article 23(5): OLG. Düsseldorf, 9.3.95 (1995) TranspR 288 (Jung, p. 154).

121. App. Paris, 18.10.73 (1973) B.T. 488, (1976) I U.L.C. 242.

122. Cass., 26.6.84 (1984) B.T. 610.

123. Loewe, para. 194.

damages payable for delay under Article 23(5). It will not be automatically awarded to the claimant who will be required to prove the extent to which he has suffered loss.[124] A clause indicating a penalty payable by the carrier for delay will be null and void under Article 41.[125]

9.43 Compensation in respect of delay can only exceed the sum of related carriage charges in two cases. First, if the sender has made a declaration of a special interest in delivery in the event of an agreed time-limit being exceeded, under the provisions of Article 26,[126] then the maximum recoverable will be the amount entered in the consignment note.[127] Secondly, where the loss results from wilful misconduct or such default as is considered by the court as equivalent to wilful misconduct,[128] then the liability of the carrier will be unlimited and he will be liable for the full loss incurred as a result of the delay. This will apply irrespective of whether or not the sender has made a declaration of a special interest in delivery under Article 26.[129]

9.44 A question that has not been satisfactorily answered to date in relation to Article 23(5) is exactly what "damage" the claimant must prove to have resulted from the delay. On the one hand this could refer to damage to the goods themselves, on the other it could refer to damage suffered by the claimant other than to the goods. As a third possibility it could cover both. If anything, the English text tends to support the first interpretation, since with the exception of Article 29, the word "damage" is used throughout the Convention with clear reference to physical damage to the goods. Turning to the French version, however, it can be seen that, whereas the word "*avarie*" is used in Articles 17(1) and 25 to refer to actual damage to the goods themselves, the word "*préjudice*" is used in Article 23(5).[130] It can therefore be assumed, bearing in mind the Latin precision of the French language as used by jurists, that a different meaning is intended in the respective provisions. Moreover, whereas the word "*avarie*" is normally used to describe damage to a physical object,[131] the word "*préjudice*" is more applicable to a general detriment or delictual injury. It would therefore appear a logical surmise that the damage referred to in Article 23(5) is intended to cover loss or damage suffered by the claimant other than physically to the goods themselves. Given the fact that Article 23(3) as applied by Article 25 gives one rate of recovery where the goods have themselves suffered damage, it would

---

124. Cass., 6.10.75 (1975) B.T. 478, App. Amiens, 19.6.80 (1980) B.T. 490, OLG. Düsseldorf, 15.12.94 (1995) TranspR 244 (Jung, p. 155).

125. Comm. Paris, 15.6.78 (1978) B.T. 420, OLG. Munich, 26.7.85 (1985) TranspR 395. As explained in the latter case, since the claimant must prove his damage up to the limit in Article 23(5), this would be defeated by a clause which provided for a penalty without proof of damage and without a fixed upper limit, see Glockner, Theunis, p. 104. A penalty would be void under English law in any event, although there might be less objection to a liquidated damages clause since this is permissible on the basis that it represents a genuine attempt by the parties to pre-estimate the loss which will result from the breach, see Treitel, *The Law of Contract*, 10th edn. p. 929 and *Dunlop Pneumatic Tyre Co. Ltd. v. New Garage & Motor Co. Ltd.* [1915] A.C. 79. However, since such clauses can override normal principles of remoteness (*Robophone Facilities Ltd. v. Blank* [1966] 1 W.L.R. 1428) and might anyway overstep the limit in CMR, they are more likely to be treated as void. For Article 41, see *infra*, Chapter 12.

126. *Infra*, para. 9.50 ff.

127. Article 23(6).

128. Article 29. Comm. Paris, 12.10.71 (1972) 1 Gaz. Pal. 132 (1973) I U.L.C. 268. See, generally, *infra*, para. 9.72 ff.

129. App. Paris, 18.12.68 (1969) B.T. 98. See also *Antwerp United Diamond BVBA v. Air Europe* [1993] 4 All E.R. 469 (affirmed Court of Appeal, *The Times*, 22 April 1995), see *infra*, para. 9.73 fn. 217.

130. See *supra*, paras 0.14–0.17 as to the propriety of having regard to the French text.

131. See *Fothergill v. Monarch Airlines Ltd.* [1980] 2 Lloyd's Rep. 295.

appear illogical to offer an alternative remedy based upon carriage charges which may be insignificant in relation to the value of the goods carried.

9.45 It is surprising therefore to find a French decision where it was held that where damage is caused to goods through delay, compensation must not exceed the carriage charges.[132] More recently, however, the opposite view has been adopted at the highest level in both France and Germany.[133] The view that Article 23(5) applies in respect of losses other than physical damage, which is separately and exclusively provided for by Article 25, is supported by several writers,[134] and it is submitted that such is the preferable approach. A similar view was supported by case law in respect of the equivalent provisions of the previous versions of CIM.[135] The latest version, however, has been amended to make it clear that damage to the goods produced by delay is subject to the limit attached to delay rather than damage which, it seems, was the actual intention of the authors of the 1961 version of CIM.[136]

9.46 Article 23(5) also raises problems in relation to remoteness of damage, which are discussed below.[137]

Damage

9.47 Article 25 provides as follows:—

> "1. In case of damage, the carrier shall be liable for the amount by which the goods have diminished in value, calculated by reference to the value of the goods fixed in accordance with article 23, paragraphs 1, 2 and 4.
> 2. The compensation may not, however, exceed:
>   (a) if the whole consignment has been damaged the amount payable in the case of total loss;
>   (b) if part only of the consignment has been damaged, the amount payable in the case of loss of the part affected."

Article 25 deals with the question of assessing the measure of damages where goods have been damaged rather than lost. Here the first point of reference is the amount of value the goods have lost. It has thus been held in a German decision that the claimant is not entitled to any higher costs he might have incurred in having the goods repaired.[138]

9.48 The basis of calculation is then the same as that contained in Article 23(1) and (2), i.e. the value of goods at the time and place of delivery, fixed by reference to commodity

---

132. App. Aix-en-Provence, 20.12.77 (1979) I U.L.R. 299.

133. Cass., 5.12.89 (1990) B.T. 310, (1990) I U.L.R. 454, BGH., 15.10.92 (1993) 28 E.T.L. 740, (1993) TranspR 137, cf. OGH., 15.2.79 (1979) II U.L.R. 229. Cf. where damage to the goods is claimed to have caused delay see OGH., 22.3.94 (1994) TranspR 439. See also OLG Cologne 14.3.97 (1998) TranspR 195 where the delay was seen merely as providing the occasion in which frost damage occurred rather than being the cause.

134. Haak, p. 230, Libouton, (1973) 8 E.T.L. 2, para. 71, Clarke, para. 59b, cf. Nickel-Lanz, No. 111.

135. See Cass., 17.4.80 (1980) B.T. 284.

136. Article 43 of the CIM Rules of the COTIF Convention 1980, see Libouton, (1982) J.T. 724, para. 110, Clarke in Yates, para. 4.1.2.43.

137. See paras. 9.55–9.59.

138. BGH., 13.2.80 (1981) I U.L.R. 260. See, however, OLG. Munich, 5.7.89 (1990) TranspR 16, (1990) I U.L.R. 438 where the goods were considered to have diminished in value by at least the sum that had been spent to repair the damage. Similarly, OLG. Hamburg, 15.1.98 (1998) TranspR 290: the court required a deduction to be made to take into account the replacement of new parts for old. Cf. further Cass., 25.11.97 (1997) B.T. 852.

exchange price, current market price, or normal value.[139] Article 23(3), containing the limit of liability in case of loss, is not expressly referred to, but is in effect applied to cases of damage by Article 25(2).[140] Under the latter provision, if the whole consignment has been damaged, the measure of damages may equal but not exceed the sum which would have been paid if the consignment was a total loss. Where only part of the consignment has been damaged, the measure of damages cannot exceed the sum payable if that part had been a total loss. Difficulties may arise in this respect where, as in the case of a shipment of machinery, damage to one or more parts may render the whole shipment unusable. In such circumstances it is not clear from the text of the Convention whether this is to be regarded as damage to the whole consignment under Article 25(2)(a), or as damage to the individual parts of the machinery, so that the limit on compensation would be based on the value of the parts in question under Article 25(2)(b). Loewe is of the view that Article 25(2)(a) applies to such cases,[141] and is generally supported by the case law.[142] In a recent German decision, however, the court took the view that the distinction between partial and total depreciation is an objective and economic one, the issue depending upon whether or not there was permanent economic loss. Consequently, where the damaged component was repaired in less than two weeks and although the claimant suffered loss from the inability to use the whole machine, since there was no permanent depreciation of the machine, Article 25(2)(b) was applied.[143] Furthermore, as to whether in appropriate cases partial damage will be sufficiently severe to amount to constructive total loss, which would then bring the claim within the provisions of Article 25(2)(a), see the discussion below in relation to Article 32(1)(a) and (b).[144]

9.49 The position in relation to consolidated shipments will be the same in relation to partial damage as in relation to partial loss, so reference should be made to the earlier discussion on this point.[145] Also, by virtue of the reference to Article 23(4) in Article 25, carriage charges, Customs duties and other charges incurred in respect of the carriage will be repayable, on a proportionate basis in the case of partial damage, and the same problems arise as discussed above[146] as to exactly what falls within the scope of those words. Apart from Article 23(4), there is scope under Article 25(1) to bring into account additional costs incurred after the goods have been damaged, as part of the process of

---

139. See *supra*, para. 9.2 ff. It has been held in Germany that the compensation payable under Article 25(1) is a standard rate compensation in the calculation of which no account should be taken of the particular situation of the damaged goods (in the particular case, the real possibilities of sale), BGH., 15.10.92 (1993) 28 E.T.L. 740, (1993) TranspR 137.

140. *William Tatton & Co. Ltd.* v. *Ferrymasters Ltd.* [1974] 1 Lloyd's Rep. 203, 205.

141. Loewe, para. 201.

142. App. Aix-en-Provence, 25.9.90 (1991) B.T. 340, cf. App. Paris, 31.1.84 (1984) B.T. 543, App. Orléans, 18.1.95 (as reported by Tilche, (1995) B.T. 123), RB. Brussel, 9.1.78 (1977–78) J.P.A. 278 (similarly where partial loss and damage affects the value of the whole consignment, OLG. Stuttgart, 15.9.93 (1994) TranspR 156). Cf. Comm. Lyon, 10.11.75 (1976) B.T. 175, criticised by Libouton, (1982) J.T. 724, para. 109, and Haak, p. 217, where the parts for a machine were divided for carriage between several lorries. The court awarded damages only for the despatch of the damaged parts and not for the fact that all the parts had to be returned for the purposes of restoration (the decision was reversed, App. Lyon, 21.7.77 (1977) B.T. 97, but on grounds not relevant to this point). Cf., further, BGH., 3.7.74 (1975) 10 E.T.L. 75.

143. BGH., 6.2.97 (1997) TranspR 335, (1997) U.L.R. 834, (1998) 33 E.T.L. 701, cf. OLG. Hamburg, 15.1.98 (1998) TranspR 290.

144. *Infra*, para. 10.59 ff.

145. *Supra*, para. 9.18 ff.

146. *Supra*, para. 9.21.

determining the extent to which the goods have diminished in value as indicated in *ICI PLC* v. *MAT Transport Ltd.*[147]

Additional compensation

9.50 In the absence of wilful misconduct,[148] higher compensation for loss, damage or delay can only be claimed[149] where the value of the goods has been declared in accordance with the procedure laid down in Article 24 or else where a special interest in delivery has been declared under Article 26.[150] In both cases an entry to that effect must be made in the consignment note.[151] The Articles provide as follows:—

"Article 24

The sender may, against payment of a surcharge to be agreed upon, declare in the consignment note a value for the goods exceeding the limit laid down in article 23, paragraph 3, and in that case the amount of the declared value shall be substituted for that limit.

Article 26

1. The sender may, against payment of a surcharge to be agreed upon, fix the amount of a special interest in delivery in the case of loss or damage or of the agreed time-limit being exceeded, by entering such amount in the consignment note.
2. If a declaration of a special interest in delivery has been made, compensation for the additional loss or damage proved may be claimed, up to the total amount of the interest declared, independently of the compensation provided for in articles 23, 24 and 25."

It might at first be thought that there is a substantial overlap between Articles 24 and 26, but in fact they serve distinct functions. Article 24 will be relevant where the limit provided by Article 23(3) is not sufficiently high for the goods in question.[152] The limit

---

147. *Supra*, para. 9.38, *per* Staughton, J., at p. 362. Cf. the view of Brown, J., in *William Tatton & Co. Ltd.* v. *Ferrymasters Ltd.* [1974] 1 Lloyd's Rep. 203, 206, and cf. further OLG. Munich (1990) TranspR 16, (1990) I U.L.R. 438, OLG. Celle, 29.10.98 (1999) TranspR 106.

148. As to which, see *infra*, para. 9.72 ff.

149. Some courts, however, do not see these provisions as imposing a limit on the power of the court to award additional damages on the basis of abuse of process, e.g. Cass., 17.6.97 (1997) B.T. 596.

150. The amount of the special interest must be declared, see App. Montpellier, 19.2.87 (1987) B.T. 578 where the contract stated that the date for delivery amounted to a declaration of special interest, which was to be equal to the value of the goods. There being no indication of value of the goods, the clause was insufficient to qualify under Article 26.

151. Whatever doubts there may be about the need for other particulars to appear in the consignment note it is thought that under Articles 24 and 26 such inclusion is necessary, being expressly provided for by both Articles. Loewe, paras. 196 and 203, is of the same view. See BGH., 14.7.93 (1993) 28 E.T.L. 917 in respect of Article 26, cf. App. Reims, 13.7.77 (1977) B.T. 406, OGH., 30.8.90 (1992) TranspR 406. See also Athens Court of Appeal, 1985 (No. 5663) (1987) 22 E.T.L. 65, 67. Certainly inclusion of a declared value on a document established unilaterally by the sender, containing neither the carrier's stamp nor signature and with an unidentifiable signature, will not be sufficient, App. Paris, 15.12.77 (1978) B.T. 53 (see Lamy, para. 1579, Jur. 4, cf. App. Versailles, 4.11.93 (1994) B.T. 19, where it was said that there must at least be an indication on a contractual document), neither will the declaration of value on Custom documents be sufficient (App. Versailles, 6.7.94 (1994) B.T. 759, Cass. (Belg.), 10.2.94 (1994) Bull. Cass. et Pas. 165 (Putzeys, "Le droit des transports en Belgique (II) 1993–1995," (1996) U.L.R. 557, 567), Krings, p. 779, nor the stipulation of a cash on delivery requirement (App. Paris, 6.2.91 (1991) B.T. 295). Inclusion on an initial consignment note is sufficient, notwithstanding the later consolidation of the goods with other goods under a groupage consignment note which fails to include it, Hof. Brussel, 16.11.77 (1980) 15 E.T.L. 319, cf. App. Versailles, 4.11.93, *supra*.

152. Or indeed the provision could be used, for the purposes of jurisdictions not party to the Protocol, as a method of avoiding the uncertainties involved in ascertaining a figure for the limit in Article 23(3) in its original form: see *supra*.

can thus be raised to an appropriate level,[153] but this does not otherwise affect the basis of compensation, which will be calculated with reference to Article 23 or 25 in the normal way. Article 26, on the other hand, is designed for the situation where there are other extraneous matters, which might otherwise be unrecoverable as consequential loss. This in turn raises the question to what extent Article 26 overrides normal rules of remoteness of damage, which is discussed below.[154]

9.51 Further, Article 24, by reason of its reference to Article 23(3) alone, can have no effect on the maximum recoverable under Article 23(5) in respect of delay, which would therefore remain restricted to the amount of the carriage charges. Article 26, however, does provide for delay, but only where there is an agreed time limit. It therefore cannot assist the claimant in respect of delay where he is relying on the carrier's failure to deliver within a reasonable time under Article 19.[155]

9.52 Both Articles 24 and 26 provide for "payment of a surcharge to be agreed upon". This raises the possible argument that in order to be able to rely on either Article, such additional charges must have been specifically agreed as such, and that further they must have been paid prior to the carriage. Loewe[156] is of the view that neither is a necessary prerequisite, and it is thought that this must be the preferable view, the wording merely making it clear that the carrier will be entitled to charge extra under either Article.[157] This will presumably in practice often not be invoiced as an identifiable item of additional charge, but will appear as part of a single, albeit larger, charge for the carriage. The burden of proving that payment of the surcharge has been made will rest on the claimant.[158]

9.53 It is also important to note in this context that the terms of the carrier's insurance policy may well exclude liability under Articles 24 and 26 unless prior notice has been given to the insurance company and an increased premium paid.[159] This can pose particular problems for the successive carrier, who may not be aware of the declaration of higher value or special interest in delivery, and thus might be dependent on his driver spotting the modification when the consignment note is handed over.

9.54 Where a special interest in delivery has been declared under Article 26, additional loss or damage, if proved, can be claimed without regard to the provisions of Articles 23 to 25. The requirement that the loss be proved makes it clear that there is no automatic entitlement to the full amount of the declared interest; what is not entirely clear, however,

---

153. An agreement to fix a lower maximum figure for a proportionately reduced carriage charge would be void under Article 41: *infra*, Chapter 12. It has been held in France that there is no duty on the carrier to advise the sender of the possibility of increasing the limit in the case of goods which are of particularly high value: Cass., 12.10.81 (1981) B.T. 576, (1982) 17 E.T.L. 294 (reversing App. Paris, 19.9.79 (1979) B.T. 481). Such a duty was, however, applied to a *commissionnaire* in App. Paris, 6.1.82 (1982) B.T. 84. For the position in relation to a forwarding agent under English law, see *Von Traubenberg* v. *Davis, Turner & Co. Ltd.* [1951] 2 Lloyd's Rep. 462.

154. It might be observed at this stage, however, that one of the reasons why the dissenting minority in *James Buchanan and Co. Ltd.* v. *Babco Forwarding and Shipping Ltd.* [1978] 1 Lloyd's Rep. 119 (*supra*, para. 9.25 ff.) thought the excise duty unrecoverable was because they thought Article 26 was in the Convention to cover just such a situation.

155. See *supra*, Chapter 7, para. 7.1 ff.

156. Loewe, paras. 198 and 203.

157. Cf. OLG. Düsseldorf, 28.10.82 (1983) VersR 749, and OLG. Hamburg, 29.5.80 (1980) VersR 950, but note the criticism of the latter case by Haak, p. 233. Cf. Lamy, paras. 528 and 529, derived implicitly from Cass., 10.7.89 (1989) B.T. 591. Agreement for a surcharge was stated to be necessary in App. Montpellier, 19.2.87 (1987) B.T. 578, App. Versailles, 6.7.94 (1994) B.T. 759 and App. Versailles, 4.11.93 (1994) B.T. 19.

158. See Cass., 10.7.89 (1989) B.T. 591.

159. See, e.g., clause 3(2) of Lloyd's CMR policy, see Appendix D.

is how far a declaration under Article 26 will override national rules as to remoteness of damage, so as to enable the cargo interests to recover in respect of items of loss which, under the national rules, would be unrecoverable. This question is dicussed in the next section.

The scope of recovery under Articles 23(5) and 26

9.55 There is one problem which is common to both the delay provisions of Article 23(5) and the provision for the declaration of a special interest in delivery under Article 26. Both provide for upper limits of compensation, and both make it clear by their wording that the sum in question is not automatically recoverable, but that loss must be proved up to that limit. Neither Article, however, provides any guidance as to what sums are recoverable within this limit, it therefore remains unclear as to whether the national courts are to apply their own rules as to remoteness.

9.56 Under English law, the basic test of remoteness in contract is that those losses are recoverable which are in the reasonable contemplation of the parties at the time the contract was made. The defendant will therefore be liable if the loss in question is to be regarded as likely to occur in the normal course of events, or because of special facts of which he has actual or constructive knowledge.[160] Any loss in respect of which cargo interests are claiming must fall within one of these two categories before it is recoverable. In carriage of goods cases under English law, a fluctuating market at the place of destination is regarded as likely to arise in the normal course of events, so that a fall in value through delay in delivery will be recoverable,[161] whereas if there are special circumstances increasing the measure of the loss, then the carrier will not be liable unless he knew or should have known of the potential loss. So if a contract at a particularly lucrative price is lost through delay, if the carrier is unaware of that particular circumstance then damages will be based on the normal market price.[162]

9.57 Loewe, with reference to Article 26, is of the view that national rules of remoteness apply to the quantification of damages,[163] and his reasoning would seem equally applicable to Article 23(5). He argues that establishing a code on remoteness would have been beyond the scope of the Convention, and difficult to achieve. He concluded that "apart from material damage (loss sustained and loss of earnings) other compensation may be payable provided that the national law applicable in situations of this kind so provides".[164] This does, of course, beg the question of what is meant by loss of earnings. On the other hand, Hardingham argues in respect of Article 23(5) that a "claimant who proves that a delay within the meaning of CMR has occurred is entitled to recover from the carrier in respect of any detriment genuinely suffered[165] and which can be quantified in monetary terms, subject to the limitation that the recovery cannot exceed the amount of

---

160. *Hadley* v. *Baxendale* (1854) 9 Exch. 341.
161. *Czarnikow* v. *Koufos* [1967] 2 Lloyd's Rep. 457.
162. *Horne* v. *Midland Railway* (1873) L.R. 8 C.P. 131.
163. Loewe, para. 204. See also Haak, p. 232 n. 198 where reference is made to W/TRANS/SCI/438, 19.4.72, Nos. 71, 72.
164. Ibid.
165. At the very least a causal link must be established between the loss suffered by the claimant and the act or omission within the responsibility of the carrier, see Cass., 26.6.84 (1984) B.T. 610, cf. App. Nimes, 29.10.80 (1981) B.T. 256, (1981) I U.L.R. 267. See also BGH., 27.6.85 (1986) 21 E.T.L. 102.

the carriage charges".[166] In other words, he is of the view that normal English rules of remoteness will not apply under Article 23(5), and again his reasoning is equally applicable to Article 26.

9.58 Hardingham's view certainly has the greater attraction, since, as he points out, in an international convention, it is in the interests of uniformity undesirable to have regard to the technical rules of national law. In so far as the English courts may feel any reluctance to depart from normal concepts of remoteness in relation to these two provisions, it would seem to be arguable that Hardingham's approach can in any event be reconciled with normal rules of remoteness. The key to such an argument lies in the aspect of Master Jacob's judgment in *James Buchanan & Co. Ltd.* v. *Babco Forwarding & Shipping (U.K.) Ltd.*[167] referred to earlier.[168] Applying that reasoning to the present circumstances, it can be argued that a carrier, because he is undertaking an international carriage subject to the Convention, has actual or constructive notice of its provisions. Returning to normal English rules of remoteness, it will be recalled that there are two possibilities: damage is recoverable if it can be regarded as in the reasonable contemplation of both parties at the time the contract was made as likely to result, either in the normal course of things or from special circumstances of which both parties have actual or constructive knowledge. It may be that, at least in relation to Article 26, the declaration of the special interest in delivery, which it will be remembered has to be agreed by the carrier, can be regarded as giving him sufficient knowledge of the potential loss. Even though he may not know the precise circumstances, he is put on notice by the declaration that loss, damage or delay could produce a loss up to that amount, and therefore he ought not to be able to object that any loss occurring within that limit is too remote. A similar argument can be made in relation to damages under Article 23(5), although in such circumstances the carrier's knowledge of his potential liability would have to be deemed to derive from the provisions of Article 23(5) itself rather than from any declaration in the consignment note.[169] Such is not, however, inconsistent with the statement of Master Jacob referred to above. Some support for this view, at least in relation to Article 23(5) where there is an agreed time limit, can be found in a French decision where there was a stipulated date for delivery which was not complied with. The carrier objected to the consignor's claim for the full amount of the carriage charges, which was based on the consignor's obligation to pay a larger sum to the consignees in respect of the delay under the terms of the contract between them, on the basis that the urgency of the situation had not been made clear. Thus, the issue of remoteness seems to have been the substance of the carrier's defence, but the date for delivery was clearly indicated in the consignment note, and it was held that the consignor, having proved his loss, was entitled to succeed.[170]

---

166. Hardingham, "The delay provisions of CMR", [1979] LMCLQ 193, 197.
167. Set out so far as material at [1978] 1 Lloyd's Rep. 119, 130.
168. The relevant passage is set out *supra*, para. 9.31.
169. Cf. Hardingham, op. cit., p. 198.
170. App. Bordeaux, 9.2.72 (1972) B.T. 114, (1974) I U.L.C. 341. Cf. BGH., 30.9.93 (1994) 29 E.T.L. 97, (1994) TranspR 16, where damages were awarded under Article 23(5) to a sender who had paid a sum to his buyer as compensation for lost production caused by the delay and to prevent the buyer from placing the seller on a black list, thus threatening future loss of profits to the sender. The court noted ((1994) 29E TL 97 at p. 102) that damages under Article 23(5) can include not only damages suffered by the recipient due to lost production but also loss suffered by the sender due to claims for compensation by his buyer. Furthermore, expenses were held to be recoverable if they are necessary to prevent threatened damage from occurring. Clearly, the court considered the expense to avert a threat to future profits as within the scope of recoverable damage.

9.59 At first sight, *Horne* v. *Midland Railway*[171] would suggest that the above view cannot be correct in relation to English law, since there it had been made clear that delivery by a particular date was essential, yet the resulting losses were in the main treated as too remote. However, that case can be distinguished on the basis that there was no equivalent provision putting the carriers on notice as to their potential liability.

Interest and conversion rate

9.60 Article 27 provides as follows:—

> "1. The claimant shall be entitled to claim interest on compensation payable.[172] Such interest, calculated at five per centum per annum, shall accrue from the date on which the claim was sent in writing to the carrier, or, if no such claim has been made, from the date on which legal proceedings were instituted.
> 2. When the amounts on which the calculation of the compensation is based are not expressed in the currency of the country in which payment is claimed, conversion shall be at the rate of exchange applicable on the date and at the place of payment of compensation."

Article 27 provides for a fixed rate of interest of 5%.[173] When rates of interest are high this may tend to favour the carrier, and it has not always proved acceptable to the individual national courts. French courts, for example, have awarded interest at a rate of 6%,[174] but this is clearly not permitted under the terms of the Convention.[175] Interest will accrue either from the date on which a written claim made under Article 30(1) has been sent to the carrier,[176] or, if none has been made in writing, from the date on which legal

---

171. (1873) L.R. 8 C.P. 131. See *supra*, para. 9.56.

172. This applies to compensation claimed under Article 21 as well as under Article 17 etc., BGH., 10.10.91 (1992) TranspR 100. It does not apply to claims for freight charges by a carrier which are, accordingly, governed by the relevant national law (see *infra*), Comm. Tongres, 22.4.76 (1976) J.C.B. 365, Hof. 's Gravenhage, 13.6.80 (1980) S. & S. No. 126, see Libouton, (1982) J.T. 724, para. 113. On one view it represents a limit on liability, the limit being subject to the application of Article 29 in cases of wilful misconduct (*infra*, para. 9.72 ff.), *B. Paradise Ltd.* v. *Islander Trucking*, Queen's Bench Division, 28.1.85, unreported except on LEXIS, and see also OLG. Hamm, 25.5.92 (1992) TranspR 410, OLG. Düsseldorf, 11.5.89 (1990) TranspR 60, Comm. Bruxelles, 2.3.90 (1992) R.D.C.B. 819. Contrary authority supports the view that it is not so subject: Cass., 19.12.95 (1996) B.T. 463, App. Paris, 5.4.95 (1995) B.T. 811, App. Versailles, 29.2.96 (1996) B.T. 385, HR., 20.11.98 (1999) 34 E.T.L. 254. The starting point to the *right* to claim interest indicated in Article 27(1), apart from the application of national law, is not affected by Article 29: OLG. Munich, 16.1.91 (1992) TranspR 181, see Thume, *infra*, para. 9.63 fn. 185, at p. 369.

173. In France the issue of whether there can be capitalisation of interest was held to be a matter of national law, and Article 114 of the Civil Code which provides for this, as applied in Cass., 17.3.92 (1992) B.T. 253. See also Cass., 25.11.97 (1999) 34 E.T.L. 248. Cf., however, Hof. Antwerpen, 30.5.90 (1991) J.P.A. 126.

174. App. Paris, 18.12.68 (1969) B.T. 98. App. Colmar, 10.7.70 (1970) B.T. 358, (1971) U.L.C. 137, criticised by Libouton, (1973) 8 E.T.L. 2, para. 72. Cf. Cass., 24.10.95 (1995) B.T. 762; cf., however, RB. Rotterdam, 12.5.78 (1979) 7 E.L.D. 180: even after proceedings have been commenced, interest is limited to 5%. Interest at 5% was applied in Cass., 3.3.98 (1998) B.T. 231, (1998) U.L.R. 897 and see further cases cited in Lamy, para. 1583.

175. A possible justification for such an approach is to read Article 27 as imposing a rate of 5% only where the claimant expressly claims interest at this rate, leaving the courts free to apply national law where he does not, e.g. App. Douai, 5.10.93 (1996) B.T. 16, 17. As noted by Tilche in the summary of the case this view is not in keeping with the spirit of the Convention and the objective of uniformity. For the application of Article 27(1) and (2) in cases of wilful misconduct, see *infra*, para. 9.75.

176. See the discussion of *William Tatton & Co. Ltd.* v. *Ferrymaster Ltd.* [1974] 1 Lloyd's Rep. 203 *infra*, para. 10.85 as to the requirements of a valid written claim.

proceedings commence.[177] An oral claim will therefore not be effective to start interest running. The Convention makes no provision for the payment of interest prior to the issue of the writ in the absence of a written claim.[178] Interest therefore cannot accrue as from the date of the delivery of the goods in question, as failure to perform a contractual obligation under the Convention does not *per se* give a right to interest on damages subsequently awarded.[179]

9.61 The English courts have a general discretionary power to award interest upon all or any part of any debt or damages, which if awarded can cover the whole or any part of the period between the date when the cause of action arose and the date of judgment.[180] There is no universal rule as to the rate of such interest, but it is awarded on the basis that the plaintiff would be entitled to borrow "so as not to be out of his money",[181] and in *Miliangos* v. *George Frank (Textiles) Ltd.*[182] interest was awarded at 1% over the average minimum lending rate calculated on a day-to-day basis over the period in question. It has also been held[183] that where a foreign currency judgment is awarded, the appropriate rate is that prevailing in the financial market of that country.

9.62 In general, therefore, English law provides for the award of interest at a rate in excess of the 5% stipulated under the Convention, and further allows recovery from the date the cause of action arose, rather than from the date of a written claim. On the strict wording of Article 27(1) it might be argued that it would be open to an English court to award interest at the higher rate in respect of the period between the accrual of the cause of action and the date of the written claim. However, it would seem clear that the intention

---

177. Although it does not form part of the decision as reported, it was held in *Sidney G. Jones Ltd.* v. *Martin Bencher Ltd.* [1986] 1 Lloyd's Rep. 54 that where there has been no written notice of the claim, interest will run from issue of the writ rather than from the date of service. Counsel for the defendants had argued that, read in conjunction with the remainder of Article 27, notice to the defendant was of central importance, and, in the absence of written notice, prior to service of the legal proceedings the defendants would not have had any notice of the claim against them. Although Popplewell, J., accepted that this argument was not without merit, he was of the view that "the institution of legal proceedings has a particular connotation, which means the issue of the writ and not the service of the writ" and he accordingly granted interest from the date of issue. However, since the decision in this case, the English courts have had to consider the significance of the distinction between issue and service in other contexts, so *quaere* whether the same result would now be reached: see *infra*, para. 10.38. Cf. also the wording of the French text, which refers to "*la demande en justice*".

178. Cf. GH. 's Hertongenbosch, 13.1.70 (1971) S. & S. No. 10, (1971) 6 E.T.L. 817. The written claim referred to is the same as that which suspends the period of prescription in Article 32(2): RB. Amsterdam, 12.4.72 (1972) S. & S. No. 102.

179. Comm. Courtrai, 25.5.72 (1972) J.P.A. 192, (1973) II U.L.C. 438.

180. Supreme Court Act 1981, Section 35A. See, generally, *McGregor on Damages*, 16th edn., Chapter 14. Separate statutory provisions govern interest on judgment debts. Such provisions cannot be used to alter the rate provided for by CMR, see HVC., 17.9.87 (1988) 23 E.T.L. 201, and the cases cited by Libouton, (1982) J.T. 724, para. 113; see also Cass., 17.3.92 (1992) B.T. 253. Nor can an award based on loss of interest due to delay in payment be awarded in derogation from Article 27: BGH., 10.10.91 (1992) TranspR 100, cf., however, App. Poitiers, 23.8.90 (1990) B.T. 586, (1991) II U.L.R. 324. Cf. further, Clarke p. 374 with Herber/Piper, Art. 27 Rdn. 8 and Koller Art. 27 Rdn. 5.

181. *Miliangos* v. *George Frank (Textiles) Ltd. (No. 2)* [1976] 3 W.L.R. 477, 481. An alternative possibility which sidesteps the 5% rule is to permit a claim, under national law, for specific damages in respect e.g. of the legal costs caused by the delay in payment, OLG. Hamm, 7.11.96 (1998) TranspR 459, (1999) U.L.R. 194 (see also Thume, Article 27, Rdn. 34, 35 and *op. cit. infra* fn. 185, cf. Herber/Piper, Article 27 Rdn. 11). In OLG. Hamburg, 22.1.98 (1998) TranspR 252, the court denied a claim for pre-trial costs by a contracting carrier against his subcontractor. This is criticised by Thume, (1998) TranspR 440. The restriction on consequential losses in Article 23(4), *supra*, Chapter 9, paras. 9.21–9.41, relates to loss consequential to the liability under Article 17, not that incurred by being required to litigate. Cf. further OLG. Hamm, 6.2.97 (1998) TranspR 34.

182. [1975] Q.B. 487.

183. *Miliangos* v. *George Frank (Textiles) Ltd. (No. 2), supra.* See also *Helmsing Schiffahrts* v. *Malta Drydocks Corp.* [1977] 2 Lloyd's Rep. 444.

of the Convention was to restrict interest to the period following the written claim, and any other interpretation would produce the illogical conclusion that the claimant might be in a better position by not submitting a written claim, thus leaving interest to accrue at the higher rate.

9.63 Article 27(1) therefore specifies when interest commences to run. It is silent, however, as to whether interest is limited to 5% from the date at the written claim right up until the date of payment, or whether it applies only up to the date of judgment. This leaves open the question as to whether interest at a higher rate might be awarded for the period between judgment and payment. The position under English law generally is that every judgment debt carries interest, the current rate being 8%,[184] and it would seem therefore that the appropriate rate after judgment is that figure.[185] Apart from this, it is clear that, unlike under the normal English rules, the courts have no discretion in respect of the period for which interest has to be awarded. In *Elektronska Industrija Oour TVA* v. *Transped Oour Kintinentalna Spedicna*[186] Hobhouse, J., said[187]:

"Article 27 is not a mere matter of procedure giving rise to a discretion. It creates an entitlement by reference to a defined period and a defined rate of interest which can only be defeated by some substantive defence or some set-off or principle of circuity."

9.64 Article 27(2) provides that where the amounts on the basis of which compensation is calculated are not expressed in the currency of the country where payment is claimed, the rate of conversion applicable will be that applicable on the day and at the place compensation is paid.[188] Thus, in a Dutch decision where tin ingots were stolen, damages under Article 23 were calculated on the basis of the world market value of tin on the day of the theft, and the sterling value so obtained was then converted into Dutch guilders at the rate prevailing on the date of payment.[189]

Extra-contractual claims

9.65 Article 28 provides as follows:—

"1. In cases where, under the law applicable, loss, damage or delay arising out of carriage under this Convention gives rise to an extra-contractual claim, the carrier may avail himself of the provisions of this Convention which exclude his liability or which fix or limit the compensation due.

2. In cases where the extra-contractual liability for loss, damage or delay of one of the persons for whom the carrier is responsible under the terms of article 3 is in issue, such person may also avail himself of the provisions of this Convention which exclude the liability of the carrier or which fix or limit the compensation due."

---

184. Judgments Act 1838, section 17, as amended by Administration of Justice Act 1970, section 44. Judgment Debts (Rate of Interest) Order 1993 (S.I. 1993 No. 564).

185. Such was the award at first instance in *James Buchanan & Co. Ltd.* v. *Babco Forwarding & Shipping (U.K.) Ltd.* (unreported) although the point does not seem to have been argued, nor was the point taken before the Court of Appeal or House of Lords. Cf., however, the view of Thume, "Article 27 CMR und Entschädigungsverzug des Frachtfuhrers" (1993) TranspR 365, 366 and Clarke, p. 347. See also, likewise contrary to the view expressed in the text, App. Bruxelles, 27.2.96 (1996) 31 E.T.L. 833, 839. Cf. further the cases cited *supra*, fn. 180.

186. [1986] 1 Lloyd's Rep. 49.

187. At p. 53.

188. See, however, Loewe, who suggests (para. 210) that where there is a fall in the currency of the award the claimant is entitled to additional payment, but no allowance is to be made for a carrier if the currency moves against him.

189. Arrond. Amsterdam, 19.1.77 (1979) 7 E.L.D. 401, cf., Hof. s' Hertogenbosch, 16.3.88 (1989) S. & S. No. 258, (1990) 18 E.L.D. No. 138, App. Paris, 24.4.97 (1997) B.T. 468.

As has been explained, the Convention is exclusively concerned with the regulation of the contract of carriage, Article 1 providing that the Convention "shall apply to every contract for the carriage of goods by road . . . ".[190] Article 28 is therefore introduced to deal with the situation where under the relevant national law a non-contractual right of action exists, which would otherwise be outside the scope of the Convention.[191] So, for example, if a shipper has employed a forwarder to arrange the contract of carriage in such circumstances that the forwarder is the sender for the purposes of CMR,[192] the shipper will have no right of action under the Convention unless the forwarder assigns his rights to him. Where such an assignment has not taken place or is not legally possible in the country in question, the only direct right of action which the shipper will have against the carrier will be in tort or delict, which, in the absence of Article 28, would not be subject to the Convention.

9.66 Under English common law, the relationship between the parties can fall into a number of distinct categories. First, of course, there is the direct contractual relationship between the consignor and the carrier. Further, where the consignor is not the owner of the consignment, but arranges the carriage as the agent for the owner, there is a contract of carriage between the owner and the carrier by virtue of the agency. Even if not the agent of the owner, if the consignor is in possession as bailee of the owner, and if the carrier accepts the goods knowing them to be the property of a customer of the consignor, this will create the relationship of bailor and bailee between the owner and the carrier respectively by virtue of sub-bailment.[193] Yet again, delivery of the goods to the carrier may in itself pass title to the goods to the consignee,[194] and in such circumstances the consignor may be treated as agent for the consignee, thus producing a direct contractual relationship between the consignee and the carrier.[195] Finally, the party interested in the goods may have rights against the carrier in tort, either in conversion, on the basis of an immediate right to possession, or in negligence.[196] However, the important point for the purposes of the present discussion is that, subject to the qualifications discussed below, it matters not how the action is framed, since under Article 28(1) the carrier, and by Article 28(2) persons for whom he is liable under Article 3, can claim the protection of the Convention.[197]

9.67 It will be seen therefore that where Article 28 applies, it overrides normal English rules of privity of contract whereby only the parties to a contract are entitled to benefit from its terms and whereby only the parties are affected by any restrictions or limitations

---

190. *Supra*, Chapter 1, paras 1.2–1.9.

191. Cf., however, Glockner, Theunis, p. 106 and also Libouton, (1982) J.T. 725, para. 115, and Lamy, paras 534 and 1586, Jur. 1, both citing App. Paris, 11.6.74 (1974) B.T. 319, see also Herber/Piper, Art. 28, Rdn. 4 Koller Art. 28, Rdn. 3; cf. Haak, pp. 239–241.

192. As to which question see *supra*, Chapter 1, paras 1.13–1.54.

193. *Morris* v. *C. W. Martin & Sons Ltd.* [1965] 2 Lloyd's Rep. 63, 73. See also *The Pioneer Container* [1994] 2 All E.R. 250, 261, [1994] 1 Lloyd's Rep. 593, 601 (sub. nom. *The K.H. Enterprise*).

194. By virtue of the Sale of Goods Act 1979, section 18, Rule 5(2).

195. See the cases cited by Brandon, J., at first instance in *The Albazero* [1974] 2 Lloyd's Rep. 38, 44. Even where the consignor has no interest in the goods when damaged, he may in limited circumstances sue in his own right: ibid., [1976] 2 Lloyd's Rep. 467. The consignee can also sue in tort where he has the property or sufficient possessory title at the time of the damage to the goods, *The Aliakmon* [1986] A.C. 785, 809.

196. For the essentials of actions in negligence and conversion, see Clerk and Lindsell, *Torts*.

197. Note that both paragraphs of Article 28 are permissive only. Presumably therefore if more favourable standard terms are potentially applicable, then it is open to the carrier and sub-contractor to rely on those terms rather than the CMR defences. There does not appear to be any conflict with Article 41 *infra*, since in such circumstances the Convention is not of compulsory application.

on their right of action. So long as the claim is caught by Article 28, third party claimants are restricted in their actions to the extent therein provided, and third party defendants protected to a like extent.[198]

9.68 As indicated above, however, there are a number of qualifications to be made to the scope of Article 28. First, it is restricted to claims for "loss, damage or delay arising out of carriage under this Convention". If, therefore, a claim is otherwise outside the Convention, Article 28 can be of no avail. So, for example, if a carrier fails to perform the contract at all, this will not be within the Convention,[199] since Articles 17 et seq. only deal with the carrier's liability for loss, damage or delay, and thus any domestic rights of action, whether in contract or tort, would be unaffected by Article 28.[200]

9.69 There is a second apparent qualification arising from the fact that Article 28 permits the carrier to avail himself of the provisions of the Convention "which exclude his liability or which fix or limit the compensation due". In other words, it stops short of incorporating the provisions of the Convention generally. It would, however, seem that the main provisions of which the carrier would wish to take advantage can be brought within the words used, although it is perhaps of significance that Article 29[201] specifically refers to provisions which shift the burden of proof,[202] whereas Article 28 does not, and it must be open to doubt whether those provisions can be properly described as excluding, fixing or limiting liability. It would also appear that the jurisdiction provisions of Article 31 cannot really be so described, nor can an agreed arbitration clause under Article 33. It would therefore seem that such provisions will only bind or protect a third party by virtue of being contractually incorporated into the arrangements with that party in one of the ways outlined above. On the other hand, it would seem that the limitation provisions in Article 32[203] can properly be so described. In particular, the latter provision refers to "an action arising out of carriage under this Convention" which would seem appropriate to include claims caught by virtue of Article 28.[204]

9.70 As a necessary corollary to the extra-contractual protection afforded to the carrier, Article 28(2) applies the same protection to "persons for whom the carrier is responsible under the terms of Article 3".[205] It follows that before such persons are entitled to the protection of Article 28(2), it is necessary that the carrier should be responsible for their

---

198. Confirmation that normal rules of privity are inapplicable can be derived from sections 1 and 14(2) of the Carriage of Goods by Road Act. Section 1 provides that the Convention "*shall have the force of law*" so far as it concerns the rights and liabilities of "persons concerned in the carriage of goods by road under a contract to which the Convention applies". Section 14(2) then defines such persons so as to include, *inter alia*, the sender, the consignee, and any person for whom the carrier is responsible under Article 3. For the effect of the words "shall have the force of law" under the Hague-Visby Rules, see *The Morviken* [1982] 1 Lloyd's Rep. 325.

199. Subject to the possibility of an action for delay, which will depend on the terms of the contract. *Supra*, para. 7.6.

200. See *Shell Chemicals U.K. Ltd.* v. *P. & O. Roadtanks Ltd.* [1993] 1 Lloyd's Rep. 114 (affirmed on different grounds in the Court of Appeal, [1995] 1 Lloyd's Rep. 297). See also, HR. 15.4.94 (1994) S. & S. No. 73, (1995) N.J. 465, (1994–5) U.L.R. 378. This is also true where, as happened in that case, by delivering the wrong goods the carrier causes some damage to the claimant other than loss, or damage or delay to the goods. Saville, J., did, however, consider that Article 32 and Article 41 were relevant to the circumstances of the case. Some provisions may therefore be applicable apart from the operation of Article 28. Whether this is true of total failure to perform the carriage as opposed to the circumstances which arose in the *Shell Chemicals* case is considered further *infra*, para. 10.56.

201. *Infra*, para. 9.72 ff.

202. I.e. Article 18(2). See *supra*, Chapter 6, paras 6.115–6.116.

203. *Infra*, Chapter 10, para. 10.51 ff.

204. OGH., 27.9.83 (1984) VersR 548, see Haak, p. 239 n. 243.

205. As to which see, generally, *supra*, Chapter 3.

acts by virtue of Article 3. So, for example, if the carrier was not so responsible because that person was acting outside the course of his employment, the protection of Article 28(2) would not arise. It should also be noted that Article 28(2) may in some circumstances operate for the protection of the carrier himself, for example, if he was contractually obliged to indemnify such other persons for any claims against them.

9.71 Apart from the situations in which Article 28 might be of no avail for the reasons discussed above, it should be noted that the distinction between claims in contract and in tort may still not be without significance even if Article 28 does apply. To the extent that national rules of remoteness may be relevant,[206] at common law there may be different amounts recoverable depending on whether the action is framed in contract or in tort.[207] Detailed discussion of this topic is beyond the scope of the present work.

Wilful misconduct

9.72 Article 29 provides as follows:—

> "1. The carrier shall not be entitled to avail himself of the provisions of this chapter which exclude or limit his liability or which shift the burden of proof if the damage was caused by his wilful misconduct or by such default on his part as, in accordance with the law of the court or tribunal seised of the case, is considered as equivalent to wilful misconduct.[208]
>
> 2. The same provision shall apply if the wilful misconduct or default is committed by the agents or servants of the carrier or by any other persons of whose services he makes use for the performance of the carriage, when such agents, servants or other persons are acting within the scope of their employment.[209] Furthermore, in such a case such agents, servants or other persons shall not be entitled to avail themselves, with regard to their personal liability, of the provisions of this chapter referred to in paragraph 1."

It will be noted that there are two ways in which Articles 28 and 29 differ. First, while Article 28 refers to any provisions of the Convention which exclude liability or fix or limit compensation, Article 29 only withdraws the protection of the provisions contained in Chapter IV.[210] This means, for example, that even if wilful misconduct has occurred, the claimant must still give reservations as prescribed by Article 30.[211] However, the time limit contained in Article 32 is specifically extended by that Article to three years in the event of wilful misconduct.

9.73 Secondly, Article 28 refers to those provisions which "exclude the liability of the carrier[212] or which fix or limit the compensation due" whereas Article 29 refers to provisions which "exclude or limit his liability or which shift the burden of proof". It will be seen that there is no reference in Article 29 to the provisions which *fix* liability, and it

---

206. See *supra*, para. 9.56.

207. See, generally, *McGregor on Damages*, 16th edn.

208. For Loewe, the wording of this provision was the only serious and manifest mistake made by the drafters of CMR: "Le CMR à 40 ans," (1996) U.L.R. 429, 438.

209. Cf. *Morris* v. *C.W. Martin & Sons Ltd.* [1965] 2 Lloyd's Rep. 63 with BGH., 27.6.85 (1985) VersR 1060, (1986) 21 E.T.L. 102, discussed by Glockner, Theunis, p. 108. See also RB. Turnhout, 12.6.97 (1998) 33 E.T.L. 124.

210. I.e. Articles 17–29.

211. *Infra*, Chapter 10, paras 10.2–10.20. In respect of Article 30(3), see BGH., 14.11.91 (1992) TranspR 135, (1993) 28 E.T.L. 265.

212. This includes Article 17(5): BGH., 27.6.85 (1985) VersR 1060, (1986) 21 E.T.L. 102, OLG. Munich, 16.1.91 (1992) TranspR 181.

was held in *Lacey's Footwear (Wholesale) Ltd.* v. *Bowler International Freight Ltd.*[213] that Article 23(1) and (2) of the Convention apply in the normal way so that the value of the consignment of stolen shoes was to be fixed by reference to their value in Spain as at the date when they were taken over for carriage rather than by reference to their value in this country. On the other hand, the effect of these provisions is to limit the recovery of most forms of consequential loss, except to the extent that they fall within Article 23(4) or are brought within Article 26.[214] Similarly, Article 23(3) and (5),[215] and Article 26[216] may operate to fix liability, but clearly their primary function is the limitation of the carrier's liability, and so they would be within Article 29.[217]

9.74 It was also held in the *Lacey's Footwear* case that the concluding words of Article 23(4) whereby "no further damages shall be payable" also amounted to a provision which excluded or limited liability, so that the carrier was not entitled to rely on it. The result was that the plaintiffs were able to recover not only the value of the shoes calculated under Article 23(1) and (2), but also their loss of profit on top of that. The effect, therefore, was that the plaintiffs were to recover the value of the shoes in this country despite the earlier finding that Article 23(1) and (2) apply in the normal way.

9.75 Two other aspects were considered in the *Lacey's Footwear* case. It was held that Article 27(1) applied in the normal way, despite the wilful misconduct, so as to limit interest to 5% p.a.[218] It was also held that, even though the relevant exchange rates had altered against the plaintiffs over the intervening period, Article 29 did not prevent reliance on Article 27(2), by which the exchange rate was to be applied as at the date of payment. Both were regarded as provisions which fix liability, so they were not excluded by Article 29 despite the wilful misconduct.[219]

---

213. [1997] 2 Lloyd's Rep. 369.

214. See *supra*, para. 9.21 and 9.50. A divergence of view can be seen in France: the decision of App. Nancy, 21.1.87 (1987) B.T. 344 supports the view that Article 23(1) is displaced, as, impliedly, does App. Versailles 4.11.93 (1994) B.T. 19. A contrary decision, App. Paris, 22.4.92 (1992) B.T. 362, is criticised in the commentary. In Austria it has been described as a provision which serves to define the objective amount of damage rather than limit it to a maximum threshold: OGH., 13.7.94 (1995) TranspR 285, 287 (Mayer, p. 174). It is hard to reconcile the non-application of Article 29 to Article 23(1) with its application to Article 26 and the consequence that no declaration will be necessary to establish a special interest in delivery, cf. OLG. Munich, 31.3.98 (1998) TranspR 353.

215. *Supra*, para. 9.12 ff. and 9.42 ff. respectively. See BGH., 27.6.85 (1985) VersR 1060, (1986) 21 E.T.L. 102, cf. App. Nimes, 11.2.81 (1981) B.T. 198.

216. *Supra*, para. 9.50 ff. See App. Besançon, 23.6.82 (1982) B.T. 394.

217. A similar issue can be raised in respect of Article 24. In French domestic law, a series of decisions have held that *faute lourde* does not enable a claimant to recover more than the value declared, but recently a change has been signalled by the Cour de Cassation in this respect. Cass: 15.12.92 (1993) B.T. 97, see further, Tilche, (1993) B.T. 88. In *Antwerp United Diamond BVBA* v. *Air Europe* (see *supra*, fn. 129) Phillips, J., held that the equivalent provision in the Warsaw-Hague Convention (Article 22(2)(a)) was a limit of liability for the purposes of the wilful misconduct provision in that Convention (Article 25). It should be noted, however, that Article 22(2)(a), in effect, combines the functions of both Articles 24 and 26 of CMR.

218. This seems to have been the result. Beldam, L.J., addressed the point referring to both Article 27(1) and 27(2), but then only answered the point with reference to Article 27(2). However, it seems that his intention was that the position was the same in relation to both provisions. An earlier unreported case to the contrary, *B. Paradise Ltd.* v. *Islander Trucking Ltd.*, Queen's Bench Division, 28.1.85, unreported except on LEXIS, where the contrary conclusion was reached and the court awarded interest at 12% p.a., does not appear to have been referred to.

219. A substantial body of Continental authority also supports the view that Article 27 is not subject to Article 29: Cass., 19.12.95 (1996) B.T. 463, Krings, p. 780 App. Paris, 5.4.95 (1995) B.T. 811, App. Versailles, 29.2.96 (1996) B.T. 385, HR., 20.11.98 (1999) 34 E.T.L. 254. The argument is that Article 27 does not fix or limit liability but provides a sanction for the failure to pay, see Tilche, (1995) B.T. 16, 17.

9.76 A further possible difficulty is that Article 29(1) refers to "damage", leading to the possible argument that it concerns only physical damage to goods, and not resulting financial loss. This question has been discussed in relation to the use of the same word in Article 23(5), where the view was expressed that the word referred only to losses arising other than from damage to the goods, due to the separate provision for damage to goods in Article 25.[220] Here, however, it is thought that the word must be given the widest meaning, including both damage to the goods and resulting financial losses.[221] Any other interpretation would be contrary to all logic.[222]

9.77 The general scheme of Article 29, therefore, is that in the event of wilful misconduct or equivalent default by the carrier, he is deprived of most of the protection which would otherwise be available to him. It should be remembered, however, that this is only in respect of damage caused by wilful misconduct. In a French decision, the fraudulent conduct of the carrier which misled the sender as to its real name meant that the carrier could not rely on the one year time limit in Article 32. This did not deprive the carrier of the right to limit liability in respect of the damage to the goods which was unconnected with any wilful misconduct on his part.[223] Further, Article 29(2) provides that if the wilful misconduct or default is committed within the scope of their employment by the "agents or servants of the carrier or by any other persons of whose services he makes use for the performance of the carriage",[224] the carrier will also be deprived of his protection. In turn, such persons will themselves similarly be deprived of the protection of the provisions of Chapter IV where they have committed such wilful misconduct or default.

Wilful misconduct or equivalent default

9.78 The carrier's right to rely upon the provisions of Chapter IV which exclude or limit his liability or which shift the burden of proof will thus be lost if he is guilty of "wilful misconduct or such default on his part as, in accordance with the law of the court seised of the case, is considered as equivalent to wilful misconduct". The French text uses the word "*dol*" instead of "wilful misconduct".

9.79 To understand the implications of Article 29 it is necessary to look briefly at the wording employed in the equivalent provisions of other international transport Conventions. In Article 44 CIM the wording used is "wilful misconduct or gross negligence" (*"dol ou de faute lourde"*).[225] This wording was not favoured by the drafters of CMR, largely on the basis that not all the national legal systems concerned distinguish between different degrees of negligence. Instead, the identical wording used in Article 25 of the original Warsaw Convention of 1929 was adopted, even though by the Hague Protocol of

---

220. *Supra*, paras 9.47–9.49.

221. See *Adamastos* v. *Anglo-Saxon Petroleum Co. Ltd.* [1958] 1 Lloyd's Rep. 73, where it was held that the words "loss or damage" in Article IV(2) of the Hague Rules was not restricted to physical loss or damage.

222. See further, Clarke, p. 388 n. 33.

223. App. Paris, 25.3.88 (1988) II U.L.R. 736, (1989) B.T. 46. Cf. BGH., 27.6.85, *supra*, fn. 215.

224. The wording is identical to that used in Article 3 to denote the persons for whom the carrier is responsible under the Convention; see *supra*, Chapter 3. In *Rustenburg Platinum Mines Ltd.* v. *South African Airways* [1979] 1 Lloyd's Rep. 19, the Court of Appeal held that the words "within the scope of his employment" in Article 25 of the Warsaw Convention include theft by the employee.

225. By a Protocol of 1990, Article 44 has been amended to reflect the wording used in Article IV, Rule 4 of the Hague-Visby Rules and Article 25 of the Hague Protocol, see *infra*, fn. 226. The Protocol is now in force in the United Kingdom, see Clarke in Yates, Part 4, para. 4.1.2.44.2.

1955 Article 25 has been completely re-drafted as it was considered unsatisfactory in its existing form.[226]

9.80 In Continental law two distinct concepts are to be found: first *"faute lourde"*, which can be equated with gross negligence, and secondly, *"dol"*, which in turn equates with wilful misconduct. Such comparisons are of course only approximate, as exact translations are virtually impossible. In relation to CMR the problem arises as to what default is considered by a national court as being equivalent to *dol*. Under French law the problem is fairly simply disposed of as, by a long series of judicial precedents, the concept of *faute lourde*[227] is treated as equivalent to *dol*.[228] In other words, no distinction is drawn between the two concepts,[229] and so the question of deciding what is equivalent to *dol* does not therefore arise in the French courts in relation to Article 29.[230] Belgian law does not, however, acknowledge the existence of a default equivalent to *dol*, and both doctrine

226. The revised Article 25 provides: "The limits of liability specified in article 22 shall not apply if it is proved that the damage resulted from an act or omission of the carrier, his servants or agents, done with intent to cause damage or recklessly and with knowledge that damage would probably result; provided that, in the case of such act or omission of a servant or agent, it is also proved that he was acting within the scope of his employment".

227. It has been suggested that the concept of *faute lourde* has been applied more narrowly in France in recent years, see Clarke, p. 394, with reference to Cass., 29.1.85 (1985) 345, see also Cass., 5.1.88 (1988) B.T. 102, App. Paris, 24.4.97 (1997) B.T. 468, App. Paris, 12.10.99 (1999) B.T. 802, App. Angers, 18.11.99 (1999) B.T. 619. Nevertheless, the concept is appealed to with considerable frequency. It is defined as *"une negligence d'une extreme gravité, confinant au dol, et dénotant l'inaptitude du transporteur, maître de son action, à l'accomplisse-ment de la mission contractuelle qu'il avait acceptée"*, see Cass., 12.12.89 (1991) 26 E.T.L. 359 (failing to follow the instruction to deliver only against a particular document and in particular without obtaining any assurance regarding the capacity of the recipient), cf. Comm. Paris, 3.2.99 (1999) B.T. 443, Cass., 3.3.98 (1998) U.L.R. 897 (driving at excessive speed, cf., App. Versailles, 29.2.96 (1996) B.T. 385, see also (1995) B.T. 139, 140). The consciousness that a particular manoeuvre could result in damage was emphasised in Cass., 19.12.95 (1996) B.T. 463. Loss of a heavy and voluminous package was *faute lourde* in App. Lyon, 25.6.99 (2000) B.T. 312.

228. Cases can be cited where the French courts have found *dol* in the context of CMR, as in App. Versailles, 14.11.96 (1997) B.T. 198, (1997) U.L.R. 630 where the carrier unloaded the goods in order to fulfil another contract and left them unattended for three days in bad weather. Since this was deliberate non-performance with full knowledge of the harmful effect, the three year time limit in Article 32(1) was applied. Another example is App. Paris, 21.9.95 (1995) B.T. 857 where deliberately violating cash on delivery instructions in circumstances suggesting collusion with the consignee was considered virtually equivalent to *dol*. Cf. in Austria OGH., 17.3.98 (1998) TranspR 361.

229. Cass., 8.1.74 (1974) 9 E.T.L. 615. App. Paris, 18.12.68 (1969) B.T. 98. Comm. Paris, 12.10.71 (1972) 1 Gaz. Pal. 132, (1973) I U.L.C. 268.

230. *Faute lourde* often arises in the context of the theft of vehicles with their contents from parking places, as in Comm. Bobigny, 29.1.93 (1993) B.T. 244 (similarly, where arson is substituted for theft, Cass., 6.1.98 (1998) B.T. 39). Here the goods were stolen from a lorry park with minimal security, it being considered *faute lourde* to leave the vehicle in such a park for three days and nights falling over a weekend. Even where the driver has been subjected to violence it can be *faute lourde* if the driver parks in open country in Italy, App. Versailles, 16.1.97 (1997) B.T. 355. Leaving a lorry on a reasonably busy road near the driver's house while he went for a meal, having taken normal precautions, was not *faute lourde* in App. Paris, 3.7.92 (1992) B.T. 559, cf. App. Versailles, 30.3.95 (1995) B.T. 834, App. Paris. 29.9.95 (1995) B.T. 836, App. Paris, 26.2.97 (1997) B.T. 355, App. Paris, 22.10.97 (1997) B.T. 779, (1998) U.L.R. 200. But leaving the vehicle, which was not equipped with the security system specified by the insurers, without surveillance even for a short time in Italy could be *faute lourde* in Cass., 9.12.97 (1998) B.T. 23. In this context, the courts take into account the degree of knowledge of the carrier as to the value of the goods, cf. Cass., 16.11.93 (1993) B.T. 874, App. Versailles, 11.6.92 (1992) B.T. 582, Cass., 13.1.81 (1981) 16 E.T.L. 686, App. Lyons, 16.10.98 (1999) B.T. 782, Cass., 1.10.97 (1998) 33 E.T.L. 49, App. Paris, 5.4.95 (1995) B.T. 811, Cass., 10.10.95 (1995) B.T. 763, Comm, Paris, 9.2.99 (1999) B.T. 619, Can. 28.3.2000 (2000) B.T. 332. (cf. also, in respect of arson, App. Versailles, 7.9.95 (1995) B.T. 659). Even with this knowledge it was not *faute lourde* where the driver parked a simply covered lorry near a service station for the night, the reason for a night journey, subject to traffic restrictions, being due to the delay in loading by the sender, App. Douai, 14.1.99 (1999) B.T. 355. Cf. also Comm. Paris, 3.5.99 (1999) B.T. 782 where parking in a hotel car park was not *faute lourde*, the sender having paid only the ordinary freight. Cf. further App. Montpellier 18.11.99 (2000) B.T. 351.

and case law condemn the assimilation of *faute lourde* to *dol*, such as arises under French law. The courts have taken the view that the purpose of introducing the concept of equivalent default in Article 29 was not to provide any additional grounds for depriving the carrier of the protection of the Convention in those legal systems which already acknowledge the concept of *dol* but instead to enable the courts in those countries which do not have such a concept to apply a meaning which the court seised of the case considers to be equivalent to *dol* with the result that wilful misconduct itself is essential before Article 29 can apply.[231]

9.81 It would appear that although there is no definite pattern on the matter[232] some European countries tend towards the French view of equating gross negligence with wilful misconduct, at least for the purposes of Article 29.[233] The difficulty that arises under the Belgian interpretation is that it is in practice difficult to prove *dol* against a carrier, as, unlike *faute lourde*, it requires proof of the state of mind of the carrier or other person responsible. Consequently, it is very difficult for the cargo interest ever successfully to invoke Article 29 against the carrier,[234] since the precise circumstances surrounding the loss will often be within the exclusive knowledge of the person responsible.[235] Where, however, there is proof that the carrier committed a deliberate breach of contract which augmented the risk, wilful misconduct is likely to be established.[236] The burden of proof is clearly on the plaintiff.[237] Until recently a similar position to that taken by the French courts was maintained by the German courts. A concept of gross negligence (*große fahrlässigkeit*) was deemed equivalent default for the purpose of Article 29.[238] The test for *große fahrlässigkeit* was whether the care necessary for

231. App. Bruxelles, 30.10.75 (1976) 11 E.T.L. 238, 244. See also Hof. Gent, 26.3.92 (1992) 27 E.T.L. 847. Cf. RB. Brussel, 25.5.92 (1993) 28 E.T.L. 762 (it is noted in the English headnote of the report that an appeal has been lodged against this judgment). Similarly, RB. Brussel, 8.6.90 (1991) 26 E.T.L. 362, confirmed on appeal, 29.6.93 (unpublished) but with an appeal for cassation lodged, see Putzeys, (1991) 1 U.L.R. 190, 200. See also Hof. Brussel, 21.1.87 (1988) 23 E.T.L. 209 where a carrier who knowingly loaded his vehicle in excess of the legally permitted height was guilty of fault equivalent to wilful misconduct. Contrast the view of Wijffels, "Art. 29.1 CMR—Belgische Störung der Uniformität", (1998) 33 E.T.L. 27. More recent confirmation of the strict approach of the Belgian courts appears, however, in HVC., 27.1.95 (1995) J.P.A. 99, (1996) 31 E.T.L. 694 (Putzeys, "Le droit des transports–1993–1995", (1996) U.L.R. 557, 566, Krings, p. 782): knowingly and voluntarily committing a fault which the carrier must know is susceptible to cause damage does not imply that the carrier has committed *dol*. Cf. Hof. Brussel, 17.10.96 (1996) 31 E.T.L. 840.

232. Some of the historical sources suggest that the narrower view was that intended by the drafters of the Convention, Tuma, "Art. 29 CMR Bestandsaufnahme und Ausblick", (1993) 28 E.T.L. 649, 655, cf. Haak, pp. 241 et seq.

233. Article 1229 of the Italian Civil Code provides that a carrier cannot exempt himself from liability resulting from "*dolo o per culpa grave*"; see Cass. (It.), 16.11.80 (1980) II U.L.C. 341 and Cass. (It.), 29.3.85 (1986) II U.L.R. 640. For Denmark see Mikkelsen "Aktuelle Fragen des dänisches Landtransportrechts", (1995) 30 E.T.L. 419, 424–427, for Norway, see Norges Høyesterett, 16.3.95 (1996) 31 E.T.L. 565. For Spain, see Audencia Provincial De Barcelona, 6.4.99 (1999) U.L.R. 1040.

234. See Wijffels, "CMR—Continental Case Law: Loss by theft from a comparative viewpoint", [1977] LMCLQ 30, 31.

235. There is no obligation on the carrier under English common law principles to offer any explanation, nor does such silence on its own raise the presumption of misconduct: *H. C. Smith Ltd.* v. *G. W. Railway* [1921] 1 A.C. 178.

236. Cf. RB. Leuven, 23.10.90 (1992) 27 E.T.L. 272: a carrier was guilty of wilful misconduct when he subjected the goods to a handling during a transhipment which was not provided for in the contract and damage resulted therefrom.

237. App. Paris, 4.7.84 (1985) B.T. 158, see further Clarke, p. 379 n. 77.

238. E.g. OLG. Munich, 27.11.68 (1971) 6 E.T.L. 115 (cf., however, in respect of Article 32, LG. Verden, 19.9.90 (1991) R.I.W. 867), and BGH., 14.7.83 (1984) VersR 134, (1985) 20 E.T.L. 95, (1986) III U.L.R. 596. See further the decisions cited by Clarke, p. 392, n. 56. This was despite the controversy in the literature (see Clarke, loc. cit., n. 62 and Jesser, "Art. 29 CMR—Welches Verschulden steht dem Vorsatz gleich?"; (1997)

transportation was offended against to a particularly severe extent and something was disregarded that, in the given case, should have been obvious to anyone.[239] More recently there was the possibility that a test of wilful negligence[240] might replace the traditional test or be included as an element of it.[241]

9.82 The position in Germany has now been radically changed. Since the coming into force of the Transport Law Reform Act on 1.7.1998 the test of *große fahrlässigkeit* has been replaced by a new test which refers to damage done with intent or recklessly with knowledge that damage would probably arise. This new test is relevant to provisions in the Commercial Code relating to liability, limits of liability and time limits and is likely to become the test applied in respect of Article 29 of CMR.[242] The full implications of the change are as yet unclear. In particular it is unclear whether there will also be a change to the burden of proof since the German courts have been prepared to ease the burden in favour of the claimant in respect of those facts within the carrier's sphere of activity.[243] This does not extend to facts external to the carrier which might suggest an increased duty of care.[244]

9.83 In Austria, from 28.7.1990, CMR has been applied to inland commercial carriage of goods by road as well as international carriage. This has meant that the view taken by the Austrian courts regarding Article 29 applies to both national and international carriage. Traditionally these courts have treated *große fahrlässigkeit* as equivalent default.[245]

TranspR 169) and was perhaps explained by the German translation which referred to fault *equal* rather than *equivalent* to wilful misconduct, see Tuma in Haak and Swart, *Road Carrier's Liability in Europe*, vol. 2, p. 17, and "Artikel 29 CMR—Die Unendliche Geschichte?) (1998) 33 E.T.L. 13.

239. BGH., 28.5.98 (1998) TranspR 454, (1999) 34 E.T.L. 109, BGH, 16.7.98 (1999) TranspR 19, (1999) 34 E.T.L. 360. The test contained, therefore, both objective and subjective elements bringing into account the circumstances in which the particular carrier was placed in the light, particularly, of practice in the trade (see Jung, pp. 160–162, Herber/Piper Art. 29 Rdn. 4, cf. BGH, 17.4.97 (1998) TranspR 65, OLG. Cologne, 4.7.95 (1996) TranspR 284 (Jung, p. 161)). Organisational fault could also amount to *große fahrlässigkeit* in light of the nature and value of the goods and risks associated with the transit: BGH, 17.4.97 (1998) TranspR 25, (1998) 33 E.T.L. 60, (1998) U.L.R. 210, BGH, 28.5.98 (*supra*), cf. however, OLG. Düsseldorf, 5.6.97 (1999) TranspR 23.

240. Where the defendant recognises the danger but omits the requisite requirements, trusting that the damage will not come to pass.

241. BGH., 17.4.97 (1998) TranspR 25, (1998) 33 E.T.L. 60, (1998) U.L.R. 210, decided that wilful negligence could not be equated with *große fahrlässigkeit*, which requires especially gross and inexcusable dereliction of duty, but left open whether wilful gross negligence could be fault equivalent to wilful misconduct.

242. Although without retrospective effect: BGH, 16.7.98 (1999) TranspR 19, (1999) 34 E.T.L. 360. See further Koller, Art. 29 Rdn. 3.

243. Once the claimant had established sufficient facts to suggest gross negligence so as to place a burden of explanation on the carrier: OLG. Munich, 12.4.90 (1990) TranspR 280, 286 (Clarke, p. 380, n. 78, see further Herber/Piper, Art. 29, Rdn. 16, Koller, Art. 29 Rdn. 7, BGH. 16.7.98 (1999) TranspR 19, 23, OLG. Nuremburg, 10.12.92 (1993) TranspR 138, OLG. Hamm, 29.6.98 (1999) TranspR 201, OLG. Hamburg, 25.5.98 (1998) TranspR 351, OLG Stuttgart, 16.9.98 (1999) TranspR 66, OLG Munich 31.3.98 (1998) TranspR 353 AG Tempelhof–Kreuzberg, 18.6.98 (1998) TranspR 403. Contrast, in France, Cass., 22.4.97 (1997) B.T. 350, App. Paris, 24.4.97 (1997) B.T. 467, App. Paris, 31.3.99 (1999) B.T. 397.

244. BGH., 16.7.98 (1999) TranspR 19, 23. Note that in this case it had not been established where on the route the loss (through theft) had occurred. The court held that the carrier must show the level of care necessary to protect the goods in respect of that part of the route which involves the greatest risk. Furthermore, it was for the carrier to present and prove the circumstances which indicate that the failure to take precautions was not the cause of the damage, since in cases of goods particularly vulnerable to theft carried through areas of risk, unsatisfactory security precautions are, as a rule, the most probable cause of the loss.

245. See OGH., 14.7.93 (1994) TranspR 189, 11.7.90 (1992) TranspR 322 and cases cited by Mayer, p. 174, n. 26 and Clarke, p. 392, n. 55, cf., however, Tuma, "Artikel 29 CMR—Die Unendliche Geschichte? [1998] 33 E.T.L. 13 and in Haak and Swart, *Road Carrier's Liability in Europe* vol. 1, at p. 37.

Austrian courts also follow the same approach as the German courts in respect of the burden of proof.[246]

9.84 Under Dutch law a fault was not to be equated with *dol* unless it was of a fraudulent nature. In so deciding it was held that the mere fact that a vehicle suffered a series of breakdowns immediately subsequent to an inspection, where the carrier was not aware that the engine was still defective, would not constitute *dol*.[247] Dutch law therefore appeared to adopt the Belgian approach to the question. Under the New Civil Code, however, the carrier may not invoke any limitation in his liability to the extent that the damage has arisen from an act or omission done either with the intent to cause that damage or recklessly and with knowledge that damage would probably result therefrom,[248] and this test is now applied to Article 29 of CMR.[249]

9.85 Turning to the position at common law,[250] the same words in the original Warsaw Convention were defined by Barry, J., in the following terms in his direction to the jury in *Horabin* v. *B.O.A.C.*:

" . . . in order to establish wilful misconduct, the plaintiff must satisfy you . . . that the person who did the act knew that he was doing something wrong, and knew it at the time, and yet did it just the same, or alternatively that the person who did the act did it quite recklessly not caring whether he was doing the right thing or the wrong thing, quite regardless of the effect of what he was doing upon the safety of the aircraft and the passengers for which and for whom he was responsible. That . . . is something . . . quite different from negligence or carelessness or errors of judgment, or even incompetence, where the wrongful intention is absent."[251]

In other words, there are two distinct categories of conduct under English law: wilful on the one hand and negligent on the other, although an element of confusion is introduced by including reckless conduct with the former, whereas in an everyday context it is perhaps more synonymous with the latter. But recklessness, at least for the purposes of the civil proceedings,[252] is "akin to intentional wrongdoing, and is something essentially different in kind from negligence or carelessness".[253] What characterises conduct as wilful as opposed to negligent is the state of mind of the defendant. If it results from forgetfulness or genuine mistake, then it will amount to negligence.[254] It will only amount to wilful misconduct where he knows and appreciates that it is wrong conduct, or where such is his disregard for the consequences that his conduct must be characterised as wilful. So, for example, to use the hypothetical illustration put by Barry, J., in *Horabin* v. *B.O.A.C.*,[255] where two drivers go through a set of red lights, one by virtue of going too fast and not keeping a proper look-out, the other, in a hurry, deciding to ignore the lights on the basis

---

246. OGH., 14.7.93 (1994) TranspR 189 (see Mayer, p. 174 and Clarke, pp. 379–380).

247. Arrond. Roermond, 15.10.70 (1971) 6 E.T.L. 839. See also Hof. Den Bosch, 2.1.79 (1979) S. & S. No. 115; see further Haak, p. 249.

248. Art. 8:1108(1).

249. Hof. Amsterdam, 24.2.94 (1995) S. & S. No. 18 (Cleton, "Digest of Recent Decisions of the Netherlands Courts concerning International Transport Law Conventions", (1996) U.L.R. 755, 758, see also Hof s' Hertogenbosch, 12.5.98 (1999) S. & S. No. 22.

250. In Greece, an interpretation corresponding to that taken in English law was applied by the Hellenic Supreme Court, 12.3.98 (1999) 34 E.T.L. 100 (see Note by Murray at pp. 104–108).

251. [1952] 2 Lloyd's Rep. 450, 459.

252. Some criminal cases have in the context of particular statutes extended recklessness to include cases where there has been a high degree of negligence.

253. *Herrington* v. *British Railways Board* [1971] 2 Q.B. 107.

254. *Gordon* v. *G. W. Railway Board* (1881) 8 Q.B.D. 144.

255. *Supra*, fn. 251, at p. 460.

that hardly any traffic comes out of the side road, the first driver is negligent only, whereas the second is guilty of wilful misconduct.[256]

9.86 To take an example from a case decided on the basis of the use of the same words in railway conditions, in *Bastable* v. *North British Railway*[257] the company regulations required a load to be measured if there was any suspicion that it might be too high. The plaintiff arranged for a large load to be carried which one of the defendant company's servants chose not to measure, even though it was clearly close to the maximum height, preferring to gauge it by eye. On the journey the load hit a bridge, and it was held that the circumstances amounted to wilful misconduct.

9.87 The first reported English case[258] to consider wilful misconduct in the context of Article 29 of the CMR Convention was *Sidney G. Jones Ltd.* v. *Martin Bencher Ltd.*,[259] where, although the defendant carrier argued that the accident which resulted in damage to the goods was caused by a tyre blow-out, it was found by the court on the basis of the available evidence that the accident was in fact a result of the defendant's driver having fallen asleep at the wheel. It was accordingly necessary to consider whether this amounted to wilful misconduct.

9.88 The substance of the plaintiff's case in this respect was the fact that the regulations as to permissible driving times had been exceeded, as was evidenced by the tachograph records. Popplewell, J., made extensive reference to the *Horabin* case and to earlier railway cases turning on the use of the same words incorporated by way of special condition. The general tenor of those dicta was as to the fundamental distinction between wilful misconduct on the one hand and mere negligence on the other, the former involving conscious disregard of a risk, the latter involving only a want of care: a lack of fault rather than a willed choice. Popplewell, J., quoted[260] from the directions to the jury given by Barry, J., in the *Horabin* case set out above.[261]

9.89 In the *Martin Bencher* case, Popplewell, J., had little difficulty in deciding that the line between negligence and wilful misconduct had been crossed:

" . . . the driver was well aware of the Regulations. He was well aware of the purpose of the Regulations. He chose to ignore them and he did so deliberately. He knew that by ignoring them, he exposed the load that he was carrying . . . to a greater risk than if he had complied with the Regulations. . . . In my judgment he appreciated that he was acting wrongfully, persisted in so acting and was wholly indifferent to the consequences."[262]

9.90 However, Popplewell, J., accepted that it is to an extent a matter of degree, and that exceeding the permitted hours by a small margin may well not amount to wilful misconduct. However, on the facts before him, he categorised the excess as substantial, being one

---

256. Cf. the position in relation to carriage by air, where under Article 25 of the Warsaw-Hague Convention the plaintiff must also show that there was an awareness of the probability of the type of damage which occurred, see *Goldman* v. *Thai Airways International* [1983] 3 All E.R. 693.

257. [1905] 2 K.B. 532.

258. There is an earlier unreported decision of Webster, J., in *B. Paradise* v. *Islander Trucking Ltd.*, Queen's Bench Division, 28.1.85, unreported except on LEXIS, where it was held that, on the evidence before the court, it was an almost inevitable inference that the driver had been either a party to or at least connived at the theft, so that quite clearly it amounted to wilful misconduct. The sub-contracted haulier was a person for whom the first carriers were responsible under Article 29(2), so they were fully liable.

259. [1986] 1 Lloyd's Rep. 54.

260. At p. 59.

261. *Supra*, para. 9.85.

262. At p. 60.

hour fifty minutes over the permissive driving time of eight hours. No doubt there is scope for more difficult borderline cases.

9.90 The question of wilful misconduct under the CMR Convention also arose in the case of *Texas Instruments Ltd.* v. *Nason (Europe) Ltd.*[263] In that case, a driver was instructed by his employer to leave a trailer containing machine calculators in a public car park to await collection by another driver. The trailer was left there in the early part of the evening, and was to be collected around midnight. The driver gave evidence that he did not think that it was a good place to leave the trailer, and that he so advised his employer. Tudor Evans, J., found that there was a high risk of loss in leaving the trailer in a car park in the East End of London and that the risk was made entirely clear by the driver to his employer. He accordingly found that the employer appreciated the risk. Without deciding whether the conduct of the driver amounted to wilful misconduct, it was held that there was ample evidence of wilful misconduct on the part of the employer, and accordingly liability was not limited by the Convention.

9.91 There is some indication in the judgment of Tudor Evans, J., in the *Texas Instruments* case that it may not always be necessary to prove as a fact that the defendant in question subjectively appreciated the risk involved. In describing the actions of the employer, Tudor Evans, J., observed that:

"He knew that there was a high risk of loss. He was told so by (the driver) and, quite apart from that, as a matter of common sense (the employer) as a haulier must have appreciated the risk just as much as (the driver) . . . ."[264]

9.92 However, insofar as Tudor Evans, J., appeared to indicate that an objective assessment may be permissible, it is submitted that this obscures the fundamental distinction between wilful misconduct on the one hand and negligence on the other. Nor was this aspect necessary to the conclusion reached by Tudor Evans, J., since, as indicated above, he was satisfied of the facts that there had been a conscious disregard of the risks involved. A wholly subjective approach appears to have been adopted in another unreported case, *M. Bardiger Ltd.* v. *Halberg Spedition APS*,[265] where a driver left his vehicle unattended with a consignment of mink skins for a short period while he sought directions for the address to which the consignment was to be delivered. A quantity of the consignment was stolen in the intervening period, and it was argued on behalf of the plaintiffs that since the driver should have appreciated the risks of leaving his vehicle unattended with that cargo in that part of London, Article 29 applied. Although Evans, J., did not directly address the question of whether the appropriate test was subjective or objective, he accepted the evidence of the driver that he had no reason to think that there was any risk of theft. The learned judge said:

"If the allegation of misconduct was made out, then it would be necessary to consider whether this was 'wilful' meaning, in short, that it was deliberate and that the risk of loss was foreseeable and was accepted by him. This accurately summarises, I hope the effect of the English and European authorities which helpfully was cited to me on this topic."[266]

---

263. [1991] 1 Lloyd's Rep. 146.
264. At pp. 153–154.
265. Queen's Bench Division, 26.10.90, unreported except on LEXIS.
266. Note that in this case, Evans, J., held that the haulier undertook to carry and deliver the goods as part of a normal groupage load which was suitable for the type of goods involved. He could not be criticised (and thereby be guilty of misconduct) for failing to treat the goods otherwise than as part of a normal groupage cargo.

9.93 In *National Semiconductors (U.K.) Ltd.* v. *UPS Ltd.*,[267] which concerned carriage of a consignment of semi-conductors to Italy, when the driver arrived at the carrier's depot in Milan he was told that he could not unload for another three hours, so he went and parked in an unnamed side street, and left the vehicle unattended while he had a meal. Both the tractor unit and trailer were stolen along with the load. The carriers paid the CMR limit of liability, and the plaintiffs claimed the balance of the value of the stolen load under Article 29.

9.94 It was held that the plaintiffs had not established wilful misconduct, since the driver did not take a risk which he knew he ought not to take. Even though specific instructions had been given by the employer to the driver not to leave vehicles unattended in Italy except in an approved parking area, it was accepted that the driver did not have those instructions in mind at the time of the theft. There was evidence that on previous occasions, if he had not felt the area was safe, then he would not leave it unattended. In the present case, in the circumstances the court was not of the view that there had been a conscious taking of risk at all.

9.95 The next English case is *Longmill Corporation PLC* v. *Andrea Merzario Ltd.*[268] which has been referred to above in relation to the application of Article 17(2).[269] So far as wilful misconduct is concerned, it was first held that the fact of having parked the vehicle in the lay-by did not amount to wilful misconduct. However, in view of the evidence that the driver (i) knew that he should not leave the vehicle unattended in Italy, (ii) agreed that Italy was notorious for such thefts and that it was particularly dangerous in the Milan and Naples areas and (iii) also knew that the carrier's insurance policy had a specific exception for unattended vehicles in Italy, having rejected the driver's evidence that he left the cab unattended as a result of threats, it was held that in these circumstances leaving the vehicle unattended would have amounted to wilful misconduct.

9.96 However, it was then rather curiously held that the wilful misconduct was not causative of the loss. It was held that the plaintiffs must prove the causative link between the wilful misconduct and the loss, and, the driver having parked in a lay-by in circumstances which he had already held did not amount in themselves to wilful misconduct, it was held by His Honour Judge Diamond that the presence in the vehicle of the driver would have been unlikely to deter criminals any more than it did in the *Silber*[270] and *Cicatiello*[271] cases. The driver's evidence that he was "no hero" was accepted by the court, so that his presence would not have made any difference and the wilful misconduct was not causative of the loss. In other words, the proximate cause of the loss was where the driver parked, which did not amount to wilful misconduct, rather than the fact that the vehicle was left unattended.

9.97 The most recent English case on wilful misconduct in relation to the CMR Convention is *Lacey's Footwear (Wholesale) Ltd.* v. *Bowler International Freight Ltd.*[272] Again this concerned the theft of a consignment, but this time in this country rather than in Italy. In this case, it was not an unattended vehicle but a theft by deception, whereby the driver was intercepted outside the delivery premises, persuaded to drive some 3km

267. [1996] 2 Lloyd's Rep. 212.
268. C.L.C.C.(B.L.), 6.6.95, unreported.
269. *Supra*, para. 6.49.
270. See *supra*, para. 6.47.
271. See *supra*, para. 6.48.
272. [1997] 2 Lloyd's Rep. 369.

further down the road where the consignment was transferred onto another vehicle in the middle of an industrial estate. The driver had been told by the thieves that his vehicle was too large to be unloaded at the original address and that the consignment therefore needed to be transferred into a smaller vehicle.

9.98 Express instructions had been given to the driver that under no circumstances was he to permit off-loading anywhere other than at the delivery address. The instruction was repeated to him in Spanish, and the court accepted that he understood the instruction. Further, it was also accepted by the court, that following the original interception, there had been two phone calls made by the driver seeking instructions. In particular, in a second disputed telephone call alleged to have been made by the driver it was alleged that he was specifically instructed to return to the original delivery address, and not to deliver the goods anywhere else.

9.99 Despite the fact that neither side called the driver to give evidence, and despite the fact that there was conflicting evidence as to whether the second telephone conversation referred to above in fact took place, the judge at first instance[273] accepted that the second conversation had occurred and held that the driver's conduct amounted to wilful misconduct. He put the position in the following terms:

"Notwithstanding that very clear and precise command the driver seems to have taken it upon himself to have followed a complete stranger who was not in possession of a copy of the CMR Note for a distance of about 3km or 4km and allowed the consignment to be off loaded from his trailer onto another or other vehicles. This he did despite having telephoned [for instructions] again and yet again having been instructed to return to the Plaintiffs' premises. It is difficult to envisage a more deliberate disregard for positive instructions."

9.100 On appeal, it was argued that this was a clear case in which the driver had been duped, so that the trial judge could not reasonably have inferred wilful misconduct. It was held by the majority that the carrier was not automatically exonerated by the deception, and, referring to the clear and express instructions which had been understood by the driver but disobeyed by him, it was held that it was open to the judge to draw the inference that by disobeying those clear instructions, the circumstances amounted to wilful misconduct.

9.101 However, in a dissenting judgment, Brooke, L.J., rejected the evidence of the second telephone call. He put the position as follows[274]:

" . . . with the benefit of hindsight it was easy to say that the men's behaviour had been suspicious and that the driver had been stupid. The driver may well have been negligent—indeed, grossly negligent—when he allowed himself to be duped. . . . But the cases show that gross negligence is not enough. I for my part consider that the evidence before the judge was not of a sufficient quality to justify a finding of reckless carelessness by a person who knows he is acting in breach of his instructions."

9.102 The most recent review of the English authorities on wilful misconduct was in a case under the unamended Warsaw Convention, *The Thomas Cook Group Ltd.* v. *Air Malta Co. Ltd.*,[275] which concerned a theft of a substantial quantity of bank notes which

273. Q.B.D., 17.3.95, unreported.
274. At pp. 382–383.
275. [1997] 2 Lloyd's Rep. 399. See also *Rolls Royce PLC* v. *Heavylift-Volga DNEPR Ltd* [2000] 1 Lloyd's Rep. 653.

were being transported for the plaintiffs to the Bank of Valletta in Malta. As a result of lax security arrangements at Malta airport, it was argued on behalf of the plaintiffs that they were entitled to avoid the limit of liability which would otherwise have applied under the Warsaw Convention by reason of wilful misconduct on the part of the defendants.

9.103 Following a review of the authorities, which included the CMR cases, Cresswell, J., summarised the position as follows[276]:

"From the above authorities and in particular *Lacey's Footwear*, I derive the following steps/propositions:
1. The starting point when considering whether in any given circumstances the acts or omissions of a person entrusted with goods of another amounted to wilful misconduct is an enquiry about the conduct ordinarily to be expected in the particular circumstances.
2. The next step is to ask whether the acts or omissions of the Defendant were so far outside the range of such conduct as to be properly regarded as "misconduct". (An important circumstance would be deliberate disregard of express instructions clearly given and understood.)
3. It is next necessary to consider whether the misconduct was wilful.
4. What does not amount to wilful misconduct? Wilful misconduct is far beyond negligence, even gross or culpable negligence.
5. What does amount to wilful misconduct? A person wilfully misconducts himself if he knows and appreciates that it is misconduct on his part in the circumstances to do or fail or omit to do something and yet (a) intentionally does or fails or omits to do it or (b) persists in the act, failure or omission regardless of the consequences or (c) acts with reckless carelessness, not caring what the results of his carelessness may be. (A person acts with reckless carelessness if, aware of a risk that goods in his care may be lost or damaged, he deliberately goes ahead and takes the risk, when it is unreasonable in all the circumstances for him to do so.)
6. The final step is to consider whether the wilful misconduct (if established) caused the loss of or damage to the goods."

Having reviewed all the circumstances, he held that the acts or omissions of the defendants were not so far outside the range of conduct ordinarily to be expected in the particular circumstances as to be properly regarded as misconduct. However, even if it was misconduct, it was not in the circumstances properly to be categorised as wilful.

9.104 Summarising the position, in England as in Belgium and Holland, wilful misconduct and negligence are treated as distinct categories: the latter, however gross, cannot come within the former. Indeed, as a term denoting precise legal consequence it has been said that gross negligence "is the same thing as 'negligence' with the addition of a vituperative adjective".[277] In the light of the *Laceys Footwear*[278] case, it is possible that there may be circumstances in which the Court will be prepared to infer the necessary mental element for wilful misconduct. However, the judge's findings in that case as to the driver's disregard of express instructions were unusually clear. The dissenting judgment of Brooke, L.J., injects a proper note of caution. It should only be in the very clearest cases that a similar inference is permissible.

---

276. At pp. 407–408.
277. *Wilson* v. *Bret* (1843) 11 M. & W. 115, 116, per Rolfe, B., cited with approval by Willes, J., in *Grill* v. *General Iron Screw Colliery Co.* (1866) L.R. 1 C.P. 600. For a more modern statement to like effect see Goddard, C.J., in *Pentecost* v. *London District Auditor* [1951] 2 All E.R. 330, 333.
278. See *supra*, para. 9.97.

Equivalent default

9.105 In *Horabin* v. *B.O.A.C.*,[279] Barry, J., dealt shortly with the words "such default on his part as . . . is considered equivalent to wilful misconduct" in the equivalent provision in the Warsaw Convention:

"You need not trouble about the latter phrase, . . . because in the law of this country a default or an omission to do something can be just as much misconduct as the doing of something which is wrong."

In other words, he was treating "wilful misconduct" as relating to acts and equivalent default as relating to omissions. It is, with respect, thought that the view of the Belgian court referred to above is more in line with the likely intentions of the authors of the Convention, and that the concept of equivalent default was introduced for such legal systems as had no concept of wilful misconduct, rather than to provide for concepts of act and omission. However, in the result the answer is the same: the concept of equivalent default has no relevance for the purposes of English law.

Remoteness of damage under Article 29

9.106 The concept of remoteness is again relevant under Article 29. Under French law, for example, where breach is due to *dol* or *faute lourde* then the normal rules of remoteness, which are very similar to the English common law rules, do not apply, and the claimant can recover for all the direct consequences of the carrier's conduct, regardless of whether such consequences can properly be regarded as within the knowledge of the parties at the time the contract was made as likely to result.[280] There is no equivalent rule at common law in relation to breach of contract, and the rules are the same whether the breach was intentional or not. In an action in tort, however, there may be a difference of approach, depending on how the action is framed. If brought in negligence, then it is clear that the test of remoteness is that of foreseeability,[281] and the same is probably true for actions in conversion based on negligence.[282] However, it seems that the appropriate test outside negligence is directness of consequence rather than forseeability.[283] Although authority on the point is limited, the case of *Doyle* v. *Olby (Ironmongers) Ltd.*[284] can be used as authority to indicate that this is the correct test. That case concerned an action in the tort of deceit, which, as with wilful misconduct, can only succeed if proved to have been committed intentionally or recklessly.[285] It was there held that in deceit all losses directly resulting are recoverable and that "[i]t does not lie in the mouth of the fraudulent person that they could not reasonably have been foreseen".[286]

---

279. *Supra*, fn. 251, at p. 458.
280. Amos and Walton's *Introduction to French Law*, 3rd edn., p. 185.
281. *The Wagon Mound* [1961] 1 Lloyd's Rep. 1.
282. Based on the assimilation between actions in trespass to the person based on negligence and actions in negligence. See *Fowler* v. *Lanning* [1959] 1 Q.B. 426 and *Letang* v. *Cooper* [1964] 2 Lloyd's Rep. 339.
283. See *McGregor on Damages*, 16th edn, para. 147.
284. [1969] 2 Q.B. 158.
285. *Derry* v. *Peek* (1889) 14 App. Cas. 337.
286. [1969] 2 Q.B. 158, 167, per Lord Denning, M.R., cited with approval by Lord Hoffmann in *South Australia Asset Management Corp.* v. *York Montague Ltd.* [1996] 3 All E.R. 365, 374.

CHAPTER 10

Claims and Actions

10.1 The procedural aspects of claims and actions brought under the Convention are dealt with by Chapter V, comprising Articles 30 to 33, which are concerned respectively with the making of reservations in support of claims, jurisdiction, periods of limitation and arbitration.

Reservations in respect of loss, damage or delay

10.2 Article 30 provides as follows:—

" 1. If the consignee takes delivery of the goods without duly checking their condition with the carrier or without sending him reservations giving a general indication of the loss or damage, not later than the time of delivery in the case of apparent loss or damage and within seven days of delivery, Sundays and public holidays excepted, in the case of loss or damage which is not apparent, the fact of his taking delivery shall be *prima facie* evidence that he has received the goods in the condition described in the consignment note. In the case of loss or damage which is not apparent the reservations referred to shall be made in writing.

  2. When the condition of the goods has been duly checked by the consignee and the carrier, evidence contradicting the result of this checking shall only be admissible in the case of loss or damage which is not apparent and provided that the consignee has duly sent reservations in writing to the carrier within seven days, Sundays and public holidays excepted, from the date of checking.

  3. No compensation shall be payable for delay in delivery unless a reservation has been sent in writing to the carrier, within twenty-one days from the time that the goods were placed at the disposal of the consignee.

  4. In calculating the time-limits provided for in this article the date of delivery, or the date of checking, or the date when the goods were placed at the disposal of the consignee, as the case may be, shall not be included.

  5. The carrier and the consignee shall give each other every reasonable facility for making the requisite investigations and checks."

10.3 The starting point in the operation of Article 30 is the delivery of the goods,[1] and its overall purpose would appear to be to ensure that the carrier is given notice of any impending claim against him at an early stage. The Article is designed to take account of the differing circumstances under which the goods may be delivered, the two variables provided for being whether or not the goods are checked with the carrier as they are delivered,[2] and whether or not the loss or damage is apparent. As will be seen, the need

---

1. However, note that Article 30(3), which relates to reservations in cases of delay, refers not to delivery but to placing the goods at the disposal of the consignee. See *supra*, para. 7.9.
2. Cf. OLG. Hamburg, 13.5.93 (1994) TranspR 195.

to take account of these variables has produced a rather awkward piece of drafting in Article 30(1). On the arrival of the goods the consignee may do one of two things. The first possibility is that the consignee will check the goods in conjunction with the carrier,[3] in which case Article 30(2) will apply. If he does not do so, then Article 30(1) will apply and the action necessary will depend on whether or not the loss or damage is apparent. If it is apparent, then the reservation must be sent at the time of delivery[4]; if it is not apparent, then the reservation must be sent[5] in writing within seven days of delivery, Sundays and public holidays excepted.[6]

10.4 If the consignee fails to give a reservation as required by Article 30(1), then taking delivery amounts to *prima facie* evidence that he has received the goods in the condition described in the consignment note.[7] But as it is only *prima facie* evidence,[8] it can be rebutted by the consignee or other party interested in the goods,[9] and so does not mean that the right of action has been lost.[10] Indeed, it has been argued in relation to the similar provision in the Hague Rules that the provision has no real legal effect, since the onus of proving loss or damage necessarily rests in any event on the person alleging it.[11]

10.5 Where the consignee has checked the goods with the carrier, again the position depends on whether or not the damage is apparent. If it is not apparent, then a written reservation must be sent within seven days of the checking,[12] Sundays and public holidays again being excepted. If no such reservation is sent within the prescribed time, then effectively any damage subsequently discovered will be unrecoverable, since evidence contradicting the result of the checking will be inadmissible.[13] Equally, where there has been a check which has failed to discover apparent loss or damage, evidence contradicting its result will be inadmissible.

10.6 The concept of apparent loss or damage is thus central to both Article 30(1) and (2), but it is not defined by the Convention. It must, however, mean something wider than loss or damage actually discovered, since there would otherwise be no point to the distinction between apparent and non-apparent. It must therefore mean such loss or

---

3. The driver would presumably have authority to bind the carrier to the check, cf. Cass., 27.5.81 (1981) B.T. 407.

4. As to the meaning of "delivery", see *supra*, para. 6.9 ff. and *infra*, para. 10.59 ff.

5. In France, sending them via *commissionnaire* will be sufficient provided that the *commissionnaire* in turn, sends them to the carrier in time; Comm. Nanterre, 22.5.98 (1998) B.T. 904. But not otherwise; Cass., 26.10.99 (1999) B.T. 784.

6. If the carrier is unable to prove the date of delivery he is unable to allege that the reservations have not been sent within the seven days: Cass., 5.4.95 (1995) B.T. 660.

7. If there is no consignment note, it would seem that Article 30(1) must be read as applying to the condition of the goods at the time when the carrier took them over for carriage. App. Limoges, 2.6.67 (1967) B.T. 273; Loewe, para. 230.

8. See Cass., 1.10.97 (1998) 33 E.T.L. 248.

9. E.g. by the carrier acknowledging liability or facts which provide adequate proof that he is liable: Cass., 7.6.74 (1975) 10 E.T.L. 68. See also App. Toulouse, 26.3.69 (1971) 6 E.T.L. 131 (affidavit by process server); KG. Delft, 13.5.65 (1966) 1 E.T.L. 722, (1966) U.L.C. 104. Also cf. App. Toulouse, 12.4.94 (1984) B.T. 714 with App. Douai, 7.9.94 (1994) B.T. 623, and see also App. Paris, 2.12.81 (1982) B.T. 73.

10. Hof. Brussel, 21.1.87 (1987) 22 E.T.L. 745. Cf. Hof. Brussel, 7.2.92 (1993) 28 E.T.L. 286.

11. *Scrutton on Charterparties*, 20th edn., p. 434.

12. Although Article 30(2) refers to the date of the checking, rather than the date of delivery, this must be seen in the context of Article 30(1) which makes it clear that para. (2) only applies where there is a mutual check made upon delivery. The date of delivery and the date of checking will therefore be the same; if not, Article 30(1) applies. See OLG. Wien, 22.6.89 (1990) TranspR 158, where the carrier sought, but failed, to establish that the loss occurred between the time of delivery and the time when the goods were checked (see Clarke, p. 237 n. 89).

13. Article 30(2). Comm. Bruxelles, 19.3.74 (1974) 9 E.T.L. 773.

damage as is reasonably capable of discovery in all circumstances of the case,[14] which would include such matters as the nature of the goods and of the packaging.[15] In respect of packaging it has been held, however, that the fact that wooden packaging arrived with traces of an impact did not necessarily mean that the machine inside was damaged and therefore that such damage was apparent.[16] A good example of damage which is not apparent is provided by a French decision, where a consignment of hot-house plants was damaged by frost. Such damage may not become apparent for some days, and thus a failure to make reservations at the time of the check was not fatal to the claim.[17] The more difficult question is to what extent the consignee is required to open up the container or other packaging in which the goods are delivered to check the contents without losing his right to make a subsequent reservation. In France, it has been held that a consignee was not required to check the contents of the container.[18]

10.7 It should be noted that Article 30 is concerned with the effects of not giving the required reservations. It makes no provision as to the effect of such reservations,[19] and clearly the reservation on its own will not constitute proof of loss or damage.[20] To this there does, however, appear to be an exception. Where the carrier has participated in the

14. The claimant cannot justify a delay in making a reservation on the basis that complete freight papers were absent, when it was perfectly possible to check whether or not there was a shortfall in delivery: OLG. Linz, 27.11.89 (1990) TranspR 154, (1990) I U.L.R. 442.

15. See, e.g., *Silver* v. *Ocean Steamship Co.* [1930] 1 K.B. 416 concerning "apparent good order and condition" under the Hague Rules, which was taken to mean "visible by reasonable inspection" by Scrutton, L.J., (at p. 425). See also the discussion of "apparent condition" in relation to carrier's reservations on taking over the goods under Article 8(2), *supra*, paras 4.26–4.27.

16. App. Bordeaux, 28.4.97 (1998) B.T. 417.

17. App. Toulouse, 26.3.69 (1971) 6 E.T.L. 131. See also RB. Antwerpen, 7.1.77 (1977) 12 E.T.L. 420 (contaminated chemicals held not to be apparent damage). Other examples are provided by App. Paris, 2.12.81 (1982) B.T. 73 (fine black dust permeating the plastic packaging which damages the goods during handling after delivery), App. Paris, 16.9.92 (1992) B.T. 651 (damage to a machine which could only be discovered once it had been put into operation by a technician), cf., further, App. Douai, 7.9.94 (1994) B.T. 623.

18. Cass., 2.6.87 (unpublished, see Lamy, para. 1561, Jur. 2), App. Reims, 3.3.80 (1980) B.T. 237, App. Lyon, 22.4.88 (1989) B.T. 176, Cass., 24.11.87 (1988) B.T. 42, App. Lyon, 16.10.98 (1999) B.T. 782. Lamy, para. 509, criticises this trend as benefitting the lazy consignee, but the narrow approach probably reflects more accurately the likely realities of delivery, and reduces the scope for dispute as to the practicalities surrounding the feasibility of examination. Where the quantity of the goods is such that the consignee cannot be expected to check them immediately, subsequent discovery of a missing quantity will, similarly, be treated as a matter of non-apparent loss, Comm. Nanterre, 22.5.98 (1998) B.T. 904 (75 out of 889 rolls of tissue). In the context of Article 30(2), more pertinent will be whether it is reasonable to delay the carrier whilst a complete check is made. If it is not, damage found subsequently will be non-apparent; App. Bordeaux, 28.4.97 (1998) B.T. 417 (the carrier could not be expected to assist in the thorough examination of a cylinder weighing some 27 tonnes).

19. It has been held in Belgium that a reservation which purports to quantify the damage monetarily did not bind the consignee. It did not amount to an agreement as to the amount of the damage when later correspondence showed that this was still an open question: Comm. Bruxelles, 2.4.90 (1991) 26 E.T.L. 541 (see commentary by De Wit, at p. 547).

20. In contrast to the position in France, whereby making a reservation within Article 30(1) shifts the burden to the carrier to prove that the damage did not occur during carriage: Cass., 2.2.82 (1982) B.T. 152, (1983) 18 E.T.L. 47 (see Brunat, (1982) B.T. 146), Cass., 15.7.86 (1986) B.T. 542, Trib. Colmar, 9.1.96 (1996) B.T. 503, cf. App. Aix-en-Provence, 6.3.91 (1991) B.T. 767 (having given a precise reservation, there was no need for a further *expertise*, Cf., however, App. Bordeaux, 28.4.97, *supra*, fn. para. 10.6, fn. 16, where the view of an expert was taken into acccount in support of the case for the claimant); compare the wording of CIM which makes clear the need for the claimant to bring proof of loss or damage in transit (Article 57). See further Clarke, p. 234 and cases cited at p. 235 n. 74. Cf. further, App. Besonçon, 27.10.99 (2000) B.T. 94. Cf. in Germany BGH., 8.6.88 (1988) 23 E.T.L. 705, (1988) II U.L.R. 717. The French view is especially pertinent where an inland carrier delivers the goods after receipt by an international carrier. The reservation made by the consignee in respect of non-apparent damage shifts the burden to the international carrier: Cass., 22.9.83 (1983) B.T. 566, (1984) 19 E.T.L. 111. See also App. Rouen, 21.2.91 (1991) B.T. 507 where the carrier's contrary proof based on an *expertise* failed because it took place too long after the carriage.

checking, Article 30(2) would apply equally to him, thus preventing him from subsequently disputing the results of the check.

10.8 A number of points of general application must now be considered.

Form of reservations

10.9 As far as the actual content of the reservation is concerned, it has been held in this country both for the purposes of Article 27(1) (written claim for interest)[21] and Article 32(2) (written claim suspending limitation period)[22] that a general intimation of intention to claim is sufficient. This must be even more the case under Article 30, where it is expressly provided that all that is required is "a general indication of the loss or damage".[23] Loewe is of the opinion that the mere words "loss" or "damage" would not be enough,[24] but it would seem that the overriding purpose of the provision is to give the carrier early notice of the impending claim so as to enable him to carry out his own investigations at an early opportunity,[25] and it is thought that such words entered on the consignment note or delivery order[26] would fulfil this function and comply with Article 30. Such a view is supported by a French decision where the words "in poor condition" entered on the consignment note were accepted by the court as sufficient for a valid reservation.[27] On the other hand, in a Belgian decision, it was held that entering the words "in bad condition" and "in bulk" on the consignment note was not sufficient, on the basis that those words could not amount to reservations in terms which would permit a court to give its decision as to the carrier's liability.[28] It may be objected, however, that such is

---

21. *Supra*, para. 9.60 ff.

22. *Infra*, para. 10.81 ff.

23. Article 30(1). The quoted wording is not repeated in respect of reservations under Article 30(2), but there is no reason why anything more precise should be required under the latter paragraph.

24. Loewe, para. 222. See further the same writer in Theunis, p. 146, who maintains his position in opposition to the view expressed in the first edition of this book.

25. See, in particular, Article 30(5) *infra*, para. 10.16. Consequently, a reservation indicating damage to the packaging would not necessarily put the carrier sufficiently on notice of a claim regarding the contents for which a further reservation would need to be made within seven days of delivery: RB. Antwerpen, 30.9.94 (1995) 30 E.T.L. 232.

26. Clearly such words on their own would make little sense if communicated in any other manner.

27. Cass., 29.4.75 (1977) I U.L.C. 334. See also App. Toulouse, 12.4.94 (1994) B.T. 714 (sufficient to state that the covering sheet was torn and that the goods were wet), cf. further Cass., 15.7.86 (1986) B.T. 542, App. Rouen, 9.2.93 (1993) B.T. 201. Cf. App. Aix-en-Provence, 22.2.79 (1979) B.T. 387, where the reservation "*sous réserve de contrôle au déballage, marchandises mouillée*" was insufficient since it did not indicate that damage definitely existed, and App. Paris, 23.3.79 (1980) J.C.P. 19372 (see Lamy, para. 1563, Jur. 6) where a reference to 25% damage was insufficient since it did not indicate the nature of the alleged damage, cf. further App. Lyon, 3.7.86 (1987) B.T. 256, App. Douai, 7.9.94 (1994) B.T. 623, App. Douai, 16.1.97 (1997) B.T. 103 (which required details of both the nature and extent of the damage to be given), App. Paris, 10.11.99 (1999) B.T. 823 (the reservation "*20 tonnes nets*" was insufficient in respect of a consignment of 22 tonnes; the reservation should also have indicated the number of missing packages) and in Austria, OLG. Wien, 22.6.89 (1990) TranspR 158, (1989) II U.L.R. 815.

28. Hof. Antwerpen, 15.11.78 (1979) 14 E.T.L. 661. See also RB. Amsterdam, 29.11.78 (1980) 8 E.L.D. 103 where it was held that a valid reservation is not made if the reservation lacks any indication as to the general nature of the damage. On the other hand, striking out the words "in good state" on a delivery note was a sufficient reservation in Comm. Bruxelles, 21.9.90 (1992) R.D.C.B. 787 (see Putzeys, "Le Droit Des Transports en Belgique", (1992) I U.L.R. 190, 201). More recently it has been pointed out that reservations precede a claim and are meant as preserving the right to maintain the claim, so that a reservation was sufficient which stated that there was much damage and that 260 out of 2100 packages were completely destroyed: App. Bruxelles, 27.2.96 (1996) 31 E.T.L. 833.

clearly not the purpose of reservations under Article 30 and it is submitted that the view expressed above is to be preferred.

10.10 Where the loss or damage is not apparent Article 30(1) and (2), depending on which is applicable, both specify that the reservation must be in writing, which would include entry on the consignment note,[29] or a separate communication sent by post,[30] telex,[31] or cable. Similarly, where a claim is made for compensation for delay in delivery, a reservation must be sent in writing otherwise no compensation will be payable.[32] In this case, however, the time limit is 21 days and since, as distinct from Article 30(1) and (2), no reference is made to the exclusion of Sundays and public holidays, it would seem that they are to be included in the calculation of this period.

10.11 On the other hand, the clear implication from the express provision in Article 30(1) that the reservation must be in writing for loss or damage which is not apparent is that it need not be in writing when, in the case of apparent loss, the reservation is made at the time of delivery. As against this, the word "sending" is used in respect of both situations, which seems inappropriate to describe an oral communication. It is thought, however, that the use of the word "sending" represents something of a mistranslation of the French text,[33] where the equivalent word used is "*addressé*", which would seem to have a broader connotation, and that "sending" in the context of apparent damage should be interpreted in the broad sense of "communicating", so as to include both oral and written reservations. This view is supported by a decision where it was held that a reservation made by telephone is valid, if the consignee shows evidence of having made it within the time specified and with sufficient clarity. Such evidence was furnished in that case by a letter recording the contents of the telephone conversation, to which no objection was made.[34] Clearly, however, from an evidential point of view it will always be in the consignee's interests to ensure that the reservation is in writing, since it will be for him to show that the reservation has been properly made,[35] although this may not be necessary where the carrier subsequently fails to dispute with reasonable promptness that an oral reservation was made at the proper time.[36]

10.12 Further, proof of an oral reservation may override the appearance created by the fact that the consignee has signed for the receipt of the goods.[37] So in *City Vintages Ltd. v. SCAC Transport International*,[38] Steyn, J., accepted that the consignment note which had been signed on delivery did not amount to *prima facie* proof of the receipt of the

---

29. Cass., 29.4.75 (1975) B.T. 298, (1977) I U.L.R. 334, App. Toulouse, 12.4.94 (1994) B.T. 714.
30. Cf. App. Paris, 22.6.66 (1966) B.T. 330.
31. Cass., 24.11.87 (1988) B.T. 42, OLG. Hamburg, 6.12.79 (1980) VersR 290 (delay), cf. Hof. Antwerpen, 30.5.79 (1979) 14 E.T.L. 924. Cf. further App. Lyon, 16.10.98 (1999) B.T. 782 (*télécopie*).
32. Article 30(3), discussed *supra*, paras 7.9–7.10. Arrond. Roermond, 30.5.68 (1969) 4 E.T.L. 1019.
33. As to the propriety of having regard to the French text, see *supra*, paras 0.14–0.17.
34. Hof. Antwerpen, 21.6.78 (1978) 13 E.T.L. 601. A broad interpretation is also supported by a French decision where it was held that entry of a reservation on the consignment note which was in the possession of and retained by the carrier was "sent" within Article 30(1) as that word is properly understood: Cass., 29.4.75 (1977) I U.L.C. 334. Loewe, para. 224 also expresses the view that oral reservations will be sufficient.
35. Hof. Antwerpen, 21.6.78 (1978) 13 E.T.L. 601. Arrond. Dordrecht, 10.5.67 (1967) S. & S. No. 70.
36. Comm. Verviers, 18.5.68 (1968) 3 E.T.L. 1240, (1969) U.L.C. 159. See also RB. Antwerpen, 1.4.80 (1980) 15 E.T.L. 461. (Oral reservation proved by fact carrier took steps in his own warehouses to find out what had happened to the missing goods.) See further Loewe, Theunis, p. 146, Clarke, p. 239 n. 12.
37. App. Liège, 6.5.70 (1970) 5 E.T.L. 716 (see Clarke, p. 238 n. 93), RB. Kortrijk, 4.6.74 (1974) 9 E.T.L. 768 (see Loewe, Theunis, p. 146) where the consignee signed a delivery note in error, which noted the good condition of the goods.
38. Queen's Bench Division, 1.12.87, unreported except on LEXIS.

goods in circumstances where the driver had been aware that it was thought that there was a considerable short delivery and the word "unexamined" had been written on the consignment note and the delivery order.

Time for reservations

10.13 The time within which reservations must be made has already been explained, but some general points remain. First, in the event of the consignee being aware of the loss or damage prior to delivery, there is nothing to prevent him from making a reservation to the carrier while the goods are still in transit,[39] although further reservations as specified by Article 30 may be necessary in the event of further loss or damage being discovered at or subsequent to the delivery.

10.14 Secondly, Article 30(4) provides that, for the purposes of the time limits specified in the Article, "the date of delivery, or the date of checking, or the date when the goods were placed at the disposal of the consignee, as the case may be, shall not be included" in the calculation of the period. This provision can presumably have no application in cases of apparent loss or damage, since in such a case, whether under Article 30(1) or (2), the Convention specifies a particular time (i.e. the time of delivery) rather than a time limit as such.[40] Thirdly, it has already been explained that, with the exception of the time limit in cases of delay, Sundays and public holidays are to be excluded.[41] It would seem that whether or not a particular day is to count as a public holiday should depend on the law of the place of delivery, since the requirement to take active steps under Article 30 is on the consignee, and his ability to to do within the time allowed will obviously depend on the public holidays at that place.

10.15 Finally, where a reservation is sent in writing, it is made clear by the wording of Article 30(1), (2) and (3) that the material time is the date of the sending rather than the date of the arrival of the letter,[42] but clearly the consignee may be required to furnish proof of posting within the time allowed. Care should therefore be taken by consignees to keep adequate evidentiary records.

Facilities for checking

10.16 Article 30(5) requires both the carrier and the consignee to give each other every reasonable facility for making any necessary investigations and checks. No provision is made for any penalty if either party fails to observe this requirement, but presumably any failure on the part of either party to co-operate may be taken into account by the court in deciding the liabilities of the parties. It should also be noted that the obligation is mutual, extending, for example, to a consignee wishing to examine the vehicle used for the carriage in order to check its condition, or the condition of any refrigerating equipment.

10.17 Such judicial consideration as there has been of this provision has demonstrated the limits of the obligation rather than how either party can benefit from it. Thus in a

---

39. But it was held in a French case that reservations under Article 30(3) (delay) were not effective if sent before the goods were actually due: App. Amiens 19.6.80 (1980) B.T. 490. See *supra*, para. 7.10.
40. Loewe, para. 224.
41. See, e.g., App. Rouen, 21.2.91 (1991) B.T. 507.
42. Cf. App. Paris, 22.6.66 (1966) B.T. 330 (see Lamy, para. 1563, Jur. 2).

Dutch case it was held that a carrier could not seek to rely on it where the cargo of meat had been destroyed and the expert's report would clearly have made no difference.[43] In a Belgian case, where the carrier had not told the consignee of his intention of appointing an expert, it was held that he could not claim that Article 30(5) had not been complied with, given that he had been informed of the nature and extent of the damage on delivery, and that the consignee's expert had arranged for the meat to be sold with a view to off-setting the cost of the damage.[44]

Practice

10.18 In general, Article 30 imposes a considerable onus on the consignee to act quickly on the receipt of goods, since under Article 30(1) his claim might to an extent be prejudiced, and under Article 30(2) and (3) he may lose his right to claim.

10.19 Further complications arise where the consignee is not the claimant, since it may be the sender who is claiming, and the sender who has taken out insurance. In such circumstances the sender is dependent on the consignee for prompt action,[45] both for the purposes of his own claim and for the purposes of his insurance. In addition, it may be that neither the consignee nor, where he is a sub-contractor, the carrier is aware that carriage is subject to the Convention, particularly if there is no consignment note.[46] In view of these difficulties, if circumstances so permit, it may be appropriate for the sender to include in his contract with the consignee a requirement that the latter comply with the requirements of Article 30. Quite apart from such a provision, it may be good commercial practice for the consignee to inform the sender, but clearly notice to the sender cannot constitute a valid reservation under Article 30.[47]

Total loss

10.20 Finally, in relation to reservations under Article 30, it will be observed that the obligation to enter reservations is dependent on delivery. Where, therefore, the goods have been lost no obligation arises under Article 30.

Jurisdiction

10.21 Article 31 provides as follows:—

"1. In legal proceedings arising out of carriage under this Convention, the plaintiff may bring an action in any court or tribunal of a contracting country designated by agreement

---

43. RB. Amsterdam, 24.11.76 (1978) 6 E.L.D. 90.
44. Hof. Antwerpen, 21.6.78 (1978) 13 E.T.L. 601.
45. At least in respect of Article 30(1) and (2) but not Article 30(3), see Loewe, Theunis, p. 147. Cf. Clarke, p. 239, seemingly in respect of Article 30(1) and (2), who cites (at n. 5) App. Limoges, 7.11.77 (1977) B.T. 537. This case, however, was concerned with Article 30(3).
46. But note the position concerning the need for a consignment note in relation to successive carriers, *infra*, Chapter 11, paras 11.21–11.48.
47. Hof. Gent, 17.11.67 (1969) 4 E.T.L. 145. Similarly, notice to a forwarder which is not transmitted to the carrier in time, App. Paris, 13.1.88 (unpublished, see Lamy, para. 1566, Jur. 2). Cf. the decision of the OGH criticised by Loewe, Theunis, p. 148 (cited as OGH DK UfR 1974, p. 365), cf. App. Paris, 16.9.92 (1992) B.T. 651 where there appears to have been proof of the transmission of the reservation from the forwarder to the carrier.

between the parties and, in addition, in the courts or tribunals of a country within whose territory:

    (a) the defendant is ordinarily resident, or has his principal place of business, or the branch or agency through which the contract of carriage was made, or

    (b) the place where the goods were taken over by the carrier or the place designated for delivery is situated,

and in no other courts or tribunals.

2. Where in respect of a claim referred to in paragraph 1 of this article an action is pending before a court or tribunal competent under that paragraph, or where in respect of such a claim a judgment has been entered by such a court or tribunal no new action shall be started between the same parties on the same grounds unless the judgment of the court or tribunal before which the first action was brought is not enforceable in the country in which the fresh proceedings are brought.

3. When a judgment entered by a court or tribunal of a contracting country in any such action as is referred to in paragraph 1 of this article has become enforceable in that country, it shall also become enforceable in each of the other contracting States, as soon as the formalities required in the country concerned have been complied with. The formalities shall not permit the merits of the case to be re-opened.

4. The provisions of paragraph 3 of this article shall apply to judgments after trial, judgments by default and settlements confirmed by an order of the court, but shall not apply to interim judgments or to awards of damages, in addition to costs against a plaintiff who wholly or partly fails in his action.

5. Security for costs shall not be required in proceedings arising out of carriage under this Convention from nationals of contracting countries resident or having their place of business in one of those countries."

General

10.22 In the event of a dispute under the Convention which the parties are unable to resolve by agreement, the first problem in relation to bringing legal proceedings is to ascertain where the action should be brought. The object of Article 31(1) is to provide an answer to this problem without resort to the normal rules of private international law. Thus in *Arctic Electronics Co. (U.K.) Ltd.* v. *McGregor Sea & Air Services Ltd.*[48] Hobhouse, J., said[49] that provisions such as this are intended to be a self-contained code within which a plaintiff must found his assertion of jurisdiction. Consequently, any bases of jurisdiction provided by national law are excluded, unless consistent with CMR.[50] However, the Convention does provide for an alternative solution in Article 33 by way of contractual provision for arbitration.[51]

10.23 It will be seen that Article 31(1) permits two possibilities: first, litigation in a jurisdiction chosen by the parties, and second, litigation in a jurisdiction designated by Article 31(1) itself. The first point to note is that the fact that the parties have agreed a jurisdiction does not exclude the alternative jurisdiction based on the provisions of Article 31(1)(a) and (b), since it is provided that the latter shall be "in addition" to the former.

---

48. [1985] 2 Lloyd's Rep. 510.

49. At p. 514.

50. See further *infra*, regarding the relationship between CMR and the Brussels and Lugano Conventions in respect of jurisdiction and the enforcement of judgments. The exclusion of rules of national law that might otherwise apply should on principle also extend to those principles of *forum non conveniens* as applied by the English courts, see Clarke, para. 46a, and Glass "CMR: Putting Practice into Theory", [1984] LMCLQ 30, 39 n. 23.

51. *Infra*, para. 10.120 ff.

10.24 If the parties have agreed a jurisdiction, Article 31(1) effectively ensures that the provisions of the Convention will be applied by in effect providing that only the courts of a contracting country can be so designated.[52] Curiously, no such limitation is contained in the provision for alternative jurisdiction in paras (a) and (b) of Article 31(1). It may well be that the places there specified as permissible jurisdictions are not within contracting countries, and it may be that the courts of a non-contracting country would not be prepared to apply CMR rules where they conflict with the national law. If expressly incorporated into the contract, then presumably CMR would be applied in any event, but whether it would be equally applied where incorporated into the contract only by virtue of the operation of the Convention itself must be open to some doubt in countries which are not parties to the Convention. In practical terms, however, it will often be in the best interests of the claimant to bring his claims before the courts of a contracting country, since the enforcement provisions of Article 31(3)[53] only apply in respect of a judgment by a court of a contracting party.

Scope

10.25 Article 31 applies to all legal proceedings arising out of carriage under the Convention. It will therefore extend to extra-contractual claims referred to in Article 28,[54] and it will also apply to legal proceedings both by the cargo interests against the carrier and by the carrier against the cargo interests.[55] However, it will not cover claims which for one reason or another are not within the scope of the Convention.[56] So, for example, it will not apply to a claim for total non-performance of the contract,[57] nor will it apply to actions arising out of third party claims against the carrier in respect of damage to other vehicles or goods. The fact that goods being carried in the defendant carrier's vehicle are totally or partially the cause of such loss or damage, even if they themselves are subject to carriage under CMR, will not affect the issue. Nor must it be forgotten that a carrier may in fact be carrying several consignments within a particular country, some of which are subject to CMR and some of which are subject to national domestic law. Proceedings in respect of the latter will not be affected by Article 31, but will be subject to national procedural rules. The question is also of importance in relation to shipments carried by combined transport which are not subject to Article 2, because they are unloaded from the road vehicle. These, too, will be subject to the national domestic law of the country or countries concerned.

---

52. Contractually stipulating the courts of a non-contracting country would be null and void under Article 41, as would any purported exclusion of Article 31(1)(a) and (b).
53. *Infra*, para. 10.43 ff.
54. *Supra*, para. 9.65 ff.
55. *Frans Maas Logistics (U.K.) Ltd.* v. *CDR Trucking* [1999] 2 Lloyd's Rep. 179. But not all of Article 31 may apply if the claim is for a negative declaration. See *infra*, paras 10.40–10.41.
56. See generally *supra*, Chapter 1.
57. App. Milano, 11.7.75 (1977) I U.L.C. 336, HG. Vienna, 3.4.84 (1984) TranspR 152, see Loewe, Theunis, p. 146, cf. Clarke, p. 180; cf. *Shell Chemicals UK Ltd.* v. *P&O Roadtanks Ltd.* [1993] 1 Lloyds' Rep. 114, where the concession to this effect by the defendants was approved by Saville, J. However, in the case before him, although he held it outside the liability provisions of the Convention, he was of the view that Article 32 would have applied, if relevant. Presumably he would have also regarded Article 31 as applicable. See, generally, the discussion of this case *supra*, paras 0.26–0.34 There might, however, be a claim for delay under the Convention if there was an agreed time limit. See *supra*, para. 7.6.

The appropriate forum: agreed jurisdiction

10.26 Turning to consider in more detail the permissible jurisdictions under Article 31(1), the initial possibility provided for, as has been seen, is that the plaintiff may bring an action in the courts of a contracting country designated by agreement between the parties.[58] A number of points arise from this provision. In the first place, there is nothing in Article 31(1) to restrict the agreement as to jurisdiction to an agreement made prior to the performance of the contract. It would therefore seem that the parties can agree a jurisdiction after the claim has arisen, as long, of course, as it is a contracting country.

10.27 In the case of agreements as to jurisdiction prior to the carriage, as between the parties to that agreement, it will be necessary for the party alleging such agreement to show that the jurisdiction clause was part of the contract, in accordance with normal rules of contractual incorporation.[59] Thus, in a French decision, where a French forwarder wished to recover from a German carrier who was partly responsible for the carriage, the latter attempted to rely on a provision of the German Forwarding Conditions (ADSP) which restricted any action to the courts of his place of business (Stuttgart). The court rejected this plea on the grounds that the jurisdiction clause was not included in the documents exchanged between the parties, nor had it been brought to the notice of the forwarder and accepted by him.[60]

10.28 Additional problems arise where the person claiming was not a party to the original contract. It would seem in principle that where the original parties have agreed a particular jurisdiction that this will be binding on all parties whose rights derive from the contract of carriage, the main example of such a person being, of course, the consignee.[61] However, in another French decision it was held that the French consignee was not bound by a clause restricting legal action to a German court, since it was not included in the consignment note handed to the consignee. Inclusion in the contract of carriage between the sender and the carrier was not regarded as sufficient on the basis that it is only through the consignment note that the consignee is brought into the contract.[62] However, it has been argued earlier that a consignment note is not generally essential, at least as between the carrier and the sender, but that it is only of evidential value as to the terms of the contract.[63] For the reasons there stated, this would seem equally applicable to the present situation, particularly in view of the fact that a jurisdiction clause is not one of the matters set out in Article 6[64] for inclusion in the consignment note. Further, given that the consignee's rights derive from the contract of carriage, i.e. from those of the consignor,[65] there would seem no reason in principle why the consignee's rights should not be restricted to the same extent as those of the consignor. Finally, it can be objected that there

---

58. E.g. App. Aix-en-Provence, 4.9.96 (1997) B.T. 277.

59. See generally *Chitty on Contracts*, 28th edn., paras 12.002–12.018 cf., Hof. Gent, 19.11.93 (1994–1995) R.W. 436 (Putzeys, "Le droit des transports en Belgique (II)—1993–1995", (1996) U.L.R. 557, 568). Clearly, the agreement must be with the carrier. A jurisdiction clause in a contract of sale to which the carrier is not privy and has no knowledge would not be binding on him. Cf. Comm. Lyon, 12.2.93 (1993) B.T. 260.

60. App. Paris, 29.3.69 (1969) B.T. 159, (1970) U.L.C. 125 but see *supra*, fn. 52 and *infra*, fn. 66.

61. Loewe, para. 242. See also Carriage of Goods by Road Act 1965, section 14(2)(b).

62. App. Paris, 14.11.69 (1969) B.T. 363, (1970) U.L.C. 133.

63. *Supra*, Chapter 4, paras 4.3–4.9 and see Cass. (It.)., 28.11.75 (1976) I U.L.C. 247 where it was held that a choice of law clause could be proved despite the fact that it was not included in the consignment note.

64. *Supra*, Chapter 4, paras 4.15–4.23.

65. Article 13, *supra*, Chapter 5, paras 5.12–5.24.

is nothing in the Convention which provides that it is through the consignment note that the consignee is brought into the contract.

10.29 The problem of jurisdiction clauses also arises in relation to combined transport bills of lading. These commonly include jurisdiction clauses, which for other modes of transport or domestic road transport will be acceptable to the courts in many countries. Where, however, either prior or subsequently to the sea transit the consignment is subject to international carriage of goods by road, CMR will be applicable. Accordingly, where there is a claim for loss, damage or delay to the goods resulting from carriage by land any jurisdiction clause in the combined transport bill of lading will be inoperative to the extent to which it conflicts with the provisions of Article 31.[66] As CMR makes no provision as to the form in which the parties must formulate the jurisdiction agreement, this will be decided by national law.[67]

Jurisdiction under Article 31(1)(a) or (b)

10.30 As has already been explained, whether or not there is an agreed jurisdiction, the claimant is entitled to bring his claim in another jurisdiction within Article 31(1)(a) or (b).[68] Under Article 31(1)(a) the plaintiff can bring an action in the courts or tribunals of a country within whose territory the defendant is ordinarily resident,[69] or has his principal place of business, or the branch or agency through which the contract of carriage was made. It is not clear whether the "principal" place of business refers to the defendant's statutory or actual place of business. It remains to be seen how the courts in the various jurisdictions interpret this provision, but ideally, it should be interpreted as covering both.

10.31 The question also arises as to whether the "branch or agency" referred to in Article 31(1)(a) refers to a branch or agency of the defendant or whether it would also cover an agency of the plaintiff or some independent agency through which the contract of carriage was made. In other words, it might be argued that the possessive pronoun, appearing as it does only in relation to "principal place of business", qualifies only that phrase and not the words "branch or agency". However, the words "the defendant . . . has" must necessarily qualify both phrases, since otherwise the second half of para. (a)

---

66. RB. Antwerpen, 23.9.75 (1976) 11 E.T.L. 279. A shipment of polyester cotton goods was carried under a combined transport bill of lading from Hong Kong to Rotterdam and thence by road to Antwerp. The bill of lading contained a clause giving exclusive jurisdiction to the English courts. In an action for loss and damage discovered in Antwerp it was held that as CMR was applicable from Rotterdam to Antwerp the jurisdiction clause in the bill was inoperative to the extent that it conflicted with Article 31. A similar conflict also arose between the nine month period of prescription in the bill and that laid down in Article 32 which rendered the former inoperative to a like degree. *Infra*, para. 10.54.

67. *Quaere* whether this question will be subject to the law of the place whose jurisdiction is specified or that of the place where the contract was made—Loewe, para. 241.

68. OLG. Düsseldorf, 18.11.71 (1973) 8 E.T.L. 510—provisions in Turkish law to the contrary cannot oust the rules relating to jurisdiction in Article 31. So, also, in the case cited in fn. 66 *supra*, the Belgian court ruled that it had jurisdiction despite the attempt to confer exclusive jurisdiction in the combined transport bill of lading. Any attempt to exclude this right would be void under Article 41, see *Dresser U.K. Ltd.* v. *Falcongate Freight Management Ltd.*, Queen's Bench Division, 6.11.90, (1991) 26 E.T.L. 798, (1991) I U.L.R. 354, reversed in part on different grounds, [1992] 2 All E.R. 450. It has already been noted that jurisdiction under Article 31(a) or (b) is apparently not restricted to the courts of a contracting country. *Supra*, para. 10.24.

69. On the meaning of "ordinarily resident" see *Rothmans of Pall Mall (Overseas) Ltd.* v. *Saudi Arabian Airlines Corporation* [1980] 3 All E.R. 359 where it was held that the same words in the Warsaw Convention as amended do not enable a foreign corporation to be sued in this country merely on the basis of having a branch office here.

makes no grammatical sense at all. Ultimately, therefore, even if the word "his" only applies to "principal place of business" the rest of the sentence necessarily restricts the relevant branch or agency to that of the defendant.[70]

10.32 A further problem is that the meaning of the word "agency" is far from clear and could be subject to widely differing interpretations in the various jurisdictions, particularly as in common law countries it can be applied loosely to various commercial activities and has no precise legal connotation. There is no requirement as to the size of the branch or agency concerned.

10.33 Alternatively, an action can be brought in the courts or tribunals of the place where the goods were taken over by the carrier or the place designated for delivery of the goods.[71] This provision presupposes that the goods have actually been taken over by the carrier,[72] but as explained earlier this Article will not in any event usually apply to the situation that arises where there is a dispute concerning the total non-performance of a contract of carriage.[73]

10.34 It might be argued that where a carrier is a successive carrier within the meaning of Article 34, the "place where the goods are taken over" must refer to the place where that carrier has taken over the goods, and not where any previous carrier has done so. On the other hand, the position concerning successive carriers is regulated by Chapter VI of the Convention,[74] which makes no reference to Article 31(1), from which it would seem to follow that no special provision was thought necessary, and that the claimant can commence proceedings in the country where the goods were originally taken over. This would also seem to make good sense, given that the claimant may not know where the goods were taken over by that carrier. A third possible approach in cases involving successive carriers would be that the claimant can at his option bring his action either where the goods were taken over by that carrier, if he knows where that was, or alternatively where the goods were taken over by the original carrier.

10.35 A related issue came before the English courts in the case of *Moto Vespa S.A.* v. *MAT (Brittania Express) Ltd.*[75] In that case, goods were to be transported from Birmingham to Madrid. The defendant carrier sub-contracted the portion of the carriage from Birmingham to Perpignan, and only took over the goods themselves at Perpignan, after which the consignment was damaged. It was held, applying the decision of the Court of Appeal in *Ulster-Swift Ltd.* v. *Taunton Meat Haulage Ltd.*,[76] that the defendant as the principal carrier was carrier for the purposes of the entire journey, and that the English court therefore had jurisdiction, the goods having been taken over at Birmingham. It will be observed that this case does not squarely cover the question of claims against successive carriers who are not principal carriers, since the defendant was the principal carrier and thus responsible for the whole journey. But it does not exclude the possibility

---

70. Loewe is of the same view, para. 243.

71. Article 31(1)(b). The place designated for delivery is normally the place so indicated in the consignment note but this place may have been changed during transit by subsequent order of the sender, as in OLG. Hamburg, 7.4.94 (1995) TranspR 115, LG. Munich, 19.7.94 (1995) TranspR 116 (see Jung, p. 151). The fact that the goods never arrive does not disqualify the place designated, App. Paris, 24.10.91 (1991) B.T. 779.

72. But not that the goods have been delivered, App. Paris, 24.10.91 (1991) B.T. 779.

73. Unless there is an agreed delivery date, in which case it could give rise to a claim for delay. See *supra*, para. 7.6.

74. Articles 34 to 40, *infra*, Chapter 11.

75. [1979] 1 Lloyd's Rep. 175.

76. [1977] 1 Lloyd's Rep. 346, discussed *supra*, paras 1.16–1.17.

that, had it so suited them, the claimants might have sued before the courts of the place where the defendant actually took over the goods. Indeed, this was the argument advanced by the defendant against English jurisdiction and the finding that the English courts had jurisdiction does not necessarily mean that the French courts would not have had jurisdiction.

10.36 Two final points require mention. First, the actual place where the loss or damage occurred is not relevant in establishing which courts have jurisdiction, except indirectly in as much as it may determine which carrier may be sued in the case of successive carriage.[77] Secondly, it would seem that the question of which court or tribunal within a particular country will have jurisdiction over a particular dispute will, of course, depend on the national law. It differs from country to country as to whether the jurisdiction of a particular court is based on locality, the value of the claim or whether it is a commercial or civil dispute.[78]

Potential duplication of actions

10.37 It will have been observed from the above that the provision of Article 31(1) can produce a situation where there is more than one permissible jurisdiction under the Convention. The purpose of Article 31(2) is therefore to avoid duplication of actions, which is achieved by providing that where a claim within Article 31(1)[79] is pending before a court or tribunal,[80] or where a judgment has already been obtained, then no new action can be started between the same parties on the same grounds unless the judgment of the first court is not enforceable[81] in the country where the subsequent proceedings are commenced. This prohibition applies to any new action "between the same parties", so it will apply equally to attempts by the defendant to counter-claim in another jurisdiction.[82] But if there is an alternative plaintiff, or an alternative defendant,[83] then further proceedings could be brought, assuming that the further claim was still within the period of limitation.

---

77. See Article 36, *infra*, Chapter 11, paras 11.62–11.74.

78. That this is a matter of national law and not CMR has been recognised in France, see Lamy, para. 556, App. Riom, 18.11.77 (1977) B.T. 560, Comm. Clermont-Ferrand, 10.6.77 (1977) B.T. 366, (1979) II U.L.R. 236, Comm. Lyon, 12.2.93 (1993) B.T. 260, (Krings, p. 785) cf. App. Aix-en-Provence, 3.3.94 (1994) B.T. 451. See also Cass., 17.1.95 (1995) B.T. 90 App. Douai, 17.2.2000 (2000) B.T. 292, Comm. Bobigny, 13.6.96 (1997) B.T. 142, App. Amiens, 7.3.97 (1997) B.T. 261, Comm. Paris, 3.2.99 (1999) B.T. 445. In App. Angers, 4.12.95 (1996) B.T. 337, however, the court pointed out that an interpretation of CMR which brought in a reference to national law and then used national law to exclude CMR was impermissible and overturned the lower court's disinclination to take jurisdiction as the place where the goods were taken over by the carrier, cf. App. Rouen, 13.1.98 (1998) B.T. 361, (1998) U.L.R. 896 and App. Orléans 9.12.99 (2000) B.T. 12. For Germany, see LG. Hamburg, 22.1.79 (1980) I U.L.C. 266: Article 31 also regulates which court or tribunal within a particular country has jurisdiction. This was, however, to deal with the difficulty that otherwise there would be no court competent in Germany to deal with the dispute. The opposite view, though, was taken in BGH., 6.2.81 (1981) VersR 633, (1982) 17 E.T.L. 50, cf., however, BGH., 9.12.82 (1983) VersR 282, see Haak, p. 289. The difficulty has been resolved by the addition of a new Article 1(a) to the German version of CMR, with effect from 5.7.1989, to enable there to be a competent German court in respect of a foreign carrier where the goods are taken over or delivered or to be delivered in Germany; see further LG. Hanover, 4.9.91 (1992) TranspR 327, OLG. Karlsruhe, 20.12.95 (1996) TranspR 203 (see Jung, p. 150). For Holland, see HR., 16.9.90 (1990) II U.L.R. 259, Hof. Amsterdam, 22.2.96 (1998) N.J. No. 139, (1998) S. & S. No. 8.

79. I.e. "arising out of carriage under this Convention". See *supra*, para. 10.25.

80. Cf. LG. Nurenberg, 26.7.94 (1995) TranspR 72.

81. As to which see *infra*, para. 10.43.

82. Cf. Hof. Amsterdam, 22.2.96 (1998) N.J. No. 139, (1998) S. & S. No. 8.

83. E.g. in the case of successive carriers, *infra*, Chapter 11.

10.38 The words "action is pending" used in Article 31(2) have not been defined. The specific reference to "before a court or tribunal" must mean that at least some step in the action has been taken. The French text makes the position clearer, stating that the subsequent claim must be suspended when "*une action est en instance devant une juridiction compétente*". In England, an action commences once the writ is issued, which must therefore be the earliest point at which an action could be considered to be pending. However, a similar question has arisen before the English courts in the context of the Brussels Convention. While not strictly directly relevant to CMR, the underlying considerations which underpin the current position of the English courts in that context would seem to have equal relevance to the proper interpretation of CMR. Thus for the purposes of Articles 21–23 of the Brussels Convention, a court may find it necessary to consider whether or not another court is "first seised" of a dispute which in turn depends on whether the action is pending before a court. In *Dresser (U.K.) Ltd.* v. *Falcongate Freight Management Ltd., The Duke of Yare*, the Court of Appeal[84] considered that while this had to be determined under English law in the context of English procedure, in determining the effect of national law for the purposes of the Convention, regard must be had to its international purposes. Whereas English authority showed that an action was pending once it was brought, and that the action was brought once the writ had been issued, the Continental practice, while varying in detail, is such that the courts on the Continent would not ordinarily regard themselves as seised of proceedings until (at the earliest) the proceedings had actually been served on, and thus brought to the attention of, the defendant. Given the limited involvement both of the court and the defendant in the action at the stage of issue of the writ, the court held that an action only became definitely pending once the writ had been served.[85]

10.39 The English version of the text prohibits further actions "on the same grounds". This would seem to suggest that another action can be commenced in another jurisdiction based on the same facts, but pleaded on different grounds. That this was not the intention is made clear by the French text, which prohibits further action "*pour la même cause*", and unless Article 31(2) is applied in this way it would open the way to possible "forum shopping" in the event of it being possible to formulate alternative grounds for the claim.[86]

10.40 A limitation on the scope of Article 31(2) arose in the case of *Frans Maas Logistics (U.K.)* v. *CDR Trucking B.V.*[87] In that case proceedings for a negative declaration had been commenced in Holland by the last carrier against the cargo interests and the other carriers. Before the conclusion of the Dutch proceedings, proceedings were commenced in this country against the last carrier by the first carrier, to whom the cargo interests had assigned their rights. It was common ground between the parties that the English proceedings raised substantially the same issues as those involved in the Dutch proceedings. The last carrier, the defendant in the English proceedings, applied for a stay of those proceedings on the basis of Article 31, or in the alternative on the basis of Articles 21 and 22 of the Brussels Convention.

---

84. [1991] 2 Lloyd's Rep. 557.
85. See also *The Sargasso* [1994] 2 Lloyd's Rep. 6. The position remains unchanged following the introduction of the CPR: *SDL International Ltd.* v. *Centre de Co-operation International, Chancery Division*, 21.7.00, unreported. Cf., in respect of Article 31(2) CMR, RB. Brussel, 28.2.75 (1975) 10 E.T.L. 419.
86. See also Loewe, Theunis, p. 152.
87. [1999] 2 Lloyd's Rep. 179.

10.41 Having held that the Convention contained nothing which prevents the increasingly common practice of proceedings for a declaration of non-liability, Colman, J., then drew a distinction between paragraphs (1) and (2) of Article 31. He held that the judgments referred to in Article 31(2) contemplated only those judgments which potentially involved enforcement proceedings. In his view, enforcement proceedings would not be commenced in another jurisdiction if the initial proceedings were concerned only to obtain a negative declaration. As he put it, although such a declaration might be recognised elsewhere, it would not be the subject of enforcement. Accordingly, in his view, Article 31(2) did not prevent a claim for an enforceable remedy in the courts of one country where proceedings for a negative declaration were pending before the courts of another country. However, he concluded by staying the English proceedings in any event under Articles 21 and 22 of the Brussels Convention.

10.42 Apart from the question of a judgment which may be unenforceable in the country of original jurisdiction, which is discussed in the next paragraph, it will be noted that there are two other situations where Article 31(2) may not operate so as to prevent further claims. First, a claim might be brought in a jurisdiction which is not competent within the meaning of Article 31(1), and secondly, a court may feel compelled to accept proceedings on the basis of doubts as to the meaning of when a claim can be said to be "pending" in another jurisdiction for the purposes of Article 31(2). In either of these situations, in this country the provisions of section 3 of the Carriage of Goods by Road Act 1965 may be of assistance. That section permits the courts of this country to take account of any other proceedings, in this country or elsewhere, which have been or are likely to be commenced to enforce a liability which is limited by Article 23.[88]

The enforcement of judgments

10.43 As will have become apparent, by virtue of the provisions of Article 31(1), as to jurisdiction the claimant's action may well be brought in a country with which the defendant has no real connection. Potentially, this could cause difficulty for the claimant in the event of it transpiring that the defendant has no assets within that jurisdiction with which to satisfy the judgment. With a view to providing for such situations, or indeed for any situation where the plaintiff wishes to enforce a judgment otherwise than in the country in which it was obtained, it is provided by Article 31(3) that once an enforceable judgment had been entered by a court of a contracting country[89] in any action within Article 31(1), that judgment will become enforceable in all other contracting countries. This is subject only to the necessary formalities specified by the national law of the country in which the judgment is to be enforced, and Article 31(3) specifically provides that such formalities are not to enable the merits of the case to be reopened.

---

88. See *supra*, para. 9.2. Comparison should be made with Article 22 of the Brussels Convention. Arguably in circumstances where related actions in contracting States to this Convention are involved, but not actions between the same parties on the same grounds, the correct rule to apply is Article 22 rather than section 3 of the 1965 Act since this section, not being part of the CMR Convention, is not preserved by Article 57 of the Brussels Convention. Since both provisions, however, are permissive, it would seem to be unlikely that the discretion of the court would be affected by such differences as may be perceived between the two provisions. Cf. RB. Brussel, *supra*, fn. 85, where Article 22 was applied since Article 31(2) of CMR was not applicable. See also *infra*, para. 10.47. Cf. *Harrison & Sons Ltd.* v. *R.T. Steward Transport Ltd.* (1993) 28 E.T.L. 747.

89. But note that contracting countries are not the only possible jurisdictions under Article 31(1). See *supra*, para. 10.24.

10.44 The enforcement of foreign judgments in this country is regulated either by the Foreign Judgments (Reciprocal Enforcement) Act 1933 or by the Civil Jurisdiction and Judgments Acts 1982 and 1991.[90] By section 4 of the Carriage of Goods by Road Act 1965, Part I of the 1933 Act applies to judgments within Article 31(1) of the Convention obtained before the courts of another contracting State which has become enforceable in that country. Detailed consideration of the 1933 Act is beyond the scope of this book,[91] but basically it provides for the registration in this country of judgments to which the Act applies, by application to the High Court. Once registered, such a judgment has the same effect as a High Court judgment. The registration can be set aside, *inter alia*, if the courts of the foreign country had no jurisdiction in the circumstances of the case.[92] If therefore the defendant can show that the judgment was obtained otherwise than before the courts specified in Article 31(1), he would be entitled to have the registration of the judgment set aside for want of jurisdiction.

Enforceable judgments

10.45 The other limitation inherent in Article 31(3) is that before a judgment becomes enforceable in other jurisdictions it must have become enforceable in the country where judgment was first given. It will be up to the courts of the country where the plaintiff seeks to enforce the judgment to decide whether the judgment has become enforceable in the country of the original judgment. In this country, the application for registration[93] under the 1933 Act must be supported by an affidavit stating *inter alia* that to the best of the deponent's knowledge and belief the judgment is enforceable in the country of original jurisdiction, and that if the judgment is registered, the registration is not liable to be set aside under section 4 of the Act.[94] That section then specifies the circumstances where, on the application of the other party, the registration of the judgment must, or may, be set aside. The judgment must be set aside where it has been registered in contravention of the Act, and a judgment which is not enforceable in the country of original jurisdiction is not one which can be legitimately registered.[95]

10.46 The onus therefore is initially on the applicant for registration of the judgment to satisfy the court on his *ex parte* application that the judgment is enforceable in the country of original jurisdiction. It is then open to the other party subsequently to apply to the court for the registration to be set aside, whereupon the onus will be on that party to satisfy the court that the judgment is not enforceable in the country of original jurisdiction. The 1933

---

90. See *infra*, para. 10.47.

91. See generally *Halsbury's Laws*, 4th edn., Vol. 8(1), paras 1028–1039.

92. Section 4(1)(a)(ii). The 1933 Act sets out circumstances in which the foreign court is deemed to have, or not to have, jurisdiction, (section 4(2) and (3)) but by virtue of section 4(2) of the Carriage of Goods by Road Act 1965 these provisions do not apply to Convention cases. The 1933 Act also provides that registration of the judgment is to be set aside where the judgment has been obtained by fraud (section 4(1)(a)(iv)), and it has been held under this provision that the defendant is entitled to have the issue of fraud tried in this country: *Re Foreign Judgments (Reciprocal Enforcements) Act 1933, Syal v. Heyward* [1948] 2 K.B. 443. *Quaere*, however, whether this is consistent with the final sentence of Article 31(3): "The formalities shall not permit the merits of the case to be re-opened."

93. As to the procedure generally, see R.S.C. Order 71 which continues to apply by virtue of CPR Part 50.1 and Schedule 1.

94. R.S.C. Order 71, Rule 3.

95. Section 2(1)(b) Foreign Judgments (Reciprocal Enforcements) Act 1933.

Act gives some guidance as to what criteria are to be taken into account in this respect. A judgment is deemed final and conclusive notwithstanding that an appeal is pending or that it may be subject to appeal,[96] although in such circumstances, on application for setting aside the registration, the court can either set aside the registration or adjourn the application to permit the applicant reasonable time to have the appeal disposed of.[97] Where the judgment is set aside, either under this provision or because the judgment was not enforceable in the country of original jurisdiction, this does not prejudice a subsequent application to register the judgment when the relevant circumstances have changed.[98]

10.47 As between member States of the EC, regard must be had to the 1968 Brussels Convention on jurisdiction and the enforcement of judgments in civil and commercial matters,[99] as amended by the 1978 Convention of Accession by virtue of which the United Kingdom became a party to the original Convention.[100] This Convention does not override the provisions of CMR as to jurisdiction,[101] since Article 57 of the Brussels Convention expressly provides that it does not affect any other conventions which govern jurisdiction or the recognition or enforcement of judgments. This provision therefore preserves the rules of CMR insofar as they are applicable to the dispute. The Brussels Convention, however, will still regulate those matters which are not regulated by CMR.[102]

10.48 In respect of the recognition and enforcement of judgments, further guidance is provided by Article 57(2)(b) of the Brussels Convention which provides for the enforcement of the judgments given in a contracting State and which provides that, where a Convention on a particular matter to which both the State of origin and the State addressed are parties lays down conditions for the recognition or enforcement of judgments, those conditions shall apply, but that the provisions of the Brussels Convention as amended which concern the procedures for recognition and enforcement of judgments may be applied in any event. Article 31(3), however, can be said to lay down conditions for the recognition of judgments in respect of CMR and should, therefore, take precedence over any contradictory rule in the Brussels Convention where the two States involved are parties to CMR.[103]

---

96. Ibid., section 1(3).

97. Ibid., section 5(1).

98. Ibid., section 5(2).

99. On the Brussels and Lugano Conventions generally, see *Halsbury's Laws*, 4th edn, Vol. 8(1), paras 1040–1059.

100. Implemented by the Civil Jurisdiction and Judgments Act 1982 which incorporates the Convention (Schedule 1) and the Protocol on Interpretation (Schedule 2), as amended by the Convention of Accession (Schedule 3), into English law. Note also the similar provisions applicable to member states of EFTA contained in the Lugano Convention 1988 and implemented by the Civil Jurisdiction and Judgments Act 1991, cf. OLG. Dresden, 24.11.98 (1999) TranspR 62.

101. App. Lyon, 21.10.76 (1977) B.T. 110, RB. Antwerpen, 25.3.76 (1976) 11 E.T.L. 691, Cass., 3.6.81 (1981) B.T. 431, App. Aix-en-Provence, 12.5.87 (1987) B.T. 400, Arrond. Rotterdam, 28.10.99 (2000) S & S No. 35.

102. E.g. in respect of provisional and protective measures, see Article 24 of the Brussels Convention, see Haak, p. 277. In respect of Article 22 of the Convention, see *supra* and RB. Brussel, 28.2.75 (1975) 10 E.T.L. 419. See also Cass., 21.6.82 (1982) B.T. 513 and criticism of Haak, p. 278. Cf. further *The Maciej Rataj* [1995] 1 Lloyd's Rep. 302.

103. In particular it is arguable that there could be no recognition of a judgment given by a court without jurisdiction under CMR. As noted *supra*, para. 10.44, this derives directly from Section 4 of the Carriage of Goods by Road Act 1965, but indirectly from CMR, whereby the conditions for the application of Article 31(3) could perhaps properly be seen as limited by Article 31(1). Such a protection against the enforcement of judgments given without proper jurisdiction is not provided for by the Brussels Convention; see further Cheshire & North's *Private International Law*, 13th edn., pp. 475–476.

10.49 The provisions of Article 31(3) as to the enforcement of judgments only apply to judgments after trial, judgments by default and settlements confirmed by court order.[104] Interim judgments are excluded from its ambit, as are "awards of damages, in addition to costs against a plaintiff who wholly or partly fails in his action".[105] This last part of Article 31(4) is strangely phrased,[106] but its meaning becomes clearer with reference to the French text which uses the words "*en sus des dépens*" (over and above) instead of the words "in addition to" in the English text.[107] In other words, the paragraph prevents awards of damages over and above costs against an unsuccessful plaintiff: costs can be awarded against him, but damages cannot be. This is presumably not intended to affect a successful counterclaim by the defendant, but is aimed at such jurisdictions as permit an award of damages against the plaintiff for bringing a claim in the first place. In this country no such awards of damages are in any event possible.

Security for costs

10.50 Since Article 31(3) effectively ensures that judgments obtained in one contracting country will be enforceable, together with any award of costs, in any other contracting country, the Convention provides that security for costs in proceedings arising out of carriage under the Convention cannot be demanded from nationals of contracting countries resident or having their place of business in one of those countries.[108] Thus, the normal rules as to security for costs in this country are necessarily excluded in such claims.[109] This would apparently equally be true of a claim between two nationals of this country, but Loewe is of the opinion that in such circumstances the rule would not apply, and the national rules as to security for costs would be applicable.[110] Although from a purposive approach this may represent the better view, the Convention makes no such provision, and in the same way as the general provisions of the Convention would prevail over any national law to the contrary in cases governed by it, so it would seem that it must it also prevail over national rules as to security for costs.

Limitation of actions

10.51 The period of limitation under the Convention is dealt with by Article 32, which provides as follows:—

"1. The period of limitation for an action arising out of carriage under this Convention shall be one year. Nevertheless, in the case of wilful misconduct, or such default as in accordance with the law of the court or tribunal seised of the case, is considered as equivalent to wilful misconduct, the period of limitation shall be three years. The period of limitation shall begin to run:
(a) in the case of partial loss, damage or delay in delivery, from the date of delivery;

---

104. Article 31(4). E.g. Article 31(3) does not apply to an order authorising the sale of goods under Article 16(5) (*supra*, Chapter 5, paras 5.43–5.47), Cass. 8.10.97 (1997) B.T. 799 Krings, p. 785.
105. Ibid.
106. In particular, the comma seems entirely superfluous.
107. As to the propriety of having regard to the French text see *supra*, paras 0.14–0.17.
108. Article 31(5).
109. For general rules as to security for costs see CPR Part 25.12–15.
110. Loewe, para. 225.

(b) in the case of total loss, from the thirtieth day after the expiry of the agreed time-limit or where there is no agreed time-limit from the sixtieth day from the date on which the goods were taken over by the carrier;

(c) in all other cases, on the expiry of a period of three months after the making of the contract of carriage.

The day on which the period of limitation begins to run shall not be included in the period.

2. A written claim shall suspend the period of limitation until such date as the carrier rejects the claim by notification in writing and returns the documents attached thereto. If a part of the claim is admitted the period of limitation shall start to run again only in respect of that part of the claim still in dispute. The burden of proof of the receipt of the claim, or of the reply and of the return of the documents, shall rest with the party relying upon these facts. The running of the period of limitation shall not be suspended by further claims having the same object.

3. Subject to the provisions of paragraph 2 above, the extension of the period of limitation shall be governed by the law of the court or tribunal seised of the case. That law shall also govern the fresh accrual of rights of action.

4. A right of action which has become barred by lapse of time may not be exercised by way of counter-claim or set-off."

(a) Scope of operation

10.52 Article 32(1) provides for a limitation period of one year which will apply to any "action arising out of carriage under [the] Convention". In other words, as with Article 31, it not only applies to actions arising out of the contract of carriage but also to any action which arises out of the actual carriage itself whether in contract or in tort.[111] Consequently, the one year period of prescription is equally applicable to actions brought by the carrier as to actions brought against the carrier by the cargo interests for loss, damage or delay.[112] It has thus been held to apply to an action by the carrier for freight charges.[113] It would also apply to a claim by the sender for the recovery of an over-payment of freight charges,[114] for damages for failure to collect the cash on delivery charge,[115] and to actions in tort or delict,[116] brought within the Convention by virtue of Article 28.[117] It would

111. Provided the claim is asserted by or against persons taking part in the carriage itself as sender, carrier or consignee and not e.g. against a forwarder entrusted only with Customs clearance and the disbursement of German import turnover tax, BGH., 23.3.95 (1997) U.L.R. 410. See also, Koller Art. 32, Rdn. 1.

112. Hof. Amsterdam, 4.6.74 (1976) 11 E.T.L. 266. Cf., App. Paris, 6.10.95 (1995) B.T. 764 concerning a claim against the carrier for failing to compel consignees to return empty pallets for redelivery to the sender.

113. RB. Brussel, 13.9.68 (1969) 4 E.T.L. 1153, Hof. Amsterdam, 4.6.74 (1976) 11 E.T.L. 266, App. Versailles, 30.11.95 (1996) B.T. 191, OLG. Düsseldorf, 26.1.95 (1995) TranspR 288 (Jung, p. 165, cf. LG. Wuppertal, 25.6.92 (1992) TranspR 362), *Muller Batavier Ltd.* v. *Laurent Transport Co.* [1977] 1 Lloyd's Rep. 411—action between successive carriers for freight charges. See also BGH., 28.2.75 (1975) II U.L.R. 373, (1975) N.J.W. 1075, cf., however, OLG. Munich, 12.4.91 (1991) TranspR 298, (1991) R.I.W. 676 (Clarke, p. 143 n. 5). Similarly claims arising out of a pooling agreement, OLG. Saarbrücken, 24.2.95 (1996) R.I.W. 605 (Jung, p. 165) and recourse claims by a carrier against a subcontract carrier where no successive carriage is included, Cass. (Belg.) 25.5.84 (1984) Pas. 1. No. 543 (Krings, p. 788).

114. BGH., 18.2.72 (1972) 7 E.T.L. 860, OLG Düsseldorf, 11.7.96 (1997) TranspR 274, (1998) U.L.R. 202. Also, claims for consequential loss due to positive breach of contract, BGH., 27.10.78 (1979) VersR 276, Herber/ Piper Art. 32, Rdn. 6.

115. Hof. 's Hertogenbosch, 13.1.70 (1971) 6 E.T.L. 817, Supreme Court of Denmark, 22.4.71 (1971) U.L.C. 307, cf. App. Paris, 2.4.91 (1991) B.T. 428.

116. Or claims for compensation in criminal proceedings, Cass., 7.4.87 (1987) B.T. 343. For the possibilities of such claims in English law, see Powers of the Criminal Courts Act 1973, section 35. See, generally, *Halsbury's Statutes*, 4th edn., Vol. 12, pp. 587–590.

117. OLG. Düsseldorf, 8.5.69 (1970) 5 E.T.L. 446, including where a servant or agent of the carrier relies on Article 28(2), OGH., 27.9.83 (1984) TranspR 191. As to Article 28, see *supra*, Chapter 9, paras 9.65–9.71.

presumably therefore apply to an action by the carrier against the cargo interests for damage to his vehicle or premises resulting from the state of the goods being carried.[118]

10.53 It has been held to apply to a carrier's claim for damages for the detention of his vehicle at a Customs post due to the failure of the sender to supply adequate documentation,[119] as well as to the claim of a forwarder against the carrier for delay caused to the forwarder's trailer.[120] In *Shell Chemicals U.K. Ltd.* v. *P. & O. Roadtanks Ltd.*,[121] Saville, J., accepted that Article 32 was relevant to the claim for damage caused to the consignee by reason of the delivery of the wrong goods, this damage including the shut-down of his refinery and the loss of chemicals and production by reason of detergent rather than oil having been pumped into the refinery even though the Convention was not applicable in other respects.[122]

10.54 On the other hand, again as with Article 31, if the Convention does not apply to the carriage for any reason,[123] then clearly Article 32 will not apply, and the claim will be regulated by the limitation period under the appropriate national law. Thus, in a Belgian case, it was held that Article 32 did not apply to an action by a trailer owner against a tractor operator who damaged it.[124] Similarly, Article 32 will not apply to a dispute between shipper and forwarder, nor to a dispute between one forwarder and another.[125] Nor will it apply to other operators performing ancillary functions such as warehousemen,[126] or to a claim by the carrier against the consignor for the costs of returning empty where the consignor had promised to arrange a cargo for the return journey.[127] But where a combined transport was arranged by sea from Hong Kong to Rotterdam and then by road to Antwerp, the limitation period in the bill of lading was inoperative in respect of damage which occurred during the road transit from Rotterdam to Antwerp, since that portion of the journey was subject to the Convention, and Article 32 applied.[128]

---

118. Rodière, para. 117. In respect of damage to the vehicle: Comm. Anvers, 30.12.74 (1974) J.P.A. 367, App. Paris, 23.12.75 (1976) B.T. 48, RB. Rotterdam, 3.9.76 (1976) S. & S. No. 56, see further Libouton, (1982) J.T. 734, para. 129.

119. BGH., 11.12.81 (1983) 18 E.T.L. 63. See also, as to claims between carriers not governed by Article 39(4), BGH., 10.5.90 (1990) TranspR 418, (1991) 26 E.T.L. 351, Cass. (Belgium), 25.5.84 (1984) J.T. 288.

120. App. Paris, 7.3.90 (1991) B.T. 118, although for other reasons the time limit was not, in fact, applied, see *infra*.

121. [1993] 1 Lloyd's Rep. 114 (affirmed on appeal on different grounds [1995] 1 Lloyd's Rep. 297, see *supra*, para. 9.4).

122. At p. 116. Action by a third party, a stranger to the contract of carriage, for damage or injury e.g. caused by the road vehicle would not be limited by the Convention. According to OGH., 10.11.81 (1982) TranspR 111, only those on whose behalf or with whose agreement the goods are carried fall within the scope of Article 32, see Loewe, Theunis, p. 153. However, a claim by the sender for indemnity against costs paid to third parties due to a road accident for which the road carrier was liable was held to be within Article 32 in Hanseatic OLG., 24.1.85 (1985) TranspR 185.

123. See generally *supra*, Chapter 1.

124. RB. Antwerpen, 27.10.71 (1972) 7 E.T.L. 1054.

125. Comm. Paris, 29.6.70 (1970) B.T. 325. See also App. Aix-en-Provence, 25.9.90 (1991) B.T. 173 (the action by the customer against the forwarder was limited by national law while the customer's action against the carrier was limited by CMR). But as to the status of a forwarder in relation to the Convention, see *supra*, Chapter 1, paras 1.13–1.54. Action taken by the forwarder as sender will be subject to Article 32, Cass., 21.6.82 (1982) B.T. 416.

126. App. Paris, 12.6.70 (1970) B.T. 228. BGH., 13.1.78 (1978) 13 E.T.L. 402.

127. App. Amiens, 19.6.80 (1980) B.T. 490.

128. RB. Antwerpen, 23.9.75 (1976) 11 E.T.L. 279.

10.55 There may also be some limit on the scope of "an action arising out of carriage under this Convention" on the basis that not all claims which might be incidental to the carriage can properly be said to arise out of it.[129] This would seem to be the basis of a French decision where it was held that a claim for Customs duty paid by the carrier in his capacity as Customs agent was not subject to Article 32, on the basis of a distinction between matters which concern delivery of the goods and those which do not.[130] Clearly, this might be a difficult distinction to apply.

10.56 A similar difficulty arises where the carriage is not performed at all. If there is no carriage, can there be an action arising from it? In the *Shell Chemicals* case,[131] Saville, J., accepted the concession of the defendants that CMR did not exclude liability where the carriers wholly failed to perform the carriage before they have taken over the goods, in the context of an argument that wherever liability is not imposed by CMR it thereby excludes it. This would seem to add some weight to the view that non-performance of the carriage is wholly outside CMR, and in consequence any claim in respect of it is not governed by Article 32.[132] On the other hand, even though Saville, J., held that the claim in question in that case, a claim for contamination to a refinery resulting from the delivery of the wrong load, fell outside the liability provisions of the Convention so that common law applied,[133] he accepted nevertheless that the limitation provisions at Article 32 were applicable.

10.57 One further general point should be mentioned. There is Belgian authority to the effect that the limitation period under national law will apply in claims arising out of domestic carriage even though the parties have voluntarily adopted the provisions of the Convention.[134] There is no reason under English law why the Convention limitation period should not apply in such cases, instead of the normal six year period.

(b) Period of limitation

10.58 Although the period of limitation is generally of one year's duration, where there is wilful misconduct on the part of one of the parties, or such default as the law of the court seised of the matter may consider as equivalent to wilful misconduct, then the period of

---

129. Such as a claim for loss of goods while being stored prior to the commencement of the carriage, BGH., 13.1.78, *supra*, fn. 126, cf. App. Versailles, 6.7.94 (1994) B.T. 759.

130. Comm. Seine, 28.6.67 (1967) B.T. 297, App. Amiens, 19.6.80 (1980) B.T. 490, App. Paris, 24.2.93 (1994) B.T. 52. According to the commentary following the first of these cases, the French courts are divided as to whether Customs formalities performed by a carrier is a service ancillary to a contract of carriage or not. These decisions would seem open to doubt on the basis that Customs duties are duties specifically regarded as "incurred in respect of the carriage" by Article 23(4). See *supra*, Chapter 9, paras 9.21–9.41. In the light of this Lamy, paras 544 and 1594, Jur. 9, considers that Customs formalities performed by the carrier are within the scope of Article 32 and is supported in this by App. Lyon, 6.7.83 (1984) B.T. 339. Cf. App. Paris, 23.9.81 (1981) B.T. 538. A claim by a *commissionnaire en douane* for indemnity by the carrier in respect of a Customs fine was held not to be within Article 32 in App. Orleans, 24.3.87 (1987) B.T. 663.

131. *Supra*, para. 10.53.

132. A view maintained by Loewe, Theunis, p. 153, despite the contrary decision of the OGH., 12.2.85 (1985/9) Stra GüV 21, (see Clarke, p. 145, n. 23, see further, Herber/Piper, Art. 32. Rdn. 3, Koller, Art. 32, Rdn. 1.). Loewe cites BGH., 27.10.78 (1979) VersR 276 in support, cf. also BGH., 18.2.72 (1972) 7 E.T.L. 860, (1973) U.L.C. 264 (discussed by Clarke, pp. 144–145). A claim for the cost involved in employing another carrier to complete the carriage was held to be within Article 32 in OLG. Hamburg, 9.2.89 (1990) TranspR 191, see further Clarke, p. 152, n. 61, see also, OLG. Hamm, 20.3.97 (1998) TranspR 297.

133. On this aspect, see the discussion *supra*, paras 0.26–0.34.

134. RB. Antwerpen, 12.9.72 (1973) 8 E.T.L. 640.

limitation will be three years.[135] Reference should be made to the earlier discussion of what is meant by wilful misconduct,[136] and if the claimant wants to rely on this extended period of limitation, it will be for him to prove that there has been wilful misconduct.[137] The moment at which time begins to run then varies with the nature of the claim.[138]

PARTIAL LOSS, DAMAGE OR DELAY

10.59 First, under Article 32(1)(a) where there has been partial loss, damage,[139] or delay in delivery of goods, it will begin to run as from the time of delivery. In order to determine whether Article 32(1)(a) applies, it is therefore of central importance to establish whether the goods have been delivered, for the purpose of calculating the limitation period. With reference to these questions, it was held in a Dutch decision that where goods are stored at the place of delivery awaiting collection by the consignee, they are not delivered until so collected.[140] If, however, the consignee refuses to accept the goods because of damage, but takes delivery and stores them on behalf of the sender, it has been held by a Swedish court that the goods are then delivered and Article 32(1)(a) will apply.[141]

10.60 Where, however, there is no actual delivery, then Article 32(1)(a) cannot apply. This point was raised before the English High Court in *Moto Vespa S.A.* v. *MAT (Brittania Express) Ltd.*,[142] where two large lathes were to be carried from Birmingham to Madrid. When nearing Madrid a serious accident occurred. The lathes were badly damaged, but were loaded on to another vehicle and the journey completed. The consignees, however, rejected the goods, which were carried back to Birmingham and, in time, replaced with others. The defendants argued that the resulting claim was time-barred under Article

135. The extended time limit can apply even in circumstances where the right to limit the compensation is unaffected by the conduct of the carrier due to the fact that the damage produced was not caused by the wilful misconduct. In App. Paris, 25.3.88 (1988) II U.L.R. 736, (1989) B.T. 46 the misconduct of the carrier in concealing his real name from the sender was held to be sufficiently prejudicial to the sender's potential claim so as to justify the extended time limit. Cf. BGH., 27.6.85 (1986) 21 E.T.L. 102. In App. Aix-en-Provence, 3.11.94 (1995) B.T. 232, however, Article 32 was held only to envisage wilful misconduct committed in the execution of the contract of transport and not later, as where the defendant displayed indifference to the destruction of the goods by being unresponsive to the claim.

136. Article 29, *supra*, Chapter 9, para. 9.72 ff. The reference to wilful misconduct is in general terms without indicating whether the reference is to the carrier only or whether it also includes reference to his servants or agents. It is presumed that, in this respect, the wilful misconduct of servants or agents will produce similar effects to those envisaged in Article 29, see Libouton, (1982) J.T. 735, para. 130.

137. HR., 17.11.78 (1979) 7 E.L.D. 279.

138. The burden of proving the event causing time to run, e.g., delivery, has been held to rest on the party raising the time bar, Supremo Tribunal Da Justiça (Portugal), 17.11.94 (1997) U.L.R. 200. The court based this on Article 342(2) of the Portuguese Civil Code which was considered as not incompatible with Article 32 CMR.

139. There is damage when a change in substance occurs which decreases the value of the goods. Thus frozen foodstuffs will be deemed to have been damaged for the purposes of Article 32(1)(a) if their quality has been so diminished in transit that the only purposes for which they can be used are less profitable that those originally envisaged: OLG. Celle, 13.1.75 (1975) 3 E.L.D. 582. See also Supreme Court of Denmark, (1971) UfR 183, (1971) U.L.C. 294, 303, OLG. Wien, 23.2.89 (1990) TranspR 156. See also OLG. Hamburg 22.7.82 (1983) VersR 63, OGH., 31.3.82 (1984) TranspR 196. See also OLG Karlsruhe, 25.2.99 (1999) TranspR 349.

140. Hof. 's Hertogenbosch, 21.12.65 (1966) 1 E.T.L. 698. See also GH. 's Gravenhage, 10.5.78 (1978) 13 E.T.L. 607: notice that the goods have arrived does not of itself amount to delivery. For further discussion of the meaning of delivery, see *supra*, para. 6.9 ff.

141. Court of Appeal for Western Sweden, 9.4.74 (1974) N.D. 86, cited by Wetter, "The time bar regulations in the CMR Convention", [1979] LMCLQ 504.

142. [1979] 1 Lloyd's Rep. 175. Hardingham, "Aspects of limitation of actions under CMR", [1979] LMCLQ 362.

32(1)(a) on the basis that the period ran from the arrival of the lathes at Madrid, such date being the date of delivery.

10.61 In considering this question, Mocatta, J., referred to Article 15,[143] which he considered as indicating that the Convention envisaged circumstances which prevented delivery of the goods after arrival at destination. Taking the view that the facts of this case were an example of such circumstances, he held that there was no delivery under Article 32(1)(a) and consequently that paragraph did not apply.[144] Thus far[145] the decision is supported by a Belgian case where part of a consignment of glass was broken in transit and the whole load returned to the sender, having been rejected by the consignee. It was held that Article 32(1)(a) did not apply.[146]

10.62 A different analysis, however, was suggested by Parker, J., in *Worldwide Carriers Ltd.* v. *Ardtran International Ltd.*[147] on similar facts to those which arose in the *Moto Vespa* case, where he put forward an alternative basis for his decision with regard to the application of the sub-paragraphs of Article 32(1).[148] He held that, if, on hearing of the damage, the sender requires the goods to be returned to him, he does no more than exercise the rights of disposal granted by Article 12.[149] Accordingly, when the carrier carries out the instructions which derive from those rights, then at least in some cases there will have been a delivery pursuant to the contract and Article 32(1)(a) will apply.[150]

TOTAL LOSS

10.63 Secondly, Article 32(1)(b) provides that in the case of total loss of the goods time will begin to run as from the thirtieth day after the expiry of the agreed time limit for delivery where one has been agreed as provided for under Article 19, and in the more common case where there is no agreed time limit, from the sixtieth day from the date on which the carrier took over the goods.

10.64 Although the point has never been considered by an English appellate court, the overwhelming balance of first instance authority is to the effect that this paragraph can have no application to cases of constructive total loss. In *ICI* v. *MAT Transport*,[151] a consignment of nylon yarn was damaged in a road accident in the course of transit, and was returned to the consignors for survey without being delivered to destination. As Staughton, J., noted, one of the surveyors involved described the goods as being "in my

---

143. *Supra*, para. 5.31 ff.

144. *Quaere* whether the reference to Article 15 really takes the matter much further, since there would in any event seem to be a clear distinction between "arrival" and "delivery".

145. For other aspects of this case, see *infra*, para. 10.77 ff.

146. App. Bruxelles, 28.6.69 (1969) 4 E.T.L. 925, (1970) U.L.C. 115.

147. [1983] 1 All E.R. 692, 698.

148. See *infra*, para. 10.70.

149. See *supra*, para. 5.2 ff.

150. See also Loewe, Theunis, p. 154, who cites a decision of the West Swedish Appeals Court, 9.4.74 and Lamy, paras 546 and 1596, Jur. 1, which cites an unpublished decision reaching a similar result: App. Paris, 1.7.86, involving circumstances where the carrier had placed the goods into storage on their return (cf. Article 16(2), *supra*, para. 5.37 ff.). See also OLG. Vienna, 10.7.97 (1997) TranspR 435. In a further case (App. Poitiers, 4.11.87) the court, in circumstances where there had been no return to the sender, applied the date from which the goods had been offered to the consignee, rather than utilising Article 32(1)(c) (see *infra*). However, this decision seems questionable. In App. Versailles, 9.12.99 (2000) B.T. 71, goods were returned to the sender for repair after being damaged in transit. They were finally delivered to the consignee some months later. The later date was held to be the date of delivery given that the later delivery was effected by the same carrier and under cover of the same consignment note.

151. [1987] 1 Lloyd's Rep. 354. Cf. Hardingham, op. cit., p. 363. Cf. further OLG. Vienna, 10.7.97 (1997) TranspR 435.

opinion a 'constructive total loss' within the meaning of that term in the insurance business. In other words an insurer who had undertaken the risk of the value of the goods would have regarded that risk as falling to be indemnified, subject only to salvage". Although Staughton, J., was prepared to accept that the goods in question did amount to a constructive total loss within the meaning of section 60(2)(iii) of the Marine Insurance Act 1906,[152] he held that a total loss within the meaning of Article 32(1)(b) includes only an actual total loss within the meaning of that Act, i.e. "where the subject matter insured is destroyed, or so damaged as to cease to be a thing of the kind insured,[153] or where the assured is irretrievably deprived thereof".[154]

10.65 In so holding, Staughton, J., followed the earlier cases of *William Tatton & Co. Ltd.* v. *Ferrymasters Ltd.*[155] and *Worldwide Carriers Ltd.* v. *Ardtran International Ltd.*[156] In the first of those cases, although the damaged goods had only a residual scrap value, Browne, J., declined to treat the case as one of loss falling within Article 23 rather than one of damage falling within Article 23. However, that case was not directly concerned with Article 32. In the later *Worldwide Carriers* case, again the consignment only had a residual scrap value. Parker, J., declined to treat the goods as a total loss. As he put it[157]:

"Had the goods in fact been re-loaded and delivered instead of being returned it would in my judgment be doing violence to the language to say that the case was one of loss and not damage."

10.66 There is one English case which can be cited in support of the contrary view. In *Thermo Engineers Ltd.* v. *Ferrymasters Ltd.*[158] the consignment was so damaged that only parts of it could be salvaged, and the value of those parts represented only a small proportion of the value of the original consignment. In considering whether the plaintiff was entitled to recover the whole of the carriage charges under Article 23(4) and Article 25, Neill, J., said that[159]:

"In my judgment, it is important to have regard to the fact that although the [consignment] had some scrap value the damage was such as to render the whole unit unacceptable. The loss of part of the consignment does not provide a satisfactory parallel. I therefore consider that if compensation were to be calculated in accordance with CMR the whole [of the freight charges] would be recoverable."

---

152. I.e. "where the cost of repairing the damage and forwarding the goods to their destination would exceed their value on arrival".

153. It may be possible to regard some damaged goods as a "total write-off", as in a case where milk cartons arrived damp and water damaged and no longer suitable for consumption, OLG. Hamm, 6.2.97 (1998) TranspR 34. In such a case the goods may well have to be destroyed to prevent them from coming into the hands of consumers.

154. Section 57(1). Irretrievable deprivation can also be considered to have occurred where goods are delivered to the wrong person in circumstances where they cannot be recovered, e.g. where they are wrongfully sold at auction by the carrier, BGH., 18.5.95 (1995) TranspR 383, (1996) 31 E.T.L. 703, (1997) U.L.R. 412, cf. BGH., 27.10.78 (1979) N.J.W. 2473, OLG. Frankfurt, 30.3.77 (1978) VersR 169 (see further Clarke, p. 220). The fact that the goods might still be recoverable by legal action should not, it is suggested, prevent them from being regarded as "lost".

155. [1974] 1 Lloyd's Rep. 203.

156. *Supra*, para. 10.62.

157. At p. 696.

158. [1981] 1 Lloyd's Rep. 200.

159. At p. 206.

10.67 However, in *ICI Fibres* v. *MAT Transport*,[160] Staughton, J., while not accepting that Neill, J., was necessarily intending to treat the claim as one of total loss,[161] in any event preferred to follow the views expressed in *William Tatton & Co. Ltd.* v. *Ferrymasters* and *Worldwide Carriers Ltd.* v. *Ardtran International Ltd.* The French case referred to in the previous edition of this book in which a badly damaged machine was held to be equivalent to total loss *"en vertu d'une jurisprudence constante"* was noted but not followed on the basis that, on its own, it did not demonstrate sufficient consensus of foreign authority as to justify departing from the view he would otherwise adopt.[162]

10.68 It would seem clear, however, that in the case of shipments consolidated by the carrier, each contract of carriage is to be regarded individually for the purposes of Article 32(1). Thus, where a consignor sent only one packing case in a consolidated shipment, and that disappeared, this was held to be a total loss to which Article 32(1)(b) applied.[163]

10.69 The starting point for the operation of the 60 day period is the date on which the goods were taken over by the carrier. Where there is only one carrier, this will present no difficulty, but in the case of an action against a successive carrier by the cargo interests,[164] it might be argued that the appropriate date is that on which the goods were taken over by that carrier. Although this might extend the period during which the claimant can bring his action, it does raise a problem of its own in that it might not be possible for him to ascertain when the goods were taken over by that carrier, and in the absence of such knowledge, to be sure of being within time, proceedings would have to be connected with regard to the date when they were originally taken over.

10.70 Apart from the case where there is an actual loss of the goods, Article 32(1)(b) is also relevant to the case where the goods are deemed to be lost in circumstances where Article 20[165] is applicable. Accordingly, if the goods have not been delivered on the expiry of the period indicated in that Article, which is the same as the period indicated in Article 32(1)(b), the time limit will commence on the second day following expiry. In the circumstances which occurred in the *Moto Vespa* case,[166] the assumption was that there had been no delivery of the goods despite the fact that they were returned to the sender. In *Worldwide Carriers Ltd.* v. *Ardtran International Ltd.*,[167] Parker, J., followed the logic of this assumption and held that Article 32(1)(b) applied in these circumstances. The plaintiffs cannot advance at one and the same time the case that there was no delivery and therefore no application of Article 32(1)(a) and the argument that it was a case of damage

---

160. [1987] 1 Lloyd's Rep. 354.

161. At p. 359: "Since Article 23(4) is, by Article 25(1), also made applicable to cases of damage, it may be that the judge decided the case to be one of the [sic] total damage, so to speak. Otherwise he need not have referred to Article 25 at all."

162. Comm. Paris, 14.3.78 (1978) 13 E.T.L. 742, similarly App. Chambery, 16.6.76 (1976) B.T. 366, cf. App. Anvers, 24.3.76 (1976) J.T. 525, Comm. Bruxelles, 9.1.78 (1977–8) J.P.A. 278. In some cases there is support for the view that where, by reason of the failure to deliver the goods to the consignee, the goods have lost any real economic value (e.g. because they were uniquely created for a sale which has been lost) they are thereby a total loss (at least for the purposes of obtaining full damages, see *supra*, fn. 161), see RB. Brussel, 6.4.84 (1984) 14 E.T.L. 431, (1986) II U.L.R. 607, cf. BGH., 3.7.74 (1974) II U.L.R. 216, contrast OLG. Düsseldorf, 16.6.92 (1993) TranspR 17 (see Clarke, p. 220).

163. Hof. Brussel, 16.11.77 (1980) 15 E.T.L. 319. See also RB. Brussel, 28.2.75 (1975) 10 E.T.L. 419.

164. There is separate provision in the Convention for claims *between* successive carriers. Article 39, *infra*, Chapter 11, paras 11.113–11.133.

165. See *supra*, para. 7.11.

166. *Supra*, para. 10.60.

167. *Supra*, para. 10.62.

and not loss.[168] In *ICI PLC* v. *MAT Transport Ltd.*[169] Staughton, J., accepted this proposition and applied Article 32(1)(b). He rejected the argument that it is implicit in Article 20 that the claimant has a choice and may choose to treat the goods as damaged rather than lost.

10.71 This argument was put forward in view of the fact that the goods were re-delivered to the sender after the period indicated in Article 20. While he was reluctant to accept that for all purposes the owner is obliged to treat the goods as lost[170] he felt that, so far as limitation and the running of time are concerned, the parties are entitled to know where they stand. Consequently, non-delivery within the period referred to is conclusive evidence of loss binding on both parties.

10.72 The opposite conclusion was reached by an appellate court in Germany. On delivery of a consignment, part of the load was found to be missing. Over a year later, the missing goods were found and returned damaged. The time limit for a claim for this damage was held to run only from the delivery of the damaged goods. Once the goods re-appeared, then conceptually there was neither full nor partial loss. As the court explained,[171] the object of the provision is to give the claimant an opportunity, once the goods are found, to accept the goods and assert claims arising from any damage. If he were to be deprived of this opportunity, he would be forced to protect his claims by always asserting a claim based on the loss of the goods in circumstances where delivery did not take place at the time agreed even though the true basis of the claim, which turns out to be damage rather than loss, was not known at that time.[172]

OTHER CASES

10.73 Article 32(1)(c) provides that in all other cases, time will begin to run on the expiry of a period of three months[173] after the making of the contract of carriage. In *ICI PLC* v. *MAT Transport Ltd.*[174] the parties agreed that the "making" of the agreement for the purposes of Article 32(1)(c) occurred on the issue of the CMR consignment note. However, the goods had been taken over a few days before that, and the carrier had agreed to the carriage a month before. Although there is much to be said in terms of certainty for

---

168. [1983] 1 All E.R. 692, 696. The decision of Comm. Paris (1978) 13 E.T.L. 742 was cited in support. The idea that goods can be considered to be "lost" when they are accessible or even re-delivered appears artificial and has been much criticised, see Glass, [1984] LMCLQ 30, 42, Haak, p. 299, Clarke, pp. 155–156.
169. [1987] 1 Lloyd's Rep. 354.
170. See *supra*, para. 7.23.
171. OLG. Düsseldorf, 23.11.89 (1990) TranspR 63, 65–66.
172. The difficulty for the claimant was acute. The goods were eventually found not to have been despatched by the carrier's agent but to have remained in his store. The carrier's assertion that the goods had been despatched may have caused the claimant to genuinely doubt that he had a claim. Only on discovery of the goods was that doubt removed. The approach of the court meant that the claimant was not thereby deprived of the opportunity to recover for it. In general, however, the courts have not been sympathetic to a claimant who is unable to fully formulate his claim within the limitation period, e.g. BGH., 11.12.81 (1982) VersR 649, (1983) 18 E.T.L. 63 (carrier's vehicles detained for several months and beyond the period of limitation), App. Paris, (1982) B.T. 434 (claimant sought reimbursement of duty where payment not made to the authorities until 15 months after the goods were stolen. On loss of the goods the claimant knew it would have to pay), cf., however, OGH., 19.5.82 (1984) TranspR 193, cited by Loewe, Theunis p. 155.
173. The "months" referred to in this paragraph are calendar months: Interpretation Act 1978, section 5 and Schedule 1; Loewe, para. 260.
174. *Supra*, fn. 151.

utilising the date of the consignment note, this must be subject to the possibility of proof of the precise date of the making of the contract of carriage rather than the commencement of performance or the confirmation of it by the making out of the note itself.[175]

10.74 This provision has been held to apply to actions for the recovery of freight charges from the sender or other party responsible for payment,[176] for the recovery of excess freight charges by the sender,[177] and to a claim by a carrier against the sender for loss due to the latter's negligence.[178] It has also been applied to an action against the carrier on the basis of failure to comply with instructions in respect of payment on delivery.[179]

10.75 As has been pointed out elsewhere,[180] potential problems arise in these situations from the fact that if there are other related claims involved, different time limits may apply. Thus, for example, if a carrier is claiming his freight charges, time will commence to run under Article 32(1)(c) from three months after the contract was formed. If the goods have been partially damaged by the carrier, time in relation to that claim begins to run under Article 32(1)(a) from the date of delivery. Clearly, the only way in which the two limitation periods will coincide is if delivery has taken place exactly three months after the contract was made. If earlier than that, the time for claiming the freight charges will expire after that in respect of the claim for damage; if later, the claim for freight charges becomes barred before the claim for damage. The dangers for claimants involved in negotiations over conflicting claims are obvious. As has been suggested[181] it would be preferable if a common time limit were applicable, but on the wording of the Convention this would not seem to be open to the courts.

10.76 The problem is further compounded by the decision in *Muller Batavier Ltd.* v. *Laurent Transport Co. Ltd.*[182] that Article 32(1)(c) applies to claims for freight charges between successive carriers, rather than the provisions of Article 39(4).[183] This produces a shorter period of limitation as between successive carriers for the recovery of freight charges than for cargo claims.

10.77 Given the apparently comprehensive wording of para.(c) of Article 32[184] another aspect of the decision in *Moto Vespa S.A.* v. *MAT (Brittania Express) Ltd.*[185] seems

---

175. Cf. RB. Dordrecht, 21.5.80 (1982) S. & S. No. 8, where, in the absence of concrete facts, the despatch date specified by the invoice was taken as the starting point, see Haak, p. 299.

176. RB. Mechelen, 14.2.73 (1973) 8 E.T.L. 630. RB. Brussel, 13.9.68 (1969) 4 E.T.L. 1153. See also the cases cited by Wetter, op. cit., pp. 505–506. A claim for freight can only be made once freight has become payable. It is possible that the parties may operate a current account whereby only on closure of the account or the end of the accounting period does the balance become payable. The possibility of postponing the operation of the time limit until the closure of the account is recognised in France, provided that there is strict proof that the account is intended to operate in this way, see App. Paris, 24.2.93 (1994) B.T. 52. This approach could be justified as an application of national law by virtue of Article 32(3), see *infra*, para. 10.113 ff. A similar difficulty arises where there is an overall contract for a series of movements. It should be possible to analyse each movement as constituting the relevant contract of carriage for the purposes of CMR within the context of an overall contract of affreightment (cf. *supra*, para. 1.3 fn. 10, and Clarke, p. 153).

177. BGH., 18.2.72 (1972) 7 E.T.L. 860.

178. KG. Rotterdam, 23.6.71 (1972) S. & S. No. 8.

179. GH. 's Hertogenbosch, 13.1.70 (1971) 6 E.T.L. 817.

180. Hardingham, op. cit.

181. Ibid.

182. [1977] 1 Lloyd's Rep. 411.

183. Discussed *infra*, para. 11.127 ff.

184. "In all other cases . . . ".

185. [1979] 1 Lloyd's Rep. 175.

surprising. Mocatta, J., there held, as has already been seen,[186] that Article 32(1)(a) did not apply. However, he went further and held that Article 32(1)(c) did not apply either. He viewed the introductory words "In all other cases" as meaning cases *other than* partial loss or damage, or total loss, referred to in the preceding paragraphs of Article 32(1), and decided that since there *had been* damage, para. (c) could not apply. The end result, therefore, was that there was no applicable period of time under the Convention, and so the six year period at common law applied. Producing, as it does, yet another possible applicable limitation period, this aspect of the decision seems unfortunate, particularly as that period will in turn vary from jurisdiction to jurisdiction. Nor was it necessary to the ultimate decision in that case, since Mocatta, J., went on to decide that in any event time had been preserved by the presentation of a written claim under Article 32(2).[187] It is thus perhaps best regarded as not part of the *ratio* of that case, and open to review in the future.

10.78 In *Worldwide Carriers Ltd.* v. *Ardtran International Ltd.*,[188] Parker, J., considered the *Moto Vespa* case, but it did not prevent him from coming to a different view.[189] As noted above, he applied Article 32(1)(b) to the circumstances before him. Although he accepted the possibility that Article 32(1)(a) might apply in the alternative, he further indicated that if neither was applicable, he would have been prepared to apply Article 32(1)(c). The opening words of Article 32 plainly show that it was intended to be comprehensive and to cover all claims under the Convention. Accordingly, in his view, Article 32(1)(c) will apply to any claim when, for whatever reason, the starting points indicated in sub-paras (a) or (b) are not applicable.[190]

10.79 Article 32(1) concludes by stating that the day on which the period of limitation begins to run is not to be included in the limitation period. This provision is not restricted to Article 32(1)(c) only, but is of general application to all the cases provided for by Article 32(1).[191]

10.80 In the absence of any provision in the Convention, the question of whether or not the period of limitation laid down under Article 32 must be specifically raised by one of the parties to the action must be determined by reference to the applicable national law.[192] Under English law, reliance on any limitation period must be specifically pleaded.[193]

(c) Suspension of period of limitation

10.81 Under Article 32(2) the Convention makes provision for the suspension of the period of limitation. Broadly, this is achieved by sending a written claim to the carrier; the period of limitation is thereafter suspended until the carrier terminates the suspension by

---

186. *Supra.*
187. *Infra*, para. 10.81 ff.
188. [1983] 1 All E.R. 692.
189. *Supra*, para. 10.62.
190. In App. Bruxelles, 28.6.69 (1969) 4 E.T.L. 925, (1970) U.L.C. 115 the court applied Article 32(1)(c) in similar circumstances. The commentator in E.T.L. argues in favour of the application of para. (a). See also Libouton, (1982) J.T. 735, para. 134, who cites a similar decision of the same court: App. Bruxelles, 16.5.80 (1980–81) R.W. col. 1013, and expresses a similar view to the commentator in E.T.L. See also, Hardingham, op. cit.
191. KG. Delft, 13.5.65 (1966) 1 E.T.L. 722, (1966) U.L.C. 104.
192. OGH., 14.1.76 (1976) Oest. Jur. 435; see further *infra*, para. 10.113 fn. 302.
193. See *Halsbury's Laws*, 4th edn, Vol. 28, para. 851. The same is true under both German and Austrian law.

rejecting the claim in writing. Again, a number of points arise which require consideration.

SCOPE

10.82 Generally, of course, this provision will apply in all cases covered by Article 32(1). There is, however, one curiosity. Whereas the period of limitation laid down in Article 32(1) applies, as has been seen, to proceedings both by and against the carrier, Article 32(2) apparently only applies to an action against the carrier, since it refers only to the period of limitation being suspended "until such date as *the carrier* rejects the claim". If this interpretation be correct, it would mean that a written claim by the carrier against the consignee would not suspend the period of limitation.[194] It would, however, apparently leave open the possibility that Article 32(2) may operate to suspend a claim for freight charges by one carrier against another, because one can then properly speak in terms of the carrier rejecting the claim. This point arose in this country in *Muller Batavier Ltd.* v. *Laurent Transport Co. Ltd.*[195] which was a claim between successive carriers for carriage charges due. The plaintiffs relied on Article 39(4),[196] but also on Article 32(2) on the basis that their invoices for the carriage charges constituted written claims, thus suspending the limitation period. This argument was rejected by May, J.: "The written claim, or the claim, referred to in Article 32(2), when one looks at the Convention and its scheme as a whole is . . . in my opinion a claim for compensation for something that has gone wrong".[197] He therefore concluded that "it is in no way directed to a straightforward request by one carrier against another, or by the carrier against the consignor for payment of the freight . . . ".[198] In his view, therefore, Article 32(2) has no application to claims for freight, regardless of whether it is a claim by the carrier against the cargo interests,[199] or by one carrier against another. Although May, J., did seem to accept that his analysis might not cover all possible cases, he gave no indication as to when the result might be different. Perhaps such would be the case if there were both a claim for damage to the goods against the carrier, and a claim for carriage charges by the carrier.[200] Since the one is to an extent

194. Loewe, para. 257, expresses the view that this interpretation is correct, and there is a French authority to this effect: App. Aix-en-Provence, 15.2.79 (1979) B.T. 352 (also App. Paris, 23.9.81 (1981) B.T. 538, and App. Paris, 28.4.87 (1987) B.T. 424) but the opposite conclusion would seem to have been reached in RB. Mechelen, 14.2.73 (1973) 8 E.T.L. 630 where it was held that a formal summons referring to an unpaid invoice for freight charges is equivalent to a written protest, so that the limitation period was suspended. It should be appreciated that national law may be applicable to provide a means of suspension to the carrier, see Article 32(3) and App. Paris, 23.9.81 (1981) B.T. 538.
195. [1977] 1 Lloyd's Rep. 411.
196. Discussed *infra*, para. 11.127 ff.
197. [1977] 1 Lloyd's Rep. 411, 416.
198. Ibid. Clearly this leaves open the possibility that Article 32(2) can apply to a recourse claim by the carrier against another carrier, see App. Nancy, 29.11.94 (1995) B.T. 17, cf. Libouton, (1982) J.T. 736, para. 137, and cf. Comm. Gand, 31.1.78 (1979) R.G.A.R. No. 10025.
199. On this point, the views of May, J., must be regarded as obiter, since that issue was not raised by the facts of the case before him.
200. In this respect it should be noted that there is no right of set-off against a claim for freight, whether the cross-claim relates to loss, damage or delay: *R. H. & D. International Ltd.* v. *I.A.S. Animal Services Ltd.* [1984] 2 All E.R. 203, applying the rule in *Aries Tanker Corp.* v. *Total Transport Ltd.* [1977] 1 Lloyd's Rep. 334 to CMR cases, followed albeit with reluctance by May, J., in *United Carriers Ltd.* v. *Heritage Food Group (U.K.) Ltd.*, *The Times*, 8 March 1995. Even if a right of set-off did exist, no restriction in imposed by Article 41 (*infra*, Chapter 12) on contractual clauses removing it: BGH., 7.3.85 (1985) 20 E.T.L. 343, OGH., 31.1.91 (1993) TranspR 237, see further Clarke, p. 176, n. 34, Herber/Piper, Art. 32 Rdn. 57.

dependent upon the other, in that the cargo interests can claim a refund of carriage charges or a proportion of them,[201] it would seem both inelegant and potentially harsh to have a differing application of Article 32(2). The same reasoning would apply to claims between successive carriers involving claims for damage to the goods on the one hand and carriage charges on the other. In Germany, in a case concerning solely a claim for freight and a counter-claim for overpayment of freight, Article 32(2) was held to be inapplicable to the counter-claim since it was not a claim for compensation or damages for some loss against the carrier.[202]

FORM OF CLAIM

10.83 There is no requirement whereby a claim under Article 32(2) has to be submitted in any particular form. It must, however, be in writing, a claim by telex being sufficient for this purpose.[203] Two difficulties arise in this connection: first, the question of how precise the claim need be to have the effect of suspending the limitation period, and second, whether there is any need for supporting documents.[204]

10.84 The *Moto Vespa* case,[205] the facts of which have already been described,[206] also discussed both these issues. The defendants in that case argued that in order to suspend the period of limitation, a written claim had to be "a semi-formal claim with documents attached". The plaintiffs argued that "a general intimation in writing that in the absence of compromise the carrier would be sued for loss of or damage to the goods" was sufficient, basing their contention on the House of Lords' decision in *A/S Rendal* v. *Arcos Ltd.*[207] and on *William Tatton & Co. Ltd.* v. *Ferrymasters Ltd.*[208]

10.85 In the former case, it was held that "notice of any claim" under a charter did not need to be precisely formulated with full details, but was sufficient if it enabled the other party to take steps to meet the claim. In the latter case, which was concerned with claims in writing under Article 27,[209] it was also held that a general intimation of an intention to hold the carrier liable would be sufficient.

10.86 Mocatta, J., was of the opinion that "the reference to the return of any documents that may be attached to a written claim (does not require) such a claim to be so documented, though it may well be".[210] Turning to the documents in that case, which consisted of clear warnings of a pending claim, it was held that these constituted sufficient to comply with Article 32(2) despite the absence of any detailed claim, and of supporting

---

201. Article 23(4), *supra*, Chapter 9, paras 9.21–9.41.
202. OLG. Düsseldorf, 11.7.96 (1997) TranspR 274, (1998) U.L.R. 202.
203. RB. Antwerpen, 3.3.76 (1977) 12 E.T.L. 437, confirmed in Hof. Antwerpen, 30.5.79 (1979) E.T.L. 924, also HVC., 12.12.80 (1981) E.T.L. 250. It has been stated in the context of claims under Art. 27, *supra* para. 9.60 that it is generally agreed that the claim can be made by teleprinter, telegram, telex, teletext and fax; Thume, 1993 TranspR 365, 366. OLG. Munich, 16.1.91 (1992) TranspR 181 is cited.
204. The written claim in Article 32(2) should be distinguished from reservations made by the consignee under Article 30. The latter may not be sufficient to suspend limitation, see BGH., 9.2.84 (1985) 20 E.T.L. 275, Trib. gde. inst. Albertville, 4.5.75 (1975) B.T. 217, Comm. Angers, 18.6.68 (1968) J.D.U. 161, App. Reims, 3.3.80 (1980) B.T. 237, Comm. Nivelles, 25.6.92 (1992) B.T. 703.
205. *Moto Vespa S.A.* v. *MAT (Brittania Express) Ltd.* [1979] 1 Lloyd's Rep. 175.
206. *Supra*, para. 10.35.
207. (1937) 58 Ll.L.Rep. 287; (1937) 43 Com. Cas. 1.
208. [1974] 1 Lloyd's Rep. 203.
209. Claims for interest. *Supra*, paras 9.60–9.64.
210. [1979] 1 Lloyd's Rep. 175, 180. This reference to the return of written documents had formed the basis of the defendant's submission.

documents. There had been no rejection of the claim by the defendants, and so they failed on the limitation issue.

10.87 In the later case of *ICI PLC.* v. *MAT Transport Ltd.*[211] two letters of claim were sent by or on behalf of the plaintiffs, the first simply referring to the accident and stating that "we must hold you responsible for any losses we incur", and the second letter spelling out the nature and extent of the claim in greater detail. The claim was then rejected. It was argued on behalf of the defendants that the first letter was not effective to suspend time on the basis that there must be a present demand rather than an indication that a claim in the future was possible. Following the *William Tatton* case and the *Moto Vespa* case, Staughton, J., held that the original letter was a sufficient written claim for the purposes of Article 32(2). Staughton, J., referred to a decision in the Rotterdam District Court which took the view that the claim should contain the clearest possible description of the nature and extent of the damage, a declaration of liability and the grounds on which it is based.[212] He contrasted, however, a decision of the Amsterdam District Court in respect of the need for documents, which indicated to him the lack of uniformity on this aspect. However, there would seem to be a tendency in Continental decisions to require the claim to include enough information to enable the carrier to decide whether or not to accept it.[213] In France, the courts require the claim to be such as to enable the carrier to take a position in relation to it.[214] This approach was adopted in one case where the letter to the carrier claimed simply that between 80% and 90% of the goods had deteriorated so that the consignment was unusable and was held at the disposition of the carrier. The court held that since the letter did not describe the damage to the goods, did not impute that it was the responsibility of the carrier and did not give the carrier sufficient information so as to be able to estimate the indemnity required from him, the claim was insufficient for the purposes of Article 32(2).[215] In another case, it was said that while the claim does not have to be calculated with exact precision, it should make clear the intention of the claimant to hold the carrier responsible[216] and give him an idea of the amount claimed.[217] In Belgium, written claims which did not quantify the claim were held sufficient where the

211. [1987] 1 Lloyd's Rep. 354.
212. See, however, Loewe, Theunis, p. 157, who says that it was the intention of the authors of CMR that the claim need not include all the details or figures relating to the claim, but an account of the damages for which compensation is required, and documents need be attached only insofar as they are essential for the carrier to be able to reply to the claim.
213. Consequently what may be sufficient as a reservation made on the consignment note, in respect of Article 30 (*supra*, Chapter 10, paras 10.2–10.20) may not be sufficient for a written claim, as in App. Bruxelles 6.11.95 (1996) J.L.M.B. 188 (Putzeys, "Le droit des transports en Belgique (II) 1993–1995", (1996) U.L.R. 557, 568).
214. Mercadal, para. 267.
215. App. Toulouse, 22.11.89 (1990) B.T. 427, cf. RB. Rotterdam, 8.12.89 (1990) S. & S. No. 434, (1991) 19 E.L.D. No. 662. See also App. Riom, 15.9.99 (1999) B.T. 746.
216. A letter which merely asked for the carrier to give his opinion in respect of an expert's report was held not to be a written claim in RB. Antwerpen, 16.4.75 (1975) 10 E.T.L. 548. In OLG. Hamm, 7.11.96 (1998) TranspR 459, (1999) U.L.R. 194 a letter which stated that "as a precaution we are making you liable for costs that arise" was sufficient. There was no doubt that the "costs" referred to the damage to the goods. The letter was not merely an announcement that a claim would be made in the future, which would not suffice, cf. Thume, Article 32 Rdn. 61.
217. App. Aix-en-Provence, 7.2.90 (1990) B.T. 699 (See also, eg. Cass. 28.3.2000 (2000) B.T. 276), cf. App. Grenoble, 4.2.92 (1992) B.T. 703 where the claim was effective to suspend time despite the fact that it did not specify the exact damage. It was not possible to calculate the damage until some time later and the carrier had not rejected the claim simply by asking to be informed of the exact amount once it was known, cf. OLG. Hamm, 7.11.96, *supra*, fn. 216.

carrier could calculate the amount of the claim by means of other information which was in his possession or came to him shortly after the claim was made.[218] A useful test may be whether the carrier should appreciate the need to involve his insurer.[219] The fact that the claimant later restricts the amount of the claim has no effect on a suspension which has already come into operation.[220] Conversely, a restricted claim which is later expanded should be suspended, in respect of the expanded claim, only from the time of the later claim.[221]

10.88 There is some suggestion in the judgment of May, J., in *Muller Batavier Ltd*. v. *Laurent Transport Co. Ltd*.[222] that the written claim under Article 32(2) must be supported by any relevant documents. But it is submitted that such suggestion must be considered in context, i.e. in considering the type of claim to which Article 32(2) applied,[223] and does not have any bearing on the present question. In *ICI PLC* v. *MAT Transport Ltd*.[224] the issue of whether documents must accompany the written claim was left open. However, Staughton, J., did observe[225]:

"One can see grounds for an argument that the written claim must be accompanied by documents, although that argument encounters the difficulty that Article 32(2) does not specify which documents."[226]

10.89 So although there is one Dutch authority where it was decided that a written claim must be accompanied by all relevant documents, otherwise the period of limitation would not be suspended,[227] there is equally Dutch,[228] Belgian,[229] and French[230] authority to the contrary, and it is submitted that the views of Mocatta, J., in the *Moto Vespa* case are to be preferred.

BY WHOM AND AGAINST WHOM A CLAIM MAY BE MADE

10.90 Article 32(2) refers to a written claim as suspending the period of limitation, but does not specify from whom the claim must emanate. Clearly, this question is part of the

---

218. HVC., 12.12.80 (1981) 16 E.T.L. 250 (value of the stolen goods indicated on the waybill), Comm. Liège, 25.11.82 (1982) 17 E.T.L. 843 (an *expertise* carried out after the claim was sent was sufficient to inform the carrier of the amount of the claim). Contrast App. Bruxelles, 20.9.95 (1995) B.T. 795, although in this case no express claim for compensation was made until it was too late, cf. App. Bruxelles, 6.11.95 (1996) B.T. 258. In Hof. Gent, 25.6.86 (1987) 22 E.T.L. 421, however, neither a formal written claim nor exact figures was required. It suffices to give an accurate description of the nature and contents of the damage and the legal grounds on which the claim is based. In OLG. Düsseldorf, 8.5.69 (1970) 5 E.T.L. 446 it was held that a written claim must "mention the damaged units in a precise and determinate manner", cf. BGH., 7.11.85 (1986) 21 E.T.L. 266 (sending the carrier a photocopy of a telegram sent for partial loss by the consignee was a sufficient written claim for the purposes of the applicable national law where its content indicated that the carrier was being called to account for the loss), cf., also, OGH., 29.8.94 (1995) TranspR 110, (1995) 30 E.T.L. 211, LG. Munich, 5.7.88 (1989) R.I.W. 573, Supreme Court, Sweden, 26.3.96 (1997) U.L.R. 202.
219. OLG. Munich, 24.4.92 (1992) TranspR 360, see Clarke, p. 168.
220. OLG. Hamm, 7.11.96, *supra*, fn. 216
221. OLG. Hamm, 7.11.96, *supra*, fn. 216, at p. 462.
222. [1977] 1 Lloyd's Rep. 411.
223. *Supra*, para. 10.82.
224. *Supra*, fn. 211.
225. At p. 360.
226. Cf. further *Impex Transport* v. *A.G. Thames Holdings Ltd*. [1981] 2 Lloyd's Rep. 566, 574.
227. Arrond. Rotterdam, 5.4.74 (1974) N.J. No. 182, (1975) I U.L.C. 274.
228. Arrond. Rotterdam, 12.4.72 (1976) I U.L.C. 253.
229. RB. Antwerpen, 24.10.67 (1969) 4 E.T.L. 1035, RB. Antwerpen, 3.3.76 (1977) 12 E.T.L. 437 and App. Bruxelles, 27.2.96 (1996) 31 E.T.L. 833.
230. App. Paris, 27.2.80 (1980) B.T. 384 and App. Paris, 25.3.82 (1982) B.T. 434, cf. Comm. Creteil, 12.12.91 (1992) B.T. 151.

larger question as to who is entitled to bring proceedings. If a party is entitled to bring proceedings, then the limitation period is running against that party, and it follows that Article 32(2) will operate for his benefit.[231]

10.91 The difficulties involved in deciding who is entitled to claim under the Convention have been discussed earlier.[232] In view of these difficulties, if there is any doubt as to who is the proper party to submit the claim, the written claim should be sent to the carrier on behalf of both the sender and the consignee,[233] and in the event of proceedings subsequently becoming necessary, both the sender and the consignee should be named as plaintiffs.[234] The fact, however, that a written claim is made by one person entitled to claim does not prevent a further written claim by another party being effective for the purpose of that party's claim.[235] So far as the written claim is concerned, on normal agency principles, it will be sufficient if sent on behalf of the person entitled to claim, and it does not need to be sent by that party personally. Thus in *Moto Vespa S.A.* v. *MAT (Brittania Express) Ltd.*[236] Mocatta, J., accepted a number of documents as collectively constituting a valid written claim under Article 32(2). Although it is not entirely clear from his judgment to what extent any one of them on its own would have been sufficient, letters from the consignee's agents and from their insurers were among those documents, and both seem to have been accepted by Mocatta, J., as effective to suspend the limitation period.[237] On the other hand, in *Sidney Jones Ltd.* v. *Martin Bencher Ltd.*,[238] it was held that a claim addressed on behalf of "interested parties" did not include the plaintiff since the claim was made by assessors appointed by the insurers of one of the other defendants in the action. As Popplewell, J., put it[239]:

"Apart from the obvious difficulty that there is no evidence that assessors appointed by insurers of the third defendants were acting or indeed authorised to act on behalf of the plaintiffs, in my judgment the phrase 'interested parties' is simply not capable of including the plaintiffs".

10.92 In France, however, a written claim sent by an insurance broker has been held to be effective to suspend the time limit. Such an agent is deemed by statute to be the agent

---

231. In Belgium, the written claim is effective for the consignee only from when he has adhered to the contract and acquired the right of disposal: App Mons, 8.11.93 (1996) J.L.M.B. 168 (Putzeys, *supra*, para 10.87, fn. 213, at p. 568). In France, a *commissionnaire* (see *supra*, para. 1.50) becomes entitled to make a written claim on his own behalf on indemnifying the victim of the damage, Cass., 3.3.98 (1998) B.T. 212.

232. See *supra*, Chapter 5, paras 5.2–5.24.

233. There is German authority to the effect that a claim by the consignee will also be effective to suspend time in respect of a claim by the sender (OLG. Düsseldorf, 13.1.73 (1973) 8 E.T.L. 620). French authority suggests that this is not possible: Trib. gde. inst. Valence, 18.11.81 (1982) B.T. 211, App. Toulouse, 7.12.89, unpublished, cited in Lamy, paras 549 and 1598, Jur. 11, cf. App. Toulouse, 22.11.89 (1990) B.T. 437.

234. The party actually submitting the claim should, of course, in either situation, first obtain the authority of the other.

235. BGH., 24.10.91 (1992) 27 E.T.L. 839, (1992) TranspR 177. The last sentence of Article 32(2) does not prohibit this possibility.

236. [1979] 1 Lloyd's Rep. 175.

237. The only reason for any doubt in this respect is because the former letter referred to the wrong principals, and the second referred to the wrong party as the insured. As indicated, Mocatta, J., did not seem to consider either defect as significant. As to the second letter, this would seem correct, since it was a floating policy, and as Mocatta, J., pointed out "in the case of declarations made under a floating policy it is common for the broker's or agent's name to appear as the assured". As to the first, however, the Continental decisions referred to in the following footnotes would indicate that rather more precision may be required elsewhere.

238. [1986] 1 Lloyd's Rep. 54.

239. At p. 61.

of the assured, and the fact that the broker's letter stated that they were acting on behalf of the insurers of the goods made no difference to the position.[240]

10.93 In relation to the commencement of proceedings, however, in this country proceedings must normally be commenced in the name of the original claimant, even if an insurance company is subrogated to his rights.[241] In Continental jurisdictions, however, once the insurance company is subrogated to the claimant's rights, the insurance company can bring proceedings in its own name.[242] This has produced a certain amount of case law as to the appropriate person to submit a written claim. Thus, in a French decision where insurers submitted a claim before they were subrogated to the rights of their insured, it was held that the period of limitation was not suspended: they were not at that stage entitled to claim in their own right and they failed to plead that they were acting on behalf of their insured.[243] To similar effect is a Dutch decision where it was held that a letter of claim sent by the plaintiff was ineffective to suspend the period of limitation as at that time the plaintiff did not have the sender's rights assigned to him.[244] Furthermore, it has been held in Germany that the fact that there is a later assignment does not confer retrospective effectiveness on the earlier written claim.[245] On the other hand, in a Belgian decision it was held that once the underwriters had indemnified the consignee, they were entitled to proceed in their own name, provided they were so authorised by the consignee.[246]

10.94 Turning to the question of to whom the written claim should be made, the claim can clearly be sent to the carrier and anyone acting on his behalf if so authorised.[247] If serving the claim on an agent it will be crucial to obtain confirmation that he is authorised to receive it, lest the defendant subsequently deny such authority. In the absence of authority, the claim would be ineffective to suspend the limitation period. Thus, in a French decision where a claimant sent his claim to the insurer acting on behalf of the carrier, it was held ineffective to suspend time as against the carrier, since the carrier had not so authorised his insurer. However, since under French law the party interested in the goods has a direct right of action against the carrier's insurer, the claim was effective to

240. App. Paris, 18.12.92 (1993) B.T. 52, cf. RB. Breda, 11.1.77 (1978) S. & S. No. 89, see further Comm. Mons, 9.11.76 (1977) 12 E.T.L. 300, RB. Utrecht, 11.4.79 (1980) S. & S. No. 12, Cass. (Belg.), 3.11.83 (1985) J.P.A. 87, BGH., 21.11.96 (1997) 32 E.T.L. 442, (1997) TranspR 164. A *commissionnaire de transport* has also been held to have authority: App. Paris, 22.1.86 (1986) B.T. 482 (cf. OLG. Düsseldorf, 16.2.82 (1983) VersR 1028), but has no independent right to claim on his own behalf until he has indemnified his customer: Cass., 4.10.82 (1982) B.T. 549, App. Paris, 18.6.97 (1997) B.T. 523, see further, *supra*, para. 10.90 fn. 231 ff, cf., however, App. Reims 22.5.96 (1997) B.T. 447.

241. The exception is where the insured has expressly assigned his rights to the insurer. The right of subrogation alone is not sufficient.

242. E.g., France, Loi 13.7.1930, Article 36, para. 1. For Germany, cf. BGH., 24.10.91 (1992) 27 E.T.L. 839, (1992) TranspR 177, LG. Cologne, 10.12.93 (1995) TranspR 71.

243. App. Aix-en-Provence, 8.11.68 (1969) 4 E.T.L. 918. Conversely, once they are subrogated, App. Aix-en-Provence, 11.3.69 (1969) B.T. 389, App. Paris, 24.9.97 (1998) B.T. 363, App. Anvers, 24.3.76 (1976) J.T. 525. Cf. the difficulty faced by a French insurer whose claim was made after the prescription period had ended, and who could not rely on an action brought by the sender after he had been indemnified by them, and in consequence no longer had title to sue: App. Versailles, 7.3.96 (1996) B.T. 633.

244. Arrond. Rotterdam, 5.4.74 (1974) N.J. No. 182, (1975) I U.L.C. 274.

245. BGH., 24.10.91 (1992) 27 E.T.L. 839, (1992) TranspR 177. See, however, App. Paris, 17.11.73 (1974) B.T. 81 where subrogation took place between the sending of the claim and its receipt by the carrier.

246. RB. Antwerpen, 17.2.74 (1974) 9 E.T.L. 504. Similarly where the insurer is armed with letters of subrogation, App. Agen, 19.3.80 (1980) B.T. 502, and see other cases cited by Lamy, paras 549 and 1598.

247. The fact that the carrier is bankrupt makes no difference as to its effectiveness especially as, by keeping the limitation period alive, action may proceed against the carrier's insurer: App. Lyon, 27.10.95 (1996) B.T. 761.

suspend time as against the insurer.[248] There is no corresponding right of action against the insurer under English law,[249] so a claim sent to an insurer here under the same circumstances would be of no effect whatsoever,[250] unless authority to accept the claim can be implied. This issue arose before the Court of Appeal in *Poclain S.A.* v. *SCAC S.A.*,[251] where it was explained that the carriers' insurers would not necessarily have authority to receive the claim if all that had happened was that they had been notified of the incident by the carriers, had been given authority to make enquiries and had been told, either expressly or by reference to a course of dealing, that the carriers either did not intend to claim on them or had not yet decided whether to do so. On the facts of the case, however, a firm claim had been made on the insurers by the carriers, and the insurers were thereby expressly or impliedly authorised to deal with the claim, so that they were authorised to receive a "written claim" for the purposes of Article 32(2). This authority was not affected by the fact that the claim had not yet been established against the carriers for the purposes of the Third Parties (Rights against Insurers) Act 1930.[252]

10.95 A particular difficulty for claimants arises in the context of successive carriage, since it was held in the case of *Worldwide Carriers Ltd.* v. *Ardtran International Ltd.*[253] that a claim submitted to one carrier[254] is not effective to suspend time for the purposes of a claim against another successive carrier, even if the former had notified the latter of the claim and indicated an intention to claim an indemnity in respect of any liability arising.[255] So also in *Sidney G. Jones Ltd.* v. *Martin Bencher Ltd.*[256] it was held that a claim by one carrier against another carrier was not capable of constituting a claim on behalf of the cargo interests, even though expressed to be lodged "on behalf of interested parties".

10.96 As to the commencement of proceedings, this is only possible in this country against the carrier himself.[257] Both as regards the commencement of proceedings and the submission of a written claim, difficult questions can arise as to who is the carrier,[258] or, in the case of successive carriers, who is the appropriate carrier,[259] to whom the claim should be sent and against whom proceedings should be commenced. If there is any doubt

---

248. App. Paris, 12.6.70 (1970) B.T. 228 (cf. however, Cass. 12.7.94 (1994) Bull. IV No. 267, p. 212, see Krings, p. 787). Whether or not the insurer is an agent, there must be a genuine claim and not merely the sending of a notice of the amount of damage to the insurer, Comm. Creteil, 12.12.91 (1992) B.T. 151.

249. Or under Belgian law: Libouton, (1973) 8 E.T.L. 2, para. 91.

250. A claim direct against the insurer is permissible in the limited circumstances specified in the Third Parties (Rights against Insurers) Act 1930 but it was held in *Post Office* v. *Norwich Union Fire Insurance Society Ltd.* [1967] 1 Lloyd's Rep. 216 that any such right only accrued once the liability of the insurer to the insured had been established by judgment, arbitration or agreement. Therefore, even in the limited circumstances therein specified, this Act will not assist the claimant for the purposes of a written claim under Article 32(2).

251. [1986] 1 Lloyd's Rep. 404.

252. See, *supra*, fn. 250.

253. [1983] 1 Lloyd's Rep. 61. See Glass, [1984] LMCLQ 30, cf. Hof. Gent, 25.6.86 (1987) 22 E.T.L. 421.

254. Cf. in France where the claim is sent to the *commissionnaire de transport* rather than the carrier, App. Paris, 23.6.76 (1976) B.T. 332, Trib. gde. inst. Metz, 10.11.81 (1982) B.T. 38. A variation on this theme is a claim sent to both the *commissionnaire* and the carrier. Such a claim was not effective to suspend the limitation period in respect of the *commissionnaire*'s claim against the carrier: App. Lyon, 16.10.98 (1999) B.T. 782.

255. A "transmission" of the claim might be distinguishable, especially where the defendant carrier has entered into a direct dialogue with the claimant, cf. Comm. Liège, 25.11.82 (1982) 17 E.T.L. 843, and see also Lamy, paras 549 and 1598, Jur. 15.

256. [1986] 1 Lloyd's Rep. 54, 60–61.

257. Subject to the Third Parties (Rights against Insurers) Act 1930, *supra*, fn. 250.

258. *Supra*, Chapter 1, paras 1.11–1.54.

259. *Infra*, Chapter 11, paras 11.62–11.74.

as to such issues the only safe course is for written claims under Article 32(2) to be submitted to all possible defendants, and if proceedings are commenced, all should be named as defendants.[260]

TIME FOR WRITTEN CLAIM

10.97 Although Article 32(2) does not expressly say so, clearly a claim will not have the effect of suspending the period of limitation unless served before the expiry of that period.[261] Rather more difficulty attaches to the question of the starting point for the possible service of a claim within the Article. Loewe is of the view that because time runs from the end of the periods specified in paras (b) and (c) of Article 32(2), claims within those periods do not operate to suspend the period of limitation until the end of the period of 30 days or 60 days, as the case may be.[262]

10.98 Logically, this would seem correct, since until the limitation period had started, it is difficult to talk in terms of suspending it. Yet if applied, it could produce hardship. If, for example, a written claim is sent early in the, say, 60 day period and rejected late within the same period, the claim would have no effect on the total limitation period, whereas had the same claim been submitted after the end of the 60 day period and rejected after the same lapse of time, the limitation period would have been extended by that amount. Further, to deprive the claim within the 60 day period of any effect would be to produce a further complication: is it open to the claimant to serve a further claim after the limitation period has commenced, which would then have a suspensory effect, or would this offend the closing words of Article 32(2)?[263] For these reasons, it is thought that a preferable approach would be to give effect in the normal way to claims sent before the limitation period has strictly speaking commenced, so that such a claim would suspend the running of time until rejected. There is Belgian authority to this effect,[264] and Wetter is of the same view.[265]

10.99 This point arose for consideration by the English courts in the case of *ICI PLC. v. MAT Transport Ltd.*[266] It was there argued by the defendants that a claim submitted during the initial period of 60 days or three months cannot suspend time. This argument was rejected by Staughton, J., in the following terms[267]:

"I can think of no rhyme or reason why the Convention should have that result, or why the owner of the goods should lose his rights if he makes a claim too early but retain them if he makes the claim later."

Nor did he feel compelled to accept the defendants' argument by virtue of the language of the Convention. In particular, reference in Article 32(2) to the period starting "to run again" did not alter his view. Staughton, J., did not express any view as to whether the 60

260. See generally Hardingham, "CMR: Safeguarding the parties' interests", [1981] LMCLQ 306.
261. OGH., 22.2.90 (1992) TranspR 357.
262. Loewe, para. 263, see also Loewe, Theunis, p. 159 (see also Lemy, para. 548, cf. App. Aix-en-Provence, 3.11.94 (1995) B.T. 232). Presumably he would apply the same reasoning to the submission of a claim prior to delivery in cases to which Article 31(1)(a) applied.
263. "The running of the period of limitation shall not be suspended by further claims having the same object."
264. App. Bruxelles, 28.6.69 (1969) 4 E.T.L. 925. This decision is, however, doubted by Libouton, (1973) 8 E.T.L. 2, para. 89.
265. Op. cit., p. 507.
266. [1987] 1 Lloyd's Rep. 354.
267. At p. 361.

day or three month period would itself be suspended, although he mentioned an un-named Belgian case which he had been referred to in argument as having held that it would.

10.100 Another related point is when the written claim takes effect. As distinct from Article 30, where it is made clear that a reservation is effective from the date of sending, Article 32(2) makes no express provision as to when the suspension takes effect. It would, however, seem, from the reference in the third sentence of the paragraph to the onus of proving receipt both in relation to the claim and its rejection, that both will only take effect when received.[268] French decisions support this view in relation to both the letter of claim[269] and the letter of reply.[270]

PERIOD OF SUSPENSION

10.101 The suspension of the limitation period will continue "until such date as the carrier rejects the claim by notification in writing and returns the documents attached thereto".[271] As in the case of the claim, no special form is required for the rejection of the claim,[272] but it must be unambiguous.[273] Thus, where the claimant received two letters from the carrier on the same day, one denying liability, the other stating that the documents were with their insurers and expressing confidence in a satisfactory outcome, the period of suspension was not ended.[274]

10.102 This aspect of Article 32(2) has been considered by the English courts in two cases. In the first, *Zerowatt SPA* v. *International Express Company Ltd.*[275] Webster, J., first accepted the principle that to be effective, a rejection of the claim must be unambiguous, indicating clearly to the plaintiff that the time has come when he must decide whether or not to pursue his claim through the courts.[276] There were three letters relied upon by the defendants; the first of these indicated that a surveyor's report was awaited and stated that the preliminary indications were that a defence would be available. The letter then concluded with the words "until we have received the final surveyor's reports and have had time to study them, we cannot indicate what the position might be. In the circumstances we must refute any liability in this matter, but will almost certainly revert again

268. Loewe is of the same view: para. 264. Section 7 of the Interpretation Act 1978, by which service is deemed to have been effected in the ordinary course of post, cannot apply since it only applies where any Act "authorises or requires any document to be served by post . . . ", which is not the case under Article 32(2).

269. Comm. Paris, 25.6.79 (1979) B.T. 403 (similarly, in Belgium, App. Bruxelles, 20.9.95 (1995) B.T. 795). See, however, Cass., 20.7.83 (1984) B.T. 236, and see further Lamy, para. 551, who criticises the implicit acceptance in this latter case of the date of sending as relevant to the starting point of the suspension. A carrier could not prevent the limitation from running, however, by evading the receipt of the claim, cf. App. Paris, 25.3.82 (1982) B.T. 434, see *infra*, fn. 295. In App. Paris, 26.2.97 (1997) B.T. 447, proof of the reception of the claim could be made by all means, and resulted from the terms of a report by an expert of the insurer. In Cass., 3.3.98 (1998) B.T. 212 it resulted from the fact that the carrier had referred the matter to his insurer.

270. Comm. Corbeil-Essones, 18.4.69 (1969) 4 E.T.L. 988; (1969) B.T. 217.

271. Article 32(2). The suspension is terminated only in respect of those claimants whose claims the carrier has rejected, OLG. Düsseldorf, 16.12.82 (1983) VersR 1132, cited by Loewe, Theunis, p. 160.

272. RB. Antwerpen, 24.10.67 (1969) 4 E.T.L. 1035.

273. Cf. Supreme Court, Sweden, 26.3.96 (1997) U.L.R. 202. A request for details and additional documents is not a clear rejection of the claim: Tribunal Supremo de España, 29.6.98 (1999) 34 E.T.L. 692.

274. RB. Rotterdam, 12.5.78 (1979) 7 E.L.D. 180. See also App. Toulouse 26.3.69 (1971) 6 E.T.L. 131, *infra* fn. 291.

275. (1991) 26 E.T.L. 530, cf. Comm. Bruxelles, 2.4.90 (1991) 26 E.T.L. 541, RB Rotterdam, 15.4.88 (1988) S. & S. No. 347, (1989) 17 E.L.D. No. 407.

276. Webster, J., suggested four categories of potential response: acceptance, non-acceptance in the sense of non-admission, rejection and a communication of some other kind. Examples of acceptance appear in unpublished French decisions referred to by Lamy, paras 552 and 1601, Jur. 4 and Jur. 5.

just as soon as possible". This letter was followed by two letters which were headed "without prejudice". Webster, J., first indicated that this had the consequence that the contents of the letters were inadmissible. However, he went on to hold that, even if the paragraph in question were to be regarded as admissible, the heading to the letter reinforced the conclusion that the letter could not be construed as containing a rejection of the claim for the purposes of the Convention. Although the letter stated that "we maintain our position in this matter and repudiate all liability", it continued to qualify this by indicating that further steps were being taken to investigate the claim. The third communication relied on was a telex stating "entirely without prejudice we would refer to our Mr. F. J. Hobb's letter dated 19th August and our standpoint must remain. We continue to disclaim all liabilities". Again, it was held that the words "without prejudice" rendered the communication inadmissible. Further, having decided that the previous two communications were themselves not sufficiently clear repudiations of liability, he held that the word "disclaimed" in the context of the previous correspondence was insufficient to start time running again.

10.103 Similarly, in the case of *Microfine Minerals & Chemicals Ltd.* v. *Transferry Shipping Co.*[277] a letter from the defendants confirmed that the claim papers had been passed to their insurance brokers in the following terms: "until further information is at hand, we at this juncture have no alternative but to decline responsibility, and will leave the matter with our brokers for their consideration and reply direct to you". It was held that this was not a sufficiently clear or unambiguous rejection as to start time running again.[278]

10.104 If, on the other hand, the carrier fails to acknowledge the claim he is penalised inasmuch as the period of limitation will remain in suspension. Indeed, there is no way under the Convention in which time can begin to run again except by his rejection of the claim in writing, although it may be that the time limit under the applicable national law would eventually operate to extinguish the claim.[279]

10.105 The reference to the need to return any documents attached to the claim produces some difficulty. According to May, J., in *Muller Batavier Ltd.* v. *Laurent Transport Co. Ltd.*[280] the *raison d'être* of this provision is that all the claimant's evidence is in the hands of the defendant, thus preventing the claimant proceeding further until it is returned to him. It is indeed difficult to see any other *rationale* for the requirement, but it would seem very doubtful that in a commercial world any claimant would be prepared to put all his evidence in the hands of the other party, without at least keeping copies for himself. What is, of course, far more likely to happen is that the claimant, if he sends anything at all with his letter of claim, will send photocopies of relevant documents, and in this context the requirement to return documents attached to the claim makes very little sense. Nevertheless, there is French authority to the effect that all documents attached to

---

277. [1991] 2 Lloyd's Rep. 630. The learned judge in this case did not consider that Webster, J., in the *Zerowatt* case had intended to mean that the letter must state what the claimant must do; rather that, as a matter of effect and construction, the claimant should understand that the time has come to decide whether or not to sue.

278. Cf. OLG. Nuremberg, 12.4.91 (1992) TranspR 63.

279. Such is assumed by Wetter, op. cit., p. 507, *sed quaere* in view of the apparently unambiguous provisions in Article 32(2).

280. [1977] 1 Lloyd's Rep. 411, 416.

the claim, be they originals or photocopies, and regardless of their probative value, must be returned before the suspension will terminate.[281] However, such a view seems logically very hard to accept, and it is perhaps implicit in May, J.'s reasoning referred to above that if only photocopies have been sent, the claimant remains in a position to proceed, and that such a situation is therefore outside the intended scope of the requirement. Loewe is of the view that photocopies need not be returned,[282] and there is a Dutch decision where it was held that there was no need to return a copy of a judgment since the claimant's lawyer had other copies.[283]

10.106 However, in *Microfine Minerals & Chemicals Ltd.* v. *Transferry Shipping Co. Ltd.*[284] Judge Kershaw, sitting as a Judge of the High Court, held that the wording of the Convention was unambiguous and that in order for a repudiation to be effective for the purposes of Article 32(2) it was necessary for the documents accompanying the claim to be returned even if they had consisted only of photocopies. He was of the view that, even if the wording of the Convention was to be regarded as ambiguous, on a purposive approach he would have come to the same conclusion. He rejected the purposive construction as suggested by May, J., in the *Muller Batavier* case referred to above, preferring the view that "the return of the documents that were sent with the claim is a very clear way of expressing the rejection of the claim, and of avoiding possible disputes about the meaning of the letter, telex or whatever which rejects the claim".[285] Further, Parker, J., appears to have been of the same view, albeit clearly obiter, in *Worldwide Carriers* v. *Ardtran*.[286] So although both decisions are only at first instance, clearly caution now requires that supporting documentation be returned with a repudiation of liability.

10.107 If the carrier admits to partial liability in respect of a claim, the period of limitation will start to run again only in respect of that part still in dispute.[287] In other words, it remains suspended in relation to that part of the claim in respect of which liability is admitted.[288] Thus, if the carrier subsequently rejects that part, it would only be at that time that the prescription period in respect of that part commenced to run again. There is also German authority that where a claim is partially admitted there is no need for the carrier to return the documents attached to the claim in order to start time running again in relation to the disputed remainder.[289]

---

281. Comm. Paris, 21.12.78 (1979) B.T. 84. Cf. App. Paris, 22.5.75 (1975) B.T. 320 where the court indicated that Article 32(2) is concerned with those documents which justify the claim (see Clarke, p. 173 n. 13).

282. Loewe, para. 266. See also, by the same author, "Le CMR à 40 ans" (1996) U.L.R. 429, 435–436.

283. Arrond. Roermond, 15.10.70 (1971) 6 E.T.L. 839. As to the general question of the need to return photocopies, Libouton comments on this case: "it does not seem as though the court had grasped the actual scope of the question with which it was dealing", Libouton (1973) 8 E.T.L. 2, para. 90 fn. 178, see also (1982) J.T. 736, para. 138, cf. Loewe, Theunis, p. 160.

284. [1991] 2 Lloyd's Rep. 630.

285. At p. 632.

286. [1983] 1 Lloyd's Rep. 61, 66. A decision of the Belgium Supreme Court also supports the position taken by Judge Kershaw, Cass. (Belg.), 27.9.84 (1984) II U.L.R. 406, but cf. The Maritime and Commercial Court of Denmark, 17.6.82 (1982) 17 E.T.L. 850. See further de Beule, (1988) 23 E.T.L. 654.

287. Article 32(2).

288. Presumably this provision applies where liability in respect of part of the claim is admitted "without prejudice", since there has been no rejection of the claim.

289. OLG. Celle, 13.1.75 (1975) 10 E.T.L. 410. This obviously makes some sense in that the carrier is likely still to require the documentation in relation to the disputed balance of the claim, but see May, J.'s analysis of the reason for the requirement that documents be returned in the *Muller Batavier* case, *supra*; in the light of the requirement for the return of any documents before time can start running again, the German decision would seem open to doubt, unless, as discussed above, the documents consist of photocopies.

10.108 There is Dutch authority to the effect that the claim can effectively be rejected by a third party if it is obvious that the claim has been rejected on behalf of the carrier.[290] On the other hand, in a French decision it was held that rejection by the Paris representative of a Ghent carrier was not effective since it did not emanate from the carrier.[291]

ONUS OF PROOF

10.109 Both in relation to the receipt of the written claim and the written rejection of the claim, Article 32(2) itself specifically places the onus of proof on the party relying on such receipt. So, for example, in a claim against the carrier for damage to his goods, it will be for the cargo interests to prove receipt of the written claim by the carrier, and for the carrier to prove receipt by the cargo interests of the written rejection and the attached documents. Proof of posting could, at most, give rise to a *prima facie* presumption of receipt by the other party,[292] so it will always be in the parties' interests to send such communications by recorded delivery.

10.110 In *Zerowatt SPA* v. *International Express Company Ltd.*,[293] in the absence of evidence before the court as to the extent to which the intention of the Post Office that first-class letters should arrive the following day was realised, it was held that on the balance of probabilities a letter of claim posted by first-class post in England to defendants in England arrived on the following day. On the other hand, in the case of *Sidney G. Jones Ltd.* v. *Martin Bencher Ltd.*,[294] although the court was prepared to accept that the letter had been posted in the normal course, the evidence of the intended recipient was that if they had received that letter, responsibility would have been denied on the basis that they were not carriers, irrespective of any other defence which may have been available. Since there was nothing in the correspondence indicative of the reaction which would have been expected on receipt of such a letter, the court was satisfied on the evidence that the letter was not in fact received.[295]

10.111 As a matter of grammatical construction, it might just be argued that the word "receipt" applies only to "the claim" and not the remainder of the sentence, so that there is no need to prove receipt of the reply, just that it was sent. Such a view is strengthened by the inelegance of referring to "proof of the receipt . . . of the return of documents . . . ". That any such distinction was intended must, however, be extremely doubtful, and there

---

290. GH.'s Gravenhage, 2.4.76 (1976) 11 E.T.L. 767. Similarly, in Germany, BGH., 21.11.96 (1997) 32 E.T.L. 442, (1997) TranspR 164.

291. App. Toulouse, 26.3.69 (1971) 6 E.T.L. 131. On the other hand, there were two other reasons given why the rejection was not effective (ambiguity and failure to return the documents attached to the claim) and it is submitted the approach of the Dutch court is to be preferred. Two further French decisions are more in line with the approach of the Dutch court, App. Aix-en-Provence, 20.12.79 (1980) B.T. 104, Comm. Paris, 23.5.79 (1979) B.T. 522 and other decisions have accepted the agency of the carrier's insurer in rejecting the claim, App. Aix en Provence, 10.2.88 (1988) B.T. 347 and 14.12.95 (1996) B.T. 521, App. Grenoble, 4.17.96, (1997) B.T. 120, but not the carrier's broker, App. Montpellier, 19.5.94 (1994) Rev. Scapel 163 (see, Lamy, para. 1602, Jur. 2).

292. As to the Interpretation Act 1978, see *supra*, fn. 268.

293. (1991) 26 E.T.L. 530.

294. [1986] 1 Lloyd's Rep. 54.

295. Where a letter is sent to the correct address, a court may accept that it is more by default that the carrier has not received it and so proceed to deal with the matter as if he had, cf. App. Paris, 25.3.82 (1982) B.T. 434.

is French authority to the effect that the letter of reply only takes effect when received,[296] which might indicate that proof of receipt is necessary.[297]

FURTHER CLAIMS

10.112 The final sentence of Article 32(2) makes it clear that once the period of suspension has been terminated by the carrier, it is not open to the claimant to suspend the limitation period again by re-submitting the claim.

NATIONAL LAW

10.113 Subject to the provisions of Article 32(2), the extension of a period of limitation will be governed by the law of the court seised of the case,[298] so any additional grounds permitted by national law will also suspend the period of limitation.[299] So, for example, in this country the normal rule, by virtue of which if a period of limitation expires on a Sunday or other day on which the Supreme Court offices are closed a claim form, as a writ is now called, issued on the next available day will be within time,[300] will apply to Convention cases. So, also, will the rules relating to the validity of claim forms after their issue and the extension of that period.[301] Since application of the limitation period in Article 32 merely bars the remedy and does not extinguish the cause of action,[302] Rules of court giving a discretion to make amendments such as to the name of a party to the proceedings can apply notwithstanding that the amendment may be made after the limitation period has expired.[303]

10.114 It follows from Article 32(3) that section 39 of the Limitation Act, which excludes the application of that Act in cases in which a period of limitation is specified by any other Act, must be read subject to this provision, which effectively incorporates any provisions of the Limitation Act which extend the period of limitation into the provisions of the Convention. However, the limits of this should be noted. Thus in the case of *Higham* v. *Stena Sealink Ltd.*,[304] a case concerning similar wording in the Athens

---

296. Comm. Corbeil-Essones, 18.4.69 (1969) 4 E.T.L. 988, (1969) B.T. 217.

297. See Loewe, Theunis, p. 159 where the French text is considered to be clearer in this respect.

298. Article 32(3).

299. See e.g. in France App. Paris, 28.10.69 (1969) B.T. 7 Cass. 1.10.91 (1991) Bull, IV No. 269, p. 187, Krings, p. 787. In Germany note e.g. s. 202 BGB, see Herber/Piper, Art. 32 Rdn. 49, Koller, Art. 32 Rdn. 8 Jung, pp. 166–167 and OLG. Munich, 31.3.98 (1998) TranspR 353, OLG. Hamm, 23.1.95 (1995) TranspR 290, OLG Hamburg, 7.4.94 (1995) TranspR 115. For a Belgian decision on the effect of service of a writ in a different country see Hof. Gent, 18.3.99 (1999) 34 E.T.L. 703.

300. *Pritam Kaur* v. *S. Russell & Sons Ltd.* [1973] Q.B. 336.

301. CPR Part 7.5–6. As to want of prosecution, see *Reads Ltd.* v. *LEP Transport Ltd.*, Court of Appeal, 18.4.85, unreported except on LEXIS.

302. Haak, p. 293, Clarke, p. 141. CMR does not refer to the liability being *discharged* as in the Hague-Visby Rules (see *Aries Tanker Corporation* v. *Total Transport Inc.* [1977] 1 Lloyd's Rep. 334, 336) or *extinguished* under Article 29 of the Warsaw Convention. Unlike Article 30(3), Article 32 does not have to be raised by the court of its own motion, see Loewe, Theunis, p. 156, OGH., 14.1.76 (1976) II U.L.R. 366, cf. BGH. 14.11.91 (1992) TranspR 135, (1993) 28 E.T.L. 265.

303. E.g., under CPR Part 17.4. Cf. the position under the Hague-Visby Rules, see *The Leni* [1992] 2 Lloyd's Rep. 48, see also *The Jay Bola* [1992] 2 Lloyd's Rep. 62, *The Kefalonia Wind* [1986] 1 Lloyd's Rep. 273. See also *International Distillers & Vinters Ltd* v. *JF Hillebrand (UK) Ltd*, Q.B.D., 17.12.99, unreported except OL Lexis.

304. [1996] 2 Lloyd's Rep. 26.

Convention,[305] it was held that the plaintiff was unable to rely on section 33 of the Limitation Act, which gives the court a discretion to exclude the operation of the time limits contained in the Act. That power operates to override the limitation period, rather than to extend it. Further, on the wording of section 39, it relates only to the limitation periods contained in designated sections of the 1980 Act itself, not limitation periods from other statutes. On the other hand, the provisions of sections 28 and 28A, dealing with cases of disability, sections 29 to 31, which deal with cases of acknowledgement or part payment, and section 32, which covers cases of fraud, concealment or mistake, would all seem to constitute provisions which extend the period of limitation, so it would appear that these would fall within Article 32(3).

10.115 Another question that arises in this respect is the extent to which the parties are free to agree between themselves to extend the time limit. Clearly, this cannot be done prior to any claim arising, since such would amount to a derogation from the Convention and hence be null and void under Article 41.[306] Such extensions of time subsequent to the claim arising are, however, widely agreed and acted upon in both marine disputes[307] and CMR cases, and clearly represent a sensible approach in terms of avoiding needless litigation.

10.116 If sufficiently clear, such agreements are binding at common law[308] and would therefore equally be effective in respect of CMR claims by virtue of Article 32(3). Such an agreement might be expressed or implied, but in the case of the latter there must be clear evidence of the necessary intention.[309] So, for example, in *Zerowatt SPA* v. *International Express Company Ltd.*[310] Webster J., held, albeit obiter, that there was a tacit agreement that time should not run pending preparation of an expert's report.

10.117 There have also been some indications over the years that an estoppel might operate to prevent a defendant from relying on a limitation period where he has by his conduct led the plaintiff to believe that he will not rely on the limitation period and in reliance on this the plaintiff has abstained from commencing proceedings, but again the conduct of the defendant must be sufficiently clear.[311] Thus in the case of *Microfine Minerals & Chemicals Ltd.* v. *Transferry Shipping Co. Ltd.*,[312] the plaintiffs sought to rely on a telephone conversation in the course of which it was alleged that the loss adjuster, acting on behalf of the defendant's insurers, agreed that time was suspended. It was held, obiter, that no estoppel arose in the plaintiff's favour. On a point of law, the court was of the view that an estoppel by convention could not arise as to the construction of the document, but it was also held that in any event it would not be equitable that the plaintiff should be able to rely on an estoppel arising from a telephone conversation in which the point at issue, namely the effect of a letter of repudiation, was not specifically addressed. However, Wetter refers to Swedish and Danish marine cases where an agreement to

---

305. The Convention Relating to the Carriage of Passengers and their Luggage by Sea, 1974.
306. *Infra*, Chapter 12.
307. See now Article III, Rule 6 of the Hague-Visby Rules which expressly permits the extension of the limitation period by agreement after the cause of action has arisen.
308. *Lubovsky* v. *Snelling* [1944] K.B. 44.
309. *The Sauria and The Thent* [1957] 1 Lloyd's Rep. 396.
310. *Supra*, fn. 293.
311. See *Wright* v. *John Bagnall & Sons Ltd.* [1900] 2 Q.B. 241, 244; *Turburville* v. *West Ham Corporation* [1950] 2 K.B. 208; and *Kammins Ballrooms Co. Ltd.* v. *Zenith Investments (Torquay) Ltd.* [1971] A.C. 850. Generally, see Goodman, (1969) 119 N.L.J. 872. See further James, *Limitation of Actions*, Chapter 3.
312. [1991] 2 Lloyd's Rep. 630.

extend the limitation period has been implied from the conduct of the parties in negotiating before and after the expiry of the time limit, and expresses the view that the courts there would adopt the same approach to CMR claims.[313] As he points out, there is a French case on the Convention where the court did adopt just such an approach.[314]

10.118 It is further provided by Article 32(3) that national law shall govern the fresh accrual of rights of action. Under English law, the only way in which a claim can revive and be enforced after the expiry of the relevant limitation period is by a fresh contract to pay, which must be supported by its own consideration.[315] A bare promise to pay will not be enforceable.[316]

COUNTERCLAIMS AND RIGHTS OF SET-OFF

10.119 The final paragraph of Article 32 states that once the limitation period has expired, the claim cannot be enforced by the device of including it in a counter-claim[317] or set-off.[318] This provision is likely to be of particular relevance to carriers who delay pressing for payment of their freight charges pending the outcome of any claims against them. Reference in this respect should be made to the discussion above as to whether in such circumstances carriers can suspend the running of time in relation to their claims for freight charges.

Arbitration

10.120 Article 33 of the Convention provides that:

"The contract of carriage may contain a clause conferring competence on an arbitration tribunal if the clause conferring competence on the tribunal[319] provides that the tribunal shall apply this Convention."

10.121 The parties are therefore permitted to provide for their disputes to be settled by arbitration if they so wish.

10.122 The reference to the contract of carriage in Article 33 does not have any precise significance in terms of formalities, since, as with any other contract, the contract of carriage can consist of a number of documents, and terms agreed orally.[320] However, quite apart from the obvious evidential advantage, it would seem that the agreement as to arbitration should be in writing, since the phrase "clause conferring competence" would seem inappropriate to describe an oral agreement.[321]

313. Op. cit., p. 508.
314. Cass., 3.2.69 (1969) 4 E.T.L. 787.
315. E.g. an account stated: *Ashby* v. *James* (1843) 11 M. & W. 542. For the general principles of consideration, see *Halsbury's Laws*, 4th edn., Vol. 9(1), paras 727–747.
316. The anomalous provision whereby an acknowledgement or part payment revived the original obligation has now been removed (Limitation Act 1980, section 29(7)) but see *supra*, para. 7.29.
317. Cass., 8.1.85 (1985) 20 E.T.L. 444.
318. Article 32(4). As to when a counter-claim is exercised, see *Impex Transport Aktieselskabet* v. *A. G. Thomas Holdings Ltd.* [1981] 2 Lloyd's Rep. 556 where it was held that a counterclaim was not exercised by an affidavit in reply to an application for summary judgment under R.S.C. Order 14. The counter-claim itself was served outside the CMR time limit and so was time barred under Article 32(4).
319. This repetition seems both inelegant and unnecessary, and is avoided by the French text: "*cette clause*".
320. On formalities generally, see *supra*, Chapter 4.
321. Also, an arbitration agreement must be in writing to be within Part 1 of the Arbitration Act 1996 (see section 5) and also to be within sections 100–104 of the Arbitration Act 1996. See *infra*, para. 10.127.

10.123 The second half of Article 33 does, however, impose an important qualification on the freedom of the parties to provide for arbitration, in that the arbitration agreement must specify that the arbitration tribunal shall apply the provisions of the Convention. If this is not so specified, the clause will be of no effect.[322] Thus in an English decision,[323] it was held that a clause in a CMR contract conferring exclusive jurisdiction on Swedish arbitrators was invalid as amounting to a derogation from the Convention contrary to Article 41, since it did not expressly provide that the arbitrators were to apply the Convention. This was held to be so despite the fact that, as the Convention is a part of Swedish national law, the arbitrators would have applied the Convention in any event. Bingham, J., acknowledged that he was required to adopt a purposive approach to the question, but found himself confronted with two conflicting arguments: (a) that the purpose of the provision is to ensure that the Convention is applied; (b) that the purpose of the provision was to ensure that the application of the Convention should be clearly stated so that, in particular, intermediate carriers would be aware of the position. Having noted these conflicting arguments, he continued:

"I find this a far from easy question . . . [b]ut I am in the end impressed by the fact that the clause does require the arbitration clause to provide that the tribunal shall apply this Convention and does not adopt a course which could have been adopted, namely, that an arbitration clause should only have effect where a tribunal would apply the Convention."[324]

10.124 He therefore concluded that express provision for the application of the Convention was required.[325] However, in the more recent case of *Inco Europe Ltd.* v. *First Choice Distribution*,[326] it was held by the Court of Appeal that a requirement in the arbitration clause in question that the arbitrators should "observe the applicable imperative legal stipulations, including the provisions of international transport treaties" was distinguishable from the *AB Bofors* case, and sufficient to comply with Article 33 of the Convention.[327]

10.125 Once the claim has arisen, on the basis of the normal contractual principle that any compromise of an action is binding on the parties,[328] the parties may dispose of it in any way they agree. There would seem no reason therefore why an agreement as to arbitration made after the dispute has arisen should be invalid if it failed to specify that the Convention is to be applied,[329] although if the arbitration tribunal subsequently improperly failed to apply the Convention rules, it would be open to the parties or any one of them to apply for judicial review of the decision.[330] As to the conduct of the arbitration,

322. Arrond. Rotterdam, 10.11.70 (1971) 6 E.T.L. 273.

323. *A.B. Bofors-UVA* v. *A.B. Skandia Transport* [1982] 1 Lloyd's Rep. 410. See also Arrond. Rotterdam, 31.1.67 (1967) S. & S. No. 56.

324. [1982] 1 Lloyd's Rep. 410, 413.

325. He also expressed the view that if the arbitration clause had been valid, although Article 31(1)(b) (*supra*) confers an independent right to proceed in this country, those proceedings would be stayed under the Arbitration Act 1996, considered *infra*, para. 10.127 ff. On this see Glass, [1984] LMCLQ 30, 38.

326. [1999] 1 All E.R. 820.

327. Cf. OLG. Hamm, 29.6.98 (1999) TranspR 201 where the same provision of the Dutch Forwarding Conditions (FENEX) was held insufficient.

328. See, e.g., *Alliance Bank Ltd.* v. *Broome* (1864) 2 Dr. & Sm. 289.

329. Loewe, para. 269, (see also Loewe, Theunis, p. 164). *Quaere*, however, whether it would then be "arbitration tribunal acting by virtue of article 33" so as to be within section 7 of the Carriage of Goods by Road Act 1965, as to which see *infra*, para. 10.128.

330. The question of judicial review of arbitrators' decisions in this country is now regulated by the Arbitration Act 1996, sections 66–71.

this will be regulated by national law. Any general consideration of the English law relating to arbitration procedures would be beyond the scope of this work,[331] but reference should be made to the point at which the arbitration is deemed to commence for limitation purposes. The Carriage of Goods by Road Act 1965 provides[332] that for the purposes of Article 32 in relation to arbitration proceedings under Article 33, section 14(3) to (5) of the Arbitration Act 1996 applies. Those sub-sections provide that the way in which the arbitration is commenced depends on the force of the arbitration agreement. If the arbitrator is specific in the agreement, then the arbitration is commenced when one party serves the offer with written notice[333] requiring them to submit the latter to the specified person.[334] Secondly, where the tribunal is to be appointed by the parties, the arbitration is commenced when one party serves written notice on the offer to appoint an arbitrator or agree to the appointment of an arbitrator.[335] Finally, where the tribunal is to be appointed by a third party, the arbitration is commenced when one party gives written notice to that third party requesting the appointment.[336]

10.126 With further reference to limitation periods in the context of arbitrations, if a court orders the arbitrators' award to be set aside or declares it to be of no effect, the court can order that the period between the date of the commencement of the arbitration and the date of the court order be disregarded for limitation purposes.[337]

10.127 At first sight, parties to an arbitration agreement would appear to be at a disadvantage under the Convention, since the provision preventing successive actions in different jurisdictions,[338] and the provision for the enforcement of judgments in other jurisdictions[339] both only apply to actions before the national courts. However, the Arbitration Act 1996 gives effect in this country to the 1958 New York Convention on the Recognition and Enforcement of Foreign Arbitral Awards,[340] to which most States party to the CMR Convention are also parties.[341] The Arbitration Act 1996, in line with the provisions of the New York Convention, applies to any arbitration agreement[342] which is not a domestic arbitration agreement.[343] If the arbitration agreement is not a domestic agreement, then if any proceedings are commenced in the courts of this country, any party to the agreement may[344] apply to the court for a stay of the proceedings, which must be granted except in certain limited circumstances.[345] In effect, therefore, in such circumstances the parties to an arbitration agreement are in a similar position with respect to

331. See generally *Russell on Arbitration*, 21st edn.
332. Section 7(2)(a), as amended by Arbitration Act 1996 section 107 and schedule 3.
333. For the manner of service of notices under the Act, see section 76.
334. Arbitration Act 1996, section 14(3).
335. Arbitration Act 1996, section 14(4).
336. Arbitration Act 1996, section 14(5).
337. Arbitration Act 1996, section 13(2).
338. Article 31(2), *supra*, paras 10.37–10.42.
339. Article 31(2), *supra*, para. 10.43 ff.
340. See generally *Russell on Arbitration*, 21st edn., p. 401 ff.
341. The exceptions are as follows: Belarus, France (withdrew 1989), Iran, Tajikstan and Turkmenistan.
342. It must, however, be in writing: section100(1)(a).
343. Defined in section 100(1) of the Act. The Act also applies to any person claiming through or under such a party, so it may include an agreement which at first sight appears to be a domestic arbitration agreement.
344. It is thus not mandatory, and it is therefore open to that party to submit to the proceedings. Indeed, if a party takes any steps in those proceedings other than entry of an appearance, the right to have the proceedings stayed will be lost: section 9(3).
345. Set out in section 9(4), namely where the court is satisfied "that the arbitration agreement is null and void, inoperative or incapable of being performed.

actions in other countries as they would be under Article 31(2) of the CMR Convention.

10.128 Turning to the enforcement of awards in other jurisdictions, in this country the effect of section 7(1) of the Carriage of Goods by Road Act 1965 would seem to be to bring an arbitration award under the CMR Convention within the scope of the Foreign Judgments (Reciprocal Enforcement) Act 1933. As has been seen,[346] section 4 of the 1965 Act applies the 1933 Act to any judgment within Article 31(1) which "has been . . . given by any court or tribunal" of a country which is party to the Convention. Then by section 7(1) of the 1965 Act: "Any reference in the preceding provisions of this Act to a court includes a reference to an arbitration tribunal acting by virtue of Article 33 . . . ". It would seem to follow that the 1933 Act applies to arbitrations under Article 33, as long as the matter arbitrated would have been within Article 31(1).

10.129 However, even if the foregoing be incorrect,[347] and in any event in relation to countries other than the United Kingdom, provision for the reciprocal enforcement of arbitration awards is contained in the New York Convention of 1958, to which, as already explained, most CMR contracting States are also party.[348] The provisions of the Convention as to reciprocal enforcement are given statutory effect in this country by the Arbitration Act 1996. Under that Act,[349] a New York Convention award[350] is enforceable, by leave of the court, in the same way as a judgment or order of the court.[351] Enforcement can only be refused for a reason specified in section 5 of the Act.

10.130 Finally, it has been noted above that in this country the courts have power to take account of proceedings which have been or are likely to be commenced elsewhere for the purpose of enforcing a liability which is limited by Article 23.[352] By virtue of section 7(1) of the Carriage of Goods by Road Act 1965, arbitrators are given the same power in relation to arbitration agreements within Article 33. By virtue of the same section, the provisions of the Act as to actions against High Contracting Parties[353] apply equally to arbitrations within Article 33.

---

346. *Supra*, para. 10.44.
347. In the first edition of this book a doubt was expressed as to the intentions of the legislature in view of the fact that the 1933 Act did not at that time apply to arbitrations. This is now resolved by section 10A of the 1933 Act, added by the Civil Jurisdiction and Judgments Act 1982, section 35(1), Schedule 10, para. 4, see *Halsbury's Statutes*, 4th edn., Vol. 22, p. 457.
348. See fn. 343, *supra*.
349. Section 101(2).
350. Defined in section 100(1) as an award made in pursuance of an arbitration agreement in a state, other than the United Kingdom, which is a party to the New York Convention.
351. As to which court, see section 105.
352. Carriage of Goods by Road Act 1965, section 3, *supra*, para. 10.42.
353. Carriage of Goods by Road Act 1965, section 6.

CHAPTER 11

Successive Carriage

Scope

11.1 The question of onward carriage arises where one carrier takes over goods from another carrier for the purpose of carrying a consignment to its destination, or else to a point in transit where he in turn hands over the goods to another carrier. The carriers involved may be carriers by different modes, in which case it will be a combined transport operation, which has been discussed in Chapter 2, but in the case of onward carriage by road the movement will either be subject to CMR or the relevant domestic law. There are several methods by which such an arrangement can be made.

11.2 First, a shipper may contract with a forwarder to arrange carriage of goods by road, who in turn will sub-contract the whole transit to a road carrier. In such circumstances the contract between the shipper and the forwarder will be subject to domestic law.[1] The contract between the forwarder and the carrier will be a contract of carriage in which the shipper will not be involved unless the forwarder is treated as the shipper's agent, so creating privity of contract between the shipper and the carrier, which he will not do in most cases.[2] The contract of carriage will then be subject either to domestic law or CMR depending upon whether it falls within the terms of Article 1.[3]

11.3 Secondly, a shipper may contract with a carrier to carry goods by road, who may in turn sub-contract all or part of the carriage. If the contract between the shipper and the carrier is for the international carriage of goods by road within the meaning of Article 1, then the contract will be governed by the Convention, at least as far as the first carrier is concerned. Whether the sub-contracting carrier employed by the first carrier is a party to the CMR contract will depend upon the relationship entered into between the two carriers. If the second carrier becomes a successive carrier for the purposes of the Convention, the relationships between the second carrier, the first carrier and shipper will all be subject to CMR. If, on the other hand, the second carrier is not a successive carrier for the purposes of CMR, the relationship between the second carrier and the first carrier, and between the second carrier and the shipper, will be subject to the relevant domestic law.[4] In such

---

1. But as to the distinction between a forwarder and a carrier see *supra*, Chapter 1, paras 1.13–1.54.
2. See Hill, *Freight Forwarders*, para. 66 ff.
3. See Chapter 1.
4. Unless the parties have voluntarily incorporated CMR into the terms of the contract of carriage. See *supra*, Chapter 1, paras 1.60–1.63. If, however, the contract between the first and second carrier is for international carriage by road, this contract will fall within Article 1 of CMR and so will be subject to it. See *infra*, para. 11.38.

circumstances, the Convention will only apply in so far as it offers the sub-contractor certain defences when sued by the cargo interests under domestic law.[5]

11.4 Where a contract is subject to CMR, Chapter VI of the Convention makes special provision for carriage which is to be performed by successive carriers. First, in Article 34 it lays down how a carrier may be constituted as a successive carrier under the Convention:

> "If carriage governed by a single contract is performed by successive road carriers, each of them shall be responsible for the performance of the whole operation, the second carrier and each succeeding carrier becoming a party to the contract of carriage, under the terms of the consignment note, by reason of his acceptance of the goods and the consignment note."

A single contract

11.5 The first point to note is that the carriage in question must be governed by a single contract before a carrier can become a successive carrier under the Convention.[6] It is not sufficient for an onward carrier to take over goods from another in a transit which at some stage spans national frontiers. It is essential that one contract is entered into for the whole transit.[7] Nevertheless, if one of the carriers in question himself carries goods across a national frontier this may in itself make his part of the operation subject to CMR, independently of whether other segments of the through movement are subject to the Convention or not.[8]

11.6 Where international carriage is performed by several carriers, one part of the transit, normally the last segment, may be performed in one single country only. The question therefore arises as to whether this part of the carriage will be subject to CMR or to the domestic laws of that country. The position appears to be that, provided the complete international transit is made subject to a single CMR contract of carriage, the last segment, although solely performed in the country of destination, will be subject to the Convention, and the last carrier will be a successive carrier under Article 34.[9] Where, however, carriage has been partly performed subject to a CMR consignment note covering the international part of the transit, and partly subject to a new contract of carriage concluded and performed in one country, that of destination, the Convention will not be applicable to the latter part of the transit, which will be subject to domestic law. A French decision on this point to the contrary must, it is thought, be incorrect. In that case, goods carried from Paris to Bretten in Germany under a CMR consignment note were then carried from Bretten to Munich under a separate consignment note which made no mention of CMR. The second carrier was held to be a successive carrier for the purposes of the Convention,[10] but, given the requirement in Article 34 that there be a single contract

5. See Article 28, discussed *supra*, Chapter 9, paras 9.65–9.71. This is also subject to the possibility that the contract between the first and the second carrier is subject to CMR as a contract for international carriage by road in respect of which the cargo interests may have contractual rights, e.g., as consignee. See *infra*, para. 11.38.
6. If the carriage is not governed by a single contract CMR will not apply: Arrond.'s Hertogenbosch, 11.12.64 (1967) S. & S. No. 5. See also Dorrestein, De Beursbengel, (1961) 323.
7. Where a contract is split at a frontier, at least one part of it will be international so that the CMR should apply to that part. However, there is Dutch authority to the contrary, HR., 16.3.79 (1980) N.J. 562, cited by Haak, p. 108, fn. 92. Cf. Cass. 25.3.97 (1997) B.T. 276, (1997) U.L.R. 852 (Krings, p. 146).
8. See *supra*, Chapter 1, para. 1.2. See, e.g., App. Reims, 2.3.94 (1994) B.T. 303.
9. RB. Antwerpen, 16.4.75 (1975) 10 E.T.L. 548. See also Hof.'s Hertogenbosch, 21.12.65 (1966) 1 E.T.L. 698. Hof. Antwerpen, 14.12.83 (1983) 18 E.T.L. 809.
10. Cass., 15.5.74 (1974) B.T. 438. The decision is criticised by the commentator on the case. Cf. the Belgian cases, e.g., Antwerp, 9.6.82 J.P.A., cited in Lamy, para. 1607, Jur. 4.

of carriage, it is thought that the decision of the lower court to the effect that the second contract was subject to domestic law is to be preferred.

11.7 It is equally possible for the first part of an international transit which is to be performed solely in the country of dispatch to be entered into as a separate contract independently of the remainder of the transit. In such circumstances the domestic law may apply to the first contract, and CMR to the remaining international part of the transit.[11] Conversely, if a single contract of international carriage is to be performed by more than one carrier, the first carrier will be subject to the Convention even though his part of the carriage may be restricted to the domestic section.[12] In all circumstances, whether CMR will apply in any particular case to the complete transit will depend upon the manner in which the contract in question has been formulated, although the problem is only likely to arise where a CMR consignment note has not been issued covering the whole transit.[13]

Groupage shipments

11.8 The question of successive carriage in relation to groupage shipments can create difficulties in practice although there has been little case law to date on the subject.[14] In a French appellate decision a shipment of two lasers, plus components, was entrusted to K. for carriage from France to Germany. He in turn employed W., who in turn employed Wetsch, a German groupage company. The latter, having no vehicle available to perform the carriage, sub-contracted the carriage to Munich of the complete groupage shipment (of which the lasers were part) to M. Wetsch took delivery of the shipment from M. in Munich, cleared the goods through Customs, degrouped the shipment and delivered the lasers to the consignee. In a dispute relating to the condition of the goods, it was held that Wetsch was a successive carrier under Article 34 in effecting delivery after deconsolidation. The court rejected the argument that M. was the sole carrier for the purposes of CMR.[15]

11.9 Each case, however, will usually turn upon its particular facts. For example, in a Belgian decision involving a groupage operator, it was decided that no contractual relationship existed between the individual shippers and the carrier employed by the groupage operator, as the carrier had neither known of nor accepted the consignment notes issued by the latter. The two contracts were quite distinct. It was not a question of the

11. This may arise in particular in relation to groupage movements. Cf. Comm. Versailles, 15.1.86 (1986) B.T. 742.

12. RB. Antwerpen, 9.12.77 (1978) 13 E.T.L. 110.

13. Cf. Hof. Gent, 2.11.95 (1995–1996) R.W. 781 (Putzeys, "Le droit des transports en Belgique (II) 1993–1995, (1996) U.L.R. 557, 568).

14. Cf. Comm. Versailles, *supra*, fn. 11, and BGH., 27.1.82 (1985) 20 E.T.L. 349.

15. Cass., 19.1.76 (1976) B.T. 149. Cf. Cass., 12.5.87 (1987) B.T. 399 where a consignment of 15 packages was received in Hamburg by a groupage operator for delivery in France. The operator drew up a CMR consignment note for 172 packages. The entire groupage load was carried by chartered carrier indicated in the note and was received by a correspondent carrier of the groupage operator who noted on the consignment note that the 15 packages destined for the consignee were missing. These were later found and delivered by the correspondent carrier, who indicated on the delivery note that the goods had come from Hamburg. These facts were held insufficient to establish the liability of this carrier for the deterioration suffered by the goods in the course of the transit, since they did not amount to sufficient proof of a single contract from Hamburg to the place of delivery in France. Cf. further Cass., 6.7.93 (1993) B.T. 621.

operator being the principal carrier and the actual carrier a substitute, nor was the latter a successive carrier for the purposes of Article 34.[16]

11.10 It has been suggested that in practice the provisions of Chapter VI are not important on the grounds that where a carrier has assumed responsibility for the whole carriage and a sub-contractor is employed to perform part of the transit, the normal practice will be for the first carrier to issue a separate consignment note for the onward carriage in which he will refer to himself as sender or consignee.[17] Whether this is in fact true in many parts of Europe is open to doubt.[18] Certainly, both in the United Kingdom and in Northern Europe it is increasingly common for a CMR consignment note covering the whole carriage to be issued by the first carrier or else by a forwarder organising the carriage. In practice, therefore, an increasing amount of litigation is coming before the courts, including those of the United Kingdom, on the question of successive carriage.

The Ulster-Swift case

11.11 Important issues as to the status of carriers for the purposes of Chapter VI of the Convention arose for consideration by the Court of Appeal in *Ulster-Swift Ltd.* v. *Taunton Meat Haulage Ltd.*[19] This case has been referred to earlier both in the context of deciding who is a carrier within the meaning of the Convention,[20] and in connection with the liability of a Convention carrier under Articles 17 and 18.[21] It will be recalled that Ulster-Swift contracted with Taunton to carry a consignment of pork carcases from Northern Ireland to Switzerland. It was not disputed that the contract was subject to CMR. Taunton then sub-contracted the complete carriage to a Dutch carrier, Fransen, and again it was not disputed that this contract, too, was subject to the Convention. On arrival of the consignment it was found to have deteriorated, and was condemned and destroyed by the Swiss authorities. Having decided the initial issue of liability in favour of the plaintiffs, the court then had to consider the position as between Taunton, the defendants, and Fransen, the third party. Fransen claimed that Taunton's claim against them was time-barred under Article 32,[22] but Taunton sought to rely on Article 39(4), which provides for an extended period of limitation in relation to claims "between carriers".[23]

11.12 Fransen's argument was twofold: first, they argued that Taunton were not carriers at all, not having performed any part of the carriage, and so this was not a claim "between carriers" for the purposes of Article 39(4); alternatively, they argued that even if Taunton were carriers, they (Fransen) were not successive carriers, and so again Article 39(4) did not apply.

11.13 As to the first argument that Taunton were not carriers, this has already been considered above,[24] and suffice it to say for present purposes that it was held that Taunton were carriers within the meaning of the Convention since they had contracted to carry the goods, even though they had sub-contracted the entire performance of the carriage.

16. Comm. Antwerp, 25.11.74 (1975/6) J.P.A. 70.
17. Loewe, para. 274.
18. This is said to be common in Benelux countries, see Clarke, p. 193 citing Putzeys at n. 47.
19. [1977] 1 Lloyd's Rep. 346.
20. *Supra*, para. 1.16 ff.
21. *Supra*, Chapter 6, paras 6.114–6.116.
22. *Supra*, Chapter 10, paras 10.51–10.80.
23. Discussed *infra*, para. 11.127 ff.
24. See paras 1.16–1.17.

11.14 Turning to Fransen's alternative argument that they were not successive carriers so that Article 39(4) did not apply, this was based on the heading to Chapter VI of the Convention, which reads "Provisions relating to Carriage performed by Successive Carriers". This, Fransen argued, meant that "claims between carriers" in Article 39(4) was to be read as meaning "claims between successive carriers within the meaning of Article 34", and that even if Taunton were carriers, they (Fransen) were not successive carriers within the meaning of Article 34. The basis of this argument was that since Taunton had sub-contracted the whole of the contract of carriage they had ceased to be carriers, even if they initially were carriers, and that therefore Fransen were not successive carriers, not having carried in succession to anyone.

11.15 The court first rejected the argument that the heading to Chapter VI in any way restricted the meaning of the plain words of Article 39(4). Further, it concluded that Taunton did not need to rely on Article 34 to be carriers, since section 14(2)(c) of the Carriage of Goods by Road Act 1965 applies the Act and the Convention scheduled to it to " . . . any carrier who, in accordance with Article 34 *or otherwise*[25] is a party to the contract of carriage", and Taunton clearly was a direct party to contracts of carriage with both Ulster-Swift and Fransen. So Fransen were admittedly carriers; Taunton were carriers by virtue of the direct contractual arrangements between the parties, and so it followed that this was a "claim between carriers" for the purposes of Article 39(4).[26]

11.16 The court nevertheless then went on to consider the position if Article 39(4) did only apply to claims between successive carriers, and in this respect adopted the reasoning of Donaldson, J., at first instance. Donaldson, J., relied on a combination of Articles 1(1) and 34. Article 1(1), as has been seen, applies the Convention to " . . . every contract for the carriage of goods . . . ", and as Taunton had contracted to carry the goods, they were carriers within Article 1. Then turning to Article 34, he described its effect as follows–

"Thus the C.M.R. Convention then sets out to create an artificial statutory contract between the actual carrier and the owner of the goods."[27]

11.17 Then referring to what he termed "the scheme of the Convention", he continued:

"This scheme reflects the common, indeed the almost universal situation where the owners of goods get in touch with a carrier and make the contract for the whole of the carriage with those carriers. It is then left to that carrier to sub-contract the successive stages of the carriage, and all those people are without doubt, and nobody would dispute it, successive carriers. The only oddity

25. Italics added by Megaw, L.J., in giving the judgment of the Court of Appeal: [1978] 1 Lloyd's Rep. 346, 360.

26. Dr Hill seems to have regarded the court as having treated section 14(2) of the 1965 Act as extending the scope of potential defendants beyond that contained in the Convention itself. ("The interpretation of CMR in the English Courts", [1977] LMCLQ 212, 216.) With respect, it is not thought that this was the court's intention. Section 14(2) was used, as it surely was intended, solely to guide the court in its interpretation of the Convention. The court considered that "a carrier" under the Convention, including Article 34, was any carrier who was a party to the contract of carriage, and was confirmed in that view by section 14(2). What the court did not do was to use section 14(2) to establish who was a party to the contract. For this reason it would not be correct, as Dr Hill has suggested (ibid.), to view section 14(2)(d) as meaning that persons for whom the carrier is responsible under Article 3 can be sued direct as parties to the contract. A similar misapprehension of the effect of section 14(2) is made by Evans, J., in *M. Bardiger Ltd.* v. *Halberg Spedition APS*, Queen's Bench Division, 26.10.90, unreported except on LEXIS, which contrasts with the view of Mance, J., in *Aqualon (U.K.) Ltd.* v. *Vallana Shipping Corporation* [1994] 1 Lloyd's Rep. 669, see *supra*, para. 1.11.

27. Quoted with approval by Megaw, L.J., in the Court of Appeal [1977] 1 Lloyd's Rep. 346, 360–361.

here is that the primary carrier, the man who contracted with the owners of the goods, did not in fact undertake any stage of the carriage himself.

Looking at art. 1, par. 1, I think that the CMR Convention must have contemplated that for this purpose the company, or individual, with whom the owner of the goods contracts is the first carrier, whether or not he himself takes possession of the goods, and that all subsequent carriers are the successive carriers within the meaning of these provisions."[28]

Fransen were therefore successive carriers, and Article 39(4) applied.

11.18 A number of principles therefore emerge. First, it is clear that Article 34 will not apply to the situation where a sender of goods has himself contracted with more than one carrier for successive carriage, since it will not be carriage governed by a single contract of carriage.

11.19 Secondly, Article 34 is not restricted to the situation where a carrier performs part of the carriage himself and then sub-contracts part to a third party. It matters not that the initial carrier has not in fact carried the goods at all.[29] Further, there is English authority to the effect that the same is true of a subsequent carrier: he can become liable as a

---

28. At p. 361.

29. This view is supported by a German decision where it was held that where a carrier in his own name and for his own account employs another carrier as sub-contractor to perform either part or the whole of the carriage, then the latter will be a successive carrier under Article 34: LG. Duisburg, 10.5.68 (1969) 4 E.T.L. 979. (See also OLG. Düsseldorf, 8.5.69 (1970) 5 E.T.L. 446.) Libouton disapproves of this decision, but gives no grounds for doing so, (1973) 8 E.T.L. 2, para. 93. More recently, the German Supreme Court has left open the issue whether the carrier who sub-contracts can be a successive carrier: 25.10.84 (1985) 20 E.T.L. 268, 273 (see further, however, *infra*, fn. 31, and note support for the wider view in Thume, p. 791, and Trappe and Gierke, " 'Aufeinander folgende' Straßenfrachtführer?", (1996) TranspR 260, 262. Loewe states that where a party enters into a contract for carriage but does not perform any part of it himself the provisions of Article 34 are not applicable—para. 276. None of his reasoning in this paragraph appears to be valid under English law in the light of the *Ulster-Swift* decision. His view, it seems, emphasises the words used in Articles 36 and 37 which refer to the carrier *performing* and carriers who have *taken part* in the carriage (see Haak, p. 111 and the Dutch cases cited at p. 112 n. 113, see also Cleton "Digest of Netherlands Case Law," (1992) II U.L.R. 186, 194–195 which indicates contrasting cases reflecting both a wide and narrow approach). The Belgian Supreme Court has applied the successive carrier provisions in the context of contracting carriers, 30.5.80 (1981) I U.L.R. 263, (1983) 18 E.T.L. 79 (explained away by Libouton, (1982) J.T. 737 para. 143, on the basis that the point was not directly raised by the plea on cassation). A different assumption might underlie decisions of the same court, 17.9.87 (1988) 23 E.T.L. 210, 25.5.84 (1984) Pas. I No. 543, 30.6.95, (1995) Pas. I No. 344 (Krings p. 148 and 791, Putzeys, "Le droit des transports en Belgique (II) 1993–1995," (1996) U.L.R. 557, 568) (1996) E.T.L. 545 (see note by Geeroms, pp. 548–557), lower court decisions have not always done so (e.g., Hof. Antwerpen, 8.10.86 (1987) 22 E.T.L. 436). The French courts have taken contrasting approaches. Cf., e.g., Cass., 15.1.72 (1972) B.T. 148 (the wider view) with Cass., 15.5.72 (1972) B.T. 438 and App. Agen, 29.6.81 (1981) B.T. 433 (the narrow view). The position in France is obscured by the fact that a decision may be based not necessarily on the narrow view, but on the classification of a contractor as a *commissionnaire de transport* (see *supra*, paras 1.50–1.51, e.g. Trib. gde. inst. Valence, 18.11.81 (1982) B.T. 211. See further Marchand, "*La pluralité de transporteurs routiers selon la CMR,*" (1995) 30 E.T.L. 577, 581–582. The Cour d'Appel de Paris, 17.11.83 (1984) B.T. 390 has implicitly rejected the narrow view, but applied Article 39(4) to a claim by a contracting carrier against the performing carrier on the basis (similar to that stated in the *Ulster-Swift* case) that this provision applies to claims between carriers whether or not they are successive carriers (see the criticism made in the commentary on the case); cf. further App. Paris, 10.3.92 (1992) B.T. 278, Cass., 3.3.98 (1998) B.T. 231. In Denmark, the Admiralty and Commercial Court of Copenhagen (17.7.86 (1987) II U.L.R. 730) also took the wider view, pointing out (at p. 733) that: "The performance of the carriage involved more than the road transport itself. In the opinion of the court the wording of Chapter VI of the law cannot be interpreted so as to restrict the application of the provisions of this chapter only to those cases where more than one carrier has taken part in the actual performance of the carriage, thus rendering those provisions inapplicable to a contracting carrier who has been involved in other important aspects of the performance of the carriage". See also Supreme Court of Sweden, 26.3.96 (1997) U.L.R. 202, Tribunal Fédéral Suisse, 22.11.83 (1995) 30 E.T.L. 675, and Marchand, *supra*, at p. 581. Many decisions, however, are based on more subtle considerations than a stark contrast between these opposing views, and the issue should be seen rather in the context of a more considered analysis of how Article 34 is to be satisfied, as to which see *infra*, para. 11.42 ff. Note, however, that Libouton repeats his disapproval in (1982) J.T. 737 para. 143, and Putzeys also considers this approach to be in error: p. 106 para. 289, n. 201.

successive carrier notwithstanding the fact that he has sub-contracted the entire portion of the carriage which he has undertaken to perform.[30]

11.20 Thirdly, although Article 34 states that a subsequent carrier becomes a party to the contract of carriage "by reason of his acceptance of the goods", it is clearly the corollary of the above that the acceptance need not be personal. It will be sufficient if possession of the goods is taken by an agent or sub-contractor on his behalf.[31] It is immaterial for this purpose whether the person actually taking delivery is himself in the physical sense a carrier or successive carrier within the meaning of the Convention.[32] So a carrier could accept goods for the purposes of Article 34 by instructing a warehouseman to take delivery on his behalf for onwards carriage by someone else, or by sub-contracting the carriage to another carrier who is not himself a successive carrier within the Convention.[33]

The need for a consignment note

11.21 Article 34 provides that the second carrier and each succeeding carrier becomes a party to the contract of carriage "under the terms of the consignment note, by reason of his acceptance of the goods and the consignment note". If the carrier does in fact take over the CMR consignment note with the goods, few problems are likely to arise. However, although the approach to documentation in the road haulage industry has improved compared with that of the shipping industry, it remains rather casual, and in practice such a consignment note may well not accompany the goods in many cases.[34]

---

30. " . . . it is consistent with [the *Ulster-Swift*] decision that a carrier who sub-contracts the whole of the part of the carriage which he has undertaken to perform may nevertheless be a successive carrier within Article 34, performing his part of the carriage through his sub-contractor", *per* Robert Goff, J., in *SGS-Ates Componenti Elettronici S.p.A.* v. *Grappo Ltd.* [1978] 1 Lloyd's Rep. 281, 284 (discussed *infra*, paras 11.26–11.32). Note also the decision of App. Paris, 10.3.92 (1992) B.T. 278, which applied Article 36, *infra*, to negate the liability of an intermediate contracting carrier. Cf. OGH., 4.6.87 (1988) 23 E.T.L. 714, *infra*, fn. 31.

31. In an Austrian decision (OGH., 4.6.87 (1988) 23 E.T.L. 714) and a German decision (BGH., 10.5.90 (1991) 26 E.T.L. 351) sub-contracted carriers who themselves sub-contracted were held not to be successive carriers because they had not accepted the goods and the consignment note. As the Austrian court explained, merely passing on an order for carriage to another carrier does not mean that they have successively adhered to the original freight contract with the sender. The consequence was that in the first case, the carrier could not rely on Article 36 to escape liability, and in the second, the carrier could not rely on the time limit in Article 39 to enforce a claim against the contracting carrier. In this country, the case of *ITT Schaub-Lorenz Vertriebgesellschaft mbH* v. *Birkart Johann Internationale Spedition GmbH* [1988] 1 Lloyd's Rep. 487 (C.A.), discussed *infra*, para. 11.38 fn. 55, involved a similar situation, but the point was not taken (the position of the first sub-contracted carrier may not have depended, however, on being a successive carrier *stricto sensu*, see *infra*). The point was also not taken by the Belgian Supreme Court in its decision of 30.5.80 (1981) I U.L.R. 263. Cf., however, Hof. Antwerpen, 8.10.86 (1987) 22 E.T.L. 436. No difficulty in the application of successive rules seems to have been seen by the Tribunal Fédéral Suisse in its decision of 22.11.83 (1995) 30 E.T.L. 675.

32. It has been questioned by Clarke (p. 198 fn. 87) whether acceptance (in contrast to *taking over*) might not mean rather the assumption of responsibility rather than "taking over", the latter approach being taken in the *SGS-Ates* case (see *infra*, paras. 11.26–11.32). In so far as this might leave room for Rodière's view on the last carrier to operate (see *infra*, para. 11.64, fn. 130), it may be objected to as contrary to the intended scheme. See, however, the discussion *infra*, at para. 11.45.

33. E.g., because there is no consignment note, as to which see *infra*.

34. As Neervoort ((1985) *Cargo Claims Analysis* 51) points out, CMR's successive carriage provisions are based directly on those of the CIM Convention, and the references to the single contract of carriage and the handing over of the consignment note are clearly rooted in the practice of rail carriage. Indeed, as he notes, international rail transport by definition involves successive carriage, and the CIM Convention makes the use of a consignment note in such cases obligatory. Clearly, the danger is that procedures designed for one mode of transport are not necessarily suitable for another.

11.22 On a strict interpretation of this wording it would appear that if no consignment note accompanies the goods the carrier will not be constituted as a successive carrier for the purposes of the Convention. This situation could cause considerable problems in respect of cross-Channel traffic, particularly in respect of export/import shipments for which a tractor is employed to move an articulated trailer to and from an English port, where the chance of the actual carrier (the tractor operator) receiving any consignment note may be even less.

11.23 Does this therefore mean that few carriers ever in fact become successive carriers under CMR? To deal with this problem it has been argued that the provisions in Article 4 that the contract will merely be "confirmed" by the making of the consignment note, and that the "absence, irregularity or loss of the consignment note shall not affect the existence or the validity of the contract of carriage which shall remain subject to the provisions of this Convention", in effect render the requirements laid down in Article 34 superfluous.

11.24 This, however, raises the fundamental question of why Article 34 should have been inserted in its existing form if it was intended that Article 4 should override this aspect of its provisions. Further, Article 4 is included in Chapter III which deals with the "Conclusion and Performance of the Contract of Carriage" and in turn this is followed by Chapter IV dealing with the "Liability of the Carrier"and Chapter V dealing with "Claims and Actions". Then a totally separate Chapter VI deals with "Provisions relating to Carriage performed by Successive Carriers". It is therefore arguable as a question of strict interpretation that Article 4 in a totally separate chapter cannot impliedly override Article 34. This is reinforced by the fact that in Article 35, which deals with the giving of a receipt for the goods by successive carriers,[35] special reference is made whereby the provisions of Article 9 are to apply to the relations between successive carriers, whereas no such reference back to Article 4 is made in Article 34. It would therefore appear that Article 4 is only applicable to the original contract of carriage entered into between the first carrier and the sender, and that it cannot affect the requirement in Article 34 that a carrier must take over both goods *and* consignment note to become a successive carrier under CMR.

11.25 Support for this approach can be found in three English decisions. First, there is an unreported decision of Mais, J., on an interlocutory application in Chambers, where he set aside leave to serve notice of a writ against carriers outside the jurisdiction on the basis that, although they had taken over the goods from the primary carriers, they had not received the consignment note as required by Article 34 and so they could not be liable as successive carriers under the Convention.[36] Mais, J., was referred to Article 4, but as he pointed out: "Article 34 provides for there to be a consignment note. If Article 4 is to govern it makes Article 34 utterly futile."

11.26 Secondly, reliance can be placed on the case of *SGS-Ates Componenti Elettronici S.p.A.* v. *Grappo Ltd.*,[37] which, although not concerned directly with the need for a consignment note, was concerned with the relevance of Article 4 to the successive carriage provisions of the Convention. In that case, Grappo, the first defendants, contracted with the sender for the carriage of a reactor and components from Heathrow, near

---

35. *Infra*, para. 11.27–11.31 and paras. 11.55–11.61.
36. *Colodense Ltd.* v. *Gelders U.K. Transport Ltd. and Another* (1980) unreported.
37. [1978] 1 Lloyd's Rep. 281. Hardingham, "Actions against successive carriers under CMR", [1978] LMCLQ 400.

London, to Sicily by road. Grappo sub-contracted the whole carriage to British Road Services, the second defendants. British Road Services carried the goods themselves from Heathrow to Rotterdam, and then sub-contracted the remainder of the transit to a Dutch haulier Furtrans, the third defendants. Furtrans did not carry the goods at all, but sub-contracted the actual carriage to Van der Vegt, a Dutch haulier, who delivered the goods to the consignee in a damaged condition. The question arose as to whether Furtrans or Van der Vegt was the last carrier. The court held that Van der Vegt was in fact the last carrier.[38]

11.27 The whole dispute turned upon the question of the construction of CMR as to what documentation is required when goods are handed over from one carrier to another. The court was therefore required to construe not only the meaning of "acceptance of the consignment note" under Article 34, but also the ancillary provision laid down in Article 35 which provides as follows–

> "1. A carrier accepting the goods from a previous carrier shall give the latter a dated and signed receipt. He shall enter his name and address on the second copy of the consignment note. Where applicable, he shall enter on the second copy of the consignment note and on the receipt reservations of the kind provided for in article 8, paragraph 2.
> 2. The provisions of article 9 shall apply to the relations between successive carriers."

11.28 A consignment note had been issued which named British Road Services (the second defendants) as "carrier" together with its address and under the heading "Successive Carriers" Furtrans and its address were entered. Vegt was nowhere referred to. A copy of the consignment note was handed over with the goods to Vegt's driver and was delivered with the goods to the consignee.

11.29 The question therefore was as to whether Vegt could be a successive carrier if his name and address was not entered in the consignment note. First, the court agreed that Article 4, which states that the absence, irregularity or loss of the consignment note will not affect the existence or validity of the contract of carriage, did not apply to overcome the absence of information in the consignment note. As Robert Goff, J., stated:

> "No reliance could, it was submitted, be placed by the third defendants [Furtrans] on the second sentence of art. 4, since, unlike art. 9 (as to which see art. 35(2)), the provisions of art. 4 were not expressly made applicable to the relations between successive carriers. I accept the submission that art. 4 does not so apply."[39]

11.30 However, the court went on to consider the meaning of the words "acceptance of the consignment note" in Article 34, and decided that "acceptance" was not dependent upon the fulfilment of the provisions of Article 35. The fact that the carrier accepting the goods fails to enter his name and address on the second copy of the consignment note or fails to give a dated and signed receipt as required by Article 35 will not restrict the operation of Article 34:

> " . . . what is meant by the expression 'acceptance of the consignment note' in art. 34? In my judgment, these words are to be given their natural and ordinary meaning unless the context otherwise requires; and the natural and ordinary meaning of the words is simply that the consignment note is, like the goods—indeed normally with the goods—accepted when it is taken over by

---

38. The question was at issue because if Vegt was the last carrier any action against him was time-barred under Article 32, and no action could proceed against Furtrans because, applying Article 36, (*infra*, para. 11.62 ff.) he was neither the first nor last carrier nor was it proved the damage occurred while the goods were carried by him. A similar decision was reached in App. Paris, 10.3.92 (1992) B.T. 278.

39. [1978] 1 Lloyd's Rep. 281, 284.

the carrier concerned, by himself or through his servant or agent, with a view to carrying out the next part of the carriage of the goods pursuant to the terms of the consignment note. I can, furthermore, see no reason to qualify this simple meaning by requiring compliance with the provisions of art. 35. . . . Suppose that a carrier takes delivery of the goods from a previous carrier together with a copy of the consignment note with a view to carrying out the next part of the carriage under the terms of the consignment note, but fails to enter his name and address, or perhaps enters his name but fails to enter his address, in the copy of the consignment note. That cannot, in my judgment, mean that there has been no acceptance of the consignment note within art. 34."[40]

11.31 As the court further observed, if the requirements of Article 35 were permitted to affect the operation of Article 34, this would merely encourage carriers intentionally to omit such information from the consignment note.[41]

11.32 Regrettably, the direct question of whether there can be a successive carrier in the total absence of a consignment note did not arise in the *SGS-Ates* case. The fact, however, that the court stated that the words "acceptance of the consignment note"in Article 34 were to be given "their natural and ordinary meaning" perhaps suggests that Robert Goff, J., considered that a consignment note must either accompany the goods or be handed over to a carrier by other means to constitute the latter a successive carrier under Article 34.

11.33 In the third decision, *Harrison & Sons Ltd.* v. *R.T. Steward Transport Ltd.*,[42] Gatehouse, J., held that a carrier who accepted, not the original consignment note, but a fresh CMR consignment note issued by a preceding carrier, was not a successive carrier. No reference was made to Article 4, and although Gatehouse, J., considered the result to be curious, he put it in the following terms:

"When Article 34 refers to 'the consignment note' I have little doubt that it means exactly what it says. It is referring to the original consignment note, not to a different note."

However, this was in the context of the question of whether jurisdiction over successive carriers could depend on the residence of that carrier for the purposes of the application of Article 39(2).[43] Since Gatehouse, J., went on to hold that this provision applied notwithstanding the fact that this carrier was not a successive carrier, his comments in this respect would appear to be obiter.

11.34 Finally, in *Dresser U.K. Ltd.* v. *Falcongate Freight Management Ltd.*[44] Hobhouse, J., took the view that Robert Goff, J., in the *SGS-Ates* case, had held that Articles 34 and 35 required that there be an actual acceptance of a pre-existing consignment note. In the case before him, this meant that the absence of a consignment note was fatal to the claim that the defendants in that case were successive carriers.[45]

11.35 Practical consideration also suggests that receipt of the consignment note is an essential prerequisite of liability under the Convention as a successive carrier, since without a consignment note a subsequent carrier may well be unaware that the carriage is subject to the Convention, particularly where he is performing a segment of the carriage

---

40. Ibid.
41. Ibid. This assumes, however, that sub-carriers are thereby unable to avoid responsibility as a party to the single contract, cf. *Dresser U.K. Ltd.* v. *Falcongate Freight Management Ltd.* (1991) 26 E.T.L. 798, *infra*, para. 11.45. The essential point is whether the carrier can negate the appearance of having accepted responsibility by having accepted the goods and the consignment note by simple failure to comply with Article 35.
42. (1993) 28 E.T.L. 747.
43. *Infra*, para. 11.119 ff.
44. Queen's Bench Division, 6.11.90, (1991) 26 E.T.L. 798.
45. This was, however, obiter since he had already found that the claim against the defendants failed on the facts.

which is entirely within one country. Given that liability under Article 34 potentially extends to responsibility for the whole carriage, it is particularly important that he is aware of his status, and, as has been pointed out elsewhere, he might in the absence of a consignment note be led to believe that he will be protected by his standard trading conditions, and his insurance may only be effective when operating under those conditions.[46]

11.36 In this connection, it will be recalled that Article 6 sets out the information which is to be contained in the consignment note,[47] and although it has been argued above that Article 4 has no application to the position as between successive carriers, clearly reference to a consignment note in Article 34 makes little sense except in the context of the information specified in Article 6, in particular the required statement that the carriage is subject to the Convention.

11.37 Similarly, it is thought that as against successive carriers, the other matters specified in Article 6 must be specifically referred to in the consignment note, otherwise they will not be binding on a successive carrier. So although it is argued elsewhere in this book that matters such as agreed delivery dates and provisions relating to the collection of C.O.D. charges need not be entered in the consignment note to bind the primary carrier, in order to bind a successive carrier it would seem that they must be so included, since otherwise he may well have no notice of such matters. On the other hand, it must be remembered that there is no mandatory format for a consignment note.[48] So as long as there is a documentary record which contains the requisite information and which is handed to the successive carrier, this will be sufficient.[49]

11.38 If a carrier does not constitute a successive carrier within the meaning of the Convention, then any action against that carrier will depend on the appropriate national law.[50] There may be a contractual right against him by the carrier who sub-contracted the carriage to him, or an extra-contractual claim by the cargo interests.[51] In the former case, this contractual liability will be fully subject to CMR if the sub-contract involves

---

46. Hardingham, op. cit., p. 501. As explained by Haak, p. 107, the *Travaux Préparatoires* show that the requirement of acceptance of the consignment note was introduced as a partial answer to the danger that a successive carrier might find himself bound by an international contract of carriage of which he is ignorant. The proposal to make the requirement indicated in Article 35 a condition of liability was rejected; see Loewe, para. 275. As the German Supreme Court has pointed out (BGH., 25.10.84 (1985) 20 E.T.L. 268, 271) acceptance of the consignment note is essential because without it the successive carrier could not be aware of essential parts of his liability, e.g. value or interest agreements in the sense of Articles 24 and 26.

47. See *supra*, para. 4.15 ff.

48. Apart from the information required by Article 6 there is also the requirement of Article 5 that the consignment note is signed by the sender and the carrier. The absence of a signature can certainly affect the quality of the document as proof (as in *City Vintages Ltd.* v. *SCAC Transport International*, Queen's Bench Division, 1.12.87, unreported except on LEXIS; see *supra*, para. 4.12. See also BGH., 16.10.86 (1987) 22 E.T.L. 56. Given the purpose of the requirement in Article 34 (see *supra*, text to fn. 46) the absence of a signature should be relevant only if the absence casts doubt on whether the successive carrier has become party to the contract relied upon by the claimant. The crucial requirement is the reference to CMR required by Article 6(1)(k), Loewe, para. 275.

49. This would seem to be supported by an Italian case which held that there could be a successive carriage under Article 34 in the absence of a consignment note if the consignor's "request for carriage" was passed on by the preceding carrier. Acceptance of the goods alone, however, was not enough: Cass (It.), 19.2.79 (1979) 7 E.L.D. 442. This is in line with the (surprisingly) less formalistic approach of the Italian courts to this issue noted by Libouton, (1982) J.T. 737 para. 145 where he cites Cass. (It.), 19.12.78, Il Foro Italiano, 1979, 1512.

50. For the problems this might cause, see Hardingham, op. cit., pp. 501–502.

51. As to the position in this country see *supra*, para. 9.66.

international carriage.[52] In the latter case, Article 28 would apply so that the carrier can take advantage of the Convention exclusions and limitations of liability, whereas in the former instance, although Article 28 would not apply, the first carrier's liability will already have been limited by the Convention, so his claim against the subsequent carrier will necessarily equally be so limited. An alternative argument might be adopted where the non-successive carrier is a sub-contractor under a contract involving international carriage by road. The recipient of the goods could well claim to be the consignee not only under the original contract with the initial carrier but also in relation to the contract with the sub-contractor, and thus entitled to exercise the right to enforce the contract granted by Article 13.[53] This argument has been firmly rejected by the Federal Supreme Court in Germany on the basis that, just as the consignor under the original contract of carriage has no contractual claim against the sub-contractor who is not a successive carrier, neither has the consignee.[54] However, this view is controversial. The point has been made that there seems to be no reason why the consignee should not acquire a contractual link against the sub-carrier through the original contracting carrier, there being no reason why the contract between those parties should not qualify as a contract of carriage. On that analysis, the contracting carrier will be the sender, and the consignee under the original contract of carriage would also occupy that position in relation to the sub-contract of carriage.[55]

11.39 In general, Continental case law supports the acceptance of a consignment note as an essential formal requirement.[56] The Belgian courts have, however, evinced a conflict

---

52. In an Austrian decision (OGH., 4.6.87 (1988) 23 E.T.L. 714), a sub-contracted carrier B responsible to his contracted carrier A for the entire transit was held liable to the latter for loss occasioned by the actions of his own sub-contracted carrier C. The central point of the case was that he could not escape liability by reliance on Article 37(a), see *infra*, p. 11.75 ff. This point might well be side-stepped in England if the approach of the courts noted at paras 11.14–11.15 is applied. (Conversely, carriers who are successive carriers have a right not to be sued on the contract of carriage with their immediate contractor if they are not caught by the provisions of Article 37, as in the *Cummins* case *infra*, and in Cass. (Belg.), 30.5.80 (1983) 18 E.T.L. 79. Apart from this, the issue of title to sue of the carrier seeking recourse may well also arise, which may depend on his establishing an immediate right to possession, see *Transcontainer Express Ltd.* v. *Custodian Security Ltd.* [1988] 1 Lloyd's Rep. 128, cf. RB. Turnhout, 12.11.81 (1983) 18 E.T.L. 105. For the Antwerp Court of Appeal, the relationship between a contracting carrier and a substitute carrier engaged to perform the whole carriage is governed by national law. The contracting carrier was neither sender nor consignee and therefore had no right to sue the substitute carrier for indemnity by means of a direct claim for the compensation paid to its principal who is named as sender on the consignment note, Hof. Antwerpen, 8.11.89 (1990) 25 E.T.L. 83 (cf. BG. Switzerland, 2.6.81, (1995) 30 E.T.L. 668), distinguished in a later decision of the same court where the claim was a direct claim by a trailer operator for damage to the trailer caused by the tractor operator driving under a low bridge, 23.6.92 (1993) 28 E.T.L. 293, cf. RB. Antwerpen, 7.11.86 (1987) 22 E.T.L. 453.

53. See *supra*, para. 5.18 ff.

54. BGH., 28.4.88 (1989) 24 E.T.L. 89.

55. Thume, (1991) TranspR 85. The court has confirmed its view, however, 10.5.90 (1991) 26 E.T.L. 351 and 24.10.91 (1992) 27 E.T.L. 839, pointing out in the latter case that the sub-carrier is really in the position of agent or servant of the contracting carrier. This might explain why in *ITT Schaub-Lorenz* v. *Birkart* [1988] 1 Lloyd's Rep. 487 the contracting but non-performing sub-carrier was viewed as having made themselves party to the contract of international carriage, see Bingham, L.J., at p. 493. They had become party to a contract to which CMR applied by virtue of Article 1. This contract, however, was with the carrier who contracted with the plaintiffs, who were the senders and consignees. Even if not successive carriers, it might have been possible to see the consignees as having a direct claim against them by virtue of being the consignees under both contracts.

56. E.g., Austria: OLG. Innsbruck, 20.6.95 (1997) TranspR 343; Germany: BGH., 25.10.84 (1985) 20 E.T.L. 268; Greece: Athens Court of Appeal, No. 56630–p. 1985 (1987) 22 E.T.L. 65 (where, in respect of Articles 34–40, the consignment note was said to have a constituting and not solely an evidential function); France: App. Paris, 27.2.80 (1980) B.T. 384; Sweden: Supreme Court, 26.3.96 (1997) U.L.R. 202. See further the cases cited by Clarke, p. 199 fn. 89 and Haak, pp. 109–111.

of views on this.[57] Some have been prepared to be less formalistic.[58] A first instance decision[59] provides a good illustration. In that case, the court stated that whilst acceptance of the consignment note is no doubt a fact from which the acceptance of the international contract by the successive carrier can be deduced, it is not the only factor. A. agreed to carry goods from Sweden to Belgium. He issued a consignment note and carried off the goods from the factory to the Swedish port, where the goods were off-loaded at the premises of B. and on the same day loaded by A. into the groupage lorry provided by C. under cover of a CMR consignment note for the next section of the transit. The lorry was then shipped to the German port and driven to the destination via Antwerp, from where other goods in the vehicle were to be shipped. On arrival, the goods were missing and it transpired that they had been mistakenly shipped at Antwerp, despite the fact that C.'s driver had supervised the transhipment at Antwerp. On receiving notice of claim, C. responded on the basis of the successive carriage provisions of CMR. This fact, and other facts surrounding the loading of the lorry and other information sent by B. to C., led to the conclusion that C. must have been aware that it was participating in an international carriage subject to CMR and not as a mere sub-contractor to B. The fact that A. had issued a separate consignment note for the previous period of internal carriage made no differ-ence.[60] If one takes the sole purpose of Article 34 to be that of ensuring that a successive carrier has knowingly accepted CMR responsibility for the whole transit, then one might justify an approach whereby this can be shown by other means than acceptance of the consignment note. However, as has been said,[61] abandonment of all formal requirements would considerably reduce scope for the application of Article 34 and would increase uncertainty and remove proper protection for the successive carrier.[62]

11.40 In respect of the application of Article 34, the existence of separate consignment notes can clearly be an indication that a sub-contracting carrier is not adhering to the whole of the carriage contract made by his contractor.[63] In *Harrison & Sons Ltd.* v. *R.T. Steward Transport Ltd.*,[64] Gatehouse, J., held that a carrier who accepted, not the original CMR consignment note, but a fresh CMR consignment note issued by the carrier who had contracted for his services was not a successive carrier.[65] This is in line with a Belgian decision where the fact that the sub-contracted carrier issued his own consignment note for

57. See further Haak, p. 110, where conflicting decisions of the courts are indicated: e.g. the decision in RB. Antwerpen, 3.4.77 (1977) 12 E.T.L. 411, which held that only a carrier to whom the consignment note has been given can be a successive carrier, is contrasted with contrary decisions of the same court: 16.4.75 (1975) 10 E.T.L. 548, and 14.4.78.

58. A note by Muller attached to the decision of Comm. Bruxelles, 11.5.87 (1988) 23 E.T.L. 720, 730 suggests a trend among the Belgian courts to take a less formalistic approach. As part of this he includes decisions which have not required strict adherence to Article 35 as in the *SGS-Ates* case.

59. RB. Brussel, 6.4.84 (1984) 19 E.T.L. 431, (1986) II U.L.R. 607.

60. Cf. Comm. Verviers, 7.4.79 (1979) 14 E.T.L. 664 where the fact that the successive carrier issued his own consignment note did not prevent him from being viewed as having become party to the original contract.

61. Haak, p. 111.

62. See *supra*, text to fn.38. See as an example of complexity the decision of App. Anvers, 15.3.89 (1989) 24 E.T.L. 574, (1989) II U.L.R. 819, discussed further, *infra*, at para. 11.48.

63. Merely entering an inland place of delivery onto the international consignment note will not necessarily prevent adherence however: Cass., 21.11.95 (1995) B.T. 831, (1996) U.L.R. 594 (Krings, p. 146).

64. (1993) 28 E.T.L. 747.

65. This was, however, in the context of an issue as to jurisdiction for the purposes of claims against successive carriers under Article 39(2), *infra*, paras 11.119–11.123. Since Gatehouse, J., went on to hold that Article 39(2) applied notwithstanding that the carrier in question was not a successive carrier. His comments in this respect would appear to be obiter. See also *Dresser U.K. Ltd.* v. *Falcongate Freight Management Ltd.* (1991) 26 E.T.L. 798 where a document purporting to be a consignment note was issued in respect of the stage of

the transit performed by him meant that he was not a successive carrier, despite the fact that he had received the original consignment note and that both documents were presented to the consignee's agent on arrival.[66]

11.41 It should be noted that Article 34 does not require that the carrier must receive the consignment note at the same time as he takes over the goods, nor that he must receive the note from the carrier who hands over the goods to him. Presumably, therefore, he could receive a copy of the note either by telex or other means from the previous carrier or some earlier carrier or from some intermediary, such as a forwarder. These points have yet to be decided by the courts, but given the increasing use of modern data transmission methods such a possibility may become commonplace in the future. At present, though, it would appear that such a situation would most likely arise where the consignment note had got lost at some stage and a replacement was needed.

11.42 A further complication that may arise is that the first carrier may not issue a consignment note if he does not physically handle the goods himself, but instead leaves his sub-contractor to make out and issue one. Such a situation would be a convenient practice as the latter's driver can then hand over the top copy to the sender when he collects the goods. In such a situation, as occurred in the *Ulster-Swift* case, difficulties obviously arise, as there is no question of the sub-contractor "accepting" a consignment note from anyone. Even so, in that case the Court of Appeal treated the sub-contractor as a successive carrier,[67] and only dealt with the fact that the first carrier had not issued a consignment note in considering the status of the first carrier.[68]

11.43 This therefore still leaves open the question of whether or not the successive carrier has "accepted" the consignment note. One possible solution to the problem is that where the first carrier fails to issue a consignment note, and the second carrier does in fact do so, he issues the note as agent for the first carrier. This issue arose in the case of *Coggins* v. *LKW Walter*[69] where a sub-contractor argued that he was not a successive carrier in circumstances in which, having further sub-contracted the carriage, he had received neither the consignment note nor the goods. It was held that, by sub-contracting the carriage, the first sub-contractor had authorised the second sub-contractor to accept both the goods and the consignment note on his behalf. Judge Hallgarten put it as follows[70]:

"As I see it, where a CMR carrier delegates performance of his responsibility to a sub-contractor, he vests the sub-contractor with authority to do such things as he would have done had he performed the contract in person."

German authorities to the contrary were rejected by the court as not representing the position in this country.

transport which included carriage by sea, and could not be treated as a consignment note relating to a single contract for the whole journey into which the plaintiff was seeking to draw the carrier by sea. See *infra*, para.11.45 and para. 11.47.

66. Comm. Bruxelles, 11.5.87 (1988) 23 E.T.L. 720, see further on this case *infra*, para. 11.45. See also Libouton, (1982) J.T. 737, No. 144, and cf. the decisions in Hof. Brussel, 26.4.83 (1983) 18 E.T.L. 511 and Comm. Verviers, 7.4.79 (1979) 14 E.T.L. 664. cf. further, App. Paris, 23.10.96 (1997) B.T. 104 where a carrier, who both signed the international consignment note for the domestic section performed by him, was held to be subject to CMR.

67. See *supra*, paras 11.11–11.17.

68. In that context, it was shortly disposed of: "The insistence of Counsel for Fransen on the fact that no consignment note was issued by Taunton to Ulster-Swift is answered by Articles 4 and 9, para. 1." [1978] 1 Lloyd's Rep. 346, 359, per Megaw, L.J. In *Coggins* v. *L.K.W. Walter International* [1999] 1 Lloyd's Rep. 255 the court rejected the submission that the *Ulster-Swift* decision was determinative of this point.

69. *Supra*.

70. At p. 259.

11.44 On the Continent, particular emphasis has been placed on the fact that a successive carrier is made a joint debtor towards the sender. Any difficulty with viewing a carrier as having accepted responsibility under the original contract between the contracting carrier and the sender will mean that he will not be characterised as a successive carrier. The issue of the consignment note by the sub-carrier may not amount to acceptance of responsibility under the original contract made with the sender. In one case[71] carrier A. sub-contracted performance of the carriage to carrier B. Carrier B. was held not to be a successive carrier since he had not taken over a consignment note issued by A. to the sender. In fact, he had issued a consignment note himself, but not to the sender of the goods.[72] Similarly, in another case[73] carrier A. contracted for carriage from Nuertingen to Basel. He carried the goods part of the way himself and then sub-contracted the remainder of the carriage to carrier B., who in turn sub-contracted to carrier C. No consignment note for the entire carriage had been made out involving the consignor. Rather, carrier A. had made out a consignment note in respect of the on-carriage. Carrier B. was therefore held not to be a successive carrier in respect of the contract between carrier A. and the sender.[74] At least in the *Ulster-Swift* case, the consignment note signed on behalf of Fransen as carrier had been made out to Ulster-Swift as consignors.[75] In Belgium, an alternative possibility has been applied, whereby a carrier in a similar position to Fransen was held to have accepted a contractual responsibility towards the sender as well as towards the carrier who had engaged him.[76] The contracting carrier was a forwarder under the responsibility of a carrier.[77] The fact of having signed the consignment note raised the presumption in Article 9, and other facts, such as that the carrier took over the goods from the sender and was to deliver to the consignee and not the contracting carrier, meant that the defendant was in reality in control of the transport so that there was insufficient evidence to displace the presumption. Furthermore, as a matter of Belgian law, the defendant, having knowledge of the original agreed operation, and having signed the consignment note as carrier, had accepted a contractual obligation towards the sender.[78] Not dissimilar reasoning may explain the characterisation of the fourth defendant in *ITT*

---

71. BGH., 28.4.88 (1989) 24 E.T.L. 89.

72. See also BGH., 9.2.84 (1985) 20 E.T.L. 275 where the consignment note issued by carrier B. was not issued to the sender and named carrier A. as consignee. Cf. Admiralty and Commercial Court of Copenhagen, 17.7.86 (1987) II U.L.R. 730 where carrier A. was the contracting carrier and named as carrier on the consignment note, and carrier B. was the performing carrier and named as successive carrier on the consignment note; the court applied Article 39 in the claim for indemnity by carrier A. against carrier B.

73. BGH., 25.10.84 (1985) 20 E.T.L. 268.

74. See also Hof. Antwerpen, 8.10.86 (1987) 22 E.T.L. 436, RB. Rotterdam, 10.6.88 (1989) S. & S. No. 272, (1990) 18 E.L.D. No. 184, RB. Rotterdam, 19.8.88 (1990) S. & S. No. 104, (1990) 18 E.L.D. 962. Clarke, p. 201 notes the point taken in BGH., 25.10.84 (*supra*, fn. 73) and BGH., 28.4.88 (1989) 24 E.T.L. 89 that a party can only be a successive carrier if a consignment note has been issued, not by himself, but by someone higher up the chain. The cases, on their facts, may still leave room, however, for the possibility that a sub-contractor, issuing a consignment note to the original sender, may be characterised as having joined as a party to the original contract between the sender and his contractor, and can thereby properly be viewed as a successive carrier.

75. Cf. *Texas Instruments Ltd.* v. *Nason (Europe) Ltd.* [1991] 1 Lloyd's Rep. 146 where the tractor operator's driver signed the consignment note indicating the trailer operator as carrier, and was held to have authority to do so in view of the trailer operator's control of the movement, thus making the trailer operator the first carrier or in the alternative last or performing carrier. An interesting question, not in issue in the case, would have been the authority of the driver in respect of his employer, the tractor operator.

76. Cass., 17.9.87 (1988) 23 E.T.L. 201 (cf. *The Texas Instruments* case, *supra*, fn. 75, with regard to the trailer operator). This does not appear to make the sub-contractor a successive carrier: Hof. Gent, 25.6.86 (1987) 22 E.T.L. 421.

77. See *supra*, para. 1.13 ff.

78. Cf. App. Anvers, 15.3.89 (1989) 24 E.T.L. 574, (1989) II U.L.R. 819.

Schaub-Lorenz[79] as within the scheme of successive carriage in having made himself party to the international contract of carriage.[80]

11.45 One final point is that acceptance of the goods and the consignment note may involve more than simple physical involvement with them. The *SGS-Ates* case[81] rejected compliance with Article 35 as an absolute requirement, but the sense of the requirement is to provide a clear means by which the adherence of the sub-carrier to the original contract of carriage can be demonstrated. In the Belgian case noted above,[82] it was accepted that the mere fact of non-compliance with Article 35 did not prevent the carrier from becoming a successive carrier. But it was also accepted that the mere fact of that carrier's driver taking physical possession of the original consignment note and presenting this at the destination did not in itself prove adherence to the original contract in the absence of any indication of acceptance in the original consignment note, having taken care to issue a separate consignment note.[83] As the court noted,[84] the burden of proving acceptance is on the claimant. It may be that there is room for evidence displacing a presumption which might otherwise be derived from the simple act of taking over the consignment note. So in *Dresser U.K. Ltd.* v. *Falcongate Freight Management Ltd.*,[85] Hobhouse, J., held that an attempt to establish that the carrier by sea was a successive carrier under CMR failed, since merely by showing that the carrier is no more than a sub-contractor of the first carrier does not, without more, make that sub-contractor a party to the "single" contract as a successive carrier. The relevant carrier must agree to become a party to the totality of the contract of carriage of the goods from origin to final destination. This conclusion was distinct from the issue raised by the absence of a consignment note,[86] although it was noted that this " . . . illustrates the evidential difficulties that will arise in proving that a given carrier had accepted responsibility for the whole of the single carriage where there is no consignment note which covers the whole of that carriage which he has acceded to". In *Aqualon (U.K.) Ltd.* v. *Vallana Shipping Corporation*[87] a freight forwarder, held to be a CMR carrier, employed a sub-contracting tractor operator to collect the goods, which were loaded into a trailer, from the sender's premises for transport to the port, whereupon the trailer was transferred to a ship. The sender made out a CMR consignment note which indicated the forwarder as the carrier, but the signature on the consignment note was that of the sub-contractor's driver. The issue before the court was the status of the forwarder,[88] but Mance, J., accepted that if the forwarder was not a CMR carrier there would be no CMR contract for the whole of the carriage. It follows that the signature of the sub-contractor's driver was not taken to bind the sub-contractor to the

---

79. *Supra*, fn. 55.
80. *Per* Bingham, L.J., at pp. 491 and 493. See, however, *infra*.
81. [1978] 1 Lloyd's Rep. 281.
82. Comm. Bruxelles, 11.5.87 (1988) 23 E.T.L. 720.
83. The court cited Libouton, (1982) J.T. 737, para. 144, where the writer wondered whether the driver of the lorry could really be seen as the agent of his employer, not merely to take over the goods, but to accept responsibility under such an onerous régime as CMR.
84. At p. 724.
85. (1991) 26 E.T.L. 798.
86. See *supra*, para. 11.34. A purported CMR consignment note was issued for the sea stage of the carriage. In the view of the court, this document did not suffice to enable the plaintiffs to succeed against the carrier by sea under CMR. Although not expressly part of the judgment, the point may be made that this was clear evidence that the carrier by sea was *not* adhering to the more extensive carriage by road.
87. [1994] 1 Lloyd's Rep. 669. Cf. App. Anvers, 15.3.89 (1989) 24 E.T.L. 574, (1989) II U.L.R. 819.
88. As to which see *supra*, paras. 1.13–1.54

whole of the carriage evidenced by the consignment note. There was, presumably, neither any authority given to the driver to make such a commitment, nor any possible apparent authority, given the indication by the sender himself as to his understanding of who was the carrier in view of the way in which he had made out the consignment note.

11.46 Thus it may be possible to distinguish from the *Ulster-Swift* case[89] where there is evidence to show that a carrier is not adhering to the original contract, certainly at the start of the carriage. If this can be the case at the start of the carriage,[90] then why not at a later stage of the transport?

11.47 In the *Texas Instruments* case,[91] the tractor operator's driver signed the consignment note, but indicated the trailer operator as the carrier. The driver was held to have the authority of the trailer operator to do this. The tractor was used for the whole of the transit, and the trailer operator was held either to be the first or last or performing carrier. The position of the tractor operator was not in issue, but the judge said that it was at least a joint effort including the trailer operator. Although not entirely clear, it is possible that the tractor operator was being viewed as a successive carrier. If so, this must have been because the tractor operator had contracted for the whole journey. At the stage of issue of the consignment note, the authority of the driver seems to be relevant to determine the extent of responsibility accepted by his employer, which will reflect the actual contract made between the employer and his contractor. Arguably this may also be true at a later stage in the transit where the agreement may be one of sub-contract rather than adherence to the entire carriage contract. The distinction may well be easier to see where the claim of the sub-contractor is that he has hired the vehicle and the driver to the contractor. For such a claim to succeed with regard to the goods owner, it must be possible for the lender of the vehicle to plead the limited authority of the driver, whether in signing or taking over the consignment note. It may therefore be possible to envisage circumstances which indicate sufficiently clearly a lack of intention to adhere to the whole of the contract in the more usual case of succeeding carriage, notwithstanding the physical transmission of the consignment note from one carrier to another. A further point should be noted from the decision in *Dresser U.K. Ltd.* v. *Falcongate Freight Management Ltd.*[92] Hobhouse, J., rejected an argument that the carrier by sea could not be a successive carrier because Article 34 does not apply to the sea leg of the journey. The argument was that Article 34 cannot apply to or include that part of the carriage which falls within Article 2 of the Convention[93] because Article 34 expressly refers to a single contract which is performed by successive road carriers, whereas Article 2 refers in contrast to "the carrier by the other means of transport". He considered this view to be mistaken, since if the carrier by sea agrees to become a party to the single contract of carriage covering the whole of the transit, he thereby agrees to be a road carrier and not merely a carrier by sea.

11.48 This contrasts with the view taken in Belgium that Article 34 does not apply to combined land-sea operations. This appears from the decision of the Antwerp Court of Appeal[94] where carrier A. sub-contracted a carriage from Belgium to Sweden to carrier B.

---

89. *Supra*, para. 11.11.

90. I.e. why was acceptance of the goods and the consignment note at that stage not sufficient to make the sub-contractor a successive carrier?

91. [1991] 1 Lloyd's Rep. 146.

92. (1991) 26 E.T.L. 798.

93. See *supra*.

94. 15.3.89 (1989) 24 E.T.L. 574, (1989) II U.L.R. 819. See also Muller, (1988) 23 E.T.L. 728, Putzeys, No. 290, but cf. No. 281. Cf. further Haak, p. 108.

The latter sub-contracted the road carriage to the Belgian port to carrier C., whereupon the trailer with the goods was shipped under cover of a sea waybill with the intention of carriage to destination by carrier D. operating within the scope of a co-operation agreement with carrier B. The goods and trailer were damaged during the sea carriage. Having made the point, noted above, that Article 34 does not apply to such operations, the court held that carrier D. could not be regarded as the last successive carrier.[95] The decision may reflect a difficulty in viewing a sub-carrier as having become a party to a contract of carriage which includes a carriage by sea.[96] However, similar reasoning as in the *Dresser* case would seem to apply. Even if the sea carrier is not a road carrier, the goods will be on board the vessel under a contract with a road carrier, who will have assumed responsibility for the carrier by sea under Article 3. There is no point at which carriage by road is not continuously involved, and accordingly no reason why any sub-carrier cannot be considered, at least not for this reason alone, to have become a party to a single contract for the carriage of goods by road.

Trailer operators

11.49 The question of articulated vehicles, with their separate tractors and trailers, together with the four-wheeled trailers towed behind a normal lorry, could create difficulties as regards successive carriage as the tractor, the trailer and the lorry are all "road vehicles" for the purposes of CMR.[97] The operator of each can therefore in theory be a carrier and therefore in turn a successive carrier. Often a tractor operator may contract with a sender and then employ a third party's trailer to move the load. Conversely, a trailer operator may contract with a sender to carry a shipment and then contract with a third party for the use of his tractor to tow the load. Equally, a series of tractor operators may tow the trailer at various stages of the transit. How, therefore, do the rules of successive carriage operate in such cases? If total confusion is not to arise the relevant rule must be carefully applied to each set of facts in question.

11.50 First, the operator who contracts with the sender for the international carriage of goods by road will be the first carrier irrespective of whether he is the tractor or the trailer operator or both or neither. Secondly, if the first carrier employs a tractor to tow his own

---

95. The assumption being that acceptance of responsibility would otherwise have made the carrier the last carrier, notwithstanding that there had been no physical acceptance of the goods. See the criticism of the first instance judgment in this respect by Muller, (1988) 23 E.T.L. 729. The court also held that, since there was no evidence that the sea carrier, a maritime transportation undertaking, had assumed any responsibility for overland transportation, it was not a CMR carrier. Furthermore carrier C. was not a principal contractor but merely a substitute sub-contractor for the carriage to the port, and could not be held liable for the damage suffered in the course of the maritime segment of the journey. It should be noted, however, that carrier B. had issued what is described in the case as a contract of carriage naming carrier C. as carrier, covering that segment of the carriage. In that document, the goods were consigned to the sea carrier, although the country of destination was indicated as the place of delivery. Consequently, as in *Aqualon*, the only part of the contract to which carrier C. had become party was the carriage up to the port.

96. Or by other means, e.g. rail, cf. App. Anvers, 16.3.93 (1994) J.P.A. 429 (Putzeys, "Le droit des transports en Belgique (II) 1993–1995", (1996) U.L.R. 557, 568).

97. Article 1(2) *supra*, para. 1.55. In an unusual decision where the owner of a container of refrigerated meat carried by car ferry from London to Antwerp and the company operating a tractor employed to move it from the harbour at Antwerp to the public street were both held to be carriers within the meaning of CMR. The latter was a successive carrier. RB. Antwerpen, 28.3.66 (1966) 1 E.T.L. 708 *sed* Libouton, (1973) 8 E.T.L. para. 95, fn. 189, points out that the carriers apparently did not dispute the question of their status.

trailer, or a trailer to tow behind his tractor, that operator will be a sub-contractor to the first carrier. Whether he becomes a successive carrier under CMR will depend upon a strict application of the rules laid down above.[98] For example, if a trailer operator is employed by the first carrier on a sub-contract, although the trailer operator may "accept the goods" under Article 34 he is unlikely to "accept the consignment note" for the purposes of that Article and will therefore, it is submitted, not become a successive carrier, but merely remain a sub-contractor under the Convention for whom the first carrier is liable under Article 3.[99] Any right of action by the cargo interest against the trailer operator would therefore depend on the national law, but subject to Article 28. Such a claim would be most likely to arise in practice where the trailer was defective in any way.[100]

11.51 Similarly, where a trailer operator employs a tractor under a sub-contract to tow the trailer, then, if the trailer operator has contracted with a sender for the international carriage of goods by road, he (the trailer operator) will be first carrier and the tractor operator, on acceptance of the goods and the consignment note, will be a successive carrier under Article 34. Even if the first carrier does not himself prepare a consignment note, but instead leaves it to the tractor operator, the latter will, it is submitted, applying the principles discussed above, still be a successive carrier on the grounds that he is in fact issuing a consignment note as an agent of the first carrier, and "accepts" the consignment note for the purposes of Article 34 when he hands the first copy to the sender and retains the second copy to accompany the goods and the third copy for his own records as laid down in Article 5(1).[101]

11.52 Where an operator obtains both tractor and trailer from sub-contractors, the latter will only become successive carriers if the criteria laid down above are fulfilled. Where a series of tractors are utilised to move a trailer through the successive stages of transit, each tractor operator in turn will become a successive carrier on accepting the goods and the consignment note from his predecessor. Finally, where a contracting carrier operates a tractor with a driver employed by the tractor operator, the courts on the Continent have also asked whether the tractor operator has contracted to carry and not merely hired his vehicle and employee to the contracting operator.[102]

---

98. *Texas Instruments Ltd.* v. *Nason (Europe) Ltd.* [1991] 1 Lloyd's Rep. 146 suggests that the evidential possibilities can be complex. In this case, the trailer operator claimed that he had merely hired the trailer to the tractor operator, with the inference that he had taken no part in the carriage. This argument failed both for failure to plead it and on the evidence, which revealed that the trailer operator was in fact the organiser of the movement (cf. *Walek & Co.* v. *Chapman & Ball (International) Ltd.* [1980] 2 Lloyd's Rep. 279). This evidence assisted Tudor Evans, J., in finding that the trailer operator was the first carrier. In the alternative, the evidence also established that the trailer operator was a successive carrier accepting the goods and the consignment note through the employee of the tractor operator, the trailer operator having organised the whole of the movement.

99. If his name appears in the consignment note, he might be held to be a joint carrier with the tractor operator, as was held in App. Paris, 12.6.70 (1970) B.T. 228, Libouton, (1973) 8 E.T.L. 2, para. 94 fn. 188.

100. It should be remembered that the first carrier may be liable for the hired trailer under Article 17(3); see *supra*, paras 6.53–6.59.

101. This may not always be the case. If the tractor operator's driver issues a consignment note on behalf of the first carrier and the sub-contractor has contracted for the tractor to be available for only part of the carriage, the tractor operator may not be considered to be a successive carrier, as appears to have been the case in *Aqualon (U.K.) Ltd.* v. *Vallana Shipping Corporation* [1994] 1 Lloyd's Rep. 669.

102. Eg., Cass. (Belg.), 2.2.90 (1990) II U.L.R. 434 where the sub-contractor, having contracted to carry the goods from the place of loading to the destination, was held to be a CMR carrier. See further, *supra*, para. 1.5.

The effect of Article 34

11.53 If a carrier is a successive carrier within Article 34 he is made responsible for the performance of the whole contract of carriage. He will therefore be liable as carrier under the Convention even if the relevant loss, damage or delay or other matter complained of occurred without fault on his part during a portion of the carriage being performed by another carrier.[103] In effect, he becomes "the carrier" for the purposes of all earlier provisions of the Convention,[104] and those provisions will apply as though that carrier had performed the whole carriage. It follows that in a claim against a carrier under Article 34, the defence in Article 17(2) in respect of "circumstances which the carrier could not avoid and the consequences of which he was unable to prevent"[105] must be read as relieving the carrier of liability only if he is able to show that neither he nor the other carriers concerned in the carriage could avoid those circumstances or prevent those consequences. It will not be enough for him to show he personally was unable to do so because he was not in possession of the goods at the material time; he remains responsible despite that fact.[106]

11.54 It also follows that Article 29 will apply so as to deprive him of the benefit of the exclusions and limits of liability contained in the Convention,[107] not only where he has been guilty of wilful misconduct, but also where another carrier has been guilty of misconduct.[108] Similarly, the period of limitation for claims against him will be extended to three years under Article 32(1) in cases of wilful misconduct by other carriers.[109] A written claim against him will obviously have the effect of suspending the limitation period,[110] but, as was confirmed in *Worldwide Carriers Ltd. v. Ardtran International Ltd.*,[111] it can take effect only against that carrier; it would not take effect to suspend time as against other carriers who may be liable.[112] It was also held in the same case that the

---

103. He can only be sued, however, if he is within Article 36, and he will have a right of recovery from the carrier actually responsible under Article 37. These provisions are discussed *infra*, para. 11.62 ff. and para. 11.75 ff. respectively.

104. As has been seen (*supra*). Article 4 does not apply in deciding whether he is a successive carrier; but once it is decided he is a successive carrier under Article 34, Article 4 would, it is thought, apply as normal.

105. *Supra*, para 6.28 ff.

106. Hardingham, op. cit., p. 503, and see *Tetroc Ltd.* v. *Cross-Con (International) Ltd.* [1981] 1 Lloyd's Rep. 192 concerning the equivalent position under Article 3.

107. *Supra*, Chapter 9, paras 9.72–9.106.

108. *Contra* Hardingham, op. cit., p. 504. His reasoning is based on the fact that Article 29(2) incorporates Article 3 without incorporating Article 34. However, Article 29(2) does not need to incorporate Article 34, since as explained above the effect of Article 34 is to make the successive carrier responsible for the entire carriage; express incorporation into Article 29(2) is therefore not necessary since he is, in effect, "the carrier" within Article 29(1) regardless of whether he performed the portion of the carriage on which the misconduct occurred.

109. Hardingham agrees with the point, albeit by means of a different approach: op. cit., p. 504.

110. Under Article 32(2), *supra*, para. 10.81 ff.

111. [1983] 1 All E.R. 692, the court citing both French and German decisions in support (App. Aix-en-Provence, 8.11.68 (1969) 4 E.T.L. 918, and OLG. Düsseldorf, 13.1.73 (1973) 8 E.T.L. 620). More recently to the same effect see Cass., 10.6.86 (1986) 21 E.T.L. 416, but cf. Hof. Gent, 25.6.86 (1987) 22 E.T.L. 421. A difficult question is whether a carrier would be bound by a claim transmitted to a sub-carrier for whom he is responsible by virtue of Article 3. This point did not arise in the *Worldwide Carriers* case, and may not have been taken in the French decision referred to above. This would seem to be a much stronger point where the sub-contractor is not a successive carrier.

112. Such is apparent from the wording of Article 32(2) itself: first, a carrier cannot be in a position to reject the claim if he has not received it; secondly, the onus of proof of the receipt of the claim is on the cargo interests, which again seems inappropriate if a successive carrier is to be effected by notice to other successive carriers. Cf., however, Hof. Gent, 21.12.95 (1996–1997) R.W. 360.

fact that the successive carrier knew that a claim had been made against the first carrier and that an indemnity claim would then be pursued against the successive carrier was not sufficient to suspend the time limit for the purposes of the claim between the claimant and the successive carrier. Article 32(2) requires a written claim by the cargo interests rather than a separate and different claim for indemnity by one carrier against another.[113]

Other documentation in relation to successive carriage

11.55 Article 35 provides as follows–

> "1. A carrier accepting the goods from a previous carrier shall give the latter a dated and signed receipt. He shall enter his name and address on the second copy of the consignment note. Where applicable, he shall enter on the second copy of the consignment note and on the receipt reservations of the kind provided for in article 8, paragraph 2.
> 2. The provisions of article 9 shall apply to the relations between the successive carriers."

Article 35 requires a successive carrier on taking over goods from a previous carrier to give the latter a receipt for the goods which must be signed and dated. This receipt is not required to be in any particular form, and may be on a copy of the consignment note or else a house receipt used by either of the two carriers for such purposes. The successive carrier is required to enter his name and address on the second copy of the consignment note although this will clearly not be necessary if it has been entered already. This requirement does not appear to have any specific legal consequences, but is merely of a useful evidentiary nature. In particular, as has already been seen, the failure of a successive carrier to enter his name and address on the consignment note does not affect his status as a successive carrier under Article 34, nor does the fact that he fails to give a signed and dated receipt to the previous carrier when he accepts the goods from him.[114] It is clear that if the requirements of Article 35 were permitted to affect the operation of Article 34, this would merely encourage carriers intentionally to omit such information from the consignment note and fail to give a receipt.

11.56 At this stage it is useful to recapitulate briefly on the pattern of documentation utilised under CMR as between the sender and the first carrier. Under Article 8(1) the carrier, on taking over the goods from the sender, is required to check the accuracy of the statements in the consignment note as to the number of packages and their marks and numbers and the apparent condition of the goods and their packing.[115] Article 8(2) further provides that where the carrier has had no reasonable means of checking the accuracy of the statements referred to in Article 8(1)(a) (as to the number of packages and their marks and numbers) he shall enter his reservations in the consignment note, together with the grounds on which they are based. He shall likewise specify the grounds for any reservations which he makes as regards the apparent condition of the goods and packing.[116] If the consignment note contains no specific reservations by the carrier, it will be presumed, unless the contrary is proved, that both goods and packing appeared to be in good condition when the carrier took them over and that the number of packages, their markings

---

113. At p. 699. See further Glass, [1984] LMCLQ 30. See, also, the point made *supra*, at para. 10.95 fn. 255.
114. *SGS-Ates Componenti Elettronici S.p.A.* v. *Grappo Ltd.* [1978] 1 Lloyd's Rep. 281, discussed *supra*, paras. 11.26–11.32.
115. See *supra*, para. 4.24.
116. Article 8(2), *supra*, para. 4.24.

and number correspond with the statements in the consignment note.[117] The consignment note itself is *prima facie* evidence of the making of the contract of carriage, the conditions of the contract and the receipt of the goods by the carrier,[118] although any reservations therein will not bind the sender unless he has expressly agreed to be bound by them on the consignment note.[119]

11.57 In general terms, therefore, Article 35 aims to apply Articles 8 and 9 to the relationship between successive carriers. It provides that when a carrier accepts goods from a previous carrier he shall enter reservations of the kind provided for in Article 8(2).[120] It will be observed, however, that Article 35 introduces a new element into this pattern of documentation, namely a receipt to be issued to the proceeding carrier, on which any reservations which the successive carrier may make on taking over the goods are also to be entered. There is clearly no need for such a document as between the sender and the first carrier, since the sender will have retained the first copy of the consignment note[121] on which the first carrier should have entered any reservations he may have had.[122] Of the two remaining copies provided for by Article 5, one is retained by the first carrier, the other travels with the goods. In effect, therefore, the provision in Article 35 for a receipt is simply to ensure that each successive carrier, as he hands over the goods and the consignment note to another carrier, is entitled to documentary evidence of the successfully (or otherwise) completed stage of the carriage. The final carrier in the chain, of course, receives a receipt from the consignee,[123] with separate provision for the entry of reservations.[124]

11.58 Article 35 does not, however, say that Article 8(2) is incorporated into relations between successive carriers in its entirety, but just that reservations of the kind therein specified are to be entered on the consignment note and the receipt. This leaves unclear the status of the concluding sentence of Article 8(2) for the purposes of Article 35: "Such reservations shall not be binding on the sender unless he has expressly agreed to be bound by them". On the one hand, the use of the word "sender" in Article 8(2) is clearly inappropriate to describe a carrier, but as against that, if a carrier has expressly agreed to be bound by any reservations made by the carrier to whom he has handed over the goods, there seems no reason for not holding him so bound. It would seem, however, that the better view is that the last sentence of Article 8(2) only applies as between sender and first carrier; not only because of the use of the word "sender", but also because the sender may have contractually agreed in the consignment note for reservations to be binding in terms which would be effective, by virtue of Article 34, automatically to deprive a successive carrier of the opportunity of challenging reservations inserted by a subsequent carrier. The fact that Article 8(2) is not incorporated *in toto* into relations between successive carriers would seem designed to prevent this effect. Nevertheless, on normal contractual principles, it would be open to carriers between themselves, independently of this part of Article

---

117. Article 9(2), *supra*, para. 4.28.
118. Article 9(1), *supra*, para. 4.24.
119. Article 8(2), *supra*, para. 4.24.
120. Although Article 8(1) is not referred to it would seem implicit that it must also apply as between successive carriers.
121. Article 5(1), *supra*, para. 4.11 ff.
122. Article 8(2).
123. Article 13(1), *supra*, para. 5.12 ff.
124. Article 30, *supra*, Chapter 10, paras 10.2–10.20.

8(2), to agree that reservations will be binding. In the absence of such agreement, any reservations made can at most be of *prima facie* evidential value.

11.59 Article 9, on the other hand, is incorporated in its entirety into relations between successive carriers.[125] So in the absence of reservations on the consignment note by a successive carrier on taking over the goods, the rebuttable presumptions contained in Article 9(2) as to the condition of the goods, their packaging, the number of packages and their marks and numbers will apply in the manner explained earlier.[126]

11.60 The presumptions in Article 9(2) apply to reservations on the consignment note. It will, however, be recalled that a successive carrier is obliged under Article 35(1) to enter reservations both on the consignment note and on the receipt. This leaves open the question of effect, if any, of reservations entered on the receipt, but not on the consignment note. Presumably, however, as under Article 9(2) the absence of a reservation on the consignment note only gives rise to a *prima facie* presumption, any reservations made only on the receipt can be put forward as evidence to rebut that presumption.

11.61 It should be noted that as no reference is made to Article 8(3) in Article 35 the carrier handing over goods to a successive carrier cannot require him to check the gross weight of the goods, their quantity, or their contents for entry in the consignment note. This right is reserved solely for the sender.[127]

Rights of action against successive carriers

11.62 The question of which carrier can be sued by the cargo interests is governed by Article 36, which provides as follows–

"Except in the case of a counter-claim or a set-off raised in an action concerning a claim based on the same contract of carriage, legal proceedings in respect of liability for loss, damage or delay may only be brought against the first carrier, the last carrier or the carrier who was performing that portion of the carriage during which the event causing the loss, damage or delay occurred; an action may be brought at the same time against several of these carriers."

As stated earlier, where successive carriers are involved in a contract of carriage which is governed by a single contract, each of the carriers will be responsible for the whole transit. However, under Article 36 legal proceedings can only be brought against the first carrier, the last carrier or the carrier performing that part of the transit during which the injurious event occurred, although the Convention permits an action to be brought against more than one of these at the same time. The question of who is the first carrier in a CMR contract has already been discussed.[128]

The last carrier

11.63 The question of who is the last carrier is not discussed in the Convention. The CIM Convention, on the other hand, does make specific provision that claims can be brought against the "railway of destination", so that, even if neither the goods nor the consignment note are received by the last carrier because they have been lost in an earlier

125. Article 35(2).
126. See para. 4.24 ff.
127. Article 8(3) is discussed *supra*, para. 4.30.
128. *Supra*, Chapter 1, paras 1.13–1.54.

part of the transit, he can still be sued.[129] This is an important question where loss or damage occurs, as in many cases the last carrier will be based, or at least have some form of representation, in the country of the consignee, who is thus assisted when the question of making claims or bringing an action arises. Furthermore, the last carrier will probably be the only carrier in the chain with whom the consignee will have any direct contact.

11.64 The absence of any such provision in CMR means that the principles outlined above must be applied. Where the goods are lost in transit the last carrier will obviously be the carrier on whose segment of the journey the loss occurred, and even if a further carrier were named in the consignment note as last carrier to perform a later part of the transit, the latter would not become a successive carrier under Article 34 as he would not have accepted either the goods or the consignment note and thus would not have become a party to the contract of carriage.[130] Where, however, the goods are merely damaged or delayed, the last carrier will be the one who makes the actual delivery of the goods to the consignee under the CMR contract. Although there may be a further onward domestic movement of the goods this will not affect the situation, since the consignee for the purposes of CMR is not necessarily the ultimate consignee of the goods.[131]

11.65 A carrier may be the last carrier even if his name and address have not been entered in the second copy of the consignment note pursuant to Article 35(1), by virtue of his acceptance of the goods and the consignment note.[132] Conversely, as has already been seen, if a carrier does not accept the consignment note with the goods he does not become a successive carrier within the meaning of the Convention,[133] so it follows that he cannot be the last carrier even if it was he who delivered the goods to the consignee. In such circumstances, since the performing carrier has not taken over both the goods and the consignment note so as not properly to be constituted a successive carrier for the purposes of Article 34, the only possible rights of action in respect of the final stage of the carriage would be either, under the domestic law, against the carrier actually performing that stage, or, under Article 3,[134] against the carrier who actually made use of his services, who is thus, in effect, the last carrier.[135] In other words, the last carrier for the purposes of the

---

129. Article 55(3) CIM.

130. *Supra*, para. 11.5 ff. Rodière is of the view (para. 131) that the last envisaged carrier can be sued, regardless of whether or not the goods ever reached him. He argues, by analogy with the CIM Convention, that such was intended for the benefit of the cargo interests. As against this, however, it may be objected that the provision in CIM on which he relies (Article 55(3) CIM, *supra*) does not appear in CMR. However, a Belgian decision has held that a party can be a successive carrier by virtue of having agreed successively to assume responsibility for the totality of the carriage: Comm. Anvers, 28.1.85 (1985) 20 E.T.L. 117 (reversed on appeal on different grounds, 15.3.89 (1989) 24 E.T.L. 574, (1989) II U.L.R. 819, see *supra*, fn. 94). This decision is, however, criticised by Muller in (1988) 23 E.T.L. 729.

131. See *supra*, para. 11.5 ff.

132. See *SGS-Ates Componenti Elettronici S.p.A.* v. *Grappo Ltd.* [1978] 1 Lloyd's Rep. 281, 284–285, discussed *supra*, para. 11.26.

133. *Supra*, para. 11.21 ff.

134. *Supra*, Chapter 3.

135. This would seem to follow from the wording of Article 3: " . . . the carrier shall be responsible for . . . persons of whose services he makes use for the performance of the carriage . . . *as if such acts or omissions were his own*" (italics added). If correct, it can be argued that this reasoning would apply equally in cases where the last carrier was a successive carrier within the meaning of the Convention, thus producing in an appropriate case four possible defendants: the first carrier, the carrier responsible for the loss, damage, etc., the last carrier, and the carrier who made use of the last carrier's services. This possibility, which does not appear to have been considered in *SGS-Ates Componenti Elettronici S.p.A.* (*supra*, fn. 132), would have produced the opposite result in that case. Nor is there any reason why this argument should be restricted to the last carrier; it could apply equally to any carrier, who would not otherwise be liable under Article 36, who makes use of the services of the carrier actually responsible for the loss, damage or delay (see Hardingham, [1978] LMCLQ 499, 503). In *Texas*

Convention is not always the carrier who actually physically delivers the goods. If that carrier is not a successive carrier within the meaning of the Convention, then the successive carrier who made use of his services must be the last carrier by virtue of Article 3.[136]

The carrier responsible for the loss, damage or delay

11.66 As regards an action against the carrier performing that part of the transit during which the injurious event occurred, in practice it is often extremely difficult to prove when the event in fact occurred.[137] Such will be the case, for example, where the carrier subsequent to the one responsible for the injurious event has failed to make relevant reservations in the consignment note. Proof by extraneous evidence is generally difficult to obtain even if a claims adjuster is employed, since often the matter complained of is only discovered after receipt of the goods by the consignee. For that reason most claims are made against the first or the last carrier, as being most accessible to the cargo interests. Much, however, will also depend upon the procedural ease of action, which will differ in each case.

Concurrent actions and jurisdiction

11.67 Although Article 36 specifies that an action can be brought concurrently against more than one of the carriers liable, it must not be forgotten that the question of jurisdiction as between the cargo interests and a carrier is regulated by Article 31(1). If reference is made to the discussion of that paragraph,[138] it will be seen that the right to bring concurrent actions does not necessarily mean that those actions can be commenced in the same jurisdiction: such would indeed normally only be possible in the country where the goods were originally taken over for carriage or the place designated for delivery,[139] which would not necessarily be the most convenient forum for the cargo interests. On the other hand, Article 31(2)[140] would not operate to prevent concurrent

---

Instruments Ltd. v. *Nason (Europe) Ltd.* [1991] 1 Lloyd's Rep. 146, Tudor Evans, J., was prepared to find that the carrier who employed the sub-carrier was involved in a joint effort with the sub-carrier, which in the circumstances made both of them the last or performing carriers. The objection to all of this, however, is the apparently exhaustive wording of Article 36: " . . . legal proceedings . . . may only be brought" against the first, last or responsible carrier. Conflicting decisions appear in France where in one decision the actually performing last carrier was held to be the last carrier for the purpose of Article 36 so that the carrier who had contracted for that section but who had sub-contracted to another carrier, who in turn sub-contracted to the performing carrier was not liable, Cass., 3.5.94 (1994) B.T. 390, (1994) D.M.F. 641, (1995) 30 E.T.L. 685, (see the criticism of Tassel, (1994) D.M.F. 642, cf., however, Marchand, "La pluralité de transporteurs routiers selon la CMR", (1995) 30 E.T.L. 557, 582–584). In a later decision, Article 3 was applied to make a contracting carrier liable in respect of the last part of the transit performed by a substitute and during which the loss occurred, Cass., 9.7.96 (1996) B.T. 536, (1996) D.M.F. 1147 (with observations by Achard), (1996) U.L.R. 784. See also, Krings pp. 789–790.

136. As Bingham, L.J., pointed out in *ITT Schaub-Lorenz* v. *Birkart* [1988] 1 Lloyd's Rep. 487, 493, in the context of Article 37 (see *infra*, para. 11.75 ff.), a CMR carrier responsible for the loss or damage cannot escape liability by showing that he has delegated or sub-contracted performance to a non-CMR carrier who was actually responsible by virtue of the operation of Articles 34 and 3. This would seem to be equally true in respect of the application of Article 36, cf. Comm. Verviers, 7.4.79 (1979) 14 E.T.L. 664.

137. In cases of delay, of course, all carriers may be partly responsible.

138. *Supra*, para. 10.21 ff.

139. Article 31(1)(b).

140. *Supra*, para. 10.37.

actions in different jurisdictions against different carriers, since that paragraph only applies to actions between the same parties. For the reason already discussed,[141] it is not thought that an agreed jurisdiction clause within Article 31(1) would be binding on a successive carrier unless included in the consignment note or otherwise communicated to him before or at the time of his taking over the goods.[142]

11.68 An action will only lie against a successive carrier who is not one of those referred to in Article 36, that is, an intermediate carrier who is not responsible for the matter complained of, if the domestic law permits an action against him. Different jurisdictions may differ on this point,[143] although it should be noted that the tendency in France to treat operators as *commissionnaires* rather than as successive carriers[144] may permit an action in contract against intermediary operators, who would not be liable under Article 36 if they were treated as successive carriers under CMR. The remaining possibility, of course, is that an action might be brought against a carrier in a jurisdiction which is not a signatory to the Convention, in which case the limitations laid down by Article 36 might be ignored completely.

11.69 Article 36 applies only to actions by the goods owner against carriers. It does not exclude a claim by a carrier, successive or otherwise *against* the goods owner for a negative declaration,[145] or presumably any other claim.

Nature and scope of liability

11.70 The Convention does not specify the exact basis of the liability of the three categories of carrier against whom actions may be brought under Article 36. However, as under Article 34 each successive carrier is responsible for the performance of the whole operation, liability would in common law terms appear to be both joint and several, and accordingly any carrier who can be sued under Article 36 can be sued in respect of the full amount of the total liability to the cargo interests, regardless of his share of responsibility for that loss, damage or delay. Such an interpretation is supported by several Continental decisions,[146] and by the provision in Article 37 for the recovery of compensation paid by one carrier from other carriers,[147] which clearly pre-supposes joint and several liability.

11.71 Article 36 refers to claims for loss, damage or delay, which would seem inappropriate to describe other claims, such as for re-payment of overpaid freight charges, which may arise out of the contract of carriage. In *Muller Batavier Ltd.* v. *Laurent Transport Co. Ltd.*[148] May, J., in considering the scope of Article 39(4) expressed the view that

---

141. *Supra*, paras 11.35–11.37.

142. Loewe, para. 282, states, without giving reasons, that the successive carrier would be bound but that he might have an action for damages under the national law against the first carrier.

143. For the position in this country, see *supra*, para. 9.65 ff.

144. See Libouton, (1973) 8 E.T.L. 2, paras. 13–16, Rodière, para. 128 fn. 76.

145. *Frans Maas Logistics (U.K.) Ltd* v. *CDR Trucking B.V.* [1999] 2 Lloyd's Rep. 179.

146. Hof. Gent, 20.11.75 (1976) 11 E.T.L. 231. RB. Antwerpen, 28.3.66 (1966) 1 E.T.L. 708. RB. Bruxelles, 28.2.75 (1975) 10 E.T.L. 419. LG. Duisburg, 10.5.68 (1964) 4 E.T.L. 979. Hof. Brussel 26.4.83 (1983) 18 E.T.L. 511. Hof. Antwerpen, 14.12.83 (1983) 18 E.T.L. 809. Rodière considers that successive carriers are liable "*in solidum*"; para. 133.

147. *Infra*.

148. [1977] 1 Lloyd's Rep. 411, discussed in relation to Article 39(4) *infra*, para. 11.127 ff.

" . . . art. 39, as with the other articles in chap. VI, is solely concerned with regulating the rights inter se of successive carriers to an overall carriage of goods by road, and to which Convention applies, where something has gone wrong . . . ",[149]

so Article 39(4) did not, in that case, apply to a claim between carriers for freight charges. It will be observed that May, J., expressed himself in terms of relations between successive carriers *inter se*. Although Article 36 is primarily relevant in determining initial liability to the cargo interests, it was one of the provisions to which May, J., referred in particular as supporting his conclusion, and there would seem no reason why the same analysis should not be applied to its wording. It would seem, therefore, that a claim by the cargo interests for re-payment of overpaid freight charges will be outside Article 36[150] and so regulated by the national law.[151]

11.72 However, it will be recalled that Article 34 provides that the successive carrier becomes "a party to the contract of carriage, *under the terms of the consignment note . . .* ". It would seem to follow, therefore, that if the item in question is included in the consignment note, he can be sued in respect of such matters regardless of whether he is the first carrier, last carrier or carrier responsible. It seems doubtful that such a result was intended, having regard in particular to the equivalent provisions of CIM, which provides that actions for the recovery of sums paid under the contract can only be brought against the railway which collected that sum or against the railway on whose behalf it was collected.[152] Nevertheless, in the absence of equivalent provisions in CMR, such would appear to be the position in the event of the item in respect of which the overpayment in question was made appearing in the consignment note.

11.73 As far as claims in respect of the failure to collect C.O.D. charges are concerned, the word "loss" appearing in the English text[153] would seem wide enough to cover such losses, so that such claims would be within Article 36.

Counterclaims and rights of set-off

11.74 Article 36 also provides that the cargo interests will not be restricted in their right of action against successive carriers where the former wish to raise a counter-claim or right of set-off against a successive carrier in an action based on the same contract of carriage. In other words, where a successive carrier sues the sender or consignee for non-payment of freight charges or any other additional expenses which may fall due in respect of a contract for the successive carriage of goods, the defendant will be able to raise any claim for compensation for loss, damage or delay to the goods, or any right of set-off which he may have against the plaintiff carrier, even though he may not otherwise have a primary right of action against the carrier in question because of the limitations laid down by Article 36. The cargo interests will, of course, have a choice as to their course of action where the first, last or responsible carrier is concerned. They can either pursue

---

149. At p. 415.
150. It follows that it would also be outside Articles 37–39.
151. The position may be different if there is both a claim for loss, damage or delay and a claim for overpayment of freight charges. May, J., suggested in the *Muller Batavier* case that Article 39(4) might have applied to the claim for freight charges in that case in such circumstances, and, although the position here is rather different, it would equally be in the interests of uniformity that Article 36 be applied so as to avoid unnecessary litigation.
152. Article 55(1) CIM.
153. The French text uses the word *"perte"* which would equally seem appropriate to cover such losses.

an original action themselves against all or any of the successive carriers, or also, if they have not paid the latter, await an action for outstanding charges and then counter-claim or set-off as the case may be.[154]

Contribution between carriers

11.75 Article 37 provides as follows–

"A carrier who has paid compensation in compliance with the provisions of this Convention, shall be entitled to recover such compensation, together with interest thereon and all costs and expenses incurred by reason of the claim, from the other carriers who have taken part in the carriage, subject to the following provisions:

    (a) the carrier responsible for the loss or damage shall be solely liable for the compensation whether paid by himself or by another carrier;

    (b) when the loss or damage has been caused by the action of two or more carriers, each of them shall pay an amount proportionate to his share of liability; should it be impossible to apportion the liability, each carrier shall be liable in proportion to the share of the payment for the carriage which is due to him;

    (c) if it cannot be ascertained to which carriers liability is attributable for the loss or damage, the amount of the compensation shall be apportioned between all the carriers as laid down in (b) above."

Scope

11.76 It is therefore the rule that where a carrier has compensated the cargo interests in accordance with the provisions of CMR, he is (subject to compliance with Article 39) entitled to recover that compensation, together with interest thereon, plus all costs and expenses which he has incurred as a result of such claim, from the other carriers who have taken part in the carriage. Three initial points arise out of this provision. First, the reference to "compensation in compliance with the provision of this Convention" means that both the amount of the compensation paid and the grounds on which liability was based must comply with the CMR provisions on the subject.[155] Where therefore a payment is made by a carrier which is in excess of that laid down by CMR, it would appear that the latter will only be able to recover under Article 37 from other carriers who have taken part in the carriage up to the limit of CMR liability.[156] Any excess will not be recoverable under the Convention. Similarly where compensation is paid on grounds other than those on which the carrier would be liable under CMR, then the carrier would not be able to recover under Article 37 from other carriers at all.[157] Given the mandatory nature of the Convention, no recovery would be possible at common law either, and any action brought on extra-contractual or delictual grounds would be defeated by the operation of

---

154. As to the limitation position, see Article 32 *supra*, in particular the discussion at paras 10.73–10.82. As to rights of set-off, note the decision in *R. H. & D. International* v. *I.A.S. Animal Services Ltd.*, [1984] 2 All E.R. 203, *supra*, para. 10.82, fn. 200. See also *United Carriers Ltd.* v. *Heritage Food Group (U.K.) Ltd.*, Queen's Bench Division, 10.2.95. *The Times*, 8 March 1995.

155. This view is supported by Rodière, (para. 135, fn. 80), and cf. Article 39(1) *infra*, para 11.113 ff.

156. See Articles 23–27, *supra*, Chapter 9.

157. In an Austrian case (OGH., 13.6.86 (1988) 23 E.T.L. 198) carrier A. had paid compensation for economic loss to sub-carrier B. due to a delay caused by sub-carrier C. Carrier A. was held able to recover from carrier C. on the basis of Austrian law since Articles 17 and 34 of CMR were irrelevant, this not being a claim for loss, damage or delay to the goods.

Article 28 which would give the other carrier the benefit of the provisions of CMR that exclude his liability or that limit the compensation payable.[158]

11.77 Secondly, it will be observed that Article 37 refers to "the other carriers" without making it clear whether it refers exclusively to successive carriers within the meaning of Article 34, or whether it refers to all carriers who have taken part in the carriage, regardless of their status under Article 34. This is, in fact, a problem of general application to the provisions of the Convention relating to the position as between carriers.[159] On the one hand, all these provisions form part of Chapter VI which is headed "Provisions relating to Carriage performed by Successive Carriers", which would seem to indicate that they are concerned solely with successive carriers within the meaning of Article 34. Further, it will be recalled that the question of a carrier's status for the purposes of Article 34 is thought to depend largely on whether or not he has received a copy of the consignment note, and the reasons discussed earlier in favour of the need for a consignment note under Article 34[160] would seem equally applicable in the present context. Finally, if Article 34 were to be disregarded altogether for the purposes of Articles 37 to 40, those provisions would be applicable even if it was not carriage governed by a single contract, which would seem contrary to the scheme of the Convention. Thus in *Arctic Electronics Co. (U.K.) Ltd.* v. *McGregor Sea & Air Services Ltd.*,[161] Hobhouse, J., considered Article 34, and the concept therein of a single contract, to be relevant to the jurisdictional limitation in Article 39(2) on the right of recovery under Article 37.[162]

11.78 In essence, the scheme of the Convention is to enable a carrier, by virtue of Article 37, to overcome the obstacle of absence of privity of contract in seeking recourse against another carrier who may have no contractual tie to him. A justification for this is the joint responsibility which is assumed by the successive carrier, but which is not assumed by a mere sub-contractor.[163] It may be seen as a bargain whereby, in return for the right of recourse granted by Article 37, the carrier loses those contractual rights that he might otherwise have had against his sub-contractor where the latter is a successive carrier,[164] rights which the sub-contractor might not have been able to negate by reason of the operation of Article 41[165] were that contract itself to be a CMR contract. On the other hand, in return for being subjected to responsibility to the goods owners, successive carriers are granted the freedom by Article 40[166] to adjust their liabilities *inter se*.

---

158. See the discussion of Article 28 *supra*, Chapter 9, paras 9.65–9.71.
159. Articles 37 to 40. There are two issues: (1) do these rules apply to any carrier who may not be a successive carrier; and (2) do they apply to the first carrier, who is clearly not a "successive carrier"? A Dutch court, relying on the words in the heading to Chapter VI, has held that Article 37 does not apply to the relations between a principal carrier and his sub-carrier even if the latter is also to be considered a successive carrier: Hof. The Hague, 22.3.94 (1995) S. & S. No. 20 (Cleton, "Digest of Recent Decisions of Netherlands Courts Concerning International Transport Law Conventions", (1996) U.L.R. 755, 759). It is suggested, however that the words used in the heading are referring to the circumstance where there is successive carriage and naturally includes the contracting/principal carrier.
160. See *supra*, paras 11.35–11.37.
161. [1985] 2 Lloyd's Rep. 510.
162. Consequently, where a consignment was split by one carrier between three on-carriers, the latter carriers could not be successive carriers as between themselves, so that none of them could be "carriers concerned" *inter se* for the purposes of Article 39(2) (see *infra*, para. 11.119 ff.). "Single contract" meant the contract for the goods carried by the relevant carrier and not the contract to which each of the three on-carriers were parties.
163. This was a point made by the German Federal Supreme Court in BGH., 25.10.84 (1985) 20 E.T.L. 268.
164. As in Cass. (Belg.), 30.5.80 (1981) I U.L.R. 263, (1983) 18 E.T.L. 79.
165. See *infra*, Chapter 12.
166. See *infra*, paras 11.134–11.135.

11.79 In favour of these arguments is the view of Bingham, L.J. in *ITT Schaub-Lorenz* v. *Birkart Johann Internationale Spedition G.m.b.H. & Co. K.G.*[167] He did not accept that the "carrier responsible" under Article 37(a)[168] can be a person who has not made himself party to the contract of international carriage, by which he meant an original contracting party or a carrier who has made himself a party under the terms of Article 34. Consequently, a CMR carrier cannot escape liability by showing that he sub-contracted performance to a non-CMR carrier who was actually responsible.[169]

11.80 As against these arguments, however, the Court of Appeal in *Ulster-Swift Ltd.* v. *Taunton Meat Haulage Ltd*[170] expressed the view that the heading to Chapter VI was not a sufficient reason for reading Article 39(4) as meaning only "claims between successive carriers within the meaning of Article 34",[171] and that view must logically equally extend to Articles 37 to 40. It was, however, expressed in the context of a rather different argument,[172] and the point at issue here was not directly raised. In any event the court went on to decide the case on another ground.[173] However, despite the obiter nature of this part of the decision, it was applied by Gatehouse, J., in *Harrison & Sons Ltd.* v. *R.T. Steward Transport*,[174] in holding that the carriers referred to in Articles 37 to 39 were not limited to successive carriers. Consequently, a carrier seeking to rely on Article 39(2) for the purposes of establishing jurisdiction[175] was able to identify a non-successive carrier as a "carrier concerned".[176]

11.81 If it be right that only successive carriers within the meaning of Article 34 can be claimed from under Article 37, then any claims for contribution against carriers who are not within Article 34 will be regulated by the applicable national law.[177] It is an argument

167. [1988] 1 Lloyd's Rep. 487, 493.
168. See *infra*, para. 11.87.
169. Bingham, L.J., considered that any other interpretation would frustrate the CMR scheme. It should be noted that Bingham, L.J., accepted at this point the legal basis of the fourth defendants' argument that Article 37(a) was relevant to them and by implication that they were successive carriers within the meaning of Article 34. Earlier in his judgment, however, at p. 491, he stated that it was by Article 1 that they become parties to a CMR contract. This must have been on the basis of the contract they had made with the third defendants who had contracted with the sender. This may provide some support for a view that the provisions of Articles 37–40 are relevant not only to successive carriers identified by means of Article 34, but also to carriers who are CMR carriers for the whole of the transit by means of a contract with the original contracting carrier and to the same extent as the carriage agreed between the contracting carrier and the sender. The decision in the *Ulster-Swift* case would be consistent with this.
170. [1977] 1 Lloyd's Rep. 346.
171. At p. 360.
172. See *supra*, para. 11.15.
173. Ibid.
174. (1993) 28 E.T.L. 747.
175. See *infra*, para. 11.119 ff.
176. This had the consequence that foreign successive carriers were liable to suit in England by reason of the English residence of a non-successive carrier. A logical consequence of the decision would be to allow a foreign carrier who was not a successive carrier to be subject to suit in England by reason of the English residence of another carrier. This is surely contrary to the scheme of the Convention in respect of successive carriers. See *supra*, fn. 169. See also, Marchand, "La pluralité de transporteur routier selon la CMR", (1995) 30 E.T.L. 577.
177. In this country, the position would then be regulated by the Civil Liability (Contribution) Act 1978, which does not otherwise apply to claims against carriers who are successive carriers under Article 34: Carriage of Goods by Road Act, section 5(1) (see *Harrison & Sons Ltd.* v. *R.T. Steward Transport, supra*). Since the restriction on the rights granted by the 1978 Act is specifically confined to Article 34 of the Convention, it follows that these rights should still be available in respect of carriers who are not successive carriers, providing further support for the argument in the text. The view in the text also has Continental support: BGH., 25.10.84 (1985) 20 E.T.L. 268.

in favour of the Court of Appeal's view referred to above that such an interpretation would favour greater uniformity, since it would obviate the need for reference to domestic law. However, claims against other sub-contractors who are not carriers will in any event be governed by the applicable national law.

11.82 Thirdly, Article 37 on a literal interpretation provides that a claim can only be brought by a carrier against another carrier who has taken part in the transit *after* the former has paid compensation to the cargo interests. The judgment of Master Elton in *William Tatton & Co. Ltd.* v. *Ferrymasters Ltd.*[178] supports this interpretation. However, the opposite conclusion was reached by the Court of Appeal in the later case of *ITT Schaub-Lorenz* v. *Birkart Johann Internationale Spedition G.m.b.H. & Co. K.G.*[179] The court accepted that section 5 of the Carriage of Goods by Road Act 1965 ousts the ordinary English rules relating to contribution, and that Article 37 stipulates that payment of compensation by one carrier is a pre-condition of the carrier's right to recover against another carrier. It was also accepted that any inconsistency between CMR and English procedural rules must be resolved in favour of CMR, but the court was nevertheless of the view that a claim by one carrier for a declaration that another carrier was liable to indemnify them in respect of any compensation properly payable by the former to the claimants is permissible under English procedural rules prior to actual payment without offending Article 37(a).[180] It should be realised that a declaration is a discretionary remedy under English law, so that it is not available as of right. However, as Bingham, L.J., put it[181]:

"Depending always on the facts of the particular case, third party proceedings may in principle be useful and desirable for all the familiar reasons. There is nothing in CMR to exclude such a claim appropriately framed."

11.83 In *ITT* v. *Birkart*, it was argued that the court ought not to exercise its discretion in favour of allowing the third party proceedings on the basis that it was alleged that distinct issues arose. However, having made it clear that each case turns on its own facts, it was concluded that even though the common issues may be limited, it was "highly undesirable" that there should be a long delay before the issue of indemnity was explored.

11.84 With reference to Bingham, L.J.'s reference in the *ITT* v. *Birkart* case to "all the familiar reasons", plainly any other result is likely to produce needless litigation since the carrier first sued by the cargo interests is thus effectively obliged to proceed to judgment

---

178. (1974) 9 E.T.L. 167. Master Elton held that compensation must first be paid under Article 37 before a claim could be made against another carrier. Consequently, where notice was served in France by the English defendant carrier to join the actual carrier, who was French, as a third party to an action by the cargo interests, the latter entered a conditional appearance to the third party notice in the High Court and applied for the third party notice and its service to be set aside. *Held*: that the third party notice and its service were to be set aside, except in so far as the notice constituted "due notice" and an opportunity to the third party to enter an appearance under Article 39(1). (That opportunity was not subsequently taken. [1974] 1 Lloyd's Rep. 203, 204–205.)

179. [1988] 1 Lloyd's Rep. 487. No reference was made to the judgment of Master Elton. However, unlike the decision in *William Tatton & Co. Ltd.* v. *Ferrymasters Ltd.* [1974] 1 Lloyd's Rep. 203 and the decision of the Court of Appeal in *Cummins Engine Co. Ltd.* v. *Davis Freight Forwarding (Hull) Ltd.* [1981] 2 Lloyd's Rep. 402, no difficulty in respect of jurisdiction was in issue, see *infra*, para. 11.119 ff.

180. Under CPR Part 20. In the view of Bingham, L.J., the same is true of the claim for indemnity against any sum properly paid to the plaintiffs. Strictly, there would be no direct conflict with Article 37 since as, appropriately framed, there will be no enforcement until payment.

181. [1988] 1 Lloyd's Rep 487, 494.

rather than settle the claim so as to be able to take advantage of Article 39(1),[182] whereas if the successive carrier can be compelled to join in the initial proceedings, a settlement between all three parties may well result.

11.85 The Court of Appeal cited with approval an obiter observation by Brandon, L.J., in the *Cummins* case.[183] Having referred to the opening words of Article 37(a), he continued:

"It seems to me, however, that where the procedure of the court in which the first and main action is brought allows claims by a defendant for contribution or indemnity to be added to the main action by way of third party proceedings, as is the situation in the present case, there is no good reason in principle why what is contemplated by CMR as the second and consequential action should not be brought by way of such third party proceedings."

11.86 Bingham, L.J., also noted[184] that if the defendant carrier did not wish to be heard on the trial of the plaintiffs' claim, they need not attend. However, Order 16, Rule 5 of the Rules of the Supreme Court applies to Order 16, Rule 8, so that the failure to give notice of intention to defend is deemed to be an admission of the claim. Bingham, L.J., may have had in mind the practice of the court not to make a declaration of right in default of defence.[185] He did note that arrangements can be made for the claim against the successive carrier to be heard after determination of the plaintiffs' claim.

Principles of apportionment

11.87 Turning to the amount recoverable under Article 37, the basic right is to full recovery of compensation paid, together with interest,[186] costs and expenses.[187] That basic right is then made subject to three rules, (a), (b) and (c), discussed below–

"(a)the carrier responsible for the loss or damage shall be solely liable for the compensation whether paid by himself or by another carrier."

11.88 The initial issue, of course, is as to the identity of "the carrier responsible". The effect of this provision will differ according to where the onus of proof is placed: if the onus is on the claimant carrier, it would be for him to show that the defendant carrier was the carrier actually responsible for the loss or damage. In practice, however, this may be very difficult for the claimant carrier to establish, so it is thought that a preferable approach would be to treat it as referring to the carrier in whose hands the goods were when the loss or damage was first discovered. In other words, if for example the following carrier makes reservations as to the state or quantity of the goods, the preceding carrier from whom he has taken over the goods will be "the carrier responsible", even if the loss

---

182. Discussed *infra*, para. 11.113 ff. This would appear to have been a relevant factor in the subsequent proceedings in the *Tatton* case itself: [1974] 1 Lloyd's Rep. 203, 204–205.

183. *Supra*, fn. 179.

184. At p. 495.

185. Now see CPR Part 20, whereby if the carrier is not a party to the main action and fails to acknowledge service or file a defence, then he will be bound by the decision in the main proceedings.

186. As to interest recoverable under the Convention, see Article 27, *supra*, para. 9.60 ff. but *quaere* whether Article 27 also applies to interest on interest, and note in particular the discussion of the position regarding interest after judgment.

187. The question of what costs and expenses are recoverable must, in the absence of any provision in the Convention, be regulated by the national law.

or damage could possibly have taken place while in the hands of an earlier carrier, unless evidence to this effect can be produced. The onus will therefore be upon the carrier *prima facie* "responsible" to disprove his liability. It is argued below[188] that Articles 17 and 18 apply to claims between carriers, and although the language of Article 17(1), which imposes the burden of proof on "the carrier", is at first sight unhelpful in the context of claims between carriers, it would seem consistent with the scheme of the Convention, in particular the provisions relating to the entry of reservations,[189] to read "the carrier" in Article 17(1) as meaning "the carrier *prima facie* responsible" in relation to claims between carriers.

11.89 In contrast to Article 36, paras (a), (b) and (c) of Article 37 all refer to responsibility for the loss or damage, omitting any reference to delay. It does not seem, however, that there is any significance to be attached to this omission, since the word "loss" would seem wide enough to cover loss caused through delay, and through failure to collect C.O.D. charges.[190]

11.90 It should be remembered that in *ITT Schaub-Lorenz* v. *Birkart Johann International Spedition G.m.b.H. Co. K.G.*,[191] Bingham, L.J., considered that a carrier cannot escape liability as "the carrier responsible" by showing that he had delegated or sub-contracted performance to a non-CMR carrier who was actually responsible[192] for the loss or damage. By Article 34, a successive carrier makes himself responsible for the whole operation and remains responsible for a non-CMR carrier or delegate under Article 3.[193]

11.91 The Court of Appeal in *Cummins Engine Co. Ltd.* v. *Davis Freight Forwarding (Hull) Ltd.*[194] found it necessary to consider Article 37(a). In that case a first carrier (C1) in this country sub-contracted the carriage of a consignment to a Dutch carrier (C2), who in turn sub-contracted the carriage to another Dutch carrier (C3), who then arranged for the carriage to be carried out by yet another Dutch carrier (C4). The goods were then damaged by C4 while in transit. It was held that third party notices issued in this country by C1 against C2 and C3 for contribution under Article 37 were to be set aside, since neither carrier was responsible for the damage within Article 37(a), the damage having been caused solely by C4. On the facts of the particular case, the decision as to the liability of C2 and C3 may have been justified, since there was no doubt that C4 was responsible for the loss or damage. There will often be circumstances, however, where at the outset it is far from clear to one carrier which of a number of successive carriers was responsible, and in such circumstances he must be allowed to proceed against them all lest his claim against any one of them becomes time-barred. Secondly, it may be objected that, even on

188. See *infra*, paras 11.98–11.100.
189. Article 35, *supra*, para. 11.55 ff.
190. The French text, which uses the word "*dommage*" in place of "loss or damage" in the English text, would equally seem wide enough to cover such losses.
191. [1988] 1 Lloyd's Rep. 487, 493.
192. He said that this would frustrate the CMR scheme. Presumably it would equally frustrate the CMR scheme to allow a successive carrier to show that, although physically responsible for the loss or damage, he was not legally responsible or not solely responsible since his contractor was responsible for his actions by virtue of Article 3. Cf. *supra*, para. 11.80.
193. This cannot extend to responsibility for other successive carriers, since it could lead to the result that more than one carrier is responsible for damage produced solely by one of them, cf. *supra*, fn. 135.
194. [1981] 2 Lloyd's Rep. 402.

the facts of this case, the fact that C4 was responsible for the damage does not automatically free C2 and C3 of liability since under Article 38,[195] in the event of C4's insolvency, each would have to pay a contribution, and again, unless proceedings have been commenced against all three at the outset, it may subsequently be too late to rely on Article 38.[296] Nor was the decision as to the effect of Article 37 necessary to the outcome of the case, since the court decided in any event that the provisions of Article 39(2) as to jurisdiction prohibited proceedings in this country against C2, C3 and C4.[197] For these reasons it is thought that this alternative basis for the court's decision is to be preferred.

11.92 The concluding words of Article 37(a), "whether paid by himself or by another carrier", simply make it clear that if the carrier who has initially paid the compensation is himself the carrier responsible for the loss or damage, he is unable to recover any contribution from the other carriers.

Shared responsibility

11.93 It is, of course, possible that the loss or damage complained of has been caused by more than one of the carriers. This possibility is covered by Article 37(b):

> "(b)when the loss[198] or damage has been caused by the action of two or more carriers, each of them shall pay an amount proportionate to his share of liability; should it be impossible to apportion the liability, each carrier shall be liable in proportion to the share of the payment for the carriage which is due to him."

11.94 Although Article 37(b) uses the words "caused by the action of" rather than "responsible for" which appears in para. (a), for the reasons already considered in relation to the latter phrase[199] it is thought that the two phrases mean the same, and that "caused by the action of" does not necessarily mean that the carrier concerned has caused the loss or damage through his own actions.[200] Such liability could therefore arise either as the result of the act or default of a sub-contracting carrier who is not a successive carrier under CMR but for whom the carrier is liable under Article 3, or else because the carrier is not in a position to prove that the damage actually occurred prior to him taking over the goods, either by his driver having made reservations in the consignment note or by extraneous evidence.[201]

11.95 If it is impossible to apportion liability among the carriers responsible for the loss then each of them will be liable *pro rata* to his share of the carriage charges for the whole transit.[202] As in practice it is often far from easy to apportion accurately the responsibility between two or more successive carriers the alternative of basing contribution upon the carriage charges is a flexible and practical alternative.[203] If any carrier can

---

195. Discussed *infra*, para. 11.101.
196. However, on the facts of the case the suggestion that, if C4 became insolvent, then C1 could have recourse against C2 and C3 under Article 38 (*infra*, para. 11.101–11.112) was rejected since there was no evidence to suggest that they were, or were likely to become, insolvent.
197. See *infra*, paras 11.119–11.123.
198. This includes loss consequent upon delay; see *supra*, para. 11.89.
199. *Supra*, para. 11.89.
200. Cf. Loewe, para. 283.
201. As to the burden of proof, see *infra*, paras 11.98–11.100.
202. Article 37(b). As to how this apportionment is to be applied in practice, see *infra*, paras 11.105–11.112.
203. Cf. the more complex causal apportionment of liability under Article 17(5), *supra*, paras 6.107–6.110.

in fact prove that neither he nor his sub-contractors were responsible for the loss or damage in question then it must be assumed that even though it might not be possible to prove which of the other carriers was responsible, that carrier would not be required to contribute.[204] On the other hand, in the absence of clear words to that effect, it might be argued that the court is to apportion between all the carriers in any event.

Unascertained responsibility

11.96 The final possibility is that the circumstances in which the loss or damage occurred are such that it is not possible to ascertain which of the carriers concerned is responsible. This possibility is covered by Article 37(c) which provides:

> "(c)if it cannot be ascertained to which carriers liability is attributable for the loss[205] or damage, the amount of the compensation shall be apportioned between all the carriers as laid down in (b) above."

11.97 This situation is most likely to face the last carrier, who, as the most convenient carrier to sue, will be forced to pay compensation to the cargo interests, even though he can prove that he was not personally liable in the matter. Under such circumstances he can recover the amount of compensation paid from the other carriers, either based upon their attributable liability or else upon a *pro rata* apportionment.[206]

11.98 Article 37 deals with the apportionment of liability between carriers, but makes no provision as to how that liability is to be established. The question therefore arises as to whether Articles 17 and 18 apply to claims between carriers so as to regulate the burden of proof and available defences, and this in turn raises the broader question of whether Chapter IV of the Convention, the chapter of which Articles 17 and 18 form part, applies to claims between carriers.

11.99 The conclusion that Chapter IV does apply is supported by the view already put forward[207] that any compensation paid to the cargo interests must comply with the Convention both as to the grounds of liability and as to quantum (matters governed by Chapter IV) before it can form the subject matter of a claim under Article 37. Further, it would be curious indeed if the burden of proof under Articles 17 and 18 was not to apply to claims between carriers, since a situation could then arise where the carriers initially sued could be held liable in circumstances where, were he to be deprived of the presumptions in Articles 17 and 18, he might not be able to pass on that liability. There is nothing, moreover, in the wording of Chapter IV which is inconsistent with its application to Article 37: its heading, "Liability of the Carrier" is clearly appropriate to cover claims under Article 37, and throughout the chapter the authors of the Convention have, where necessary, been careful to use the word "claimant"[208] so that there is no

---

204. This view is supported by Rodière, para. 135, and Nickel-Lanz, No. 203. See further Haak, p. 115.
205. Again, this will include loss caused through delay; *supra*, para. 11.89.
206. It should be noted that Article 40 gives successive carriers complete freedom to make any other arrangements they may wish *inter se* instead of relying on those laid down in Article 37. See *infra*, paras 11.134–11.135. Again, on the wording of the Convention it should make no difference in principle if one of the carriers can demonstrate that he was not responsible.
207. *Supra*, para. 11.76.
208. Articles 17(2), 18(2), 23(5) and 27(1).

conflict of wording which would prevent the application of the provisions of Chapter IV to claims under Article 37.[209]

11.100 So although it has been held in relation to Article 4 that, because it was not expressly incorporated into the provisions relating to successive carriers, it was therefore not relevant for the purposes of Article 35,[210] for the above reasons it is thought that such cannot represent a general rule, and that the provisions of Chapter IV in general, and of Articles 17 and 18 in particular,[211] will apply to claims between carriers under Article 37.

Insolvency of a carrier

11.101 Article 38 provides as follows–

"If one of the carriers is insolvent, the share of the compensation due from him and unpaid by him shall be divided among the other carriers in proportion to the share of the payment for the carriage due to them."

11.102 In the event of the insolvency of a carrier, his unpaid liability will thus be apportioned between the remaining carriers[212] in proportion to the amount of the carriage charges due to each of them. Clearly, therefore, it will be in the interests of carriers under the Convention to check as to the insurance arrangements,[213] or as to the ability in general to pay possible claims, of carriers performing other segments of the carriage.[214]

11.103 The Convention provides no definition of what is meant by insolvency so this will depend on the applicable national law. In this country, insolvency has no general technical connotation other than an inability to pay one's debts.[215] However, the statutory definition in the Sale of Goods Act 1979 reflects closely various judicial statements as to the meaning of insolvency in other contexts[216]: "A person is deemed to be insolvent within the meaning of this Act who has either ceased to pay his debts in the ordinary course of business, or cannot pay his debts as they become due, whether he has committed an act of bankruptcy or not . . .".[217] Such an interpretation could clearly produce some uncertainty as to whether Article 38 is to be applied as between the remaining carriers, and in practice it may well be that the carrier initially sued under Article 36 will in the first instance only be able to recover from the other carriers the contributions due under Article

209. The only exception to this is the provision in Article 22 relating to dangerous goods (*supra*, Chapter 8), which is not entirely appropriate to cover the situation where, for example, the first carrier has been told by the sender of the dangerous nature of the cargo, but the first carrier has failed to pass on that information to a successive carrier. Common sense dictates, however, that in such circumstances the successive carrier should take the benefit of Article 22 as against the first carrier.

210. *SGS-Ates Componenti Elettronici S.p.A.* v. *Grappo Ltd.* [1978] 1 Lloyd's Rep. 281, discussed *supra*, paras. 11.26–11.32.

211. It was assumed that Article 17 does so apply in the case of *Walek & Co.* v. *Chapman and Ball (International) Ltd.* [1980] 2 Lloyd's Rep. 279. On the application of Article 17(3) in this case cf., however, Glass, [1981] LMCLQ 384 and Cass., 17.6.97 (1997) B.T. 596.

212. As to whether this provision applies to carriers who are not successive carriers within Article 34, see *supra*, paras. 11.77–11.81.

213. *Quaere* whether Article 38 applies where a carrier is insolvent but has insurance cover in relation to the claim.

214. In this connection it should be noted that not all U.K. insurance policies cover the carrier's liability under Article 38.

215. *Re Muggeridge's Trusts* (1860) 29 L.J. Ch. 288.

216. E.g. *London & Counties Assets Co. Ltd.* v. *Brighton Grand Concert Hall & Picture Palace Ltd.* [1915] 2 K.B. 493.

217. Section 61(4).

37, while awaiting clarification of the position as regards a carrier claiming to be insolvent. In such circumstances he should take care to ensure that time is protected in relation to the other carriers, lest his claim against them under Article 38 become time-barred under Article 39(4).[218]

11.104 Again, it should be noted that Article 40 gives successive carriers complete freedom to make any other arrangements they may wish *inter se* instead of relying on the provisions of Article 38.

Calculation of contributions under Articles 37 and 38

11.105 Articles 37 and 38 both make provision for the apportionment of liability between carriers on the basis of "the share of the payment for carriage due to them". This formula, however, is not entirely clear as to its meaning, since it admits of at least two possible interpretations, which produce very different results when applied to the facts of individual cases. The first possible meaning is that the apportionment is to be based on the carriage charges received by each carrier after deduction of the amount he has been charged by the next carrier. The alternative possibility is that it is based on the gross amount charged by each carrier.

11.106 It will be recalled that apportionment of compensation based on the carriage charges can become necessary in two contexts: under Article 37(b) and (c) where it is impossible to apportion liability on the basis of responsibility for the loss or damage, and under Article 38 where a carrier liable to make a contribution is insolvent. The following hypothetical examples illustrate the differing effects of the two interpretations in each situation.

Example A

11.107 A sender engages a carrier (C1) to effect carriage from England to Italy. C1 carries the goods a short distance within England, then sub-contracts the remaining carriage to C2. It is impossible to ascertain which carrier was responsible for damage which had been discovered on delivery. C2 has invoiced C1 for £900, and C1 has invoiced the sender for £1,000.

> First interpretation:
> Total carriage charges invoiced = £1,000 of which
> £100 = 10% = C1's share of the payment for the carriage
> £900 = 90% = C2's share of the payment for the carriage

> Second interpretation:
> Total carriage charges invoiced = £1,000 = £900 = £1,900 of which
> £1,000 = 52.63% = C1's share of the payment for the carriage
> £900 = 47.37% = C2's share of the payment for the carriage

Example B

11.108 A sender engages a carrier (C1) to perform a carriage which C1 sub-contracts in its entirety to C2, who performs the bulk of the carriage, sub-contracting the final stage

---

218. *Infra*, para. 11.127 ff.

to C3. C3, who damaged the goods in transit, is now insolvent. C3 invoiced C2 for £150; C2 invoiced C1 for £450; C1 invoiced the sender for £480.

> First interpretation:
>
> Carriage charges to be received by C1 = £480 − £450 = £ 30
> Carriage charges to be received by C2 = £450 − £150 = £300
>
> $$\text{Total} = £330 \text{ of which}$$
>
> £30 = 9.09% = C1's share of the payment for the carriage
> £300 = 90.91% = C2's share of the payment for the carriage
>
> Second interpretation:
>
> Total carriage charges invoiced by C1 and C2 = £480 = £450
>
> $$= £930 \text{ of which}$$
>
> £480 = 51.61% = C1's share of the payment for the carriage
> £450 = 48.39% = C2's share of the payment for the carriage

11.109 It will be observed from these examples that the first interpretation will always mean that the carrier who performs the bulk of the carriage, as opposed in particular to those contracting to perform it then sub-contracting its performance, will always bear the most substantial part of the apportionment. Under the second interpretation, on the other hand, the first carrier will necessarily bear the highest proportion, since it will be the first carrier who presents the sender with the global invoice covering the whole carriage.

11.110 This last point apart, the arguments in favour of the first interpretation would seem to be twofold. First, it can be argued that the carrier who has carried the goods farthest stands a proportionately greater chance of being the party responsible for the loss or damage, so that it is right that he should bear the greater part of the apportionment. Secondly, it can be argued that the carrier who receives the greater part of the carriage charges should equally bear the greater part of the apportionment.

11.111 The first of these arguments can be shortly disposed of. Quite apart from the doubtful justice of it, it can only be applicable to an apportionment under Article 37 where liability cannot be established between two or more carriers, and is clearly of no application to Article 38, where no question of fault is at issue. The second argument, it is submitted, is equally invalid because it ignores the commercial realities of the situation, in that it fails to take account of the fact that the carrier who has received the greater part of the carriage charges may well have incurred the bulk of the expenditure in the performance of the carriage. The first interpretation is in this way illogical in that it takes account of the expense incurred by carriers in sub-contracting the carriage, but fails to take account of expenses incurred by a carrier in performing the carriage.

11.112 It must be conceded that there is an element of unfairness in the second interpretation in that, as already noted, the first carrier will always be paying the largest contribution, but the degree of unfairness is less: the first carrier pays a slightly larger contribution, instead of the carrier performing the carriage, or the largest part of the carriage, paying a substantially larger contribution. The only truly equitable basis for the apportionment would be to utilise the first interpretation but on the basis of each carrier's net profit on the carriage. Although this might be an overly complex solution, and there does not seem to be any reason in principle why it should not be applied, the burden being on individual carriers to justify expenses incurred. The only English decision which has had to consider the apportionment provisions seems to have assumed without discussion that the first interpretation applied without qualification.[219] However, since that case was, in the event, decided on other grounds, the judge's comments as to the apportionment provisions can be regarded as obiter.[220]

Procedural aspects of claims between carriers

11.113 Article 39 provides as follows–

"1. No carrier against whom a claim is made under articles 37 and 38 shall be entitled to dispute the validity of the payment made by the carrier making the claim if the amount of the compensation was determined by judicial authority after the first mentioned carrier had been given due notice of the proceedings and afforded an opportunity of entering an appearance.

2. A carrier wishing to take proceedings to enforce his right of recovery may make his claim before the competent court or tribunal of the country in which one of the carriers concerned is ordinarily resident, or has his principal place of business or the branch or agency through which the contract of carriage was made. All the carriers concerned may be made defendants in the same action.

3. The provisions of article 31, paragraphs 3 and 4, shall apply to judgments entered in the proceedings referred to in articles 37 and 38.

4. The provisions of article 32 shall apply to claims between carriers. The period of limitation shall, however, begin to run either on the date of the final judicial decision fixing the amount of compensation payable under the provisions of this Convention, or, if there is no such judicial decision, from the actual date of payment."

Effect of judgment against claimant carrier

11.114 In accordance with the overall concern of the Convention to avoid unnecessary duplication of actions,[221] Article 39 starts by providing that a carrier[222] from whom a contribution is claimed under Articles 37 and/or[223] 38 cannot challenge the validity of the payment made by the claimant carrier to the cargo interests, provided that the amount so paid was determined by "judicial authority".[224] Such provision on its own could operate unjustly against the defendant carrier, for he may have in his possession information which, if available at the first trial, might have had a bearing on liability to the cargo interests. Article 39(1) therefore concludes with the proviso that before it can be relied on

---

219. *Walek & Co.* v. *Chapman and Ball (International) Ltd.* [1980] 2 Lloyd's Rep. 279.

220. See also Tilche, (1999) B.T. 85.

221. See Article 31(2) and (3), *supra* and Article 39(3), *infra*.

222. As to whether Article 39 applies to carriers who are not successive carriers within the meaning of Article 34, see *supra*, paras 11.77–11.81.

223. Article 39(1) in fact uses the conjunctive alone, which would not seem appropriate since only on rare occasions will a carrier be liable under both Articles 37 and 38, and, on a literal interpretation, it would only be in those cases that Article 39(1) could apply. Such, clearly, cannot have been the intention of the authors of the Convention, but neither is the disjunctive appropriate, since it could then be argued that Article 39(1) would not apply in those cases, albeit rare, where there was liability under both Articles 37 and 38. The words "and" and "or" have troubled the courts over the years (see, e.g., the differing views of the House of Lords in *Federal Steam Navigation Co. Ltd.* v. *Department of Trade and Industry* [1974] 1 Lloyd's Rep. 520 and the cases therein cited) but this would seem an occasion for adopting the "new approach" to the construction of international conventions (see the comments of Lord Denning in *James Buchanan & Co. Ltd.* v. *Babco Forwarding & Shipping (U.K.) Ltd.* [1977] 1 Lloyd's Rep. 234, 236–237, and the House of Lords' decision in *Fothergill* v. *Monarch Airlines Ltd.* [1980] 2 Lloyd's Rep. 295) and regard should be had to the intent, purposes and scheme of the Convention. So viewed, it is submitted that Article 39(1) should be read as indicated in the text above.

224. The meaning of this is discussed *infra*.

by the claimant carrier, the defendant carrier must have been given notice of the proceedings and an opportunity to enter an appearance in the first action. In other words, the defendant carrier must have had the opportunity of taking part in the initial proceedings, and if he has not taken advantage of that opportunity he cannot then object to the court's award.[225] If, however, the defendant carrier does not receive notice of the initial proceedings, or only receives notice in such circumstances that he has no opportunity of entering an appearance, then Article 39(1) will not apply and he will be free to dispute the validity of any payment made pursuant to those proceedings. For this reason it is essential that if the cargo interests do not join all the carriers together in the initial action, then the carrier sued must ensure that all other carriers are given due notice of the proceedings.

11.115 The Convention gives no guidance as to the procedure or period of notice required, so these matters will depend on the national law.[226] Obviously, however, it will be in the interests of the claimant carrier to notify other carriers of the cargo interest's claim at the earliest opportunity, so that any information they may have as to the merits of the claim can be put forward at an early date. If, at that stage, proceedings have not been commenced, further notification will be necessary to comply with Article 39(1) as the claim proceeds to litigation. It is worth noting that Article 39(1) does not require the notice and opportunity of appearing to be given by the claimant carrier. It would seem to follow that if it could be shown that the defendant carrier had notice of the proceedings from whatever source, and that under the applicable law he had the opportunity of entering an appearance in those proceedings, then this aspect of Article 39(1) would be satisfied.[227] Clearly, however, the onus of proof in this respect would be on the claimant carrier, and equally clearly it would not be in his best interests to have to rely on such communication.

11.116 The Convention does not define what is meant by a "judicial authority", but this would clearly include a court of any contracting State which properly had jurisdiction under the Convention.[228] It will also be recalled that Article 31(1) apparently permits proceedings in countries which are not contracting States,[229] so it would seem that such a judgment can constitute "judicial authority" for the purposes of Article 39(1). It has further been suggested that a judgment of a court which is not within a contracting State will be binding under Article 39(1) even if it has not applied the provisions of the Convention.[230] However, this view ignores the fact that Article 39(1) applies only to claims under Articles 37 and 38. Article 37 expressly, and Article 38 by necessary

225. If he has participated in the proceedings it follows from Article 39(1) that he will be bound by the outcome.

226. "You cannot envisage any wording which would be exactly suitable for the procedures of all the countries concerned, but at least you can see what the draftsmen had in mind here", per Master Elton in *William Tatton & Co. Ltd.* v. *Ferrymasters Ltd.* (1974) 9 E.T.L. 167, 199–200. In that case Master Elton held that although the third party notice against successive carriers was of no effect in terms of compelling them to join in the action brought by the cargo interests, it was effective as "notice of the proceedings and . . . an opportunity of entering an appearance". However, this was in the context of the successive carrier having entered a conditional appearance, and Master Elton indicated that he would otherwise have thought it the wrong form of notice (ibid., at p. 200).

227. However, in *William Tatton & Co. Ltd.* v. *Ferrymasters Ltd.* (1974) 9 E.T.L. 167, 200 Master Elton was of the opinion that Article 39(1) meant that the successive carriers must be "served with some notice to which they can enter an appearance in the action". This view is not supported by the wording of Article 39(1), but, if correct, means that the possibility suggested in the text is incorrect.

228. As to which see Article 31(1), *supra*, Chapter 10, para. 10.21 ff.

229. *Supra*, para. 10.24.

230. Loewe, para. 287.

implication,[231] apply only to compensation paid "in compliance with the provisions of the Convention", so it follows that if any court, wherever it might be, failed to apply the Convention, the award would not be within Articles 37 or 38, and would therefore not be within Article 39(1). One can conclude that the better view is that Article 39(1) does apply to judgments of courts within States not party to the Convention,[232] but only if the provisions of the Convention have been applied to the question at issue.

11.117 So far as arbitration proceedings are concerned, it will be recalled that Article 33 permits arbitration where provided for by the contract of carriage. Since successive carriers become parties to the contract of carriage by virtue of Article 34, it would seem that an arbitration award will equally be binding on a successive carrier under Article 39(1) provided that the arbitration agreement was included in the consignment note or otherwise communicated to that carrier prior to or at the time of his taking over the goods.[233]

11.118 It is important to note the limits of Article 39(1). First, since it applies only to compensation "determined by judicial authority", if the claimant carrier has settled with the cargo interests without such judicial determination, then the defendant carrier will be free to dispute the validity of the payment.[234] The fact that this can produce litigation which might otherwise be needless has already been noted.[235] Secondly, Article 39(1) only provides that the defendant carrier shall not be permitted to challenge the validity of the payment. This does not prevent that carrier from disputing his liability for that payment. He would therefore be able to argue, for example, that he was not responsible for the loss or damage which has occurred.

Jurisdiction

11.119 Where a carrier who has compensated a third party wishes to enforce his right of recovery against another carrier he can make his claim before the competent courts of the country in which one of the carriers concerned is ordinarily resident, or has his principal place of business, or else where the branch or agency through which the contract of carriage was made is situated.[236] This differs from the equivalent provision in Article 31(1), which offers a wider choice of jurisdiction for an action by the cargo interests against a carrier.[237] English authority suggests that this difference involves a restriction on the jurisdictions available to a carrier seeking contribution. In *Cummins Engine Co. Ltd.* v. *Davis Freight Forwarding (Hull) Ltd.*,[238] Brandon, L.J.,[239] considered that in the

---

231. This is because Article 38 refers to the share of compensation due from a carrier, which can only be established by reference back to Article 37.

232. But note that Article 31(3) (*supra*, para. 10.43 ff.), relating to the enforcement of judgments in other jurisdictions, which is applied to Articles 37 and 38 by Article 39(3), only applies to judgments obtained in contracting countries. However, see *supra*, paras 10.47–10.48 as to the 1968 Brussels Convention on Jurisdiction and the Enforcement of Judgments in Civil and Commercial Matters.

233. See *supra*, paras. 10.120–10.130.

234. Clearly it does not prohibit the claimant carrier from making his claim for reimbursement merely because he has settled the claimant's claim: App. Paris, 21.9.95 (1995) B.T. 857.

235. *Supra*, paras 11.82–11.86.

236. Article 39(2).

237. See *supra*, Chapter 10, para. 10.21.

238. [1981] 2 Lloyd's Rep. 402.

239. [1981] 2 Lloyd's Rep. 402, 408.

context of this provision the word "may" must be interpreted as meaning "must".[240] Furthermore, he considered that there is nothing else in Article 39 or any other part of CMR which permits actions to be brought for an indemnity by one carrier against other carriers in the courts of any country other than as specified in Article 39. Eveleigh, L.J.,[241] was not prepared to go as far as to say that the word "may" is equivalent of "must".[242] Consequently, he left open the question whether contribution could be claimed against a party who had entered an appearance for the purpose of disputing the validity of the payment after receiving notice under Article 39.[243]

11.120 In the later case of *Arctic Electronics Co. (U.K.) Ltd.* v. *McGregor*,[244] Hobhouse, J., expressed a preference for the view of Eveleigh, L.J., taking the view that to give Article 39(2) a compulsory and exclusive effect would not be consistent with the proceedings contemplated and authorised by Article 39(1), which admitted the possibility of a submission to jurisdiction by a successive carrier. In his view, Article 39(2) defines the scope of the extra-territorial jurisdiction which a State which is party to the Convention is permitted to assume. As Eveleigh, L.J., had pointed out,[245] Order 11, Rule 1(1) of the Rules of the Supreme Court (now see CPR Part 6), which permit service out of the jurisdiction for CMR claims, envisages only claims which could be made according to the procedure specifically permitted or envisaged by Article 39(2). In *Harrison & Sons Ltd.* v. *R.T. Steward Transport Ltd.*,[246] Gatehouse, J., preferred the view of Brandon, J., but nevertheless felt that the difference in view was a narrow one and could be reconciled by the proposition that a contribution claim by a carrier who has paid compensation under Article 37 can be brought only against another carrier concerned who is liable to be sued in accordance with Article 39(2), or who has voluntarily entered appearance under Article 39(1). On either view therefore, it would appear that no extra-territorial jurisdiction may be taken by an English court except under Article 39(2) or resulting from voluntary submissions. Consequently, given the mandatory nature of the Convention as laid down by Article 41, it does not appear possible for the carriers to agree among themselves for the designation of an alternative jurisdiction so as to exclude one of those listed in Article 39(2), as the freedom given to carriers by Article 40 to derogate from the provisions of the Convention is restricted to that laid down by Articles 37 and 38. It does not extend to Article 39.[247]

240. Cf. the non-CMR Scottish case of *Morrison* v. *Panic Link Ltd.* (1994) S.L.T. 232, where the permissive sense indicated by means of the use of the word "may" in a jurisdiction clause could only be outweighed by a clear contrary indicator in the context in which it was used.

241. At p. 409.

242. As he pointed out, Article 39(2) does not contain the equivalent words to those in Article 31: "and in no other courts or tribunals." Cf. *William Tatton & Co. Ltd.* v. *Ferrymasters Ltd.* (1974) 9 E.T.L. 167, 198.

243. O'Connor, L.J., at p. 409 expressed himself in agreement with the reasons for the decision given by Brandon, L.J.

244. [1985] 2 Lloyd's Rep. 510.

245. In the *Cummins Engine* case at p. 409.

246. (1993) 28 E.T.L. 747.

247. Loewe, para. 288, takes the view, on Article 40, that the parties can designate an alternative jurisdiction, but for the reason given above it is not thought that this is correct. In support of Loewe's view (also Nickel-Lanz, para. 208) it is possible that the purpose of Article 39(2) is to guarantee the claimant carrier with a place of suit, which cannot be excluded by the defendant carrier, without necessarily depriving him of jurisdictional rights provided either by contract or national law. On this view there would be room, in particular, for the provisions of the Civil Jurisdiction and Judgments Act 1982 to operate in providing other jurisdictional possibilities. It has been held in France (Cass., 21.6.82 (1983) 18 E.T.L. 207, (1982) B.T. 513) that CMR does not exclude a claim in guarantee before the court seised of the original claim in accordance with Article 6 of the Brussels Convention. This case has been criticised by Haak, p. 278, and see further Clarke, p. 212.

11.121 Another aspect of Article 39(2) was also considered by the Court of Appeal in *Cummins Engine Co. Ltd.* v. *Davis Freight Forwarding (Hull) Ltd.*[248] In that case it was also held that the words "carriers concerned" in the first sentence of Article 39(2) meant only carriers from whom a contribution was claimed, not a carrier seeking contribution. It was therefore not open to a first carrier, sued by the cargo interests in this country, to seek to join three Dutch successive carriers in the proceedings; the action against them had to be brought in Holland, where each of the successive carriers had its place of business. Third party notices issued in this country by the first carriers were therefore set aside.

11.122 Where an action is brought against a successive carrier in a jurisdiction permitted by Article 39(2), that paragraph goes on to provide that all the carriers concerned may be joined as defendants in the same action. Thus, if the carrier first sued by the cargo interests claims from another carrier in the country where that carrier has, for example, his principal place of business all other successive carriers can be joined in the same action, regardless of whether an action against them in that country would otherwise have been permitted by Article 39(2). Thus in *Harrison & Sons* v. *R.T. Steward Transport Ltd.*[249] defendant carrier A. was held to be able to bring third party proceedings in England against carrier B. on the basis of the English residence of carrier C. Again, in this provision, we see the concern of the Convention to avoid unnecessary duplication of actions.

11.123 However, the "carriers concerned" for the purposes of Article 39(2) must be concerned in respect of carriage of the same goods under a single consignment note. In the case of *Arctic Electronics Co. (U.K.) Ltd.* v. *McGregor*,[250] a consignment of goods carried by air to Luxembourg under cover of a single air waybill was split by the contracting carrier into three different lots for on-carriage by road from Luxembourg to London by separate road carriers under separate consignment notes. These latter road carriers were not "carriers concerned" in relation to each other, so that the residence of one could not be used by the contracting carrier to found the jurisdiction of the court in respect of another.[251]

The enforcement of judgments

11.124 It will be recalled that in relation to claims by the cargo interests against a carrier, Article 31(3) and (4) make provision for the enforcement in one contracting country of a judgment entered in another contracting country.[252] Article 39(3) provides that those provisions apply equally to judgments under Articles 37 and 38. Subject to what follows, there do not appear to be any special problems in applying Article 31(3) and (4) to claims between carriers, so reference should be made to the earlier discussion of those provisions.[253]

---

248. [1981] 2 Lloyd's Rep. 402.
249. (1993) 28 E.T.L. 747.
250. [1985] 2 Lloyd's Rep. 510.
251. A fatal objection was the provision in Article 38 for the sharing of compensation where one of the carriers is insolvent. It was obvious that none of the performing road carriers could incur the responsibility to cover the possible insolvency of any of the others (at p. 513).
252. See generally *supra*, para. 10.43 ff.
253. Since only paras (3) and (4) of Article 32 are incorporated into relations between successive carriers, it follows that paras (1), (2) and (5) are not so incorporated. Clearly para. (1) would be inappropriate, since alternative provision as to jurisdiction is made in Article 39(2), but it is not so clear why paras (2) (pending

11.125 The only difficulty that does exist is that a simple reference back to Article 31(3) is not an entirely satisfactory method of incorporation due to the fact that Article 31(3) in its turn refers back to Article 31(1), which is clearly inapposite in relation to claims between carriers. However, bearing in mind that strict adherence to traditional rules of statutory interpretation is not always appropriate in the context of international conventions,[254] there would seem no difficulty in reading the reference to Article 31(3) in relation to successive carriage as it was clearly intended, i.e. as applying to judgments entered in respect of claims under Articles 37 and 38 in jurisdictions permitted under Article 39(2).

11.126 It should be remembered that Article 31(3) applies only to judgments in contracting countries, so that if a carrier has obtained judgment otherwise than in a contracting country he will not be able to rely on this provision for the enforcement of his judgment elsewhere. The position would then be governed by the normal rules of private international law or, in respect of States parties to the Brussels and Lugano Conventions,[255] by those Conventions as law supplemental to CMR in respect of an issue not governed by it.[256] National law will also govern the procedural matters in respect of enforcement, including limitation.

The limitation of actions

11.127 Article 39(4) provides that the provisions of Article 32,[257] which deal with the limitation of actions, shall apply to claims between carriers, except that, replacing the equivalent provision in Article 32(1), the period of limitation[258] will begin to run either on the date of the final judgment fixing the amount of compensation payable under the Convention by the carrier sued by the cargo interests, or, if there is no such judgment, from the actual date of payment.[259] This provision gives rise to a number of points. First, it should be remembered that if the carrier was paid compensation without it being determined by a court judgment, then the second carrier or carriers claimed against will not be precluded by Article 39(1) from disputing the validity of the payment. Secondly, it will be recalled that there is some authority in this country that proceedings under Article 37 cannot be commenced by one carrier against another carrier until compensation

---

actions) and (5) (security for costs) were not considered appropriate, either by way of direct incorporation or by way of equivalent provision.

254. *Fothergill* v. *Monarch Airlines Ltd.* [1980] 2 Lloyd's Rep. 295.

255. The 1968 Brussels Convention on Jurisdiction and the Enforcement of Judgments in Civil and Commercial Matters, given the force of law by the Civil Jurisdiction and Judgments Act 1982 in respect of EC States and the 1988 Lugano Convention, given the force of law by the Civil Jurisdiction and Judgments Act 1991 in respect of EFTA States.

256. See *supra*, paras 10.47–10.48.

257. See generally *supra*, Chapter 10, para. 10.51 ff.

258. The duration of the limitation period is the same as under Article 32(1), i.e., one year, or three years in cases of wilful misconduct.

259. Where there is such a judgment, the claimant carrier cannot rely on a later date of payment, BGH., 10.5.90 (1991) 26 E.T.L. 351, 355. It was also held in this case that where a judicial decision has incorrectly characterised a party as a successive carrier and required him to pay compensation to the sender, this does not alter the principle that the relevant time limit is that in Article 32 and not that in Article 39(4). The court left open the possibility, however, that since the effect of the decision was to make this carrier a co-debtor with the defendant carrier, an exceptional basis for the application of Article 37 could exist. On any analysis of the facts, however, the claimant carrier was out of time.

has actually been paid by the former to the cargo interests.[260] It would seem to follow that even where the claimant carrier under Article 37 has judgment against him, unless he has actually paid that compensation at the time proceedings are commenced against the defendant carrier, there may be some risk that that defendant carrier might apply for the proceedings to be struck out: a subsequent attempt to re-commence proceedings after payment to the cargo interests may by then be out of time.

11.128 Another question that arises in relation to Article 39(4) is as to whether it is concerned only with claims involving loss, damage or delay, or whether it equally applies to claims between carriers in respect of other matters, such as claims between them for freight charges. This question came before the English courts in *Muller Batavier Ltd.* v. *Laurent Transport Co. Ltd.*,[261] which involved a claim by one successive carrier against another in respect of unpaid freight charges. The claimant carrier's claim was time-barred under Article 32(1)(c), but it argued that since this was a claim between carriers, Article 39(4) applied, and since there had been neither judicial decision nor payment, time had not yet started to run. This argument was rejected by May, J. He started by pointing out the illogicality of trying to apply Article 39(4) to such a claim:

" . . . if, in a given case, the limitation period as between carriers under art. 39, par. 4, is only to run from the actual date of payment of whatever is due, there having been no judicial decision, no limitation period is relevant, because ex hypothesi payment has been made and consequently there is no need for any limitation period to start."[262]

In other words, the reference in Article 39(4) to the payment as starting the limitation period running cannot be appropriate to claims for unpaid freight charges between carriers, since if they have been paid there is no question of a limitation period being relevant. The relevant payment would therefore have to be by the claimant carrier to a third party in order for this part of Article 39(4) to make any logical sense. As for the reference to a judicial decision as starting the limitation period running, it is plainly nonsense to talk in terms of the judicial decision in respect of the carriage charges as starting the limitation period in respect of that very claim.

11.129 However, May, J., went on to base his decision on this aspect of the case on his finding that the word "claims" in Article 39(4), having regard in particular to the words "fixing the amount of the compensation" and in general to the overall provisions of Articles 36 to 39, meant claims by one carrier against another in respect of the latter's responsibility for something that has gone wrong during the carriage. Article 39(4), he concluded, was not applicable to a claim for freight charges, and the plaintiff's claim was therefore time-barred.

11.130 If Article 39(4) has no application whatsoever to claims between carriers for freight charges, an unpaid successive carrier is in some difficulty where there has been loss, damage or delay, because his claim for freight charges will become time-barred before any claims against him under Article 39(4). At the same time he may well feel hesitant at pressing for payment of his charges while issues relating to loss, damage or delay remain outstanding. However, on the basis of May, J.'s decision in the *Muller Batavier* case, if he is unable to obtain an agreed extension of time[263] from the carrier owing him

---

260. Discussed *supra*, paras 11.82–11.86.
261. [1977] 1 Lloyd's Rep. 411. See Hardingham, "Aspects of the limitation of actions under CMR", [1979] LMCLQ 362.
262. [1977] 1 Lloyd's Rep. 411, 414.
263. As to which, see *supra*, paras 10.115–10.116.

his charges, he will be obliged to commence proceedings lest he find his claim time-barred.

11.131 Another possibility is, however, suggested in May, J.'s judgment. Having reached the conclusion that Article 39(4) did not apply to the facts of the case before him, he continued:

"Were there claims to compensation for damage to the goods or delay to the goods, or injury to the goods, or loss of the goods, in the course of the carriage, then art. 39, par. 4, might well be relevant."[264]

Presumably he cannot have been referring simply to a situation where there was damage, delay or loss, since then plainly Article 39(4) would apply. It would therefore seem that May, J., was referring to situations where there was such a claim *and* a claim for freight charges. Such an interpretation makes sense in the context of Article 23(4)[265] whereby a refund of freight charges may have to be made to the cargo interests by the carrier initially claimed against, who may have therefore been withholding them from the carrier to whom they were due, and would in such circumstances avoid a differing period of limitation as between the respective claims.

11.132 Such an interpretation does, however, produce difficulties of its own. It may happen that the carrier first claimed from under Article 36 does not proceed under Article 37 against another carrier until that carrier's claim for freight charges has become time-barred under Article 32. Is the claim for freight charges then revived under Article 39(4)? On the one hand, this would appear to be expressly precluded by Article 32(4) which provides that once a right of action has become time-barred, it cannot be exercised by way of counter-claim or set-off. On the other hand, however, it can be argued that the effect of the claim under Article 37 is to bring the defendant carrier's claim for carriage charges within Article 39(4), and that Article 32(4) is no longer applicable so as to prevent the counter-claim. If the claim for freight charges is revived, the Convention thus produces a curious limbo for such claims: even though time-barred under Article 32, they may subsequently cease to be so by virtue of the carrier owing the charges claiming a contribution under Article 37.

11.133 If Article 39(4) does not operate to revive the claim in such circumstances, it does at least seem from May, J.'s suggestion that, if a carrier is claimed against under Article 37 in respect of loss, damage or delay before the limitation period under Article 32 has expired in respect of his carriage charges, the longer period in Article 39(4) can then apply to his counter-claim for those charges; Article 39(4) will have come into operation before the claim was time-barred, so there is no conflict with Article 32(4). This more limited view is unlikely to be of much assistance, however, since a carrier thus sued would presumably in most cases issue his counter-claim at the stage he is sued, thus bringing himself within the Article 32 limitation period in any event.

Alternative arrangements between carriers

11.134 Article 40 has already been mentioned in discussing Articles 37 and 38. It provides as follows:—

---

264. [1977] 1 Lloyd's Rep. 411, 415.
265. *Supra*, Chapter 9, 9.21 ff.

"Carriers shall be free to agree among themselves on provisions other than those laid down in articles 37 and 38."

In effect, therefore, carriers are permitted to make whatever contractual arrangements they may deem fit as regards the division of liability among themselves. Clearly, it matters not whether Article 40 applies only to successive carriers within the meaning of Article 34, since any such agreement would in any event be binding on them under normal contractual principles. However, such an agreement can only operate as between the carriers themselves, since any attempt to affect the rights of the cargo interests would be inoperative by virtue of Article 41.[266]

11.135 Carriers may wish to take advantage of Article 40, for example, where there is a course of trading between them or where they are associated companies, either of which situations may warrant different arrangements for the division of liability between them than those contained in Articles 37 and 38. In the absence of such circumstances, there would be no advantage to one carrier in restricting the liability of another, as he would merely preclude himself from the rights of recovery he would otherwise have, and may affect the validity of his insurance cover. Nevertheless, he may, either through a weaker economic position or through mere inadvertence to another carrier's standard trading conditions, contractually agree to a restriction on another carrier's liability which affords the latter greater protection than he would otherwise be entitled to under Articles 37 and 38. Thus in *Walek & Co.* v. *Chapman and Ball (International) Ltd.*,[267] the carriers sued by the cargo interests (C1) were claiming indemnity from successive carriers (C2) on the alternative grounds of either Article 37 or their standard terms of business. In the event, it was held that C1 were entitled to a full indemnity from C2 under Article 37(a). Mocatta, J., went on, however, to hold that even if an apportionment under Article 37(b) would not have been appropriate, C1 were nevertheless entitled to a full indemnity under their standard trading terms which were, he also held, incorporated into the contract, Article 40 was not in fact referred to in the judgment, but there was no need for reference to it since C2 do not appear to have raised Article 41 by way of objection to the standard trading terms; had they done so, Article 40 would clearly have applied.

266. *Infra*, Chapter 12. An interesting question is whether Article 40 only applies where there is an agreement between all the successive carriers, or whether it can still apply even though only some of the successive carriers involved are party to the agreement (cf. LG. Duisburg, 18.3.75 (1975) E.T.L. 527, (1976) E.T.L. 290, Libouton, (1982) J.T. para. 149). The more natural assumption would seem to be the latter. However, this could produce some difficulty where a carrier is sued by a non-contracting carrier or where there has been an insolvency and a carrier is asked to contribute even though he has excluded his liability. In English law, the concept of bailment on terms would probably provide relief in most cases. Otherwise, the position would have to be resolved by means of express or implied indemnities.

267. [1980] 2 Lloyd's Rep. 279.

CHAPTER 12

Derogation from the Convention

12.1 Article 41 provides as follows:—

"1. Subject to the provisions of article 40, any stipulation which would directly or indirectly derogate from the provisions of this Convention shall be null and void. The nullity of such a stipulation shall not involve the nullity of the other provisions of the contract.

2. In particular, a benefit of insurance in favour of the carrier or any other similar clause, or any clause shifting the burden of proof shall be null and void."

12.2 Article 41 is of fundamental importance, since in effect it means that it is not possible for parties to either decrease or increase the rights and liabilities of parties under a contract for the international carriage of goods by road which is subject to CMR.[1] Any attempt to do so will be struck out, although the remainder of the terms of the contract will remain unaffected. This means that both the powerful consumer and the powerful carrier are prevented from dictating onerous terms to the weaker party in the bargain. Equally, it would seem that the terms of Article 41 prevent a consumer or carrier from offering terms more favourable than those contained in the Convention, so that unfair competition by large enterprises is thereby avoided.[2] However, there are exceptions where the parties are given freedom of contract within the framework of the Convention. First, Articles 37 and 38, which govern relationships between carriers, are subject to amendment if the parties so desire.[3] Secondly, under Article 31 the parties have a limited freedom to select the forum in which disputes are to be adjudicated upon.[4] Thirdly, Article 33 permits the parties to settle their disputes by arbitration if they so wish instead of by formal court action, provided that the arbitration agreement specifies that the rules of the Convention be applied.[5]

12.3 Apart from these exceptions, Article 41 makes it impossible to avoid the operation of the Convention in any contract which falls within the scope of Article 1. The parties cannot therefore increase the liability of the carrier, and thus, by way of example, a contractual provision making the carrier liable for all loss resulting from defective loading or stowage which has been performed by the sender will be null and void.[6] Similarly, where a carrier has undertaken to provide the documentation necessary for Customs

---

1. KG. Rotterdam, 23.6.71 (1972) S. & S. No. 8.
2. Cf. Loewe, para. 292.
3. Article 40, *supra*, Chapter 11, para. 11.134 ff.
4. *Supra*, Chapter 10, para. 10.21 ff.
5. *Supra*, Chapter 10, para. 10.120 ff. Thus in *A.B. Bofors-UVA* v. *A.B. Skandia Transport* [1982] 1 Lloyd's Rep. 410 (discussed *supra*, paras. 10.123–10.124) it was held that, because the arbitration clause did not expressly provide for the application of CMR, it failed to satisfy Article 33 and was therefore invalid under Article 41.
6. Comm. Verviers, 27.11.69 (1969–70) Jur. Liège 174.

clearance of goods, and delay results from its inadequacy, he will not be liable in negligence for this, as it would constitute a derogation from Article 11(2).[7] Conversely, any attempt by a carrier to restrict his liability will equally be null and void.[8] However, any matter which is not provided for in CMR will be subject to the relevant domestic law.[9]

12.4 One area where problems relating to Article 41 are likely to arise is where a forwarder is involved in international road haulage and wishes to rely upon his standard trading conditions to restrict or limit his liability.[10] Before the advent of CMR in most countries the forwarder was free to restrict his liability, but under CMR it is necessary first to consider the precise status of the forwarder under the contract.[11] If he is properly to be regarded not as forwarder but as carrier for the purposes of the Convention, then the Convention will mandatorily apply, and any attempt to introduce forwarding conditions into the contract will be null and void to the extent that they conflict with the Convention. There are decisions in most Continental countries on this point,[12] and in this country an attempt by the carrier to rely on the arbitration clause contained in the Nordic Conditions was rejected by the court on the basis of the failure of that clause to comply with Article 33.[13] Article 41 also precludes a carrier from relying on forwarding conditions for any ancillary operations which are covered by the Convention, such as selecting an operator to unload goods where delivery is not possible. Under Article 16(2) the carrier will be responsible for the unloading of the goods in such circumstances and cannot avoid CMR liability.[14]

12.5 In *Eastern Kayam Carpets Ltd.* v. *Eastern United Freight Ltd.*,[15] forwarders sought to rely on their trading conditions to restrict their liability for having failed to carry out an instruction to deliver the goods only against certain documents. Since the forwarders were held to have contracted as carriers they were therefore subject to the CMR Convention. However, having held that the instruction in question did not amount to an instruction to collect a "cash on delivery" charge within Article 21, this raised the question of whether the forwarders were therefore entitled to rely on their standard trading conditions, which would have limited their liability in such circumstances. However, Hirst, J., held that

---

7. App. Paris, 27.1.71 (1971) B.T. 115, (1971) U.L.C. 151.

8. Hof. Brussel, 13.1.72 (1972) 7 E.T.L. 585. See also Arrond. Rotterdam, 12.4.72 (1972) S. & S. No. 103, (1976) I U.L.C. 253 where it was held that any clauses purporting to relieve a carrier from liability or to modify the rates of interest fixed by CMR are null and void, so that even where a carrier does not oppose a claim by the cargo interest for interest at the normal 9%, the latter could still only obtain 5% under Article 27(1).

9. LG. Bremen, 6.5.65 (1966) 1 E.T.L. 691. See also Arrond. Breda, 16.12.69 (1970) 5 E.T.L. 67.

10. E.g. the BIFA standard trading conditions (forwarding terms approved by the British International Freight Association), see generally Yates, Part 7.

11. See generally *supra*, Chapter 1, para. 1.13 ff.

12. E.g. where a forwarder is a CMR carrier he cannot rely on the Dutch FENEX Conditions—GH. 's Gravenhage, 17.5.68 (1968) 3 E.T.L. 1227. See also RB. Amsterdam, 12.4.72 (1972) S. & S. No. 102. Nor can he rely on the Belgian *Conditions Generales des Expéditeurs de Belgique*—Civ. Charleroi, 1.10.68 (1969) J.P.A. 129; see also App. Bruxelles, 25.5.72 (1972) J.P.A. 219 and Comm. Bruxelles, 26.10.72 (1973) 8 E.T.L. 516. In Hof. Brussel, 23.12.71 (1972) 7 E.T.L. 865 a forwarder who attempted to rely on French domestic law to limit his liability was unable to do so, being treated as a carrier and therefore subject to CMR. Nor can he rely on the Austrian AÖSp. (where a six-month period of limitation is provided)—Cass., 5.6.72 (1972) B.T. 484, or the German ADSp. (which provides that where the forwarder is assimilated to a carrier by Article 413(1) HGB, he only incurs the liability of a *spediteur*)—BGH., 18.2.72 (1972) 7 E.T.L. 860, (1973) I U.L.C. 264. See also, Comm. Bruxelles, 12.2.77 (1978) 13 E.T.L. 285.

13. *A.B. Bofors-UVA* v. *A.B. Skandia Transport* [1982] 1 Lloyd's Rep. 410.

14. RB. Amsterdam, 12.4.72 (1972) S. & S. No. 102.

15. Queen's Bench Division, 6.12.83, unreported except on LEXIS.

Article 41 rendered the conditions null and void for all purposes, including the case in issue. This may have been expressed a little too widely, since Article 41 does not prevent the parties from agreeing on terms dealing with matters not regulated by CMR. The decision would, however, seem correct insofar as any trading condition which *ex facie* offended[16] the provisions of CMR would automatically be rendered null and void regardless of the issue before the court.[17] On the other hand, in *Noble v. The R.H. Group*[18] it was held by the Court of Appeal that Article 41 did not render null and void a clause which enabled a carrier to claim an indemnity in respect of its liability for personal injury to its own employee during the course of unloading the vehicle due to its own failure to properly secure the load. It was said that Article 41 does not invalidate provisions of a contract concerning matters which are not regulated by the Convention and here the clause, on the facts of the case, did not derogate from the carrier's responsibilities whether under Article 17 or Article 3. The position would have been otherwise if the claim had been in respect of damage to the claimant's goods.

12.6 Most recently, in the case of *Lacey's Footwear (Wholesale) Ltd. v. Bowler International Freight Ltd.*,[19] no less than three different conclusions were reached in the three judgments with regard to the defendants' standard terms as they applied to a failure on their part to arrange insurance cover. It was held at first instance that the terms were not incorporated, then on appeal it was held by the majority that they did not apply as a matter of construction and by the minority that inadequate notice had been given of the particular term relied upon. It does not appear to have even been argued that Article 41 applied so as to invalidate the conditions in any event.[20]

12.7 Article 41(2) specifically provides that any clause giving the carrier the benefit of insurance will be void.[21] This would in any event seem to fall within the general prohibition in Article 41(1) since if the cargo interest assigns his right to the insurance moneys to the carrier the carrier is in effect relieved of all liability. However, presumably it was thought necessary to make separate provision because insurance is in many ways a distinct matter from that of the contract of carriage, and there may be attempts to evade the effect of Article 41(1) by way of provision in the contract of insurance between the cargo interest and their insurers, aimed at preventing the insurers from exercising their rights of subrogation. Thus in a Belgian case, a carrier inserted in his conditions of trading a clause which provided that he was not required to pay compensation if the cargo interests had not taken out an insurance policy and that such a policy must provide that the insurer abandoned any right of subrogation against the carrier. Either way, therefore, the carrier could not be claimed from, and it was held that this amounted to a derogation from the Convention under Article 41(2).[22] There is also early Belgian authority to the effect that the result is the same even if the carrier does not attempt to obtain such a benefit through his general conditions of contract but where the insurer voluntarily abandons his

16. See *The Morviken* [1982] 1 Lloyd's Rep. 325 in respect of the Hague-Visby Rules.
17. See also *Shell Chemicals U.K. Ltd. v. P. & O. Roadtanks Ltd.* [1993] 1 Lloyd's Rep. 114, 119 (at first instance), cf. BGH., 9.2.79 (1980) 15 E.T.L. 84.
18. 5.2.93, unreported except on LEXIS.
19. [1997] 2 Lloyd's Rep. 369.
20. The first instance decision in *Gefco (U.K.) Ltd. v. John Mason*, Q.B.D., (Birmingham District Registry), 21.2.2000 (unreported) may render this issue of limited continuing importance. See the discussion of this case, *supra*, paras 0.27–0.36.
21. See OGH., 17.3.98 (1998) TranspR 361.
22. Comm. Bruxelles, 10.11.71 (1971) J.P.A. 227.

rights of subrogation in the contract of insurance between the insurer and the cargo interests.[23] Although such an interpretation is consistent with the wording of Article 41(2), which is not restricted to clauses in the contract of carriage, there is French authority to the contrary based on the view that the contract of insurance is not within the field of application of the Convention. It was therefore held that since the carriers had not influenced the agreement between the cargo interests and their insurers, they would be entitled to benefit from it.[24] Given that Article 1 restricts the scope of the Convention to contracts for the carriage of goods, it would seem that the better view is that a provision in a contract of insurance which is wholly independent of the contract of carriage is outside the scope of the application, and that the French decision is therefore to be preferred. Such indeed would now appear to be the position in Belgium, where it has more recently been held by the Cour de Cassation that a clause contained in the contract of insurance between the consignee and his insurer whereby the insurer waived any right of recovery against the carrier was unaffected by Article 41.[25]

12.8 Article 41(2) also specifically provides that any clause shifting the burden of proof shall be null and void. This has two aspects. First, where CMR makes provision as to the burden of proof, obviously any attempt to alter this would in any case derogate from the Convention and would be covered by the general provision in Article 41(1). Secondly, it is possible that Article 41(2) is relevant where CMR does not make any provision as to the burden of proof. In such a case it is arguable that the mere fact that a burden of proof is placed on a party where none existed before would constitute a derogation from the Convention. The question does not yet appear to have come before the courts in relation to the burden of proof,[26] but in terms of general principle, it would seem that if the Convention makes no provision on any particular matter, contractual provision between the parties so as to cover that matter should not amount to derogation from the Convention.[27]

---

23. Comm. Bruxelles, 21.1.73 (1973) J.C.B. 201. Libouton disagrees with the result of this case: (1974) B.T. 528, 530.

24. App. Paris, 26.6.81 (1981) B.T. 395. In the event, the clause did not protect the carriers since it expressly excluded cases of *faute grave*, and the court held that the circumstances of the damage came within that exclusion.

25. Cass. (Belg)., 9.4.81 (1983) 11 E.L.D. 31. The report does not specify whether or not the carrier was instrumental in the insertion of the clause. The clauses covered by Article 41(2) are subject to the objection that the carrier profits since an element of insurance cost can be taken to be included in the freight. A court found less objection to a clause separating out liability insurance which is paid for by the sender with a corresponding reduction in freight, the sender himself being a transport operator who has assumed liability to the goods owner: BGH., 10.12.98 (1999) TranspR 155.

26. Loewe, para. 296, supports the view that to establish a burden of proof where none exists under CMR would be a derogation, but gives no reasons.

27. See Binding Advice, 3.8.76 (1977) S. & S. No. 241, (1978) 6 E.L.D. 91, *supra*, Chapter 6, fn. 281.

APPENDIX A

Carriage of Goods by Road Act 1965

An Act to give effect to the Convention on the Contract for the International Carriage of Goods by Road signed at Geneva on 19th May 1956; and for purposes connected therewith.

[5th August 1965]

1. Subject to the following provisions of this Act, the provisions of the Convention on the Contract for the International Carriage of Goods by Road (in this Act referred to as "the Convention"), as set out in the Schedule to this Act, shall have the force of law in the United Kingdom so far as they relate to the rights and liabilities of persons concerned in the carriage of goods by road under a contract to which the Convention applies.

2.—(1) Her Majesty may by Order in Council from time to time certify who are the High Contracting Parties to the Convention and in respect of what territories they are respectively parties.

(2) An Order in Council under this section shall, except so far as it has been superseded by a subsequent Order, be conclusive evidence of the matters so certified.

3.—(1) A court before which proceedings are brought to enforce a liability which is limited by article 23 in the Schedule to this Act may at any stage of the proceedings make any such order as appears to the court to be just and equitable in view of the provisions of the said article 23 and of any other proceedings which have been, or are likely to be, commenced in the United Kingdom or elsewhere to enforce the liability in whole or in part.

(2) Without prejudice to the preceding subsection, a court before which proceedings are brought to enforce a liability which is limited in the said article 23 shall, where the liability is, or may be, partly enforceable in other proceedings in the United Kingdom or elsewhere, have jurisdiction to award an amount less than the court would have awarded if the limitation applied solely to the proceedings before the court, or to make any part of its award conditional on the result of any other proceedings.

4.—(1) Subject to the next following subsection, Part I of the Foreign Judgments (Reciprocal Enforcement) Act 1933 (in this section referred to as "the Act of 1933") shall apply, whether or not it would otherwise have so applied, to any judgment which—

(a) has been given in any such action as is referred to in paragraph 1 of article 31 in the Schedule to this Act, and

(b) has been so given by any court or tribunal of a territory in respect of which one of the High Contracting Parties, other than the United Kingdom, is a party to the Convention, and

(c) has become enforceable in that territory.

(2) In the application of Part I of the Act of 1933 in relation to any such judgment as is referred to in the preceding subsection, section 4 of that Act shall have effect with the omission of subsections (2) and (3).

(3) The registration, in accordance with Part I of the Act of 1933, of any such judgment as is referred to in subsection (1) of this section shall constitute, in relation to that judgment, compliance with the formalities for the purposes of paragraph 3 of article 31 in the Schedule to this Act.

5.—(1) Where a carrier under a contract to which the Convention applies is liable in respect of any loss or damage for which compensation is payable under the Convention, nothing in [section 1 of the Civil Liability (Contribution) Act 1978[1]], or section 3(2) of the Law Reform (Miscellaneous Provisions) (Scotland) Act 1940 shall confer on him any right to recover contribution in respect of that loss or damage from any other carrier who, in accordance with article 34 in the Schedule to this Act, is a party to the contract of carriage.

(2) The preceding subsection shall be without prejudice to the operation of article 37 in the Schedule to this Act.

6. Every High Contracting Party to the Convention shall, for the purposes of any proceedings brought in a court in the United Kingdom in accordance with the provisions of article 31 in the Schedule to this Act to enforce a claim in respect of carriage undertaken by that Party, be deemed to have submitted to the jurisdiction of that court, and accordingly rules of court may provide for the manner in which any such action is to be commenced and carried on; but nothing in this section shall authorise the issue of execution, or in Scotland the execution of diligence, against the property of any High Contracting Party.

7.—(1) Any reference in the preceding provisions of this Act to a court includes a reference to an arbitration tribunal acting by virtue of article 33 in the Schedule to this Act.

(2) For the purposes of article 32 in the Schedule to this Act, as it has effect (by virtue of the said article 33) in relation to arbitrations,—

    (a) as respects England and Wales, subsections (3) to (5) of [section 34 of the Limitation Act 1980[2]] (which determine the time at which an arbitration is deemed to be commenced) shall apply;

    (b) as respects Northern Ireland, subsections, (2) to (4) of section 72 of the Statute of Limitations (Northern Ireland) 1958 (which make similar provision) shall apply; and

    (c) as respects Scotland, an arbitration shall be deemed to be commenced when one party to the arbitration serves on the other party or parties a notice requiring him or them to appoint an arbiter or to agree to the appointment of an arbiter or, where the arbitration agreement provides that the reference shall be to a person named or designated in the agreement, requiring him or them to submit the dispute to the person so named or designated.

8.—(1) If it appears to Her Majesty in Council that there is any conflict between the provisions of this Act (including the provisions of the Convention as set out in the Schedule to this Act) and any provisions relating to the carriage of goods for reward by land, sea or air contained in—

    (a) any other Convention which has been signed or ratified by or on behalf of Her Majesty's Government in the United Kingdom before the passing of this Act, or

    (b) any enactment of the Parliament of the United Kingdom giving effect to such a Convention,

Her Majesty may by Order in Council make such provision as may seem to Her to be appropriate for resolving that conflict by amending or modifying this Act or any such enactment.[3]

(2) Any statutory instrument made by virtue of this section shall be subject to annulment in pursuance of a resolution of either House of Parliament.

8A[4].—(1) If at any time it appears to Her Majesty in Council that Her Majesty's Government in the United Kingdom have agreed to any revision of the Convention, Her Majesty may by Order in Council[5] make such amendment of—

1. Words substituted by Civil Liability (Contribution) Act 1978, Sch. 1 para. 7.
2. Words substituted by Limitation Act 1980, Sch. 3 para. 6.
3. By s. 6(1)(b) of the Carriage by Air and Road Act 1979 the powers to make Orders in Council conferred by ss. 8, 8A and 9 included power to make Orders in Council in respect of the 1965 Act as amended by the 1979 Act.
4. Inserted by the Carriage by Air and Road Act 1979, s. 3(3).
5. See fn. 3, *supra*.

(a) [the provisions set out in the Schedule to][6] this Act; and

[(b) the definition of, and references in this Act to, or to particular provisions of, the Convention; and

(c) section 5(1) of the Carriage by Air and Road Act 1979,][6]

as appear to Her to be appropriate in consequence of the revision.

(2) In the preceding subsection "revision" means an omission from, addition to or alteration of the Convention and includes replacement of the Convention or part of it by another convention.

(3) An Order in Council under this section shall not be made unless a draft of the Order has been laid before Parliament and approved by a resolution of each House of Parliament.

9. Her Majesty may by Order in Council[7] direct that this Act shall extend, subject to such exceptions, adaptations and modifications as may be specified in the Order, to—

(a) the Isle of Man;

(b) any of the Channel Islands;

(c) any colony;

(d) any state or territory which is for the time being a protectorate or protected state for the purposes of the British Nationality Act 1948.

10. In its application to Scotland, the Schedule to this Act shall have effect as if—

(a) any reference therein to a plaintiff included a reference to a pursuer;

(b) any reference therein to a defendant included a reference to a defender; and

(c) any reference to security for costs included a reference to caution for expenses.

11.—(1) In the application of this Act to Northern Ireland, any reference to an enactment of the Parliament of Northern Ireland shall be construed as a reference to that enactment as amended by any Act of that Parliament, whether passed before or after this Act, and to any enactment of that Parliament passed after this Act and re-enacting the said enactment with or without modification.

(2) In the application of section 4 of this Act to Northern Ireland, any reference to the Foreign Judgments (Reciprocal Enforcement) Act 1933 is a reference to that Act as it applies in Northern Ireland.

(3) [Repealed by Northern Ireland Constitution Act 1973, Schedule 6 Part I]

12. An Order in Council made under any of the preceding provisions of this Act may contain such transitional and supplementary provisions as appear to Her Majesty to be expedient and may be varied or revoked by a subsequent Order in Council made under that provision.

13. This Act shall bind the Crown.

14.—(1) This Act may be cited as the Carriage of Goods by Road Act 1965.

(2) The persons who, for the purposes of this Act, are persons concerned in the carriage of goods by road under a contract to which the Convention applies are—

(a) the sender,

(b) the consignee,

(c) any carrier who, in accordance with article 34 in the Schedule to this Act or otherwise, is a party to the contract of carriage,

(d) any person for whom such a carrier is responsible by virtue of article 3 in the Schedule to this Act,

(e) any person to whom the rights and liabilities of any of the persons referred to in paragraphs (a) to (d) of this subsection have passed (whether by assignment or assignation or by operation of law).

---

6. Deleted by the International Transport Conventions Act 1983, s. 9 and Sch. 2.
7. See fn. 3, *supra.*

(3) Except in so far as the context otherwise requires, any reference in this Act to an enactment shall be construed as a reference to that enactment as amended or extended by or under any other enactment.

(4) This Act shall come into operation on such day as Her Majesty may by Order in Council appoint; but nothing in this Act shall apply in relation to any contract for the carriage of goods by road made before the day so appointed.

SCHEDULE

CONVENTION ON THE CONTRACT FOR THE INTERNATIONAL CARRIAGE OF
GOODS BY ROAD

CHAPTER I
SCOPE OF APPLICATION

Article 1

1. This Convention shall apply to every contract for the carriage of goods by road in vehicles for reward, when the place of taking over the goods and the place designated for delivery, as specified in the contract, are situated in two different countries, of which at least one is a Contracting country, irrespective of the place of residence and the nationality of the parties.

2. For the purposes of this Convention, "vehicles" means motor vehicles, articulated vehicles, trailers and semi-trailers as defined in article 4[8] of the Convention on Road Traffic dated 19th September 1949.

3. This Convention shall apply also where carriage coming within its scope is carried out by States or by governmental institutions or organizations.

4. This Convention shall not apply:

    (a) to carriage performed under the terms of any international postal convention;

    (b) to funeral consignments;

    (c) to furniture removal.

5. The Contracting Parties agree not to vary any of the provisions of this Convention by special agreements between two or more of them, except to make it inapplicable to their frontier traffic or to authorise the use in transport operations entirely confined to their territory of consignment notes representing a title to the goods.

Article 2

1. Where the vehicle containing the goods is carried over part of the journey by sea, rail, inland waterways or air, and, except where the provisions of article 14 are applicable, the goods are not unloaded from the vehicle, this Convention shall nevertheless apply to the whole of the carriage. Provided that to the extent that it is proved that any loss, damage or delay in delivery of the goods which occurs during the carriage by the other means of transport was not caused by an act or omission of the carrier by road, but by some event which could only have occurred in the course of and by reason of the carriage by that other means of transport, the liability of the carrier by road shall be determined not by this Convention but in the manner in which the liability of the carrier by the

---

8. The definitions set out in this article are, so far as material, as follows:—

"Motor vehicle" means any self-propelled vehicle normally used for the transport of persons or goods upon a road, other than vehicles running on rails or connected to electric conductors.

"Articulated vehicle" means any motor vehicle with a trailer having no front axle and so attached that part of the trailer is superimposed upon the motor vehicle and a substantial part of the weight of the trailer and of its load is borne by the motor vehicle. Such a trailer shall be called a "semi-trailer".

"Trailer" means any vehicle designed to be drawn by a motor vehicle.

other means of transport would have been determined if a contract for the carriage of the goods alone had been made by the sender with the carrier by the other means of transport in accordance with the conditions prescribed by law for the carriage of goods by that means of transport. If, however, there are no such prescribed conditions, the liability of the carrier by road shall be determined by this Convention.

2. If the carrier by road is also himself the carrier by the other means of transport, his liability shall also be determined in accordance with the provisions of paragraph 1 of this article, but as if, in his capacities as carrier by road and as carrier by the other means of transport, he were two separate persons.

CHAPTER II

PERSONS FOR WHOM THE CARRIER IS RESPONSIBLE

Article 3

For the purposes of this Convention the carrier shall be responsible for the acts and omissions of his agents and servants and of any other persons of whose services he makes use for the performance of the carriage, when such agents, servants or other persons are acting within the scope of their employment, as if such acts or omissions were his own.

CHAPTER III

CONCLUSION AND PERFORMANCE OF THE CONTRACT OF CARRIAGE

Article 4

The contract of carriage shall be confirmed by the making out of a consignment note. The absence, irregularity or loss of the consignment note shall not affect the existence or the validity of the contract of carriage which shall remain subject to the provisions of this Convention.

Article 5

1. The consignment note shall be made out in three original copies signed by the sender and by the carrier. These signatures may be printed or replaced by the stamps of the sender and the carrier if the law of the country in which the consignment note has been made out so permits. The first copy shall be handed to the sender, the second shall accompany the goods and the third shall be retained by the carrier.

2. When the goods which are to be carried have to be loaded in different vehicles, or are of different kinds or are divided into different lots, the sender or the carrier shall have the right to require a separate consignment note to be made out for each vehicle used, or for each kind or lot of goods.

Article 6

1. The consignment note shall contain the following particulars:

(a) the date of the consignment note and the place at which it is made out;
(b) the name and address of the sender;
(c) the name and address of the carrier;
(d) the place and the date of taking over of the goods and the place designated for delivery;
(e) the name and address of the consignee;
(f) the description in common use of the nature of the goods and the method of packing, and, in the case of dangerous goods, their generally recognised description;
(g) the number of packages and their special marks and numbers;
(h) the gross weight of the goods or their quantity otherwise expressed;

337

   (i) charges relating to the carriage (carriage charges, supplementary charges, Customs duties
        and other charges incurred from the making of the contract to the time of delivery);
   (j) the requisite instructions for Customs and other formalities;
   (k) a statement that the carriage is subject, notwithstanding any clause to the contrary, to the
        provisions of this Convention.

2. Where applicable, the consignment note shall also contain the following particulars:

   (a) a statement that transhipment is not allowed;
   (b) the charges which the sender undertakes to pay;
   (c) the amount of "cash on delivery" charges;
   (d) a declaration of the value of the goods and the amount representing special interest in
        delivery;
   (e) the sender's instructions to the carrier regarding insurance of the goods;
   (f) the agreed time-limit within which the carriage is to be carried out;
   (g) a list of the documents handed to the carrier.

3. The parties may enter in the consignment note any other particulars which they may deem
useful.

Article 7

1. The sender shall be responsible for all expenses, loss and damage sustained by the carrier by
reason of the inaccuracy or inadequacy of:

   (a) the particulars specified in article 6, paragraph 1, (b), (d), (e), (f), (g), (h), and (j);
   (b) the particulars specified in article 6, paragraph 2;
   (c) any other particulars or instructions given by him to enable the consignment note to be
        made out or for the purpose of their being entered therein.

2. If, at the request of the sender, the carrier enters in the consignment note the particulars
referred to in paragraph 1 of this article, he shall be deemed, unless the contrary is proved, to have
done so on behalf of the sender.

3. If the consignment note does not contain the statement specified in article 6, paragraph 1(k),
the carrier shall be liable for all expenses, loss and damage sustained through such omission by the
person entitled to dispose of the goods.

Article 8

1. On taking over the goods, the carrier shall check:

   (a) the accuracy of the statements in the consignment note as to the number of packages and
        their marks and numbers, and
   (b) the apparent condition of the goods and their packaging.

2. Where the carrier has no reasonable means of checking the accuracy of the statements referred
to in paragraph 1(a) of this article, he shall enter his reservations in the consignment note together
with the grounds on which they are based. He shall likewise specify the grounds for any reservations
which he makes with regard to the apparent condition of the goods and their packaging. Such
reservations shall not bind the sender unless he has expressly agreed to be bound by them in the
consignment note.

3. The sender shall be entitled to require the carrier to check the gross weight of the goods or their
quantity otherwise expressed. He may also require the contents of the packages to be checked. The
carrier shall be entitled to claim the cost of such checking. The result of the checks shall be entered
in the consignment note.

Article 9

1. The consignment note shall be *prima facie* evidence of the making of the contract of carriage, the conditions of the contract and the receipt of the goods by the carrier.

2. If the consignment note contains no specific reservations by the carrier, it shall be presumed, unless the contrary is proved, that the goods and their packaging appeared to be in good condition when the carrier took them over and that the number of packages, their marks and numbers corresponded with the statements in the consignment note.

Article 10

The sender shall be liable to the carrier for damage to persons, equipment or other goods, and for any expenses due to defective packing of the goods, unless the defect was apparent or known to the carrier at the time when he took over the goods and he made no reservations concerning it.

Article 11

1. For the purposes of the Customs or other formalities which have to be completed before delivery of the goods, the sender shall attach the necessary documents to the consignment note or place them at the disposal of the carrier and shall furnish him with all the information which he requires.

2. The carrier shall not be under any duty to enquire into either the accuracy or the adequacy of such documents and information. The sender shall be liable to the carrier for any damage caused by the absence, inadequacy or irregularity of such documents and information, except in the case of some wrongful act or neglect on the part of the carrier.

3. The liability of the carrier for the consequences arising from the loss or incorrect use of the documents specified in and accompanying the consignment note or deposited with the carrier shall be that of an agent, provided that the compensation payable by the carrier shall not exceed that payable in the event of loss of the goods.

Article 12

1. The sender has the right to dispose of the goods, in particular by asking the carrier to stop the goods in transit, to change the place at which delivery is to take place or to deliver the goods to a consignee other than the consignee indicated in the consignment note.

2. This right shall cease to exist when the second copy of the consignment note is handed to the consignee or when the consignee exercises his right under article 13, paragraph 1; from that time onwards the carrier shall obey the orders of the consignee.

3. The consignee shall, however, have the right of disposal from the time when the consignment note is drawn up, if the sender makes an entry to that effect in the consignment note.

4. If in exercising his right of disposal the consignee has ordered the delivery of the goods to another person, that other person shall not be entitled to name other consignees.

5. The exercise of the right of disposal shall be subject to the following conditions:

(a) that the sender or, in the case referred to in paragraph 3 of this article, the consignee who wishes to exercise the right produces the first copy of the consignment note on which the new instructions to the carrier have been entered and indemnifies the carrier against all expenses, loss and damage involved in carrying out such instructions;

(b) that the carrying out of such instructions is possible at the time when the instructions reach the person who is to carry them out and does not either interfere with the normal working of the carrier's undertaking or prejudice the senders or consignees of other consignments;

(c) that the instructions do not result in a division of the consignment.

6. When, by reason of the provisions of paragraph 5(b) of this article, the carrier cannot carry out the instructions which he receives, he shall immediately notify the person who gave him such instructions.

7. A carrier who has not carried out the instructions given under the conditions provided for in this article, or who has carried them out without requiring the first copy of the consignment note to be produced, shall be liable to the person entitled to make a claim for any loss or damage caused thereby.

Article 13

1. After arrival of the goods at the place designated for delivery, the consignee shall be entitled to require the carrier to deliver to him, against a receipt, the second copy of the consignment note and the goods. If the loss of the goods is established or if the goods have not arrived after the expiry of the period provided for in article 19, the consignee shall be entitled to enforce in his own name against the carrier any rights arising from the contract of carriage.

2. The consignee who avails himself of the rights granted to him under paragraph 1 of this article shall pay the charges shown to be due on the consignment note, but in the event of dispute on this matter the carrier shall not be required to deliver the goods unless security has been furnished by the consignee.

Article 14

1. If for any reason it is or becomes impossible to carry out the contract in accordance with the terms laid down in the consignment note before the goods reach the place designated for delivery, the carrier shall ask for instructions from the person entitled to dispose of the goods in accordance with the provisions of article 12.

2. Nevertheless, if circumstances are such as to allow the carriage to be carried out under conditions differing from those laid down in the consignment note and if the carrier has been unable to obtain instructions in reasonable time from the person entitled to dispose of the goods in accordance with the provisions of article 12, he shall take such steps as seem to him to be in the best interests of the person entitled to dispose of the goods.

Article 15

1. Where circumstances prevent delivery of the goods after their arrival at the place designated for delivery, the carrier shall ask the sender for his instructions. If the consignee refuses the goods the sender shall be entitled to dispose of them without being obliged to produce the first copy of the consignment note.

2. Even if he has refused the goods, the consignee may nevertheless require delivery so long as the carrier has not received instructions to the contrary from the sender.

3. When circumstances preventing delivery of the goods arise after the consignee, in exercise of his rights under article 12, paragraph 3, has given an order for the goods to be delivered to another person, paragraphs 1 and 2 of this article shall apply as if the consignee were the sender and that other person were the consignee.

Article 16

1. The carrier shall be entitled to recover the cost of his request for instructions and any expenses entailed in carrying out such instructions, unless such expenses were caused by the wrongful act or neglect of the carrier.

2. In the cases referred to in article 14, paragraph 1, and in article 15, the carrier may immediately unload the goods for account of the person entitled to dispose of them and thereupon the carriage shall be deemed to be at an end. The carrier shall then hold the goods on behalf of the person so entitled. He may however entrust them to a third party, and in that case he shall not be under any

liability except for the exercise of reasonable care in the choice of such third party. The charges due under the consignment note and all other expenses shall remain chargeable against the goods.

3. The carrier may sell the goods, without awaiting instructions from the person entitled to dispose of them, if the goods are perishable or their condition warrants such a course, or when the storage expenses would be out of proportion to the value of the goods. He may also proceed to the sale of the goods in other cases if after the expiry of a reasonable period he has not received from the person entitled to dispose of the goods instructions to the contrary which he may reasonably be required to carry out.

4. If the goods have been sold pursuant to this article, the proceeds of sale, after deduction of the expenses chargeable against the goods, shall be placed at the disposal of the person entitled to dispose of the goods. If these charges exceed the proceeds of sale, the carrier shall be entitled to the difference.

5. The procedure in the case of sale shall be determined by the law or custom of the place where the goods are situated.

CHAPTER IV

LIABILITY OF THE CARRIER

Article 17

1. The carrier shall be liable for the total or partial loss of the goods and for damage thereto occurring between the time when he takes over the goods and the time of delivery, as well as for any delay in delivery.

2. The carrier shall however be relieved of liability if the loss, damage or delay was caused by the wrongful act or neglect of the claimant, by the instructions of the claimant given otherwise than as the result of a wrongful act or neglect on the part of the carrier, by inherent vice of the goods or through circumstances which the carrier could not avoid and the consequences of which he was unable to prevent.

3. The carrier shall not be relieved of liability by reason of the defective condition of the vehicle used by him in order to perform the carriage, or by reason of the wrongful act or neglect of the person from whom he may have hired the vehicle or of the agents or servants of the latter.

4. Subject to article 18, paragraphs 2 to 5, the carrier shall be relieved of liability when the loss or damage arises from the special risks inherent in one or more of the following circumstances:

    (a) Use of open unsheeted vehicles, when their use has been expressly agreed and specified in the consignment note;

    (b) the lack of, or defective condition of packing in the case of goods which, by their nature, are liable to wastage or to be damaged when not packed or when not properly packed;

    (c) handling, loading, stowage or unloading of the goods by the sender, the consignee or persons acting on behalf of the sender or the consignee;

    (d) the nature of certain kinds of goods which particularly exposes them to total or partial loss or to damage, especially through breakage, rust, decay, desiccation, leakage, normal wastage, or the action of moth or vermin;

    (e) insufficiency or inadequacy of marks or numbers on the packages;

    (f) the carriage of livestock.

5. Where under this article the carrier is not under any liability in respect of some of the factors causing the loss, damage or delay, he shall only be liable to the extent that those factors for which he is liable under this article have contributed to the loss, damage or delay.

Article 18

1. The burden of proving that loss, damage or delay was due to one of the causes specified in article 17, paragraph 2, shall rest upon the carrier.

341

2. When the carrier establishes that in the circumstances of the case, the loss or damage could be attributed to one or more of the special risks referred to in article 17, paragraph 4, it shall be presumed that it was so caused. The claimant shall however be entitled to prove that the loss or damage was not, in fact, attributable either wholly or partly to one of these risks.

3. This presumption shall not apply in the circumstances set out in article 17, paragraph 4(a), if there has been an abnormal shortage, or a loss of any package.

4. If the carriage is performed in vehicles specially equipped to protect the goods from the effects of heat, cold, variations in temperature or the humidity of the air, the carrier shall not be entitled to claim the benefit of article 17, paragraph 4(d), unless he proves that all steps incumbent on him in the circumstances with respect to the choice, maintenance and use of such equipment were taken and that he complied with any special instructions issued to him.

5. The carrier shall not be entitled to claim the benefit of article 17, paragraph 4(f), unless he proves that all steps normally incumbent on him in the circumstances were taken and that he complied with any special instructions issued to him.

Article 19

Delay in delivery shall be said to occur when the goods have not been delivered within the agreed time-limit or when, failing an agreed time-limit, the actual duration of the carriage having regard to the circumstances of the case, and in particular, in the case of partial loads, the time required for making up a complete load in the normal way, exceeds the time it would be reasonable to allow a diligent carrier.

Article 20

1. The fact that goods have not been delivered within thirty days following the expiry of the agreed time-limit, or, if there is no agreed time-limit, within sixty days from the time when the carrier took over the goods, shall be conclusive evidence of the loss of the goods, and the person entitled to make a claim may thereupon treat them as lost.

2. The person so entitled may, on receipt of compensation for the missing goods, request in writing that he shall be notified immediately should the goods be recovered in the course of the year following the payment of compensation. He shall be given a written acknowledgment of such request.

3. Within the thirty days following receipt of such notification, the person entitled as aforesaid may require the goods to be delivered to him against payment of the charges shown to be due on the consignment note and also against refund of the compensation he received less any charges included therein but without prejudice to any claims to compensation for delay in delivery under article 23 and, where applicable, article 26.

4. In the absence of the request mentioned in paragraph 2 or of any instructions given within the period of thirty days specified in paragraph 3, or if the goods are not recovered until more than one year after the payment of compensation, the carrier shall be entitled to deal with them in accordance with the law of the place where the goods are situated.

Article 21

Should the goods have been delivered to the consignee without collection of the "cash on delivery" charge which should have been collected by the carrier under the terms of the contract of carriage, the carrier shall be liable to the sender for compensation not exceeding the amount of such charge without prejudice to his right of action against the consignee.

Article 22

1. When the sender hands goods of a dangerous nature to the carrier, he shall inform the carrier of the exact nature of the danger and indicate, if necessary, the precautions to be taken. If this

information has not been entered in the consignment note, the burden of proving, by some other means, that the carrier knew the exact nature of the danger constituted by the carriage of the said goods shall rest upon the sender or the consignee.

2. Goods of a dangerous nature which, in the circumstances referred to in paragraph 1 of this article, the carrier did not know were dangerous, may, at any time or place, be unloaded, destroyed or rendered harmless by the carrier without compensation; further, the sender shall be liable for all expenses, loss or damage arising out of their handing over for carriage or of their carriage.

Article 23

1. When, under the provisions of this Convention, a carrier is liable for compensation in respect of total or partial loss of goods, such compensation shall be calculated by reference to the value of the goods at the place and time at which they were accepted for carriage.

2. The value of the goods shall be fixed accordingly to the commodity exchange price or, if there is no such price, according to the current market price or, if there is no commodity exchange price or current market price, by reference to the normal value of goods of the same kind and quality.

3.[9] Compensation shall not, however, exceed 8.33 units of account per kilogram of gross weight short.

[3 Compensation shall not, however, exceed 25 francs per kilogram of gross weight short. "Franc" means the gold franc weighing 10/31 of a gramme and being of millesimal fineness 900.]

4. In addition, the carriage charges, Customs duties and other charges incurred in respect of the carriage of the goods shall be refunded in full in case of total loss and in proportion to the loss sustained in case of partial loss, but no further damages shall be payable.

5. In the case of delay, if the claimant proves that damage has resulted therefrom the carrier shall pay compensation for such damage not exceeding the carriage charges.

6. Higher compensation may only be claimed where the value of the goods or a special interest in delivery has been declared in accordance with articles 24 and 26.

7.[10] The unit of account mentioned in this Convention is the Special Drawing Right as defined by the International Monetary Fund. The amount mentioned in paragraph 3 of this article shall be converted into the national currency of the State of the Court seised of the case on the basis of the value of that currency on the date of the judgment or the date agreed upon by the Parties.

Article 24

The sender may, against payment of a surcharge to be agreed upon, declare in the consignment note a value for the goods exceeding the limit laid down in article 23, paragraph 3, and in that case the amount of the declared value shall be substituted for that limit.

Article 25

1. In case of damage, the carrier shall be liable for the amount by which the goods have diminished in value, calculated by reference to the value of the goods fixed in accordance with article 23, paragraphs 1, 2 and 4.

2. The compensation may not, however, exceed:

    (a) if the whole consignment has been damaged the amount payable in the case of total loss;

    (b) if part only of the consignment has been damaged, the amount payable in the case of loss of the part affected.

---

9. Substituted for following para. by Carriage by Air and Road Act 1979, s. 4(2)(a).
10. Para. 7 inserted by Carriage by Air and Road Act 1979, s. 4(2)(b).

Article 26

1. The sender may, against payment of a surcharge to be agreed upon, fix the amount of a special interest in delivery in the case of loss or damage or of the agreed time-limit being exceeded, by entering such amount in the consignment note.

2. If a declaration of a special interest in delivery has been made, compensation for the additional loss or damage proved may be claimed, up to the total amount of the interest declared, independently of the compensation provided for in articles 23, 24 and 25.

Article 27

1. The claimant shall be entitled to claim interest on compensation payable. Such interest, calculated at five per centum per annum, shall accrue from the date on which the claim was sent in writing to the carrier or, if no such claim has been made, from the date on which legal proceedings were instituted.

2. When the amounts on which the calculation of the compensation is based are not expressed in the currency of the country in which payment is claimed, conversion shall be at the rate of exchange applicable on the day and at the place of payment of compensation.

Article 28

1. In cases where, under the law applicable, loss, damage or delay arising out of carriage under this Convention gives rise to an extra-contractual claim, the carrier may avail himself of the provisions of this Convention which exclude his liability or which fix or limit the compensation due.

2. In cases where the extra-contractual liability for loss, damage or delay of one of the persons for whom the carrier is responsible under the terms of article 3 is in issue, such person may also avail himself of the provisions of this Convention which exclude the liability of the carrier or which fix or limit the compensation due.

Article 29

1. The carrier shall not be entitled to avail himself of the provisions of this chapter which exclude or limit his liability or which shift the burden of proof if the damage was caused by his wilful misconduct or by such default on his part as, in accordance with the law of the court or tribunal seised of the case, is considered as equivalent to wilful misconduct.

2. The same provision shall apply if the wilful misconduct or default is committed by the agents or servants of the carrier or by any other persons of whose services he makes use for the performance of the carriage, when such agents, servants or other persons are acting within the scope of their employment. Furthermore, in such a case such agents, servants or other persons shall not be entitled to avail themselves, with regard to their personal liability, of the provisions of this chapter referred to in paragraph 1.

CHAPTER V

CLAIMS AND ACTIONS

Article 30

1. If the consignee takes delivery of the goods without duly checking their condition with the carrier or without sending him reservations giving a general indication of the loss or damage, not later than the time of delivery in the case of apparent loss or damage and within seven days of delivery, Sundays and public holidays excepted, in the case of loss or damage which is not apparent, the fact of his taking delivery shall be *prima facie* evidence that he has received the goods in the

condition described in the consignment note. In the case of loss or damage which is not apparent the reservations referred to shall be made in writing.

2. When the condition of the goods has been duly checked by the consignee and the carrier, evidence contradicting the result of this checking shall only be admissible in the case of loss or damage which is not apparent and provided that the consignee has duly sent reservations in writing to the carrier within seven days, Sundays and public holidays excepted, from the date of checking.

3. No compensation shall be payable for delay in delivery unless a reservation has been sent in writing to the carrier, within twenty-one days from the time that the goods were placed at the disposal of the consignee.

4. In calculating the time-limits provided for in this article the date of delivery, or the date of checking, or the date when the goods were placed at the disposal of the consignee, as the case may be, shall not be included.

5. The carrier and the consignee shall give each other every reasonable facility for making the requisite investigations and checks.

Article 31

1. In legal proceedings arising out of carriage under this Convention, the plaintiff may bring an action in any court or tribunal of a contracting country designated by agreement between the parties and, in addition, in the courts or tribunals of a country within whose territory:

(a) the defendant is ordinarily resident, or has his principal place of business, or the branch or agency through which the contract of carriage was made, or

(b) the place where the goods were taken over by the carrier or the place designated for delivery is situated,

and in no other courts or tribunals.

2. Where in respect of a claim referred to in paragraph 1 of this article an action is pending before a court or tribunal competent under that paragraph, or where in respect of such a claim a judgment has been entered by such a court or tribunal no new action shall be started between the same parties on the same grounds unless the judgment of the court or tribunal before which the first action was brought is not enforceable in the country in which the fresh proceedings are brought.

3. When a judgment entered by a court or tribunal of a contracting country in any such action as is referred to in paragraph 1 of this article has become enforceable in that country, it shall also become enforceable in each of the other contracting States, as soon as the formalities required in the country concerned have been complied with. The formalities shall not permit the merits of the case to be re-opened.

4. The provisions of paragraph 3 of this article shall apply to judgments after trial, judgments by default and settlements confirmed by an order of the court, but shall not apply to interim judgments or to awards of damages, in addition to costs against a plaintiff who wholly or partly fails in his action.

5. Security for costs shall not be required in proceedings arising out of carriage under this Convention from nationals of contracting countries resident or having their place of business in one of those countries.

Article 32

1. The period of limitation for an action arising out of carriage under this Convention shall be one year. Nevertheless, in the case of wilful misconduct, or such default as in accordance with the law of the court or tribunal seised of the case, is considered as equivalent to wilful misconduct, the period of limitation shall be three years. The period of limitation shall begin to run:

(a) in the case of partial loss, damage or delay in delivery, from the date of delivery;

(b) in the case of total loss, from the thirtieth day after the expiry of the agreed time-limit or where there is no agreed time-limit from the sixtieth day from the date on which the goods were taken over by the carrier;

(c) in all other cases, on the expiry of a period of three months after the making of the contract of carriage.

The day on which the period of limitation begins to run shall not be included in the period.

2. A written claim shall suspend the period of limitation until such date as the carrier rejects the claim by notification in writing and returns the documents attached thereto. If a part of the claim is admitted the period of limitation shall start to run again only in respect of that part of the claim still in dispute. The burden of proof of the receipt of the claim, or of the reply and of the return of the documents, shall rest with the party relying upon these facts. The running of the period of limitation shall not be suspended by further claims having the same object.

3. Subject to the provisions of paragraph 2 above, the extension of the period of limitation shall be governed by the law of the court or tribunal seised of the case. That law shall also govern the fresh accrual of rights of action.

4. A right of action which has become barred by lapse of time may not be exercised by way of counter-claim or set-off.

Article 33

The contract of carriage may contain a clause conferring competence on an arbitration tribunal if the clause conferring competence on the tribunal provides that the tribunal shall apply this Convention.

CHAPTER VI

PROVISIONS RELATING TO CARRIAGE PERFORMED BY SUCCESSIVE CARRIERS

Article 34

If carriage governed by a single contract is performed by successive road carriers, each of them shall be responsible for the performance of the whole operation, the second carrier and each succeeding carrier becoming a party to the contract of carriage, under the terms of the consignment note, by reason of his acceptance of the goods and the consignment note.

Article 35

1. A carrier accepting the goods from a previous carrier shall give the latter a dated and signed receipt. He shall enter his name and address on the second copy of the consignment note. Where applicable, he shall enter on the second copy of the consignment note and on the receipt reservations of the kind provided for in article 8, paragraph 2.

2. The provisions of article 9 shall apply to the relations between successive carriers.

Article 36

Except in the case of a counter-claim or a set-off raised in an action concerning a claim based on the same contract of carriage, legal proceedings in respect of liability for loss, damage or delay may only be brought against the first carrier, the last carrier or the carrier who was performing that portion of the carriage during which the event causing the loss, damage or delay occurred; an action may be brought at the same time against several of these carriers.

Article 37

A carrier who has paid compensation in compliance with the provisions of this Convention, shall be entitled to recover such compensation, together with interest thereon and all costs and expenses incurred by reason of the claim, from the other carriers who have taken part in the carriage, subject to the following provisions:

(a) the carrier responsible for the loss or damage shall be solely liable for the compensation whether paid by himself or by another carrier;

(b) when the loss or damage has been caused by the action of two or more carriers, each of them shall pay an amount proportionate to his share of liability; should it be impossible to apportion the liability, each carrier shall be liable in proportion to the share of the payment for the carriage which is due to him;

(c) if it cannot be ascertained to which carriers liability is attributable for the loss or damage, the amount of the compensation shall be apportioned between all the carriers as laid down in (b) above.

Article 38

If one of the carriers is insolvent, the share of the compensation due from him and unpaid by him shall be divided among the other carriers in proportion to the share of the payment for the carriage due to them.

Article 39

1. No carrier against whom a claim is made under articles 37 and 38 shall be entitled to dispute the validity of the payment made by the carrier making the claim if the amount of the compensation was determined by judicial authority after the first mentioned carrier had been given due notice of the proceedings and afforded an opportunity of entering an appearance.

2. A carrier wishing to take proceedings to enforce his right of recovery may make his claim before the competent court or tribunal of the country in which one of the carriers concerned is ordinarily resident, or has his principal place of business or the branch or agency through which the contract of carriage was made. All the carriers concerned may be made defendants in the same action.

3. The provisions of article 31, paragraphs 3 and 4, shall apply to judgments entered in the proceedings referred to in articles 37 and 38.

4. The provisions of article 32 shall apply to claims between carriers. The period of limitation shall, however, begin to run either on the date of the final judicial decision fixing the amount of compensation payable under the provisions of this Convention, or, if there is no such judicial decision, from the actual date of payment.

Article 40

Carriers shall be free to agree among themselves on provisions other than those laid down in articles 37 and 38.

CHAPTER VII

NULLITY OF STIPULATIONS CONTRARY TO THE CONVENTION

Article 41

1. Subject to the provisions of article 40, any stipulation which would directly or indirectly derogate from the provisions of this Convention shall be null and void. The nullity of such a stipulation shall not involve the nullity of the other provisions of the contract.

2. In particular, a benefit of insurance in favour of the carrier or any other similar clause, or any clause shifting the burden of proof shall be null and void.

[*Chapter VIII of the Convention is not reproduced. This deals with the coming into force of the Convention, the settlement of disputes between the High Contracting Parties and related matters.*]

PROTOCOL OF SIGNATURE

1. This Convention shall not apply to traffic between the United Kingdom of Great Britain and Northern Ireland and the Republic of Ireland.

French text of the Convention on the Contract for the International Carriage of Goods by Road

PREAMBULE

Les parties contractantes,

Ayant reconnu l'utilité de régler d'une manière uniforme les conditions du contrat de transport international de marchandises par route, particulièrement en ce qui concerne les documents utilisés pour ce transport et la responsabilité du transporteur,

Son convenues de ce qui suit:

CHAPITRE PREMIER

CHAMP D'APPLICATION

Article premier

1. La présente Convention s'applique à tout contrat de transport de marchandises par route à titre onéreux au moyen de véhicules, lorsque le lieu de la prise en charge de la marchandise et le lieu prévu pour la livraison, tels qu'ils sont indiqués au contrat, sont situés dans deux pays différents dont l'un au moins est un pays contractant. Il en est ainsi quels que soient le domicile et la nationalité des parties.

2. Pour l'application de la présente Convention, il faut entendre par «véhicules» les automobiles, les véhicules articulés, les remorques et les semi-remorques, tels qu'ils sont définis par l'article 4 de la Convention sur la circulation routière en date du 19 septembre 1949.

3. La présente Convention s'applique même si les transports rentrant dans son champ d'application sont effectués par des Etats ou par les institutions ou organisations gouvernementales.

4. La présente Convention ne s'applique pas

a) Aux transports effectués sous l'empire de conventions postales internationales;

b) Aux transports funéraires;

c) Aux transports de déménagement.

5. Les parties contractantes s'interdisent d'apporter par voie d'accords particuliers conclus entre deux ou plusieurs d'entre elles toute modification à la présente Convention, sauf pour soustraire à son empire leur trafic frontalier ou pour autoriser dans les transports empruntant exclusivement leur territoire l'emploi de la lettre de voiture représentative de la marchandise.

Article 2

1. Si le véhicule contenant les marchandises est transporté par mer, chemin de fer, voie navigable intérieure ou air sur une partie du parcours, sans rupture de charge sauf, éventuellement, pour l'application des dispositions de l'article 14, la présente Convention s'applique, néan-moins, pour l'ensemble du transport. Cependant, dans la mesure où il est prouvé qu'une perte, une avarie ou un retard à la livraison de la marchandise qui est survenu au cours du transport par l'un des modes de

transport autre que la route n'a pas été causé par un acte ou une omission du transporteur routier et qu'il provient d'un fait qui n'a pu se produire qu'au cours et en raison du transport non routier, la responsabilité du transporteur routier est déterminée non par la présente Convention, mais de la façon dont la responsabilité du transporteur non routier eût été déterminée si un contrat de transport avait été conclu entre l'expéditeur et le transporteur non routier pour le seul transport de la marchandise conformément aux dispositions impératives de la loi concernant le transport de marchandises par le mode de transport autre que la route. Toutefois, en l'absence de telles dispositions, la responsabilité du transporteur par route sera déterminée par la présente Convention.

2. Si le transporteur routier est en même temps le transporteur non routier, sa responsabilité est également déterminée par le paragraphe 1 comme si sa fonction de transporteur routier et sa fonction de transporteur non routier étaient exercées par deux personnes différentes.

CHAPITRE II

PERSONNES DONT RÉPOND LE TRANSPORTEUR

Article 3

Pour l'application de la présente Convention, le transporteur répond, comme de ses propres actes et omissions, des acts et omissions de ses préposés et de toutes autres personnes aux services desquelles il recourt pour l'exécution du transport lorsque ces préposés ou ces personnes agissent dans l'exercice de leurs fonctions.

CHAPITRE III

CONCLUSION ET EXÉCUTION DU CONTRAT DE TRANSPORT

Article 4

Le contrat de transport est constaté par une lettre de voiture. L'absence, l'irrégularité ou la perte de la lettre de voiture n'affectent ni l'existence ni la validité du contrat de transport qui reste soumis aux dispositions de la présente Convention.

Article 5

1. La lettre de voiture est établie en trois exemplaires originaux signés par l'expéditeur et par le transporteur, ces signatures pouvant être imprimées ou remplacées par les timbres de l'expéditeur et du transporteur si la législation du pays où la lettre de voiture est établie le permet. Le premier exemplaire est remis à l'expéditeur, le deuxième accompagne la marchandise et le troisième est retenu par le transporteur.

2. Lorsque la marchandise à transporter doit être chargée dans des véhicules différents, ou lorsqu'il s'agit de différentes espèces de marchandises ou de lots distincts, l'expéditeur ou le transporteur a le droit d'exiger l'établissement d'autant de lettres de voiture qu'il doit être utilisé de véhicules ou qu'il y a d'espèces ou de lots de marchandises.

Article 6

1. La lettre de voiture doit contenir les indications suivantes:

a) Le lieu et la date de son établissement;
b) Le nom et l'adresse de l'expéditeur;
c) Le nom et l'adresse du transporteur;
d) Le lieu et la date de la prise en charge de la marchandise et le lieu prévu pour la livraison;
e) Le nom et l'adresse du destinataire;

f) La dénomination courante de la nature de la marchandise et le mode d'emballage, et, pour les marchandises dangereuses, leur dénomination généralement reconnue;

g) Le nombre des colis, leurs marques particulières et leurs numéros;

h) Le poids brut ou la quantité autrement exprimée de la marchandise;

i) Les frais afférents au transport (prix de transport, frais accessoires, droits de douane et autres frais survenant à partir de la conclusion du contrat jusqu'à la livraison);

j) Les instructions requises pour les formalités de douane et autres;

k) L'indication que le transport est soumis, nonobstant toute clause contraire, au régime établi par la présente Convention.

2. Les cas échéant, la lettre de voiture doit contenir, en outre, les indications suivantes:

a) L'interdiction de transbordement;

b) Les frais que l'expéditeur prend à sa charge;

c) Le montant du remboursement à percevoir lors de la livraison de la marchandise;

d) La valeur déclarée de la marchandise et la somme représentant l'intérêt spécial à la livraison;

e) Les instructions de l'expéditeur au transporteur en ce qui concerne l'assurance de la marchandise;

f) Le délai convenu dans lequel le transport doit être effectué;

g) La liste des documents remis au transporteur.

3. Les parties peuvent porter sur la lettre de voiture toute autre indication qu'elles jugent utile.

Article 7

1. L'expéditeur répond de tous frais et dommages que supporterait le transporteur en raison de l'inexactitude ou de l'insuffisance;

a) Des indications mentionnées à l'article 6, paragraphe 1, b, d, e, f, g, h et j;

b) Des indications mentionnées à l'article 6, paragraphe 2;

c) De toutes autres indications ou instructions qu'il donne pour l'établissement de la lettre de voiture ou pour y être reportées.

2. Si, à la demande de l'expéditeur, le transporteur inscrit sur la lettre de voiture les mentions visées au paragraphe 1 du présent article, il est considéré, jusqu'à preuve du contraire, comme agissant pour le compte de l'expéditeur.

3. Si la lettre de voiture ne contient pas la mention prévue à l'article 6, paragraphe 1, k, le transporteur est responsable de tous frais et dommages que subirait l'ayant droit à la marchandise en raison de cette omission.

Article 8

1. Lors de la prise en charge de la marchandise, le transporteur est tenu de vérifier:

a) L'exactitude des mentions de la lettre de voiture relative au nombre de colis, ainsi qu'à leurs marques et numéros;

b) L'état apparent de la marchandise et de son emballage.

2. Si le transporteur n'a pas de moyens raisonnables de vérifier l'exactitude des mentions visées au paragraphe 1, a, du présent article, il inscrit sur la lettre de voiture des réserves qui doivent être motivées. Il doit de même motiver toutes les réserves qu'il fait au sujet de l'état apparent de la marchandise et de son emballage. Ces réserves n'engagent pas l'expéditeur, si celui-ci ne les a pas expressément acceptées sur la lettre de voiture.

3. L'expéditeur a le droit d'exiger la vérification par le transporteur du poids brut ou de la quantité autrement exprimée de la marchandise. Il peut aussi exiger la vérification du contenu des colis. Le

transporteur peut réclamer le paiement des frais de vérification. Le résultat des vérifications est consigné sur la lettre de voiture.

Article 9

1. La lettre de voiture fait foi, jusqu'à preuve du contraire, des conditions du contrat et de la réception de la marchandise par le transporteur.

2. En l'absence d'inscription sur la lettre de voiture de réserves motivées du transporteur, il y a présomption que la marchandise et son emballage étaient en bon état apparent au moment de la prise en charge par le transporteur et que le nombre des colis ainsi que leurs marques et numéros étaient conformes aux énonciations de la lettre de voiture.

Article 10

L'expéditeur est responsable envers le transporteur des dommages aux personnes, au matériel ou à d'autres marchandises, ainsi que des frais, qui auraient pour origine la défectuosité de l'emballage de la marchandise, à moins que, la défectuosité étant apparente ou connue du transporteur au moment de la prise en charge, le transporteur n'ait pas fait de réserves à son sujet.

Article 11

1. En vue de l'accomplissement des formalités de douane et autres à remplir avant la livraison de la marchandise, l'expéditeur doit joindre à la lettre de voiture ou mettre à la disposition du transporteur les documents nécessaires et lui fournir tous renseignements voulus.

2. Le transporteur n'est pas tenu d'examiner si ces documents et renseignements sont exacts ou suffisants. L'expéditeur est responsable envers le transporteur de tous dommages qui pourraient résulter de l'absence, de l'insuffisance ou de l'irrégularité de ces documents et renseignements, sauf en cas de faute du transporteur.

3. Le transporteur est responsable au même titre qu'un commissionnaire des conséquences de la perte ou de l'utilisation inexacte des documents mentionnés sur la lettre de voiture et qui accompagnent celle-ci ou qui sont déposés entre ses mains; toutefois, l'indemnité à sa charge ne dépassera pas celle qui serait due en cas de perte de la marchandise.

Article 12

1. L'expéditeur a le droit de disposer de la marchandise, notamment en demandant au transporteur d'en arrêter le transport, de modifier le lieu prévu pour la livraison ou de livrer la marchandise à un destinataire différent de celui indiqué sur la lettre de voiture.

2. Ce droit s'éteint lorsque le deuxième exemplaire de la lettre de voiture est remis au destinataire ou que celui-ci fait valoir le droit prévu à l'article 13, paragraphe 1; à partir de ce moment, le transporteur doit se conformer aux ordres du destinataire.

3. Le droit de disposition appartient toutefois au destinataire dès l'établissement de la lettre de voiture si une mention dans ce sens est faite par l'expéditeur sur cette lettre.

4. Si, en exerçant son droit de disposition, le destinataire ordonne de livrer la marchandise à une autre personne, celle-ci ne peut pas désigner d'autres destinataires.

5. L'exercice du droit de disposition est subordonné aux conditions suivants:

a) L'expéditeur ou, dans le cas visé au paragraphe 3 du présent article, le destinataire qui veut exercer ce droit doit présenter le premier exemplaire de la lettre de voiture, sur lequel doivent être inscrites les nouvelles instructions données au transporteur, et dédommager le transporteur des frais et du préjudice qu'entraîne l'exécution de ces instructions;

b) Cette exécution doit être possible au moment où les instructions parviennent à la personne qui doit les exécuter et elle ne doit ni entraver l'exploitation normale de l'entreprise du transporteur, ni porter préjudice aux expéditeurs ou destinataires d'autres envois;

c) Les instructions ne doivent jamais avoir pour effet de diviser l'envoi.

6. Lorsque, en raison des dispositions prévues au paragraphe 5, b, du présent article, le transporteur ne peut exécuter les instructions qu'il reçoit, il doit en aviser immédiatement la personne dont émanent ces instructions.

7. Le transporteur qui n'aura pas exécuté les instructions données dans les conditions prévues au présent article ou qui se sera conformé à de telles instructions sans avoir exigé la présentation du premier exemplaire de la lettre de voiture sera responsable envers l'ayant droit du préjudice causé par ce fait.

Article 13

1. Après l'arrivée de la marchandise au lieu prévu pour la livraison, le destinataire a le droit de demander que le deuxième exemplaire de la lettre de voiture lui soit remis et que la marchandise lui soit livrée, le tout contre décharge. Si la perte de la marchandise est établie, ou si la marchandise n'est pas arrivée à l'expiration du délai prévu à l'article 19, le destinataire est authorisé à faire valoir en son propre nom vis-à-vis du transporteur les droits qui résultent du contrat de transport.

2. Le destinataire qui se prévaut des droits qui lui sont accordés aux termes du paragraphe 1 du présent article est tenu de payer le montant des créances résultant de la lettre de voiture. En cas de contestation à ce sujet, le transporteur n'est obligé d'effectuer la livraison de la marchandise que si une caution lui est fournie par le destinataire.

Article 14

1. Si, pour un motif quelconque, l'exécution du contrat dans les conditions prévues à la lettre de voiture est ou devient impossible avant l'arrivée de la marchandise au lieu prévu pour la livraison, le transporteur est tenu de demander des instructions à la personne qui a le droit de disposer de la marchandise conformément à l'article 12.

2. Toutefois, si les circonstances permettent l'exécution du transport dans des conditions différentes de celles prévues à la lettre de voiture et si le transporteur n'a pu obtenir en temps utile les instructions de la personne qui a le droit de disposer de la marchandise conformément à l'article 12, il prend les mesures qui lui paraissent les meilleures dans l'intérêt de la personne ayant le droit de disposer de la marchandise.

Article 15

1. Lorsque, après l'arrivée de la marchandise au lieu de destination, il se présente des empêchements à la livraison, le transporteur demande des instructions à l'expéditeur. Si le destinataire refuse la marchandise, l'expéditeur a le droit de disposer de celle-ci sans avoir à produire le premier exemplaire de la lettre de voiture.

2. Même s'il a refusé la marchandise, le destinataire peut toujours en demander la livraison tant que le transporteur n'a pas reçu d'instructions contraires de l'expéditeur.

3. Si l'empêchement à la livraison se présente après que, conformément au droit qu'il détient en vertu de l'article 12, paragraphe 3, le destinataire a donné l'ordre de livrer la marchandise à une autre personne, le destinataire est substitué à l'expéditeur, et cette autre personne au destinataire, pour l'application des paragraphes 1 et 2 ci-dessus.

Article 16

1. Le transporteur a droit au remboursement des frais que lui cause sa demande d'instructions, ou qu'entraîne pour lui l'exécution des instructions reçues, à moins que ces frais ne soient la conséquence de sa faute.

2. Dans les cas visés à l'article 14, paragraphe 1, et à l'article 15, le transporteur peut décharger immédiatement la marchandise pour le compte de l'ayant droit; après ce déchargement, le transport est réputé terminé. Le transporteur assume alors la garde de la marchandise. Il peut toutefois confier la marchandise à un tiers et n'est alors responsable que du choix judicieux de ce tiers. La marchandise reste grevée des créances résultant de la lettre de voiture et de tous autres frais.

3. Le transporteur peut faire procéder à la vente de la marchandise sans attendre d'instructions de l'ayant droit lorsque la nature périssable ou l'état de la marchandise le justifie ou lorsque les frais de garde sont hors de proportion avec la valeur de la marchandise. Dans les autres cas, il peut également faire procéder à la vente lorsque, dans un délai raisonnable, il n'a pas reçu de l'ayant droit d'instructions contraires dont l'exécution puisse équitablement être exigée.

4. Si la marchandise a été vendue en application du présent article, le produit de la vente doit être mis à la disposition de l'ayant droit, déduction faite des frais grevant la marchandise. Si ces frais sont supérieurs au produit de la vente, le transporteur a droit à la différence.

5. La façon de procéder en cas de vente est déterminée par la loi ou les usages du lieu où se trouve la marchandise.

CHAPITRE IV

RESPONSABILITÉ DU TRANSPORTEUR

Article 17

1. Le transporteur est responsable de la perte totale ou partielle ou de l'avarie, qui se produit entre le moment de la prise en charge de la marchandise et celui de la livraison, ainsi que du retard à la livraison.

2. Le transporteur est déchargé de cette responsabilité si la perte, l'avarie ou le retard a eu pour cause une faute de l'ayant droit, un ordre de celui-ci ne résultant pas d'une faute du transporteur, un vice propre de la marchandise, ou des circonstances que le transporteur ne pouvait pas éviter et aux conséquences desquelles il ne pouvait pas obvier.

3. Le transporteur ne peut exciper, pour se décharger de sa responsabilité, ni des défectuosités du véhicule dont il se sert pour effectuer le transport, ni de fautes de la personne dont il aurait loué le véhicule ou des préposés de celle-ci.

4. Compte tenu de l'article 18, paragraphes 2 à 5, le transporteur est déchargé de sa responsabilité lorsque la perte ou l'avarie résulte des risques particuliers inhérents à l'un des faits suivants ou à plusieurs d'entre eux:

a) Emploi de véhicules ouverts et non bâchés, lorsque cet emploi a été convenu d'une manière expresse et mentionné dans la lettre de voiture;

b) Absence ou défectuosité de l'emballage pour les marchandise exposées par leur nature à des déchets ou avaries quand elles ne sont pas emballées ou sont mal emballées;

c) Manutention, chargement, arrimage ou déchargement de la marchandise par l'expéditeur ou le destinataire ou des personnes agissant pour le compte de l'expéditeur ou du destinataire;

d) Nature de certaines marchandises exposées, par des causes inhérentes à cette nature même, soit à perte totale ou partielle, soit à avarie notamment par bris, rouille, détérioration interne et spontanée, dessication, coulage, déchet normal ou action de la vermine et des rongeurs;

e) Insuffisance ou imperfection des marques ou des numéros de colis;

f) Transport d'animaux vivants.

5. Si, en vertu du présent article, le transporteur ne répond pas de certains des facteurs qui ont causé le dommage, sa responsabilité n'est engagée que dans la proportion où les facteurs dont il répond en vertu du présent article ont contribué au dommage.

354

Article 18

1. La preuve que la perte, l'avarie ou le retard a eu pour cause un des faits prévus à l'article 17, paragraphe 2, incombe au transporteur.

2. Lorsque le transporteur établit que, eu égard aux circonstances de fait, la perte ou l'avarie a pu résulter d'un ou de plusieurs des risques particuliers prévus à l'article 17, paragraphe 4, il y a présomption qu'elle en résulte. L'ayant droit peut toutefois faire la preuve que le dommage n'a pas eu l'un de ces risques pour cause totale ou partielle.

3. La présomption visée ci-dessus n'est pas applicable dans le cas prévu à l'article 17, paragraphe 4, a, s'il y a manquant d'une importance anormale ou perte de colis.

4. Si le transport est effectué au moyen d'un véhicule aménagé en vue de soustraire les marchandises à l'influence de la chaleur, du froid, des variations de température ou de l'humidité de l'air, le transporteur ne peut invoquer le bénéfice de l'article 17, paragraphe 4, d, que s'il fournit la preuve que toutes les mesures lui incombant, compte tenu des circonstances, ont été prises en ce qui concerne le choix, l'entretien et l'emploi de ces aménagements et qu'il s'est conformé aux instructions spéciales qui ont pu lui être données.

5. Le transporteur ne peut invoquer le bénéfice de l'article 17, paragraphe 4, f, que s'il fournit la preuve que toutes les mesures lui incombant normalement, compte tenu des circonstances, ont été prises et qu'il s'est conformé aux instructions spéciales qui ont pu lui être données.

Article 19

Il y a retard à la livraison lorsque la marchandise n'a pas été livrée dans le délai convenu ou, s'il n'a pas été convenu de délai, lorsque la durée effective du transport dépasse, compte tenu des circonstances et, notamment, dans le cas d'un chargement partiel, du temps voulu pour assembler un chargement complet dans des conditions normales, le temps qu'il est raisonnable d'allouer à des transporteurs diligents.

Article 20

1. L'ayant droit peut, sans avoir à fournir d'autres preuves, considérer la marchandise comme perdue quand elle n'a pas été livrée dans les trente jours qui suivent l'expiration du délai convenu ou, s'il n'a pas été convenu de délai, dans les soixante jours qui suivent la prise en charge de la marchandise par le transporteur.

2. L'ayant droit peut, en recevant le paiement de l'indemnité pour la marchandise perdue, demander, par écrit, à être avisé immédiatement dans le cas où la marchandise serait retrouvée au cours de l'année qui suivra le paiement de l'indemnité. Il lui est donné par écrit acte de cette demande.

3. Dans les trente jours qui suivent la réception de cet avis, l'ayant droit peut exiger que la marchandise lui soit livrée contre paiement des créances résultant de la lettre de voiture et contre restitution de l'indemnité qu'il a reçue, déduction faite éventuellement des frais qui auraient été compris dans cette indemnité, et sous réserve de tous droits à l'indemnité pour retard à la livraison prévue à l'article 23 et, s'il y a lieu, à l'article 26.

4. A défaut soit de la demande prévue au paragraphe 2, soit d'instructions données dans le délai de trente jours prévu au paragraphe 3, ou encore si la marchandise n'a été retrouvée que plus d'un an après le paiement de l'indemnité, le transporteur en dispose conformément à la loi du lieu où se trouve la marchandise.

Article 21

Si la marchandise est livrée au destinataire sans encaissement du remboursement qui aurait dú être perçu par le transporteur en vertu des dispositions du contrat de transport, le transporteur est tenu d'indemniser l'expéditeur à concurrence du montant du remboursement, sauf son recours contre le destinataire.

Article 22

1. Si l'expéditeur remet au transporteur des marchandises dangereuses, il lui signale la nature exacte du danger qu'elles présentent et lui indique éventuellement les précautions à prendre. Au cas où cet avis n'a pas été consigné sur la lettre de voiture, il appartient à l'expéditeur ou au destinataire de faire la preuve, par tous autres moyens, que le transporteur a eu connaissance de la nature exacte du danger que présentait le transport desdites marchandises.

2. Les marchandises dangereuses qui n'auraient pas été connues comme telles par le transporteur dans les conditions prévues au paragraphe 1 du présent article peuvent à tout moment et en tout lieu être déchargées, détruites ou rendues inoffensives par le transporteur, et ce sans aucune indemnité; l'expéditeur est en outre responsable de tous frais et dommages résultant de leur remise au transport ou de leur transport.

Article 23

1. Quand, en vertu des dispositions de la présente Convention, une indemnité pour perte totale ou partielle de la marchandise est mise à la charge du transporteur, cette indemnité est calculée d'après la valeur de la marchandise au lieu et à l'époque de la prise en charge.

2. La valeur de la marchandise est déterminée d'après le cours en bourse ou, à défaut, d'après le prix courant sur le marché ou, à défaut de l'un et de l'autre, d'aprés la valeur usuelle des marchandises de même nature et qualité.

3. Toutefois, l'indemnité ne peut dépasser 25 francs par kilogramme du poids brut manquant. Le franc s'entend du franc-or, d'un poids 10/31 de gramme au titre de 0,900.

4. Sont en outre remboursés le prix du transport, les droits de douane et les autres frais encourus à l'occasion du transport de la marchandise, en totalité en cas de perte totale, et au prorata en cas de perte pertielle; d'autres dommages-intérêts ne sont pas dus.

5. En cas de retard, si l'ayant droit prouve qu'un préjudice en est résulté, le transporteur est tenu de payer pour ce préjudice une indemnité qui ne peut pas dépasser le prix du transport.

6. Des indemnités plus élevées ne peuvent être réclamées qu'en cas de déclaration de la valeur de la marchandise ou de déclaration d'intérêt spécial à la livraison, conformément aux articles 24 et 26.

Article 24

L'expéditeur peut déclarer dans la lettre de voiture, contre paiement d'un supplément de prix à convenir, une valeur de la marchandise excédant la limite mentionnée au paragraphe 3 de l'article 23 et, dans ce cas, le montant déclaré se substitue à cette limite.

Article 25

1. En cas d'avarie, le transporteur paie le montant de la dépréciation calculée d'après la valeur de la marchandise fixée conformément à l'article 23, paragraphes 1, 2 et 4.

2. Toutefois, l'indemnité ne peut dépasser:

a) Si la totalité de l'expédition est dépréciée par l'avarie, le chiffre qu'elle aurait atteint en cas de perte totale;

b) Si une partie seulement de l'expédition est dépréciée par l'avarie, le chiffre qu'elle aurait atteint en cas de perte de la partie dépréciée.

Article 26

1. L'expéditeur peut fixer, en l'inscrivant à la lettre de voiture, et contre paiement d'un supplément de prix à convenir, le montant d'un intérêt spécial à la livraison, pour le cas de perte ou d'avarie et pour celui de dépassement du délai convenu.

2. S'il y a eu déclaration d'intérêt spécial à la livraison, il peut être réclamé, indépendamment des indemnités prévues aux articles 23, 24 et 25, et à concurrence du montant de l'intérêt déclaré, une indemnité égale au dommage supplémentaire dont la preuve est apportée.

Article 27

1. L'ayant droit peut demander les intérêts de l'indemnité. Ces intérêts, calculés à raison de 5 pour 100 l'an, courent du jour de la réclamation adressée par écrit au transporteur ou, s'il n'y a pas eu de réclamation, du jour de la demande en justice.

2. Lorsque les éléments qui servent de base au calcul de l'indemnité ne sont pas exprimés dans la monnaie du pays où le paiement est réclamé, la conversion est faite d'après le cours du jour et du lieu du paiement de l'indemnité.

Article 28

1. Lorsque, d'après la loi applicable, la perte, l'avarie ou le retard survenu au cours d'un transport soumis à la présente Convention peut donner lieu à une réclamation extra-contractuelle, le transporteur peut se prévaloir des dispositions de la présente Convention qui excluent sa responsabilité ou qui déterminent ou limitent les indemnités dues.

2. Lorsque la responsabilité extra-contractuelle pour perte, avarie ou retard d'une des personnes dont le transporteur répond aux terms de l'article 3 est mise en cause, cette personne peut également se prévaloir des dispositions de la présente Convention qui excluent la responsabilité du transporteur ou qui déterminent ou limitent les indemnités dues.

Article 29

1. Le transporteur n'a pas le droit de se prévaloir des dispositions du présent chapitre qui excluent ou limitent sa responsabilité ou qui renversent le fardeau de la preuve, si le dommage provient de son dol ou d'une faute qui lui est imputable et qui, d'après la loi de la juridiction saisie, est considérée comme équivalente au dol.

2. Il en est de même si le dol ou la faute est le fait des préposés du transporteur ou de toutes autres personnes aux services desquelles il recourt pour l'exécution du transport lorsque ces préposés ou ces autres personnes agissent dans l'exercice de leurs fonctions. Dans ce cas, ces préposés ou ces autres personnes n'ont pas d'avantage le droit de se prévaloir, en ce qui concerne leur responsabilité personnelle, des dispositions du présent chapitre visées au paragraphe 1.

CHAPITRE V

RÉCLAMATIONS ET ACTIONS

Article 30

1. Si le destinataire a pris livraison de la marchandise sans qu'il en ait constaté l'état contradictoirement avec le transporteur ou sans qu'il ait, au plus tard au moment de la livraison s'il s'agit de pertes ou avaries apparentes, ou dans les sept jours à dater de la livraison, dimanche et jours fériés non compris, lorsqu'il s'agit de pertes ou avaries non apparentes, adressé des réserves au transporteur indiquant la nature générale de la perte ou de l'avarie, il est présumé, jusqu'à preuve contraire, avoir reçu la marchandise dans l'état décrit dans la lettre de voiture. Les réserves visées ci-dessus doivent être faites par écrit lorsqu'il s'agit de pertes ou avaries non apparentes.

2. Lorsque l'état de la marchandise a été constaté contradictoirement par le destinataire et le transporteur, la preuve contraire au résultat de cette constatation ne peut être faite que s'il s'agit de pertes ou avaries non apparentes et si le destinataire a adressé des réserves écrites au transporteur dans les sept jours, dimanche et jours fériés non compris, à dater de cette constatation.

3. Un retard à la livraison ne peut donner lieu à indemnité que si une réserve a été adressée par écrit dans le délai de 21 jours, à dater de la mise de la marchandise à la disposition du destinataire.

4. La date de livraison ou, selon le cas, celle de la constatation ou celle de la mise à disposition n'est pas comptée dans les délais prévus au présent article.

5. Le transporteur et le destinataire se donnent réciproquement toutes facilités raisonnables pour les constatations et vérifications utiles.

Article 31

1. Pour tous litiges auxquels donnent lieu les transports soumis à la présente Convention, le demandeur peut saisir, en dehors des juridictions des pays contractants désignées d'un commun accord par les parties, les juridictions du pays sur le territoire duquel:

a) Le défendeur a sa résidence habituelle, son siège principal ou la succursale ou l'agence par l'intermédiaire de laquelle le contrat de transport a été conclu, ou

b) Le lieu de la prise en charge de la marchandise ou celui prévu pour la livraison est situé,

et ne peut saisir que ces juridictions.

2. Lorsque dans un litige visé au paragraphe 1 du présent article une action est en instance devant une juridiction compétente aux termes de ce paragraphe, ou lorsque dans un tel litige un jugement a été prononcé par une telle juridiction, il ne peut être intenté aucune nouvelle action pour la même cause entre les mêmes parties à moins que la décision de la juridiction devant laquelle la première action a été intentée ne soit pas susceptible d'être exécutée dans le pays où la nouvelle action est intentée.

3. Lorsque dans un litige visé au paragraphe 1 du présent article un jugement rendu par une juridiction d'un pays contractant est devenu exécutoire dans ce pays, il devient également exécutoire dans chacun des autres pays contractants aussitôt après accomplissement des formalités prescrites à cet effet dans le pays intéressé. Ces formalités ne peuvent comporter aucune revision de l'affaire.

4. Les dispositions du paragraphe 3 du présent article s'appliquent aux jugements contradictoires, aux jugements par défaut et aux transactions judiciaries, mais ne s'appliquent ni aux jugements qui ne sont exécutoires que par provision, ni aux condamnations en dommages et intérêts qui seraient prononcés en sus des dépens contre un demandeur en raison du rejet total ou partiel de sa demande.

5. Il ne peut être exigé de caution de ressortissants de pays contractants, ayant leur domicile ou un établissement dans un de ces pays, pour assurer le paiement des dépens à l'occasion des actions en justice auxquelles donnent lieu les transports soumis à la présente Convention.

Article 32

1. Les actions auxquelles peuvent donner lieu les transports soumis à la présente Convention sont prescrites dans le délai d'un an. Toutefois, dans le cas de dol ou de faute considérée, d'après la loi de la juridiction saisie, comme équivalente au dol, la prescription est de trois ans. La prescription court:

a) Dans le cas de perte partielle, d'avarie ou de retard, à partir du jour où la marchandise a été livrée;

b) Dans le cas de perte totale, à partir du trentième jour après l'expiration du délai convenu ou, s'il n'a pas été convenu de délai, à partir du soixantième jour après la prise en charge de la marchandise par le transporteur;

c) Dans tous les autres cas, à partir de l'expiration d'un délai de trois mois à dater de la conclusion du contrat de transport.

Le jour indiqué ci-dessus comme point de départ de la prescription n'est pas compris dans le délai.

2. Une réclamation écrite suspend la prescription justqu'au jour où le transporteur repousse la réclamation par écrit et restitue les pièces qui y étaient jointes. En cas d'acceptation partielle de la réclamation, la prescription ne reprend son cours que pour la partie de la réclamation qui reste litigieuse. La preuve de la réception de la réclamation ou de la réponse et de la restitution des pièces est à la charge de la partie qui invoque ce fait. Les réclamations ultérieures ayant le même objet ne suspendent pas la prescription.

3. Sous réserve des dispositions du paragraphe 2 ci-dessus, la suspension de la prescription est régie par la loi de la juridiction saisie. Il en est de même en ce qui concerne l'interruption de la prescription.

4. L'action prescrite ne peut plus être exercée, même sous forme de demande reconventionnelle ou d'exception.

Article 33

Le contrat de transport peut contenir une clause attribuant compétence à un tribunal arbitral à condition que cette clause prévoie que le tribunal arbitral appliquera la présente Convention.

CHAPITRE VI

DISPOSITIONS RELATIVES AU TRANSPORT EFFECTUÉ PAR TRANSPORTEURS SUCCESSIFS

Article 34

Si un transport régi par un contrat unique est exécuté par des transporteurs routiers successifs, chacun de ceux-ci assume la responsabilité de l'exécution du transport total, le second transporteur et chacun des transporteurs suivants devenant, de par leur acceptation de la marchandise et de la lettre de voiture, parties au contrat, aux conditions de la lettre de voiture.

Article 35

1. Le transporteur qui accepte la marchandise du transporteur précédent remet à celui-ci un reçu daté et signé. Il doit porter son nom et son adresse sur le deuxième exemplaire de la lettre de voiture. S'il y a lieu, il appose sur cet exemplaire, ainsi que sur le reçu, des réserves analogues à celles qui sont prévues à l'article 8, paragraphe 2.

2. Les dispositions de l'article 9 s'appliquent aux relations entre transporteurs successifs.

Article 36

A moins qu'il ne s'agisse d'une demande reconventionnelle ou d'une exception formulée dans une instance relative à une demande fondée sur le même contrat de transport, l'action en responsabilité pour perte, avarie ou retard ne peut être dirigée que contre le premier transporteur, le dernier transporteur ou le transporteur qui exécutait la partie du transport au cours de laquelle s'est produit le fait ayant causé la perte, l'avarie ou le retard; l'action peut être dirigée à la fois contre plusieurs de ces transporteurs.

Article 37

Le transporteur qui a payé une indemnité en vertu des dispositions de la présente Convention a le droit d'exercer un recours en principal, intérêts et frais contre les transporteurs qui ont participé à l'exécution du contrat de transport, conformément aux dispositions suivantes:

a) Le transporteur par le fait duquel le dommage a été causé doit seul supporter l'indemnité, qu'il l'ait payée lui-même ou qu'elle ait été payée par un autre transporteur;

b) Lorsque le dommage a été causé par le fait de deux ou plusieurs transporteurs, chacun d'eux doit payer un montant proportionnel à sa part de responsabilité; si l'évaluation des parts de responsabilité est impossible, chacun d'eux est responsable proportionnellement à la part de rémunération du transport qui lui revient;

c) Si l'on ne peut déterminer quels sont ceux des transporteurs auxquels la responsabilité est imputable, la charge de l'indemnité due est répartie, dans la proportion fixée en b, entre tous les transporteurs.

Article 38

Si l'un des transporteurs est insolvable, la part lui incombant et qu'il n'a pas payée est répartie entre tous les autres transporteurs proportionnellement à leur rémunération.

Article 39

1. Le transporteur contre lequel est exercé un des recours prévus aux articles 37 et 38 n'est pas recevable à contester le bien-fondé du paiement efectué par le transporteur exerçant le recours, lorsque l'indemnité a été fixée par décision de justice, pourvu qu'il ait été dûment informé du procès et qu'il ait été à même d'y intervenir.

2. Le transporteur qui veut exercer son recours peut le former devant le tribunal compétent du pays dans lequel l'un des transporteurs intéressés a sa résidence habituelle, son siège principal ou la succursale ou l'agence par l'entremise de laquelle le contrat de transport a été conclu. Le recours peut être dirigé dans une seule et même instance contre tous les transporteurs intéressés.

3. Les dispositions de l'article 31, paragraphes 3 et 4, s'appliquent aux jugements rendus sur les recours prévus aux articles 37 et 38.

4. Les dispositions de l'article 32 sont applicables aux recours entre transporteurs. La prescription court, toutefois, soit à partir du jour d'une décision de justice définitive fixant l'indemnité à payer en vertu des dispositions de la présente. Convention, soit, au cas ou il n'y aurait pas eu de telle décision, à partir du jour du paiement effectif.

Article 40

Les transporteurs sont libres de convenir entre eux de dispositions dérogeant aux articles 37 et 38.

CHAPITRE VII

NULLITÉ DES STIPULATIONS CONTRAIRES À LA CONVENTION

Article 41

1. Sous réserve des dispositions de l'article 40, est nulle et de nul effet toute stipulation qui, directement ou indirectement, dérogerait aux dispositions de la présente Convention. La nullité de telles stipulations n'entraîne pas la nullité des autres dispositions du contrat.

2. En particulier, seraient nulles toute clause par laquelle le transporteur se ferait céder le bénéfice de l'assurance de la marchandise ou toute autre clause analogue, ainsi que toute clause déplaçant le fardeau de la preuve.

CHAPITRE VIII

DISPOSITIONS FINALES

Article 42

1. La présente Convention est ouverte à la signature ou à l'adhésion des pays membres de la Commission économique pour l'Europe et des pays admis à la Commission à titre consultatif conformément au paragraphe 8 du mandat de cette commission.

2. Les pays susceptibles de participer a certains travaux de la Commission économique pour l'Europe en application du paragraphe 11 du mandat de cette commission peuvent devenir parties contractantes à la présente Convention en y adhérant après son entrée en vigueur.

3. La Convention sera ouverte à la signature jusqu'au 31 août 1956 inclus. Après cette date, elle sera ouverte à l'adhésion.

4. La présente Convention sera ratifiée.

5. La ratification ou l'adhésion sera effectuée par le dépôt d'un instrument auprès du Secrétaire de l'Organisation des Nations Unies.

Article 43

1. La présente Convention entrera en vigueur le quatre-vingt-dixième jour après que cinq des pays mentionnés au paragraphe 1 de l'article 42 auront déposé leur instrument de ratification ou d'adhésion.

2. Pour chaque pays qui la ratifiera ou y adhérera après que cinq pays auront déposé leur instrument de ratification ou d'adhésion, la présente Convention entrera en vigueur le quatre-vingt-dixième jour qui suivra le dépôt de l'instrument de ratification ou d'adhésion dudit pays.

Article 44

1. Chaque partie contractante pourra dénoncer la présente Convention par notification adressée au Secrétaire général de l'Organisation des Nations Unies.

2. La dénonciation prendra effet douze mois après la date à laquelle le Secrétaire général en aura reçu notification.

Article 45

Si, après l'entrée en vigueur de la présente Convention, le nombre de parties contractantes se trouve, par suite de dénonciations, ramené à moins de cinq, la présente Convention cessera d'être en vigueur à partir de la date à laquelle la dernière de ces dénonciations prendra effet.

Article 46

1. Tout pays pourra, lors du dépôt de son instrument de ratification ou d'adhésion ou à tout moment ultérieur, déclarer, par notification adressée au Secrétaire général de l'Organisation des Nations Unies, que la présente Convention sera applicable à tout ou partie des territoires qu'il représente sur le plan international. La Convention sera applicable au territoire ou aux territoires mentionnées dans la notification à dater du quatre-vingt-dixième jour après réception de cette notification par le Secrétaire général ou, si à ce jour la Convention n'est pas encore entrée en vigueur, à dater de son entrée en vigueur.

2. Tout pays qui aura fait, conformément au paragraphe précédent, une déclaration ayant pour effet de rendre la présente Convention applicable à un territoire qu'il représente sur le plan international pourra, conformément à l'article 44, dénoncer la Convention en ce qui concerne ledit territoire.

Article 47

Tout différent entre deux ou plusieurs parties contractantes touchant l'interprétation ou l'application de la présente Convention que les parties n'auraient pu régler par voie de négociation ou par un autre mode de règlement pourra être porté, à la requête d'une quelconque des parties contractantes intéressées, devant la Cour internationale de Justice, pour être tranché par elle.

Article 48

1. Chaque partie contractante pourra, au moment où elle signera ou ratifiera la présente Convention ou y adhérera, déclarer qu'elle ne se considère pas liée par l'article 47 de la Convention. Les autres parties contractantes ne seront pas liées par l'article 47 envers toute partie contractante qui aura formulé une telle réserve.

2. Toute partie contractante qui aura formulé une réserve conformément au paragraphe 1 pourra a tout moment lever cette réserve par une notification adressée au Secrétaire général de l'Organisation des Nations Unies.

3. Aucune autre réserve à la présente Convention ne sera admise.

Article 49

1. Après que la présente Convention aura éte en vigueur pendant trois ans, toute partie contractante pourra, par notification adressée au Secrétaire général de l'Organisation des Nations Unies, demander la convocation d'une conférence à l'effet de reviser la présente Convention. Le Secrétaire général notifiera cette demande à toutes les parties contractantes et convoquera une conférence de revision si, dans un délai de quatre mois à dater de la notification adressé par lui, le quart au moins des parties contractantes lui signifient leur assentiment à cette demande.

2. Si une conférence est convoquée conformément au paragraphe précédent, le Secrétaire général en avisera toutes les parties contractantes et les invitera à présenter, dans un délai de trois mois, les propositions qu'elles souhaiteraient voir examiner par la conférence. Le Secrétaire général communiquera à toutes les parties contractantes l'ordre du jour provisoire de la conférence, ainsi que le texte de ces propositions, trois mois au moins avant la date d'ouverture de la conférence.

3. Le Secrétaire général invitera à toute conférence convoquée conformément au présent article tous les pays visés au paragraphe 1 de l'article 42, ainsi que les pays devenus parties contractantes en application du paragraphe 2 de l'article 42.

Article 50

Outre les notifications prévues à l'article 49, le Secrétaire général de l'Organisation des Nations Unies notifiera aux pays visés au paragraphe 1 de l'article 42, ainsi qu'aux pays devenus parties contractantes en application du paragraphe 2 de l'article 42:

a) Les ratifications et adhésions en vertu de l'article 42;
b) Les dates auxquelles la présente Convention entrera en vigueur conformément à l'article 43;
c) Les dénonciations en vertu de l'article 44;
d) L'abrogation de la présente Convention conformément à l'article 45;
e) Les notifications reçues conformément à l'article 46;
f) Les déclarations et notifications reçues conformément aux paragraphes 1 et 2 de l'article 48.

Article 51

Après le 31 août 1956, l'original de la présente Convention sera déposé auprès du Secrétaire général de l'Organisation des Nations Unies, qui en transmettra des copies certifiées conformes à chacun des pays visés aux paragraphes 1 et 2 de l'article 42.

EN FOI DE QUOI, les soussignés, à ce dûment autorisés, ont signé la présente Convention.

FAIT à Genève, le dix-neuf mai mil neuf cent cinquante-six, en un seul exemplaire, en langues anglaise et française, les deux textes faisant également foi.

1 Expéditeur (nom, adresse, pays) Sender (name, address, country)	**LETTRE DE VOITURE INTERNATIONALE** **INTERNATIONAL CONSIGNMENT NOTE** No 24382 **CMR** Ce transport est soumis, nonobstant toute clause contraire, à la Convention relative au contrat de transport international de marchandises par route (CMR). This carriage is subject, notwithstanding any clause to the contrary, to the Convention on the Contract for the International Carriage of goods by road (CMR).
2 Destinataire (nom, adresse, pays) Consignee (name, address, country)	**16** Transporteur (nom, adresse, pays) Carrier (name, address, country)
3 Lieu prévu pour la livraison de la marchandise (lieu, pays) Place of delivery of the goods (place, country)	**17** Transporteurs successifs (nom, adresse, pays) Successive carriers (name, address, country)
4 Lieu et date de la prise en charge de la marchandise (lieu, pays, date) Place and date of taking over the goods (place, country, date)	**18** Réserves et observations du transporteur Carrier's reservations and observations
5 Documents annexés Documents attached	

6 Marques et numéros / Marks and Nos   **7** Nombre des colis / Number of packages   **8** Mode d'emballage / Method of packing   **9** Nature de la marchandise / Nature of the goods   **10** No statistique / Statistical number   **11** Poids brut, kg / Gross weight in kg   **12** Cubage m3 / Volume in m3

Classe / Class   Chiffre / Number   Lettre / Letter   (ADR ')

13 Instructions de l'expéditeur
Sender's instructions

19 Conventions particulières
Special agreements

20 A payer par To be paid by :	Expéditeur Senders		Monnaie/Currency	Destinataire Consignee
Prix de transport Carriage charges				
Réductions Deductions —				
Solde / Balance				
Suppléments Supplém charges :				
Frais accessoires Other charges ›				
TOTAL :				

14 Prescriptions d'affranchissement
Instructions as to payment for carriage

☐ Franco / Carriage paid

☐ Non franco / Carriage forward

21 Etabli à / Established in   le / on   19

15 Remboursement / Cash on delivery

22

23

24 Marchandises reçues / Goods received

Lieu / Place   le / on   19

Signature et timbre de l'expéditeur
Signature and stamp of the sender

Signature et timbre du transporteur
Signature and stamp of the carrier

Signature et timbre du destinataire
Signature and stamp of the consignee

Modèle IRU 1976

LETTRE DE VOITURE INTERNATIONALE   **CMR**   **INTERNATIONAL CONSIGNMENT NOTE**

COPY 1 SENDER
COPY 2 CONSIGNEE
COPY 3 CARRIER

Approved by FTA/RHA/SITPRO UK 1981/1987/1993
*NB FOR DANGEROUS GOODS SPECIFY 1. Substance identification number (if applicable)
2. Substance description 3. Class 4. Item number and letter (if any) 5. The initials 'ADR' or 'RID'
6. Other statements as required by ADR or RID.

TATE FREIGHT FORMS (0908) 221162
(1.93)
SITPRO APPROVED Licensee NO. 20

Sender (name, address, country) Expediteur (nom, adresse, pays) 1	Sender's/agent's reference Reference de l'expediteur/de l'agent 2/3
Consignee (name, address, country) Destinataire (nom, adresse, pays) 4	Carrier (name, address, country) Transporteur (nom, adresse, pays) 5
Place & date of taking over the goods (place, country, date) Lieu et date de la prise en charge des marchandises (lieu, pays, date) 6	Successive carriers Transporteurs successifs 7
Place designated for delivery of goods (place, country) Lieu prevu pour la livraison des marchandises (lieu, pays) 8	This carriage is subject, notwithstanding any clause to the contrary, to the Convention on the Contract for the International Carriage of Goods by Road (CMR) Ce transport est soumis nonobstant toute clause contraire a la Convention Relative au Contrat de Transport International de Marchandises par Route (CMR)

Shipping marks; no. & kind of packages, description of goods* Marques et nos; no et nature des colis, designation des marchandises* 9	Gross weight (kg) 10 Poids brut (kg)	Volume (m3) 11 Cubage (m3)

Carriage charges Prix de transport 12	Senders instructions for customs, etc. Instructions de l'expediteur (optional) 13
Reservations Reserves 14	Documents attached Documents annexes (optional) 15
	Special agreements Conventions particulieres (optional) 16
Goods received Marchandises recues 17	Signature of carrier Signature du transporteur 18	Company completing this note Societe emettrice 19
		Place and date, signature Lieu et date, signature 20

LLOYD'S GOODS IN TRANSIT (C.M.R.) POLICY

Any person not an Underwriting Member of Lloyd's subscribing this Policy, or any person uttering the same if so subscribed, will be liable to be proceeded against under Lloyd's Acts.

Printed at Lloyd's, London, England.

(Subscribed only by Underwriting Members of Lloyd's all of whom have complied with the requirements of the Insurance Companies Act, 1958, as to security and otherwise.)

1. CONTRACT

Whereas the Insured named in the Schedule herein has made to us, the Underwriters who have hereunto subscribed our Names, a written proposal bearing the date specified in the Schedule, and any amendments thereto, which is hereby agreed to be the basis of the contract and is deemed to be incorporated herein, and has paid to us the premium specified in the Schedule,

2. COVER

The Underwriters agree, subject to the terms, conditions, limits of liability and exceptions contained herein, or endorsed or otherwise expressed hereon, to indemnify the Insured against the Insured's legal liability as a carrier under the provisions of the Convention on the Contract for the International Carriage of Goods by Road (C.M.R.), hereinafter called the Convention, for loss of or destruction of or damage to goods and/or merchandise whilst in transit per Insured's vehicles and/or trailers and/or containers (including all such conveyances in the Insured's care, custody or control) to or from any address in the United Kingdom, Continent of Europe and Eire, including any loading and unloading and packing and unpacking; and whilst in the normal course of transit the goods and/or merchandise are temporarily housed on or off the vehicles and/or trailers or in containers that are on or off the vehicles and/or trailers.

3. EXCLUSIONS

This Policy does not cover liability in respect of or arising from or caused by
 (1) Inadequate documentation.
 (2) Liability under the Convention for Articles 7 (3), 21, 24, 26, 38 and 40, but liability under Articles 24 and 26 may be included provided prior agreement be given by Underwriters and additional premiums agreed.
 (3) Mildew, vermin, inherent vice, deterioration, insufficiency of insulation, electrical and/or mechanical derangement unless caused by external means; contamination, shortage in weight, evaporation, taint, consequential loss, loss of market; or delay except in so far as Article 23(5) of the Convention applies.
 (4) The conveyance of livestock, jewellery, furs, watches, precious metals and stones, Treasury notes, bullion, cash, bonds, deeds, stamps, securities, manuscripts, documents and plans, safety razor blades, non ferrous metal and scrap, perfumery, tobacco, cigars, cigarettes, wines, spirits and the like, explosives, radios, television sets, tape recorders and similar articles, unless permission is obtained from the Underwriters and rates and terms agreed prior to the carriage of such goods and/or merchandise.
 (5) Theft of or from vehicles and/or trailers and/or containers when left unattended whilst in the actual care and custody of the Insured unless,
 (i) such vehicle and/or trailer and/or container is garaged in a building or parked in a fully enclosed yard, which is securely closed and locked or under constant surveillance, and
 (ii) such vehicle and/or trailer and/or container has all the openings closed and securely locked and all keys removed, so far as local fire regulations permit.
 Nevertheless between 0600 hours and 1800 hours only, where the Insured is unable to comply with the requirement of (i) and (ii) above, theft of or from the vehicle and/or trailer and/or container is covered unless consequent upon the failure of the Insured to take all reasonable safety and security precautions.
Further this Policy does not cover liability
 (6) Directly or indirectly occasioned by, happening through, or in consequence of war, invasion, acts of foreign enemies, hostilities (whether war be declared or not), civil war, rebellion, revolution, insurrection, military or usurped power or confiscation or nationalisation or requisition or destruction of or damage to property by or under the order of any government or public or local authority.
 (7) Directly or indirectly occasioned by, happening through, or in consequence of strikes, lockouts, labour disturbances, riots and civil commotions unless previously agreed to delete and an additional premium agreed.
 (8) Directly or indirectly caused by or contributed to by or arising from ionising radiations or contamination by radioactivity from any nuclear fuel or from any nuclear waste from the combustion of nuclear fuel.

4. DANGEROUS DRUGS

" It is understood and agreed that no claim under this Policy will be paid in respect of drugs to which the various International Conventions relating to Opium and other dangerous drugs apply unless
 (1) the drugs shall be expressly declared as such in the consignment note and the name of the country from which, and the name of the country to which they are consigned shall be specifically stated in the consignment note, and
 (2) the proof of loss is accompanied either by a licence, certificate or authorization issued by the Government of the country to which the drugs are consigned showing that the importation of that consignment into that country has been approved by that Government, or, alternatively, by a licence, certificate or authorization issued by the Government of the country from which the drugs are consigned showing that the export of the consignment to the destination stated has been approved by that Government, and
 (3) the route by which the drugs were conveyed was usual and customary."

5. DUE DILIGENCE

The Insured shall take all reasonable precautions for the protection and safeguarding of the goods and/or merchandise and use such protective appliances as may be specified in the Policy and all vehicles and protective devices shall be maintained in good order. Such devices shall be used at all times and shall not be varied or withdrawn without written consent by the Underwriters. It is the duty of the Insured in the event that any property of the Insured, or for which they are responsible, be lost or damaged, to take all reasonable steps to effect its recovery and/or preservation.

6. NON-CONTRIBUTION

This Policy does not cover any liability or loss or damage which at the time of the attachment of liability or the happening of such loss or damage is insured by or would, but for the existence of this Policy, be insured by any other existing Policy or Policies except in respect of any excess beyond the amount which would have been payable under such other Policy or Policies had this Insurance not been effected.

7. CANCELLATION

This Policy may be cancelled at any time by either party giving 15 days' notice (7 days in respect of Exclusion (7) being deleted) in writing by prepaid letter post properly addressed to the last known address of the Insured or to the Brokers. Underwriters receiving notice will receive and retain the customary short term or earned premium whichever is the greater or when giving notice receive and retain pro rata or earned premium whichever is the greater.

8. CLAIMS PROCEDURE

In the event of a happening likely to give rise to a claim against the Insured in respect of the goods and/or merchandise the insured shall as soon as possible give notice thereof to the Underwriters and furnish full particulars. Every letter, notice, writ, summons and process relating thereto shall be notified or forwarded to the Underwriters immediately on receipt. In respect of any such claim no admission, offer, promise, payment or indemnity shall be made or given by or on behalf of the Insured without the written consent of the Underwriters. The Underwriters shall be entitled to take over and conduct in the name of the Insured the defence or settlement of any such claim or to prosecute in the name of the Insured for their own benefit any claim for indemnity or damages or otherwise and shall have full discretion in the conduct of any proceedings or in the settlement of any claim and the Insured shall give all such information and assistance as the Underwriters may require. In connection with any such claim the Underwriters will pay costs and expenses of litigation recoverable by any claimant from the Insured and costs and expenses incurred by the Underwriters on behalf of the Insured.

9. SUBROGATION

The Insured shall at the request and at the expense of the Underwriters do and concur in doing and permit to be done all such acts and things as may be necessary or reasonably required by the Underwriters for the purposes of enforcing any rights and remedies or of obtaining relief or indemnity from other parties to which the Underwriters shall be or would become entitled or subrogated upon paying any claim arising under this Policy whether such acts and things shall be or become necessary or required before or after his Indemnification by the Underwriters.

10. SUB-CONTRACTORS and/or SUCCESSIVE CARRIERS

The expression "in transit" is extended to cover transit per Sub-Contractors' and/or Successive Carriers' vehicles and/or trailers and/or containers subject to the following additional terms and conditions :—

(1) The Insured shall ascertain that all Sub-Contractors and/or Successive Carriers carry valid insurance under which the vehicle and/or trailer and/or container limit is sufficient to cover the value of the sub-contracted load (including all Customs or Excise Duties and/or any other charges) and that in all other respects the cover provided by such insurance is not less extensive than the cover provided by this Policy, and the Insured shall obtain written confirmation of such cover if required by the Underwriters.

(2) The benefit of this Policy shall in no circumstances whatsoever pass to Sub-Contractors and/or Successive Carriers or the Insurers of Sub-Contractors and/or Successive Carriers.

(3) It is a condition precedent to Underwriters' liability under this Policy in respect of any claim arising out of transit per Sub-Contractors' and/or Successive Carriers' vehicles and/or trailers and/or containers that the property shall not be delivered by the Insured or his servants or agents, and that the Insured shall take all reasonable precautions to ensure that the property is not delivered by the owner or his servants or agents to any person purporting to be a Sub-Contractor and/or Successive Carrier or a representative of a Sub-Contractor and/or Successive Carrier unless the bona-fides of the said person shall have been established beyond reasonable doubt.

11. EXCESS

12. LIMITS

13. PREMIUM

See Schedule.

If the Insured shall make any claim knowing the same to be false or fraudulent, as regards the amount or otherwise, this Policy shall become void and all claims hereunder shall be forfeited.

Now know Ye that We, the Underwriters, members of the Syndicate(s) whose definitive Number(s) in the attached list are set out in the Table opposite, or attached opposite, hereby bind Ourselves, each for his own part and not one for Another, our Heirs, Executors and Administrators, and in respect of his due proportion only, to pay or make good to the Insured or to the Insured's Executors or Administrators, all such liability as aforesaid arising from an occurrence during the continuance of this Policy; not exceeding in all the Total Sum Insured specified in the said Table, such payment to be made within Seven Days after such liability is proved, and so that the due proportion for which each of Us the Underwriters is liable shall be ascertained by reference to his proportion as ascertained according to the said list of the Amount, Percentage or Proportion of the Total Sum Insured which is in the said Table set opposite the definitive Number of the Syndicate of which such Underwriter is a member.

In Witness whereof the Manager of Lloyd's Policy Signing Office has subscribed his Name on behalf of each of Us.

LLOYD'S POLICY SIGNING OFFICE,

MANAGER.

Definitive Numbers of Syndicates and Amount, Percentage or Proportion of the Total Sum Insured shared between the Members of those Syndicates.

LLOYD'S GOODS IN TRANSIT (CMR) POLICY

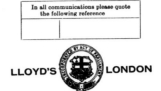

LLOYD'S **LONDON**

Lloyd's Goods in Transit (C.M.R.) Policy

Insured

Premium

Policy and Stamp

Date of Expiry

The Insured is requested to read this Policy and, if it is incorrect, return it immediately for alteration.

In the event of any occurrence likely to result in a claim under this Policy, immediate notice should be given to:—

SCHEDULE

The Policy No.	*1*

The Name and Address of the Insured

The period of insurance is from   to   ,
both days inclusive, and for such further period or periods as may be mutually agreed upon.

The date of the written proposal

EXCESS — The Insured shall bear the first £   of each and every claim except claims in respect of or arising from loss or damage caused by fire.

LIMITS — £   per vehicle or vehicle and/or trailer and/or container combined.
£   any one loss which is the maximum sum payable for any claim or series of claims arising out of any one happening or event.

PREMIUM — The deposit premium of £   is based on £   being the estimated gross annual haulage charges which includes shipping charges and any other charges including those of Sub-Contractors and/or Successive Carriers.
The Policy is subject to a minimum premium of £   . The premium to be adjusted at   per cent. on the certified gross annual haulage charges (as described above) on expiry.

Dated in London, the

370

Index